AF388925

Advances in Computational Intelligence Applications in the Mining Industry

Advances in Computational Intelligence Applications in the Mining Industry

Editors

Rajive Ganguli
Sean Dessureault
Pratt Rogers

MDPI • Basel • Beijing • Wuhan • Barcelona • Belgrade • Manchester • Tokyo • Cluj • Tianjin

Editors

Rajive Ganguli
Mining Engineering
University of Utah
Salt Lake City, UT
United States

Sean Dessureault
Technology and Innovation
Mosaic Company
Tampa, FL
United States

Pratt Rogers
Mining Engineering
University of Utah
Salt Lake City, UT
United States

Editorial Office
MDPI
St. Alban-Anlage 66
4052 Basel, Switzerland

This is a reprint of articles from the Special Issue published online in the open access journal *Minerals* (ISSN 2075-163X) (available at: www.mdpi.com/journal/minerals/special_issues/ACIAMI).

For citation purposes, cite each article independently as indicated on the article page online and as indicated below:

LastName, A.A.; LastName, B.B.; LastName, C.C. Article Title. *Journal Name* **Year**, *Volume Number*, Page Range.

ISBN 978-3-0365-3159-5 (Hbk)
ISBN 978-3-0365-3158-8 (PDF)

Contents

About the Editors

Rajive Ganguli

Dr. Rajive Ganguli is the Malcolm McKinnon Professor of Mining Engineering and Associate Dean of Assessment in the College of Mines and Earth Sciences at the University of Utah. He was previously affiliated with the University of Alaska Fairbanks, Jim Walter Resources (Alabama) and Hindustan Copper Limited (India). He has three degrees in Mining Engineering, and is a registered Professional Engineer (Alaska) and a mine foreman (Alabama). He is very interested in the grand challenges of mining, from exploiting Big Data to tackling the various issues that make mine operations difficult. Over the last two decades, among other things, he has been working on computational intelligence applications in the mining industry. He has partnered with mines in the United States, Mongolia, and Mexico in developing the industrial applications of computational intelligence.

Sean Dessureault

Dr. Sean Dessureault is the Vice President of Technology and Innovation for The Mosaic Company, a leading vertically integrated fertilizer producer with mines throughout the Americas and a global distribution network. From 2002 to 2018, Dessureault was a professor at the University of Arizona, where he engaged in research related to the integration and effective use of mining information systems and sustainability. In 2004, he founded MISOM Technologies, an IT services company with expertise in data integration, control rooms, and automation. MISOM eventually developed the FARA platform, which brings together Internet of Things (IoT) sensors, a tablet-based mobile app, and modern approaches such as Big Data and gamification to create innovative form digitization, worker engagement, and fleet management. In late 2017, MISOM was acquired by MST Global, where Dr. Dessureault served as Chief Innovation Evangelist until joining Mosaic in 2019.

Pratt Rogers

Dr. Pratt Rogers is a mining professional with extensive research and business experience in data management systems, automation, mining technology, and health and safety management systems. Dr. Rogers started in a mining faculty at the University of Utah in 2016 as an Assistant Professor. Prior to joining the University of Utah, Dr. Rogers worked as the VP of Product Development at MISOM Technologies. He has Bachelor's (2008), Master's (2012), and Doctoral (2015) degrees in Mining Engineering from The University of Arizona. He has worked as a site engineer for Luminant Mining and as a mining analytics consultant across the United States, Canada, and Mexico. Dr. Rogers continues to provide data analytics consultation for various mining companies. His research interests concentrate on the reliability of data management systems in mining organizations for a variety of objectives, such as fatigue management, short-range planning, and operational excellence. Recently, Dr. Rogers has been working on a research project funded by NIOSH focused on fatigue management modeling and monitoring tools.

Editorial

Introduction to the Special Issue "Advances in Computational Intelligence Applications in the Mining Industry"

Rajive Ganguli [1,*], Sean Dessureault [2] and Pratt Rogers [1]

1 Department of Mining Engineering, University of Utah, Salt Lake City, UT 84112, USA; pratt.rogers@utah.edu
2 The Mosaic Company, Tampa, FL 33605, USA; Sean.Dessureault@mosaicco.com
* Correspondence: rajive.ganguli@utah.edu

Citation: Ganguli, R.; Dessureault, S.; Rogers, P. Introduction to the Special Issue "Advances in Computational Intelligence Applications in the Mining Industry". *Minerals* **2022**, *12*, 67. https://doi.org/10.3390/min12010067

Received: 14 December 2021
Accepted: 28 December 2021
Published: 5 January 2022

Publisher's Note: MDPI stays neutral with regard to jurisdictional claims in published maps and institutional affiliations.

This is an exciting time for the mining industry, as it is on the cusp of a change in efficiency as it gets better at leveraging data. After decades of focusing on collecting data, the industry has developed to where the focus now is on utilizing the data. The utilization of data typically involves developing models that are used to better understand mining processes, with a variety of computational intelligence (CI) techniques being at the forefront of methods used to develop models. Modeling and data collection add value by presenting analytics so that humans, from frontline workers to corporate executives, can respond as quickly as possible to changing conditions.

CI is often defined as a class of techniques, which includes neural networks, fuzzy systems, and evolutionary computing. Many papers in this issue make excellent use of these techniques to advance the state of the industry. However, given the broad nature of the mining industry, we also chose to include other data-driven computational techniques that are advancing the state of the art, regardless of whether they fall directly under CI. Our focus was more on capturing the advances than maintaining the purity of the techniques.

The papers in this issue advance the state of the art in four broad categories: mine operations, mine planning, mine safety, and advancing the sciences, primarily in image processing applications. In the field of mining operations, Both and Dimitrakopolous [1] utilize drill hole penetration rates to predict ball mill throughput. They combine a variety of techniques, including neural networks, in their work. Young and Rogers [2] acknowledge the important role stockpiles play in managing ore that is supplied to the mill, and the industry's struggle in understanding the grade distribution within the stockpiles. They demonstrate that data from mine dispatch systems can be combined with traditional interpolation techniques to obtain the grade distribution of stockpiles. Olivier and Aldrich [3] similarly show the value of combining simplicity with operational data. They extract control rules from semi-autogenous grinding (SAG) mill operational data using decision trees. In controlling the mill for power draw, the decision trees identify the same factors as important as random forests.

In the field of mine planning, authors have either leveraged existing mine plans or offered methods to improve mine plans. At a Mongolian mine, Sarantsatsral et al. [4] use random forests to predict rock types in various mine planning scenarios. They determine that rock types could be predicted relatively well for some mine planning scenarios. de Carvalho and Dimitrakopoulos [5] improve real-time truck dispatch decisions by basing them on a deep Q-learning reinforcement neural network model. The reinforcement model is trained based on a continuous real-time discrete event simulation (DES) model, which simulates short-term mine plans. Wilson et al. [6] utilize partial least squares (PLS) regressions to model the geological uncertainty in oil-sands. They combine the PLS models with DES methods to stabilize plant throughput, despite uncertainties in geology and processing methods. Park et al. [7] leverage the Internet of Things to collect truck travel times and environmental data from the transportation systems at a limestone mine. They apply various machine learning models to identify when the transportation system suffered bottlenecks. The models are then used to anticipate problems in the transportation system, aiding production planning.

Three diverse papers related to mining health and safety are included in this Special Issue. Talebi et al. [8] explore the complex problem of mine operator fatigue using computational intelligence. The authors use a random forest model with operational technology data and a PERCLOS fatigue monitoring system. The model identifies some interesting leading indicators of fatigue found in operational technology. Many health and safety management systems (HSMSs) are dependent upon qualitative/narrative datasets. Ganguli et al. [9] explore using natural language processing (NLP) to contextualize these datasets. The authors use large US-based MSHA datasets to train NLP models. These models can then be used to improve the analysis of HSMS data at mine sites. Mining companies strive to reduce risk to their operations and surrounding stakeholders. Chomacki et al. [10] explore methods to improve the understating of mining impacts on local stakeholders. Two models are created to assess risk to surface buildings from underground mining units of operation. These tools will help manage the complex risks of mining impacts on proximity stakeholders.

Five papers were included in the Special Issue that utilizes computational intelligence tools to advance fundamental science in the areas of prospectivity mapping, rock/ore classification, and rock fragmentation. First, Lachaud et al. [11] present a data-driven mineral prospectivity model to identify areas with higher discovery potential. They use existing geological datasets to train random forest machine learning models to improve exploration decisions. Next, Sinaice et al. [12] present a model to help mining companies more quickly classify rock masses using hyperspectral imaging, neighborhood component analysis, and machine learning. By integrating these computational tools, the authors present a rock mass classification model that can quickly and accurately predict geological properties. Advanced imaging technologies are changing geological sampling and analysis. Iwaszenko and Róg [13] provide an image analysis model to segment important geological features of coal. The modeling can speed up the analysis, thereby influencing key mineral processing decisions and earlier capturing valuable time and energy.

In addition to the image analysis discussed, Tungul et al. [14] provide an updated approach to simplifying fragmentation analysis using smartphones and GNSS technology. The authors showcase a methodology that can reduce the inherent error of GNSS. The methodology can reduce the cost of fragmentation analysis and improve the speed of analysis. This has the potential to allow smaller operations access to this critical mining and mineral processing variable. Along the lines of rock fragmentation computational intelligence, Dumakor-Dupey et al. [15] provide a review of computational intelligence and blast-induced impacts. The authors explore various blast-impact empirical and machine learning models. The paper provides a guide for future research in this area.

The editors are pleased with the results of the Special Issue and appreciate the contributions of the authors, which include important contributions to computational intelligence and operational excellence. In addition, the contributions to advancing fundamental science in the mining domain will yield important results in the future. Digital transformation's benefits rest on computational intelligence and a culture of process change around analytics. The mining and minerals industry, academia, and governments need to continue to invest in research and development in this area. The research presented in this Special Issue is an important, albeit small, contribution to this endeavor.

Conflicts of Interest: The authors declare no conflict of interest.

References

1. Both, C.; Dimitrakopoulos, R. Applied Machine Learning for Geometallurgical Throughput Prediction—A Case Study Using Production Data at the Tropicana Gold Mining Complex. *Minerals* **2021**, *11*, 1257. [CrossRef]
2. Young, A.; Rogers, W.P. Modelling Large Heaped Fill Stockpiles Using FMS Data. *Minerals* **2021**, *11*, 636. [CrossRef]
3. Olivier, J.; Aldrich, C. Use of Decision Trees for the Development of Decision Support Systems for the Control of Grinding Circuits. *Minerals* **2021**, *11*, 595. [CrossRef]

4. Sarantsatsral, N.; Ganguli, R.; Pothina, R.; Tumen-Ayush, B. A Case Study of Rock Type Prediction Using Random Forests: Erdenet Copper Mine, Mongolia. *Minerals* **2021**, *11*, 1059. [CrossRef]
5. de Carvalho, J.P.; Dimitrakopoulos, R. Integrating Production Planning with Truck-Dispatching Decisions through Reinforcement Learning While Managing Uncertainty. *Minerals* **2021**, *11*, 587. [CrossRef]
6. Wilson, R.; Mercier, P.H.J.; Patarachao, B.; Navarra, A. Partial Least Squares Regression of Oil Sands Processing Variables within Discrete Event Simulation Digital Twin. *Minerals* **2021**, *11*, 689. [CrossRef]
7. Park, S.; Jung, D.; Nguyen, H.; Choi, Y. Diagnosis of Problems in Truck Ore Transport Operations in Underground Mines Using Various Machine Learning Models and Data Collected by Internet of Things Systems. *Minerals* **2021**, *11*, 1128. [CrossRef]
8. Talebi, E.; Rogers, W.P.; Morgan, T.; Drews, F.A. Modeling Mine Workforce Fatigue: Finding Leading Indicators of Fatigue in Operational Data Sets. *Minerals* **2021**, *11*, 621. [CrossRef]
9. Ganguli, R.; Miller, P.; Pothina, R. Effectiveness of Natural Language Processing Based Machine Learning in Analyzing Incident Narratives at a Mine. *Minerals* **2021**, *11*, 776. [CrossRef]
10. Chomacki, L.; Rusek, J.; Słowik, L. Selected Artificial Intelligence Methods in the Risk Analysis of Damage to Masonry Buildings Subject to Long-Term Underground Mining Exploitation. *Minerals* **2021**, *11*, 958. [CrossRef]
11. Lachaud, A.; Marcus, A.; Vučetić, S.; Mišković, I. Study of the Influence of Non-Deposit Locations in Data-Driven Mineral Prospectivity Mapping: A Case Study on the Iskut Project in Northwestern British Columbia, Canada. *Minerals* **2021**, *11*, 597. [CrossRef]
12. Sinaice, B.B.; Owada, N.; Saadat, M.; Toriya, H.; Inagaki, F.; Bagai, Z.; Kawamura, Y. Coupling NCA Dimensionality Reduction with Machine Learning in Multispectral Rock Classification Problems. *Minerals* **2021**, *11*, 846. [CrossRef]
13. Iwaszenko, S.; Róg, L. Application of Deep Learning in Petrographic Coal Images Segmentation. *Minerals* **2021**, *11*, 1265. [CrossRef]
14. Tungol, Z.P.L.; Toriya, H.; Owada, N.; Kitahara, I.; Inagaki, F.; Saadat, M.; Jang, H.D.; Kawamura, Y. Model Scaling in Smartphone GNSS-Aided Photogrammetry for Fragmentation Size Distribution Estimation. *Minerals* **2021**, *11*, 1301. [CrossRef]
15. Dumakor-Dupey, N.K.; Arya, S.; Jha, A. Advances in Blast-Induced Impact Prediction—A Review of Machine Learning Applications. *Minerals* **2021**, *11*, 601. [CrossRef]

Article

Applied Machine Learning for Geometallurgical Throughput Prediction—A Case Study Using Production Data at the Tropicana Gold Mining Complex

Christian Both * and **Roussos Dimitrakopoulos ***

COSMO—Stochastic Mine Planning Laboratory, Department of Mining and Materials Engineering, McGill University, 3450 University Street, Montreal, QC H3A 0E8, Canada
* Correspondence: christian.both@mail.mcgill.ca (C.B.); roussos.dimitrakopoulos@mcgill.ca (R.D.)

Abstract: With the increased use of digital technologies in the mining industry, the amount of centrally stored production data is continuously growing. However, datasets in mines and processing plants are not fully utilized to build links between extracted materials and metallurgical plant performances. This article shows a case study at the Tropicana Gold mining complex that utilizes penetration rates from blasthole drilling and measurements of the comminution circuit to construct a data-driven, geometallurgical throughput prediction model of the ball mill. Several improvements over a previous publication are shown. First, the recorded power draw, feed particle and product particle size are newly considered. Second, a machine learning model in the form of a neural network is used and compared to a linear model. The article also shows that hardness proportions perform 6.3% better than averages of penetration rates for throughput prediction, underlining the importance of compositional approaches for non-additive geometallurgical variables. When adding ball mill power and product particle size, the prediction error (RMSE) decreases by another 10.6%. This result can only be achieved with the neural network, whereas the linear regression shows improvements of 4.2%. Finally, it is discussed how the throughput prediction model can be integrated into production scheduling.

Keywords: tactical geometallurgy; data analytics in mining; ball mill throughput; measurement while drilling; non-additivity

Citation: Both, C.; Dimitrakopoulos, R. Applied Machine Learning for Geometallurgical Throughput Prediction—A Case Study Using Production Data at the Tropicana Gold Mining Complex. *Minerals* **2021**, *11*, 1257. https://doi.org/10.3390/min11111257

Academic Editors: Rajive Ganguli, Sean Dessureault and Pratt Rogers

Received: 11 October 2021
Accepted: 10 November 2021
Published: 12 November 2021

1. Introduction

In recent years, the amount of collected and centrally stored production data in the mining industry has increased massively with the implementation of digital technologies. Some examples of centrally stored datasets in operating mines are records of fleet management systems [1], measurement while drilling (MWD) [2], measurements of material characteristics using sensor techniques [3], and other key performance indicators at the processing plants. While potentially all mine planning activities can benefit from the analysis of production data (data analytics), interdisciplinary fields such as geometallurgy can particularly gain from this growing data. Geometallurgy aims to capture the relationships between spatially distributed rock characteristics and its metallurgical behavior when the mined materials are processed and transformed into sellable products. One pertinent part of geometallurgy is the optimization of comminution circuits and the prediction of comminution performance indicators such as throughput in the mineral processing facilities [4–6]. However, value is only added to the operation when the gained geometallurgical knowledge is integrated into decision-making processes, whereas appropriate methods are still mostly lacking for the tactical or short-term production planning horizon [7]. Another current limitation is the cost-intensive sampling and laboratory testing of rock hardness and grindability [8]. The present article shows a case study at the Tropicana Gold mining complex that demonstrates how production data combined with machine learning can be used to construct a data-driven geometallurgical throughput prediction model and how such a model can subsequently be utilized for short-term mine production scheduling.

The optimization of comminution circuits has traditionally relied on well-accepted comminution laws and ore hardness and grindability indices for ball/rod mills [9,10] and SAG mills [11–13]. These comminution models are routinely used for optimized grinding circuit design, using averages or ranges of ore hardness tests of the mineral deposits to be extracted. Instead of using constant values representing whole deposits, geometallurgical programs account for the heterogeneity of geometallurgical variables within the mineral reserve and their effect on downstream processes over time [14]. A typical geometallurgical workflow includes a spatial model, which comprises geostatistically simulated or estimated variables (e.g., grindability). Several case studies have demonstrated how throughput rates of a comminution circuit can be predicted using spatial geometallurgical models of hardness and grindability indices in combination with comminution theory [15–18]. Although some of these throughput models have demonstrated high accuracy in reconciliation studies, there are notable challenges when using and integrating them into decision-making processes such as short-term production scheduling. First, the geometallurgical sampling program requires cost-intensive laboratory testing to obtain the abovementioned hardness and grindability indices [8,13]. The high associated costs spent in early project stages can be prohibitively large and typically result in very sparse sampling, although research is being conducted to increase the number of samples by using alternate data measurement tools and small-scale processing tests [19]. Second, the throughput prediction models are built to evaluate the weekly or monthly performance of mine production schedules a posteriori, instead of integrating them into short-term production scheduling. Third, none of the models account for the inherent uncertainty of the geometallurgical variables stemming from the imperfect knowledge of the orebody.

There have been efforts to incorporate geometallurgical hardness properties and their associated geological uncertainty into mine production scheduling in single, open-pit mines [20] and in mining complexes [21]. The stochastic optimization models are developed for long-term production scheduling and require that hardness and grindability indices are geostatistically simulated for volumes of selective mining units (mining block). However, most of the frequently utilized hardness and grindability indices are non-additive [13,22,23]. Geometallurgical samples are also collected on large support scales [24,25] and are typically very sparse, as mentioned earlier. These complicating factors make the joint spatial interpolation of geometallurgical variables and their change of support from point measurements to mining blocks challenging [25–28]. Morales et al. [20] optimize the mine production schedule using precalculated mill throughputs and economic values for each block independently. The method thus ignores that extracted materials are blended in stockpiles and in processing facilities; consequently, the non-additive comminution behavior of blended materials and resulting metal production cannot be correctly assessed. Kumar and Dimitrakopoulos [29] optimize a mining complex while including predefined ratios of hard and soft rock, to achieve a consistent throughput in processing streams. However, these ratios are defined arbitrarily, and details of short-term planning are not addressed.

Both and Dimitrakopoulos [30] present a new approach that integrates a geometallurgical throughput prediction model into short-term stochastic production scheduling for mining complexes. The stochastic production-scheduling formulation builds upon simultaneous stochastic optimization of mining complexes [31,32] which optimizes pertinent components of a mining complex in a single mathematical model and incorporates geological uncertainty to minimize technical risk. Instead of using block throughput rates, the production-scheduling formulation calculates the throughput of blended materials using an empirically created throughput prediction model, learning from previously observed throughput rates at the ball mill [30]. One limitation of this work is that the integrated throughput prediction model so far has only considered rock hardness, density, lithology, and weathering degree of the mineral reserve. This ignores that mill throughput rates also depend on operating factors of the processing plant, such as power draw, utilization rates, and particle size distributions. Second, a multiple linear regression (MLR) has been used for throughput predictions, which is unable to capture potential nonlinear relationships among input variables and geometallurgical response.

The case study at the Tropicana Gold mining complex shown in this article expands the method presented in Both and Dimitrakopoulos [30] in multiple ways. First, the recorded plant measurements power draw, feed particle size, and product particle size of the ball mill are newly considered to improve the prediction of ball mill throughput rates. Second, a more powerful supervised learning method in the form of an artificial neural network is tested and compared to MLR, since the addition of the new comminution-related features increases the possibilities of nonlinear interactions between predictive and response variables. The plant measurements, including the observed ball mill throughput, are retrieved from the comminution circuit at the Tropicana Gold mining complex. The other dataset used in this case study to predict ball mill throughput comprises penetration rates from measurement while drilling (MWD). The use of this dataset is motivated by its ability to indicate the strength and hardness of the intact rock [2,33,34]. The penetration rates are converted into a set of hardness proportions per selective mining unit (SMU) which has recently been proposed to build a link between intact rock hardness and comminution performance of the rock in milling and grinding circuits [30]. The present article also compares the prediction capabilities of hardness proportions to averages of penetration rates. In this way, the effect of ignoring non-additivity of hardness-related geometallurgical variables can be quantified, an issue that has had little attention in the literature thus far.

In the following sections, the components of the Tropicana Gold mining complex are introduced first, together with all utilized production data that are used for the prediction of ball mill throughput. The supervised machine learning model is discussed next, including a statistical analysis of the present dataset and a hyperparameter calibration. Analysis of results, discussion, and conclusions follow.

2. The Tropicana Gold Mining Complex and Utilized Production Data for Ball Mill Throughput Prediction

The Tropicana Gold mining complex is located in western Australia in the west of the Great Victoria Desert. The gold deposit is mined from four pits, Boston Shaker, Tropicana, Havana, and Havana South (from north to south), as can be seen in the aerial view in Figure 1. In addition, the mining complex contains a processing plant, stockpiles, a tailings facility, and multiple waste dumps. Gold is produced onsite in a single processing stream, consisting of a comminution circuit and a carbon-in-leach (CIL) plant.

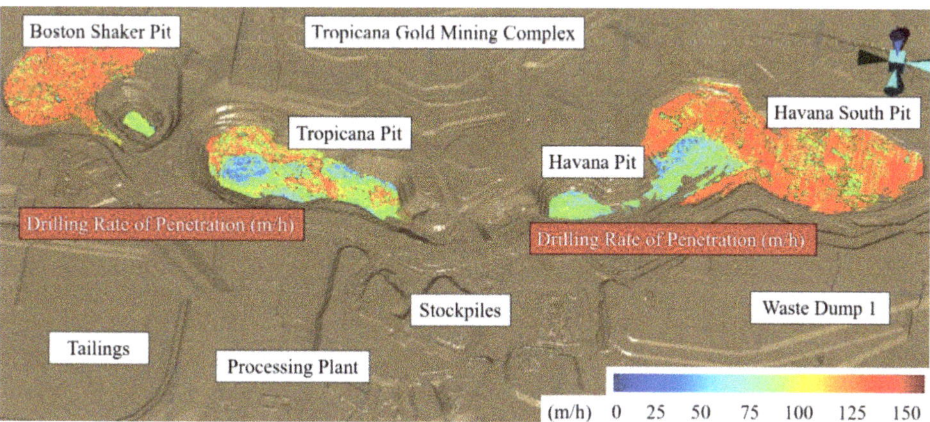

Figure 1. Components of the Tropicana Gold mining complex and a heat map of drilling rate of penetration (ROP) retrieved from measurement while drilling (MWD).

The displayed dataset in the four pits in Figure 1 shows the drilling rate of penetration (ROP) from production drilling (blastholes), which is part of the measurement while drilling (MWD) dataset collected at the Tropicana Gold mining complex. It is clearly visible how ROP reflects the heterogeneity of the rock and decreases with depth. Exemplary, easy-to-drill (softer) rock is found towards the surface (red colors at Havana South Pit and Boston Shaker Pit), whereas difficult-to-drill (harder) rock is located deeper in the pits (green–blue colors in Havana Pit, Tropicana Pit, and deeper cutback of Boston Shaker Pit). Both and Dimitrakopoulos [30] demonstrate strong correlations between the rate of penetration (ROP) of drilled rock and ball mill throughput when these rock parcels are sent to the processing plant. They subsequently present a method that predicts ball mill throughput using ROP. This article extends this work by utilizing additional measurements in the processing plant related to ball mill throughput.

The relevant material flow in the mining complex is shown together with all utilized production data in Figure 2. Detailed material tracking in daily intervals is performed using truck cycle data, starting from the material extraction in the pits and ending at the crusher. Crucially, material tracking includes all dumping and rehandling activities at run-of-mine (ROM) stockpiles, since rehandled material accounts for 80–90% of processed ore in the Tropicana Gold mining complex. In this way, ROP entries recorded in the pits can be successfully linked to observed measurements in the processing plant, including the observed throughput of the ball mill. Details of successful implementations of material tracking that include stockpiles can be found in Wambeke et al. [35] and Both and Dimitrakopoulos [30].

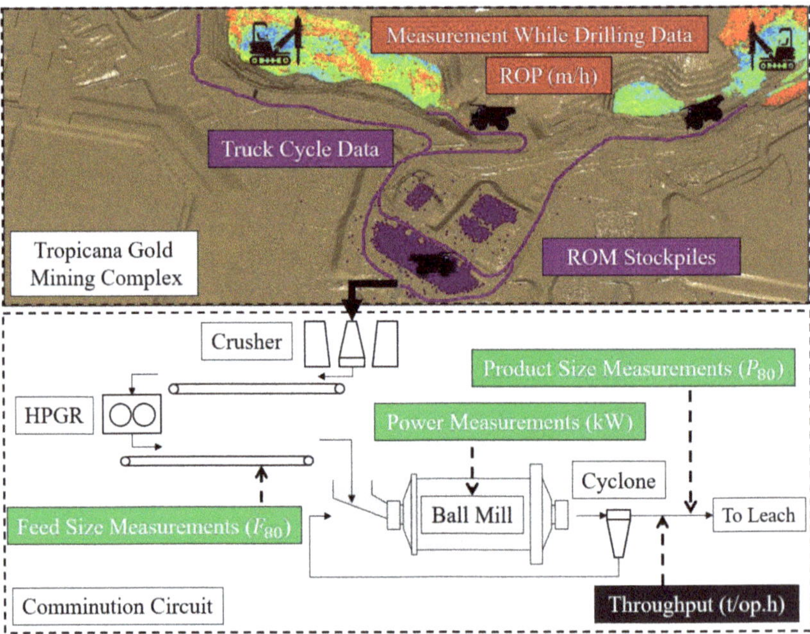

Figure 2. Material flow and utilized production data for ball mill throughput prediction in the Tropicana Gold mining complex.

The comminution circuit at Tropicana Gold mining complex comprises three stages: crushing (primary and secondary crusher), grinding (high-pressure grinding roll, HPGR), and milling (ball mill). The cyclone overflow is sent to the CIL plant to extract the gold. The recorded average power draw of the ball mill and the particle size distributions entering and leaving the ball mill are of particular interest for throughput prediction. Note that the feed and product particle size distributions are subsequently defined by their 80% passing diameters in μm. The feed particle size measurements (F_{80}) are performed using image analyzers on the conveyor belt of the HPGR product. Shift composites of cyclone overflow samples are used for product particle size measurements (P_{80}).

The relevance of all presented measurements above can be derived from comminution theory, such as Bond's law of comminution [9,10]. The Bond equation (Equation (1)) calculates the specific energy of the ball mill (W in kWh/t) required to grind the ore from a known feed size (F_{80}) to a required product size (P_{80}).

$$W = Wi * \left(\frac{10}{\sqrt{P_{80}}} - \frac{10}{\sqrt{F_{80}}} \right) \tag{1}$$

The Work index (Wi in kWh/t) is a measure of the ore's resistance to crushing and grinding [9]. In this article, it is useful to substitute the specific energy of the ball mill (energy delivered per ton of ore in kWh/t) by the quotient of mill power draw (kW) and mill throughput (processed tons per operating hour), as shown in Equation (2).

$$\frac{Power}{TPH} = Wi * \left(\frac{10}{\sqrt{P_{80}}} - \frac{10}{\sqrt{F_{80}}} \right) \tag{2}$$

Equation (3) is obtained by rearranging Equation (2) for ball mill throughput (TPH).

$$TPH = \frac{Power}{Wi * \left(\frac{10}{\sqrt{P_{80}}} - \frac{10}{\sqrt{F_{80}}} \right)} \tag{3}$$

Next to the measured power draw and particle size distributions, it is clear that throughput predictions of the ball mill must include some kind of information about ore hardness. Generally, the harder the material, the higher its resistance against comminution, thus needing to reside longer in the ball mill to reach the desired product size, given constant power draw and particle feed size. In Bond's equation, TPH is inversely proportional to Wi, as shown in Equation (4).

$$TPH \propto \frac{1}{Wi} \tag{4}$$

As introduced above, the role of informing ore hardness is taken over by ROP measurements in this article. By utilizing cost-effective and easily accessible production data (MWD information generated by drilling machines), costly and time-consuming laboratory tests spent for Wi estimates of the geological reserve can be replaced. Mwanga et al. [8] report that the typical sample volume required for Bond tests is relatively large (2–10 kg, depending on test modification), and requires crushed ore smaller than 3.35 mm (passing a 6-mesh sieve). Furthermore, several grinding cycles are necessary to reach the steady state of the simulated closed circuit. The alternative utilization of ROP is especially promising as a substitute for Wi because of its demonstrated ability to indicate rock type, strength, and alteration [34,36–38]. In general, high ROP (in m/h) indicates less competent rock, bearing lower Wi. In turn, TPH is expected to increase, as shown in Equation (5).

$$ROP \left(\frac{m}{h} \right) \nearrow \implies Wi \left(\frac{kWh}{t} \right) \searrow \implies TPH \nearrow \tag{5}$$

Note that the dependencies in Equation (5) may be nonlinear. Rather, potentially nonlinear dependencies call for more sophisticated prediction models for TPH prediction, which are subsequently discussed in Section 3.

3. Application of Supervised Machine Learning for Throughput Prediction

This section discusses the use of supervised machine learning to create a throughput-prediction model at the Tropicana Gold mining complex. Supervised machine learning models require labelled datasets for training, consisting of data pairs $\{x_i, y_i\}, i = 1, \ldots, N,$

whereas x_i is a vector of predictor variables, and y_i is the known response. In this article, the known response (label) is the observed ball mill throughput, and the M predictor variables (features) comprise of the geological attributes of the ore and measured variables in the comminution circuit. Throughput responses are recorded on a continuous scale, rendering the supervised learning problem a regression task ($y_i \in \mathbb{R}$).

3.1. Neural Networks

A feed-forward neural network is chosen as a supervised learning model for the potentially nonlinear task of ball mill throughput prediction. In its essence, feed-forward neural networks are fully connected, layered combinations of neurons that find their origins in the perceptron model [39]. A single neuron (perceptron) calculates the inner product between its internal weight vector, w^T, and the input vector, x. After adding a bias term, $b \in \mathbb{R}$, the resulting value is passed through a nonlinear activation function, $g(\cdot)$, creating a scalar output $z = g(w^T x + b)$. Several connected neurons to x form the so-called first hidden layer of the neural network. If the outputs of the first hidden layer are passed through another layer of neurons, a multilayer neural network is built [40]. The output layer comprises a single neuron that receives as input the vector of hidden outputs, z and provides an estimate, $\hat{y} \in \mathbb{R}$. Neural networks are the method of choice in this article because they have the proven advantage of being capable of approximating every arbitrary function using either one hidden layer of exponentially many neurons, or multiple consecutive neural layers consisting of fewer neurons [41]. This gives neural networks theoretical advantages over linear prediction models, such as multiple linear regression, which has been tested in previous work for throughput prediction [30]. Univariate statistics and correlations in the present dataset, including potential nonlinearities, are discussed next, followed by the discussion of the utilized neural network architecture, and tuning of its hyperparameters.

3.2. Dataset and Statistical Analysis

The dataset for throughput prediction contains the hardness-related rate of penetration (ROP) of the ore, which has been tracked in the Tropicana Gold mining complex, as presented in Figure 2. The power draw, F_{80}, and P_{80} measurements, as well as a ball mill utilization factor reflecting ball mill up- and down-time, are also included. A 7-day moving average of the data is calculated for an observed time horizon of six months (February–August 2018), which reduces noise in the dataset and helps recognize trends of higher and lower throughput rates that are more likely connected to rock properties of the material processed. In the six-month interval, extraction mainly occurs in two pits, the Tropicana and Havana Pit, and material is continuously stockpiled at the ROM stockpiles. Univariate statistics of the predictive variables and the response variable (throughput) are shown in Table 1.

Table 1. Univariate statistics of predictive variables (features) and ball mill throughput (response).

	Average ROP (m/h)	Ball Mill Power (kW)	Ball Mill Utilization (%)	P_{80} (μm)	F_{80} (mm)	Ball Mill Throughput (t/op.h)
Minimum	35.0	9996	0.7	76.5	10.3	796.4
Mean	41.4	13,002	1.0	83.3	13.1	926.5
Maximum	53.6	13,435	1.0	93.2	15.0	1007.9
Std. Dev.	3.45	685.9	0.052	3.12	1.00	34.8
Coeff. of Var. (CV)	0.083	0.053	0.053	0.037	0.077	0.038
Skewness	0.88	−3.01	−3.03	0.57	−0.36	−1.07
Kurtosis	1.20	9.26	9.30	0.94	−0.13	3.05
Count	181	181	181	176	153	181

Table 2 shows linear correlations between pertinent features and observed TPH using Pearson's correlation coefficient, in Equation (6) below, with x_i and y_i representing individual sample points and $\bar{x}$, $\bar{y}$ indicating sample means. Note that correlations in Table 2 can be inflated because they are calculated after applying the moving average.

$$r = \frac{\sum_{i=1}^{n}(x_i - \bar{x})(y_i - \bar{y})}{\sqrt{\sum_{i=1}^{n}(x_i - \bar{x})^2} * \sqrt{\sum_{i=1}^{n}(y_i - \bar{y})^2}} \tag{6}$$

Table 2. Pearson's correlation coefficient between predictor variables (geological, comminution-related) and ball mill throughput.

No.	Category	Feature	Unit	Pearson's Correlation Coefficient to Ball Mill Throughput
A1	Ore hardness using average pen-rate values	Average ROP	(m/h)	0.44
B1	(Harder material)	<26 m/h	(%)	−0.274
B2		26–29 m/h	(%)	−0.370
B3		29–32 m/h	(%)	−0.525
B4	Ore hardness expressed by	32–35 m/h	(%)	−0.361
B5	proportions of	35–38 m/h	(%)	−0.416
B6	penetration rate	38–42 m/h	(%)	−0.318
B7	intervals	42–46 m/h	(%)	0.028
B8		46–53 m/h	(%)	0.356
B9		53–62 m/h	(%)	0.386
B10	(Softer Material)	>62 m/h	(%)	0.498
C1		Feed size F80	(mm)	0.046
C2	Measurements in the	Product size P80	(μm)	0.063
C3	comminution circuit	Power	(kW)	0.382
C4		Mill Utilization	(%)	0.374

The tracked ROP entries are henceforth used in two different ways to inform material hardness. The feature 'Average ROP' comprises weighted averages of continuous ROP values linked to the materials that are transported to the crusher in the same observed time interval. In contrast, Both and Dimitrakopoulos [30] propose a compositional approach, which partitions ROP into easier-to-drill (softer rock) and difficult-to-drill (harder rock) categories, using a set of ROP intervals. The split in multiple intervals results in proportions of harder or softer materials sent to the comminution circuit in a given time interval. A detailed explanation of how to calculate these hardness proportions is given in Both and Dimitrakopoulos [30]. The listed features in Table 2 can broadly be distinguished into three categories, whereas the first two categories are related to ore hardness. Average ROP comprises the first category (A1), and hardness proportions built by intervals of penetration rates comprise the second category (B1–B10). The third feature category reflects measurements at the comminution circuit (C1–C4).

By comparing the Pearson correlation coefficients in Table 2, it can be seen that some variables correlate more strongly with TPH, whereas other variables do not. A stronger positive correlation of TPH for 'Average ROP' (in m/h) gives the first evidence of the usefulness of this feature (A1). The compositional approach effectively partitions the distribution of penetration rates into multiple hardness categories. Here, a higher percentage of difficult-to-penetrate material in the processed ore blend (B1–B6) indicates harder material, thus lowering TPH, which is confirmed by the negative correlation in Table 2. Conversely, a higher fraction of easier-to-penetrate material in the blend is expected to increase TPH, which is equally confirmed in Table 2 through positive correlations of categories B8–B10. Interestingly, some hardness categories show a stronger correlation (positive and negative) than the average ROP feature (A1). This indicates that additional information may be conveyed through the creation of hardness categories. The prediction potential of average penetration rates and hardness proportions is compared in detail in Section 4.1.

According to Equation (5), the relationship between ball mill power and TPH is directly proportional. This theoretical relationship is empirically well reflected in Table 2, showing a stronger positive correlation between ball mill power (C3) and TPH. The power measurements thus comprise an important part of throughput prediction, subsequently performed in Section 4 of this article. Although the ball mill utilization (C4) is not part of Bond's equation, it is not surprising to see a stronger correlation to TPH. Events of planned and unplanned ball mill downtime, i.e., utilization < 100 percent, ramp-up and ramp-down processes, are among the effects that also lower the effective throughput per operating hour. A redundancy between ball mill utilization and ball mill power is observed, confirmed by similar statistics of power and utilization in Table 1, which explains similar correlation in Table 2. Relationships between TPH and particle sizes of the ore that result from Bond's law (Equation (1)) are shown in Equations (7) and (8).

$$P_{80} \nearrow \implies TPH \nearrow \tag{7}$$

$$F_{80} \searrow \implies TPH \nearrow \tag{8}$$

On the one hand, a coarser product particle size (larger P_{80} value) results in higher TPH (Equation (7)), given that ore characteristics, energy input and feed particle size stay the same. On the other hand, a finer-grained feed size (smaller F_{80} value) can also lead to an increased TPH because less grinding work needs to be applied to reach the desired product size (Equation (8)). In the present dataset, the particle size measurements (C1–C2) show very little correlation in Table 2. This can be for several reasons. Contrary to power draw, the relationships in Equations (7) and (8) are nonlinear, and the particle size measurements are incomplete for some periods, as indicated in Table 1. Additionally, one must consider that particle size measurements over running belts are error-prone, especially when using image analyzers for F_{80}. It is analyzed in Section 4 whether particle size measurements can enhance throughput prediction in practice. Note that all comminution variables are scaled before usage by dividing by their maximum value. Compositional data naturally comprises fractional values in [0,1] and thus does not have to be scaled.

3.3. Network Architecture and Hyperparameter Search

In its implementation, the architecture of a feed-forward neural network requires the calibration of several hyperparameters. The hyperparameter setting is relevant to the evaluation process and robustness of the approach. Therefore, it becomes obvious to explore the hyperparameter space in order to find a stable region of this space [42]. However, due to the small size of the dataset (181 data points) and the need to test on the entire horizon (181 days) to extrapolate the overall performance of the proposed approach, the dataset cannot be split. Instead, k-fold cross-validation is used to measure the configuration quality, thus minimizing the information loss [43]. Different periods are used for different folds (20 folds) to simulate the more realistic scenario where a prediction is made over a new period. The network architecture is implemented in Python using the scikit-learn package [44]. The squared error between the observed throughput, y_i, $i = 1, \ldots, N$, and predicted throughput, $\hat{y}_i$, $i = 1, \ldots, N$, is chosen as the loss function to be minimized during training, and the rectified linear unit is chosen as activation function. The quasi-Newtonian L-BFGS algorithm [45] is used to minimize the loss function, which proved to converge more quickly on the small dataset compared to stochastic gradient methods. Finally, the root-mean-squared error (RMSE) is used for comparisons.

Early stopping of training is important to prevent overfitting in neural networks, and therefore, the number of training iterations is a hyperparameter that needs to be calibrated [46]. It was found that the validation error was minimal after five iterations.

L2 regularization was tested but did not significantly increase generalization potential in this application.

Number of Layers and Neurons

Figure 3 shows a sensitivity analysis of the number of neurons for two selected feature sets. In Figure 3a, only hardness-related features are used, whereas Figure 3b includes more features. Given the stochastic processes involved during training, each network configuration is repeated 20 times using random initializations of weights. This procedure results in a sample of errors that are shown by boxplots.

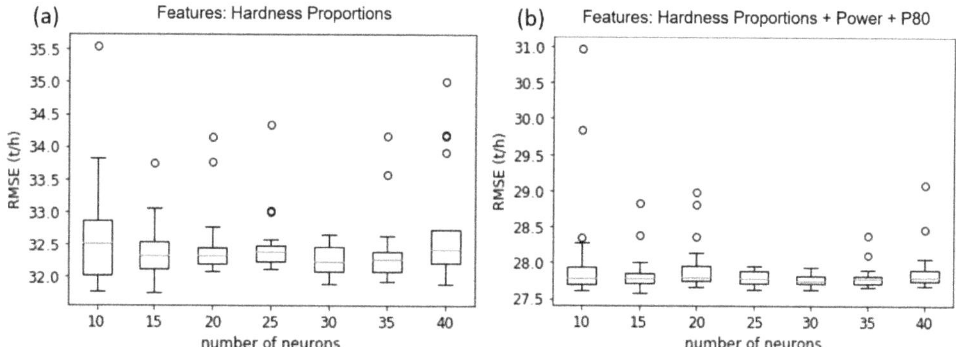

Figure 3. Comparison of the number of neurons for two selected feature sets: (**a**) hardness proportions and (**b**) hardness proportions, ball mill power and product particle size (P80).

Figure 3 shows that the average error and error variance reduce for both feature sets as the number of neurons increases. A plateau is reached at 25 to 30 neurons. This is expected since a too small number of neurons is not able to adequately map the underlying function. Note that this behavior can be observed independently of the number of layers. Two fully connected hidden layers are used in Figure 3a, whereas a single connected hidden layer was used for the sensitivity analysis in Figure 3b. For the best choice of layers, another sensitivity analysis is performed by varying the number of hidden layers from one to four. Figure 4 shows the results performed on the same selected feature sets.

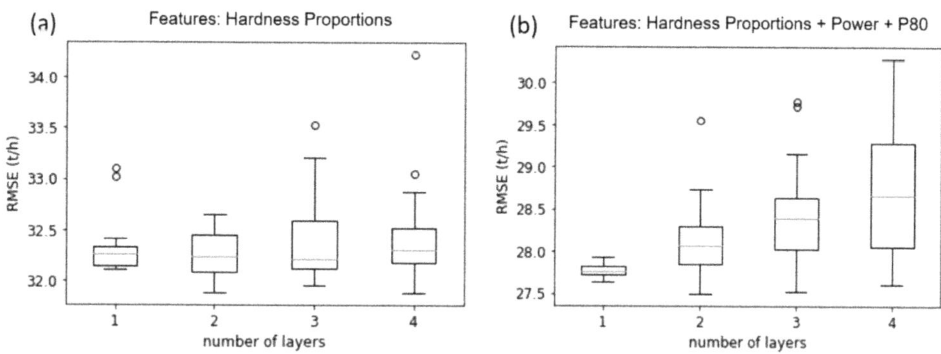

Figure 4. Comparison of the number of hidden layers for two selected feature sets: (**a**) hardness proportions and (**b**) hardness proportions, ball mill power and product particle size (P80).

Figure 4 indicates that one hidden layer delivers the most stable results on all tested feature sets. Although the addition of more layers can reduce the error in individual runs, as seen in Figure 4a, the network appears more prone to overfitting and the error variance increases. For larger feature sets (Figure 4b), overfitting appears to be exacerbated the more layers are used. The obtained results demonstrate the strength of parsimony of parameters (POP), as the model with the smallest size (i.e., one hidden layer) performs best.

4. Results and Analysis

Section 4 is subdivided into two separate parts that aim to analyze the effects of different feature sets on throughput prediction, and then benchmark the presented neural network against a multiple regression model. Section 4.1 addresses the prediction of ball mill throughput using hardness-related variables only. In Section 4.2, pertinent comminution variables are added individually, and their effect on throughput prediction is evaluated.

4.1. Hardness-Related Variables (Effect of Non-Additivity)

This subsection aims to answer how different ways of informing about the hardness and grindability of the geological reserve using penetration rates from blasthole drilling perform for throughput prediction. Specifically, the prediction potential of the average rate of penetration (ROP) is compared to the prediction behavior of hardness proportions created using penetration rate intervals. Figure 5a shows a graphical comparison of ball mill throughput (left axis) and average ROP of the processed ore (right axis). Figure 5b,c illustrates the evolution in time of two distinct hardness proportions compared to throughput, and are discussed subsequently.

It can be seen in Figure 5a that average ROP follows ball mill throughput well in many periods of the observed time horizon. Together with the strong positive correlation reported in Table 2, the similar behavior of both variables in Figure 5a confirms the hypothesis that penetration rates recorded by drilling machines can contribute to informing the comminution performance and grindability of the processed ore. Next, this feature is tested using 20-fold cross-validation. The performance of average ROP as a single feature for throughput prediction is shown in Figure 6a (neural network) and Figure 6b (multiple regression).

When comparing Figure 6a,b, there appears to be no obvious advantages of the neural network compared to multiple regression, which can be explained by the fact that only one single feature is used. Although following the general trends of throughput in most of the observed time intervals, the results reveal weaknesses in predicting the right magnitudes of low and high throughputs. A possible explanation for this weakness can be found when considering penetration rates as a non-additive variable. Non-additivity is present if linear averages of a variable, for instance penetration rates of two separate rock entities, are different from the expected value of the combined (blended) sample. Thus, taking mathematical averages can be detrimental to such variables. Other well-known examples are metal recovery [47] and other variables representing product quality [48].

In fact, the feature 'average ROP' has gone through an averaging process twice. First, penetration rates are averaged within a mining block when changing the support from simulated grid nodes (point support) to mineable volumes (SMU) to reflect mine selectivity. This standard process is only innocuous for additive variables such as metal grades (at constant density). Second, a weighted average by tonnage of each truckload is calculated per day, accounting for all sources of material that are blended. For the alternative feature set of hardness proportions, penetration rates in point support are split into several categories using penetration rate intervals. This procedure avoids the averaging of harder and softer parts within the geological reserve. Instead, proportions of softer and harder material are preserved in the ore blends that are processed in the mill (compositional approach). A discussion of how to build hardness proportions and how many hardness categories are needed can be found in Both and Dimitrakopoulos [30].

Figure 5b,c illustrates the evolution of two distinct hardness proportions compared to TPH. Figure 5b shows the proportions of soft material arriving at the mill, informed by the percentage of high penetration rates (greater than 62 m/h) in the ore blend. Here, higher throughputs are expected to occur when more of this soft material arrives at the mill. Indeed, large proportions of softer material in Figure 5b coincide with high mill throughput, which is most visible for days 1–10 as well as for days 170–181 of the observed period. Figure 5c shows the proportions of harder material, which is reflected by penetration rates that fall in the interval of 29 to 32 m/h. Larger proportions of this material category should have a negative effect on throughput. Interestingly, Figure 5c shows that the lowest mill throughput (days 128–133) coincides with the peaking of the fractions of harder material. Conversely, the highest throughput is achieved when the proportions of this harder-to-penetrate material are the smallest.

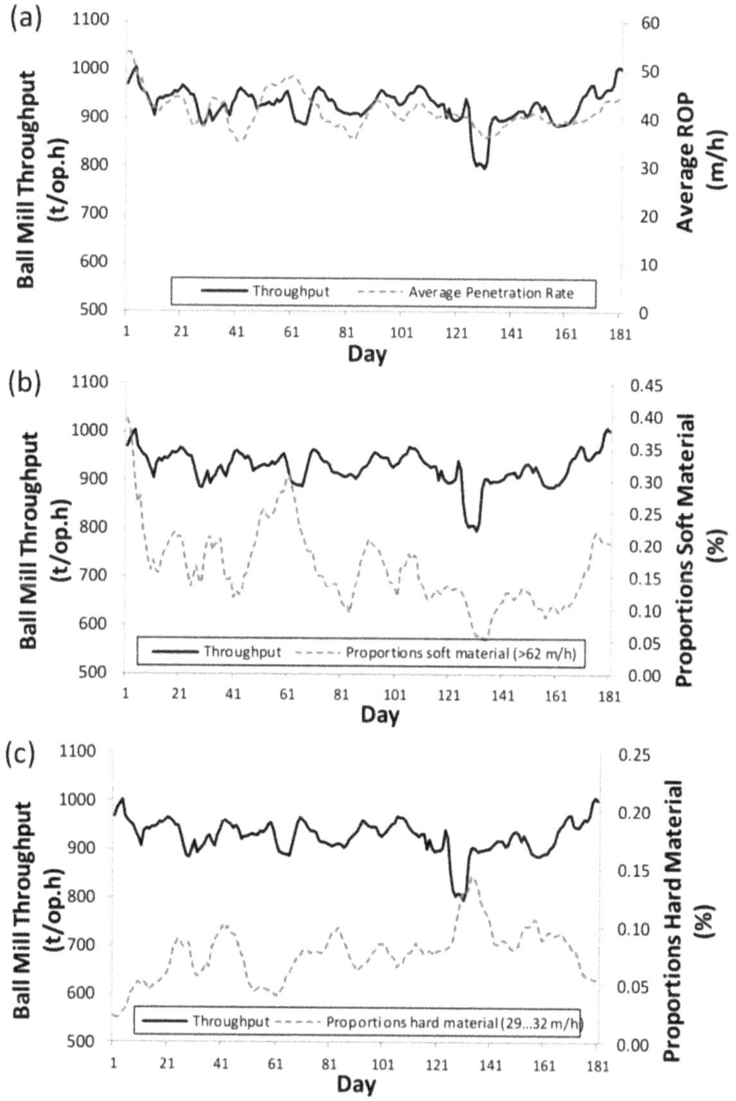

Figure 5. Moving average of ball mill throughput compared to moving average of (**a**) average rate of penetration (ROP), (**b**) proportions of softer material (high penetration rates in the interval >62 m/h), and (**c**) proportions of harder material (low penetration rates falling in the interval of 29–32 m/h).

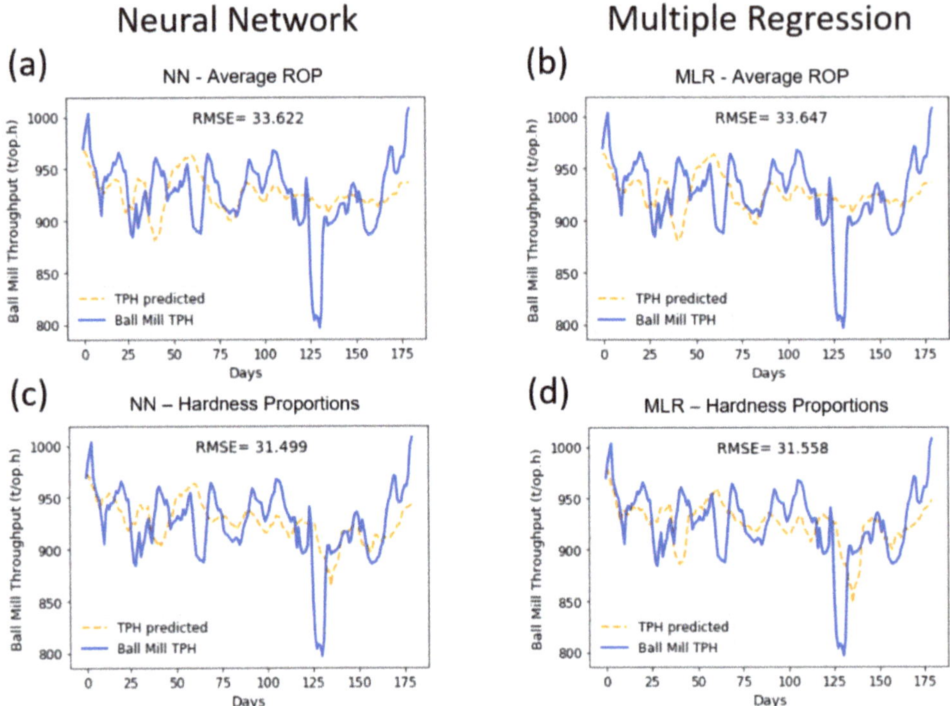

Figure 6. Ball mill throughput prediction (20-fold cross-validation) using (**a**) average ROP (NN), (**b**) average ROP (MLR), (**c**) hardness proportions (NN), and (**d**) penetration rate categories (MLR).

The performance of hardness proportions for throughput prediction is shown in Figure 6c (neural network) and Figure 6d (multiple regression). The highs and lows of throughput are more closely predicted, leading to a reduction in the prediction error by 6.3% for both prediction models. This indicates that classification into hardness proportions is advantageous over using a single, continuous hardness variable. The difference between the neural network and the multiple regression model is relatively small.

4.2. Effect of Comminution Variables on Prediction

Several comminution variables were identified as potential candidates to improve throughput prediction in Sections 2 and 3. In this subsection, the hardness feature set comprising hardness categories is enhanced by one additional comminution variable at a time. To analyze the effects of the neural network, a comparison to a multiple linear regression model is provided for each experiment.

4.2.1. Ball Mill Power

The ball mill power measurements showed the potential to improve the prediction of ball mill throughput due to its proportional relationship to TPH in Bond's law (Equation (1)) and its strong correlation in the present dataset shown in Table 2. Figure 7a shows a graphical comparison between the daily average power draw of the ball mill and TPH. Power draw stays mostly constant for the observed time horizon, including some distinctive drops in power in the second half of the observed time horizon. These power drops tend to occur at times when the mill throughput decreases as well. It is thus not surprising that adding ball mill power as a feature for throughput prediction especially enhances the periods of sharp throughput decrease, as shown in Figure 8a.

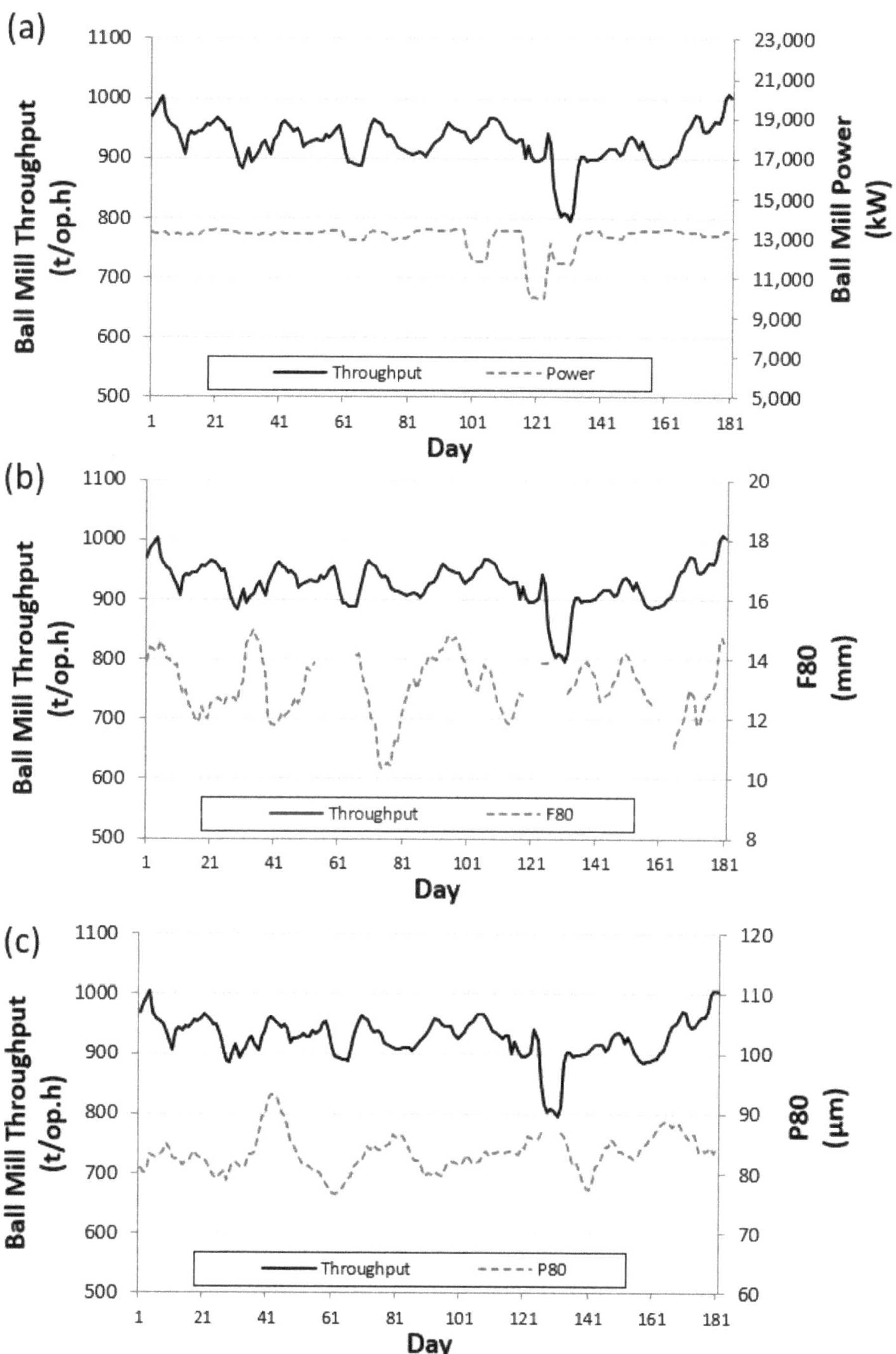

Figure 7. Moving average (7 days) of ball mill throughput compared to moving average of (**a**) ball mill power (**b**) feed particle size: (F80), and (**c**) product particle size (P80).

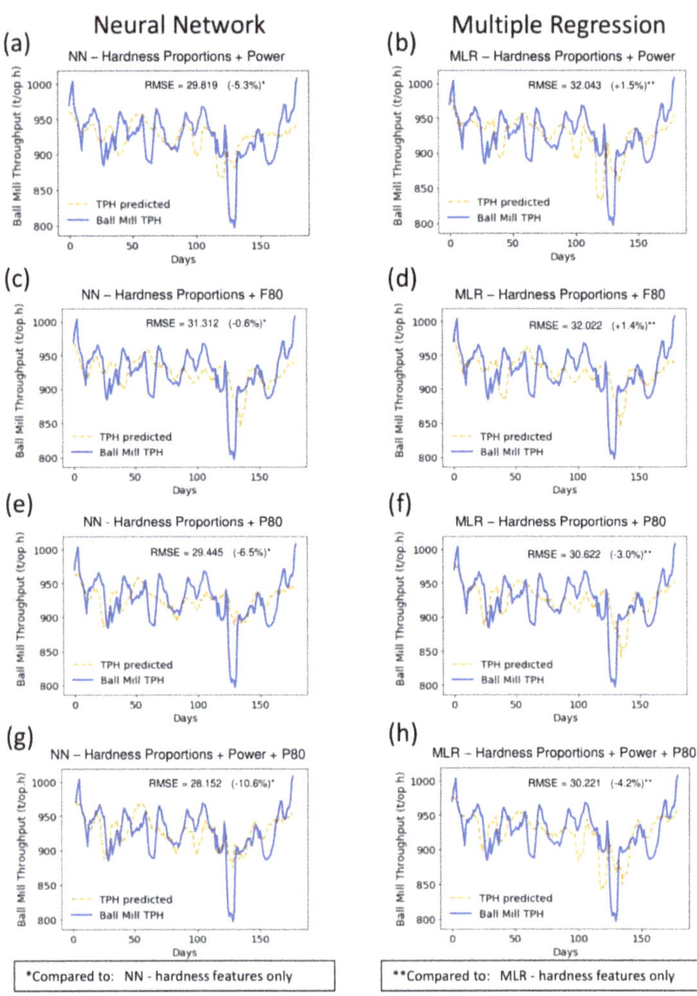

Figure 8. Ball mill throughput prediction (20-fold cross-validation) using as additional features: (**a**) ball mill power (NN), (**b**) ball mill power (MLR), (**c**) feed particle sizes (NN), (**d**) feed particle sizes (MLR) (**e**) product particle sizes (NN), (**f**) product particle sizes (MLR) (**g**) power and P80 (NN), (**h**) power and P80 (MLR)–RSME is compared in brackets to respective model predictions (neural network/multiple regression) using hardness features only.

By comparing the predictive performance of the neural network (NN) with the performance of the multiple linear regression model (MLR) in Figure 8b, the superiority of the neural network becomes apparent. MLR overestimates the influence of ball mill power, seen in the sharp decrease in days 120–125. The neural network predicts closer to the true throughput, which can be noticed visually and statistically. Compared to the sole utilization of hardness proportions (Section 4.1), the RMSE decreases by 5.3% when using the neural network, whereas the error for MLR rises by 1.5%.

4.2.2. Particle Sizes

Compared to ball mill power measurements, particle size measurements indicate a low empirical correlation in the present dataset between particle sizes and TPH (Table 2).

The theoretical relations to throughput (Equations (5), (8) and (9)) cannot be confirmed by visual analysis in Figure 7b,c alone. The graphs also show a large amount of missing data, especially for feed particle size (F80) measurements. No visible trends are recognizable.

By comparing the prediction behavior when adding particle sizes in Figure 8c–f, the following conclusions may be drawn. Adding F80 measurements seems to not significantly enhance throughput prediction in this case study since the RMSE decreases only marginally when using the NN (-0.6%, Figure 8c) and increases when using MLR ($+1.4\%$, Figure 8d). The addition of product size measurements (P80) seems to have a positive effect on throughput prediction in this case study, which is noticeable for both prediction models. However, the NN prediction error (-6.5%) in Figure 8e reduces notably more than the MLR prediction error (-3.0%) in Figure 8f, showing the superiority of the NN when dealing with nonlinear features. The biggest gain in prediction accuracy can be obtained when using both well-performing features, power draw and P80, together. Here, the strengths of the neural network become most apparent, showing the lowest error in Figure 8g and a 10.6% error reduction compared to ore hardness only. The MLR also shows the lowest recorded error (-4.2%, Figure 8h), but the error decreases much less than the NN. To summarize, the more features are added, the better their interdependencies can be interpreted by NN.

5. Discussion

Next to the superior performance of hardness proportions combined with power draw and product size measurements, the results obtained above show that the use of neural networks can decrease the ball mill throughput-prediction error compared to using multiple regression. Short-term decision making, such as short-term mine production scheduling, can benefit from the demonstrated improvements in throughput prediction presented in this article. A conventional short-term production schedule for the Tropicana Gold mining complex is shown in Figure 9.

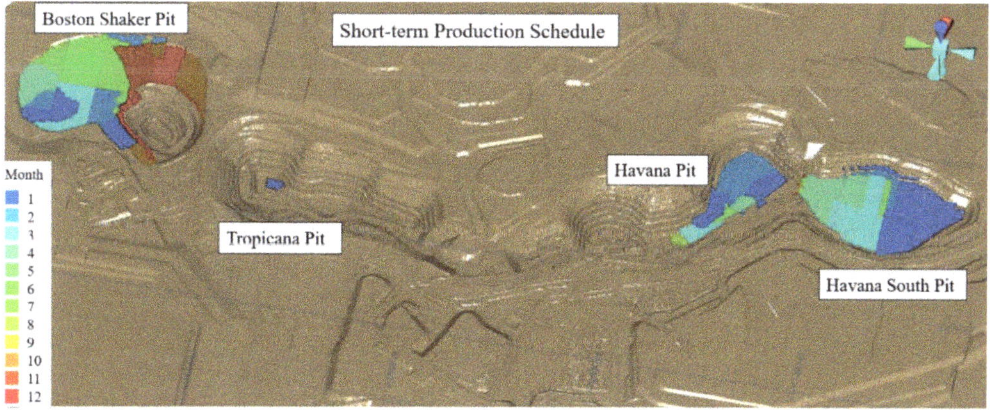

Figure 9. Example of a monthly short-term production schedule in the Tropicana Gold mining complex.

As can be seen in Figure 9, short-term extraction can take place in multiple pits and different mining areas within the pits in the same period of extraction, leading to blended material streams at the processing plant(s). As a recent development in short-term mine planning, the incorporation of a geometallurgical throughput-prediction model into short-term production scheduling has been demonstrated in Both and Dimitrakopoulos [30]. Instead of building predefined throughput estimates per mining block, the authors predict the ball mill throughput of blended materials using a multiple regression model, and use these predictions for short-term production scheduling in a stochastic optimization model. Figure 10 illustrates how the trained neural network in this article, together with comminution variables at the ball mill, can replace the multiple regression model for production scheduling optimization.

Key Constraint per Period *t* and Orebody Scenario *s* :

$$Schedule\_Tonnes_{s,t} - Throughput_{s,t} \cdot Hours_{Mill} - d_{s,t}^+ + d_{s,t}^- = 0$$

Multiple Linear Regression (Both and Dimitrakopoulos [30]) :

$$w_0^* + \sum_j w_j^* \cdot Hardness\_proportion_{j,s,t} = Throughput_{s,t}$$

Neural Network proposed in this Article :

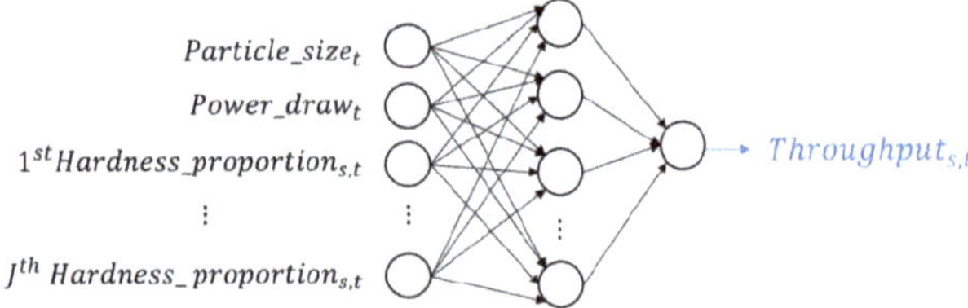

Figure 10. Comparison of models for ball mill throughput prediction and integration into short-term production scheduling.

The stochastic constraint shown in Figure 10 ensures that for every period and simulated orebody scenario, the scheduled ore tonnage equals the tonnage resulting from the predicted hourly throughput and available mill hours. The deviation variables, $d_{,t,s}^+$ and $d_{,t,s}^-$, penalize deviations between the scheduled tonnage and realizable mill tonnage in the objective function of the mathematical program, which is discussed in detail in Both and Dimitrakopoulos [30]. The hardness proportions serving as input to the neural network represent the weighted hardness proportions of the materials to be scheduled together in a single short-term period. Furthermore, the planned power draw, as well as the planned feed and product particle sizes for the future scheduled materials, can now serve as input to the production scheduling optimization, since the neural network has been trained on these attributes. Note that nonlinear production-scheduling formulations combined with a metaheuristic solution method, such as simulated annealing, can handle these internal nonlinear computations in the optimization process, which have been developed for long-term and short-term planning [31,49].

6. Conclusions

This article shows a case study at the Tropicana Gold mining complex that demonstrates improvements of a geometallurgical throughput-prediction model using collected production data in mines and processing plants, combined with supervised machine learning. The key improvements over a previous publication are: (i) including and testing the influence of measurements in the comminution circuit that likely affect ball mill throughput rates in a nonlinear way, (ii) utilizing a supervised learning model in the form of a neural network to approximate nonlinear relationships between predictor and response variables, and (iii) testing if compositional approaches can account for non-additive geometallurgical variables better than average-type information. Finally, recommendations are given on how to integrate the prediction model into short-term production scheduling.

Results show that adding ball mill power draw and product particle size measurements can decrease the prediction error of throughput by 10.6% compared to throughput prediction using geological hardness variables only. This result can only be achieved with the trained neural network, whereas the linear regression model shows improvements of up to 4.2%. Available feed size measurements in the presented case study appear too imprecise to positively affect the throughput prediction. A neural network structure of one hidden layer comprising 30 neurons delivers the most stable predictions and shows the lowest error variance. However, the advantages of the neural network are partly offset by the more time-intensive hyperparameter search compared to the linear model, which is easy to apply and shows comparative performance in some cases.

Finally, hardness proportions decrease the prediction error compared to the use of averages of penetration rates. This underlines the importance of compositional approaches for non-additive geometallurgical variables. A key takeaway is that the shown compositional approach is not limited to ore hardness variables. Instead, it is conceivable to utilize compositional approaches for other non-additive (geometallurgical) variables as well.

Future work aims to create more data-driven prediction models of metallurgical responses in mining complexes using production data generated in the mines and processing plants. Next to the demonstrated prediction of comminution performance, the data-driven prediction of metal recovery, consumption of reagents, and other revenue and cost factors should be considered. The integration of these prediction models into decision-making processes, such as short-term production scheduling, is pertinent for meeting key production targets in mineral value chains.

Author Contributions: Conceptualization, C.B. and R.D.; methodology, C.B.; software, C.B.; validation, C.B. and R.D.; formal analysis, C.B.; investigation, C.B.; resources, R.D.; data curation, C.B.; writing—original draft preparation, C.B.; writing—review and editing, C.B. and R.D.; visualization, C.B.; supervision, R.D.; project administration, R.D.; funding acquisition, R.D. All authors have read and agreed to the published version of the manuscript.

Funding: This research was funded by the National Science and Engineering Research Council of Canada (NSERC) CRD Grant CRDPJ 500414-16, NSERC Discovery Grant 239019, and the COSMO mining industry consortium (AngloGold Ashanti, AngloAmerican, BHP, De Beers, IAMGOLD, Kinross, Newmont Mining and Vale).

Acknowledgments: Special thanks are in order to AngloGold Ashanti Limited and the Tropicana Gold Mine (AngloGold Ashanti Australia Ltd., 70% and manager, IGO Ltd., 30%), in particular Mark Kent, Tom Wambeke, Aaron Caswell, Johan Viljoen and Louis Cloete for providing the data used in this study and long-standing collaboration.

Conflicts of Interest: The authors declare no conflict of interest.

References

1. Moradi Afrapoli, A.; Askari-Nasab, H. Mining Fleet Management Systems: A Review of Models and Algorithms. *Int. J. Min. Reclam. Environ.* **2017**, *33*, 42–60. [CrossRef]
2. Rai, P.; Schunnesson, H.; Lindqvist, P.A.; Kumar, U. An Overview on Measurement-While-Drilling Technique and Its Scope in Excavation Industry. *J. Inst. Eng. Ser. D* **2015**, *96*, 57–66. [CrossRef]
3. Lessard, J.; De Bakker, J.; McHugh, L. Development of Ore Sorting and Its Impact on Mineral Processing Economics. *Miner. Eng.* **2014**, *65*, 88–97. [CrossRef]
4. Williams, S.R. A Historical Perspective of the Application and Success of Geometallurgical Methodologies. In Proceedings of the Second AusIMM International Geometallurgy Conference, Brisbane, Australia, 30 September—2 October 2013; AusIMM: Carlton, Australia, 2013; pp. 37–47.
5. Dobby, G.; Bennett, C.; Kosick, G. Advances in SAG Circuit Design and Simulation Applied: The Mine Block Model. In Proceedings of the International Autogenous and Semi-Autogenous Grinding Technology—SAG 2001, Vancouver, BC, Canada, 30 September—3 October 2001; Volume 4, pp. 221–234.
6. Bueno, M.; Foggiatto, B.; Lane, G. Geometallurgy Applied in Comminution to Minimize Design Risks. In Proceedings of the 6th International Conference Autogenous and Semi-Autogenous Grinding and High Pressure Grinding Roll Technology, Vancouver, BC, Canada, September 2015; University of British Columbia (UBC): Vancouver, BC, Canada, 2015; Volume 40, pp. 1–19.
7. McKay, N.; Vann, J.; Ware, W.; Morley, W.; Hodkiewicz, P. Strategic and Tactical Geometallurgy—A Systematic Process to Add and Sustain Resource Value. In Proceedings of the Third AUSIMM International Geometallurgy Conference, Perth, Australia, 15–16 June 2016; AusIMM: Carlton, Australia; pp. 29–36.
8. Mwanga, A.; Rosenkranz, J.; Lamberg, P. Testing of Ore Comminution Behavior in the Geometallurgical Context—A Review. *Minerals* **2015**, *5*, 276–297. [CrossRef]
9. Bond, F.C. The Third Theory of Comminution. *Trans. AIME Min. Eng.* **1952**, *193*, 484–494.
10. Bond, F.C. Crushing & Grinding Calculations Part 1. *Br. Chem. Eng.* **1961**, *6*, 378–385.

11. Starkey, J.; Dobby, G. Application of the MinnovEX SAG Power Index at Five Canadian SAG Plants. *Autogenous Semi Autogenous Grind.* **1996**, 345–360.

12. Morrell, S. Predicting the Specific Energy of Autogenous and Semi-Autogenous Mills from Small Diameter Drill Core Samples. *Miner. Eng.* **2004**, *17*, 447–451. [CrossRef]

13. Amelunxen, P.; Berrios, P.; Rodriguez, E. The SAG Grindability Index Test. *Miner. Eng.* **2014**, *55*, 42–51. [CrossRef]

14. Dominy, S.C.; O'Connor, L.; Parbhakar-Fox, A.; Glass, H.J.; Purevgerel, S. Geometallurgy—A Route to More Resilient Mine Operations. *Minerals* **2018**, *8*, 560. [CrossRef]

15. Flores, L. Hardness Model and Reconciliation of Throughput Models to Plant Results at Minera Escondida Ltda., Chile. *SGS Miner. Serv. Tech. Bull.* **2005**, *5*, 1–14.

16. Bulled, D.; Leriche, T.; Blake, M.; Thompson, J.; Wilkie, T. Improved Production Forecasting through Geometallurgical Modeling at Iron Ore Company of Canada. In Proceedings of the 41st Annual Meeting of Canadian Mineral Processors, Ottawa, ON, Canada, 20–22 January 2009; pp. 279–295.

17. Keeney, L.; Walters, S.G.; Kojovic, T. Geometallurgical Mapping and Modelling of Comminution Performance at the Cadia East Porphyry Deposit. In Proceedings of the First AusIMM International Geometallurgy Conference, Brisbane, Australia, 5–7 September 2011; AusIMM: Carlton, Australia, 2011; pp. 5–7.

18. Alruiz, O.M.; Morrell, S.; Suazo, C.J.; Naranjo, A. A Novel Approach to the Geometallurgical Modelling of the Collahuasi Grinding Circuit. *Miner. Eng.* **2009**, *22*, 1060–1067. [CrossRef]

19. Keeney, L.; Walters, S.G. A Methodology for Geometallurgical Mapping and Orebody Modelling. In Proceedings of the First AusIMM International Geometallurgy Conference, Brisbane, Australia, 5–7 September 2011; AusIMM: Carlton, Australia, 2011; pp. 217–225.

20. Morales, N.; Seguel, S.; Cáceres, A.; Jélvez, E.; Alarcón, M. Incorporation of Geometallurgical Attributes and Geological Uncertainty into Long-Term Open-Pit Mine Planning. *Minerals* **2019**, *9*, 108. [CrossRef]

21. Kumar, A.; Dimitrakopoulos, R. Application of Simultaneous Stochastic Optimization with Geometallurgical Decisions at a Copper–Gold Mining Complex. *Min. Technol. Trans. Inst. Min. Metall.* **2019**, *128*, 88–105. [CrossRef]

22. Yan, D.; Eaton, R. Breakage Properties of Ore Blends. *Miner. Eng.* **1994**, *7*, 185–199. [CrossRef]

23. Amelunxen, P. The Application of the SAG Power Index to Ore Body Hardness Characterization for the Design and Optimization of Autogenous Grinding Circuits. Master's Thesis, McGill University, Montreal, QC, Canada, 2003.

24. Garrido, M.; Ortiz, J.M.; Villaseca, F.; Kracht, W.; Townley, B.; Miranda, R. Change of Support Using Non-Additive Variables with Gibbs Sampler: Application to Metallurgical Recovery of Sulphide Ores. *Comput. Geosci.* **2019**, *122*, 68–76. [CrossRef]

25. Deutsch, J.L.; Palmer, K.; Deutsch, C.V.; Szymanski, J.; Etsell, T.H. Spatial Modeling of Geometallurgical Properties: Techniques and a Case Study. *Nat. Resour. Res.* **2016**, *25*, 161–181. [CrossRef]

26. Deutsch, C.V. Geostatistical Modelling of Geometallurgical Variables—Problems and Solutions. In Proceedings of the Second AusIMM International Geometallurgy Conference, Brisbane, Australia, 30 September–2 October 2013; AusIMM: Carlton, Australia, 2013; pp. 7–15.

27. van den Boogaart, K.G.; Konsulke, S.; Tolosana-Delgado, R. Non-Linear Geostatistics for Geometallurgical Optimisation. In Proceedings of the Second AusIMM International Geometallurgy Conference, Brisbane, Australia, 30 September–2 October 2013; AusIMM: Carlton, Australia, 2013; pp. 253–257.

28. Ortiz, J.M.; Kracht, W.; Pamparana, G.; Haas, J. Optimization of a SAG Mill Energy System: Integrating Rock Hardness, Solar Irradiation, Climate Change, and Demand-Side Management. *Math. Geosci.* **2020**, *52*, 355–379. [CrossRef]

29. Kumar, A.; Dimitrakopoulos, R. Production Scheduling in Industrial Mining Complexes with Incoming New Information Using Tree Search and Deep Reinforcement Learning. *Appl. Soft Comput.* **2021**, *110*, 107644. [CrossRef]

30. Both, C.; Dimitrakopoulos, R. Integrating Geometallurgical Ball Mill Throughput Predictions into Short-Term Stochastic Production Scheduling in Mining Complexes. *Int. J. Min. Sci. Technol.* **2021**, 1–37, accepted.

31. Goodfellow, R.C.; Dimitrakopoulos, R. Global Optimization of Open Pit Mining Complexes with Uncertainty. *Appl. Soft. Comput. J.* **2016**, *40*, 292–304. [CrossRef]

32. Montiel, L.; Dimitrakopoulos, R. Optimizing Mining Complexes with Multiple Processing and Transportation Alternatives: An Uncertainty-Based Approach. *Eur. J. Oper. Res.* **2015**, *247*, 166–178. [CrossRef]

33. Park, J.; Kim, K. Use of Drilling Performance to Improve Rock-Breakage Efficiencies: A Part of Mine-to-Mill Optimization Studies in a Hard-Rock Mine. *Int. J. Min. Sci. Technol.* **2020**, *30*, 179–188. [CrossRef]

34. Vezhapparambu, V.; Eidsvik, J.; Ellefmo, S. Rock Classification Using Multivariate Analysis of Measurement While Drilling Data: Towards a Better Sampling Strategy. *Minerals* **2018**, *8*, 384. [CrossRef]

35. Wambeke, T.; Elder, D.; Miller, A.; Benndorf, J.; Peattie, R. Real-Time Reconciliation of a Geometallurgical Model Based on Ball Mill Performance Measurements—A Pilot Study at the Tropicana Gold Mine. *Min. Technol.* **2018**, *127*, 1–16. [CrossRef]

36. Horner, P.C.; Sherrell, F.W. The Application of Air-Flush Rotary Percussion Drilling Techniques in Site Investigation. *Q. J. Eng. Geol.* **1977**, *10*, 207–220. [CrossRef]

37. Sugawara, J.; Yue, Z.; Tham, L.; Law, K.; Lee, C. Weathered Rock Characterization Using Drilling Parameters. *Can. Geotech. J.* **2003**, *40*, 661–668. [CrossRef]

38. Yue, Z.Q.; Lee, C.F.; Law, K.T.; Tham, L.G. Automatic Monitoring of Rotary-Percussive Drilling for Ground Characterization-Illustrated by a Case Example in Hong Kong. *Int. J. Rock Mech. Min. Sci.* **2004**, *41*, 573–612. [CrossRef]

39. Rosenblatt, F. The Perceptron: A Probabilistic Model for Information Storage and Organization in the Brain. *Psychol. Rev.* **1958**, *65*, 386–408. [CrossRef]
40. Hornik, K.; Stichcombe, M.; White, H. Multilayer Feedforward Networks. *Neural Netw.* **1989**, *2*, 359–366. [CrossRef]
41. Hornik, K. Approximation Capabilities of Multilayer Feedforward Networks. *Neural Netw.* **1991**, *4*, 251–257. [CrossRef]
42. Bengio, Y.; Grandvalet, Y. No Unbiased Estimator of the Variance of K-Fold Cross-Validation. *J. Mach. Learn. Res.* **2004**, *5*, 1089–1105. [CrossRef]
43. Hastie, T.; Tibshirani, R.; Friedman, J. The Elements of Statistical Learning. *Bayesian Forecast. Dyn. Model.* **2009**, *1*, 1–694. [CrossRef]
44. Pedregosa, F.; Varoquaux, G.V.; Gramfort, A.; Michel, V.; Thirion, B.; Grisel, O.; Blondel, M.; Prettenhofer, P.; Weiss, R.; Dubourg, V.; et al. Journal of Machine Learning Research: Preface. *J. Mach. Learn. Res.* **2011**, *12*, 2825–2830.
45. Liu, D.C.; Nocedal, J. On the Limited Memory BFGS Method for Large Scale Optimization. *Math. Program.* **1989**, *45*, 503–528. [CrossRef]
46. Goodfellow, I.; Bengio, Y.; Courville, A. *Deep Learning*; MIT Press: Cambridge, MA, USA, 2016.
47. Carrasco, P.; Chilès, J.-P.; Séguret, S. Additivity, Metallurgical Recovery, and Grade. In Proceedings of the 8th International Geostatistics Congress; Santiago, Chile, 1–5 December 2008; p. 10.
48. Coward, S.; Vann, J.; Dunham, S.; Steward, M. The Primary-Response Framework for Geometallurgical Variables. In Proceedings of the Seventh International Mining Geology Conference Proceedings, Perth, Australia, 17–19 August 2009; AusIMM: Carlton, Australia, 2009; pp. 109–113.
49. Both, C.; Dimitrakopoulos, R. Joint Stochastic Short-Term Production Scheduling and Fleet Management Optimization for Mining Complexes. *Optim. Eng.* **2020**, *21*, 1717–1743. [CrossRef]

 *minerals*

Article

Modelling Large Heaped Fill Stockpiles Using FMS Data

Aaron Young * and **William Pratt Rogers**

Department of Mining Engineering, University of Utah, Salt Lake City, UT 84112, USA; pratt.rogers@utah.edu
* Correspondence: Aaron.S.Young@utah.edu

Abstract: The frequent best practice for managing large low-grade run-of-mine (ROM) stockpiles is to average the entire stockpile to only one grade. Modern ore control and mineral processing procedures need better precision. Low-precision models hinder the ability to create a digital mine-to-mill model and optimize the holistic mining process. Prior to processing, poorly characterized stockpiles are often drilled and sampled, despite there being no geological reason for relationships between samples to exist. Stockpile management is also influenced by reserve accounting and lacks a common operational workflow. This paper provides a review of base and precious metal run-of-mine (ROM) pre-crusher stockpiles in the mining industry, and demonstrates how to build a spatial model of a large long-term stockpile using fleet management system (FMS) data and geostatistical code in Python and R Studio. We demonstrate a framework for modelling a stockpile believed to be readily workable for most modern mines through use of established geostatistical modelling techniques applied to the type of data generated in a FMS. In the method presented, each bench of the stockpile is modeled as its own geological domain. Size of dump loads is assumed to contain the same volume of material and grade values that match those of the grade data tracked in the FMS. Despite the limitations of these inputs, existing interpolation techniques can lead to increased understanding of the grade distribution within stockpiles. Using the framework demonstrated in this paper, engineers and stockpile managers will be able to leverage operational data into valuable insight for empowered decision making and smoother operations.

Keywords: stockpiles; operational data; mine-to-mill; geostatistics; ore control; mine optimization

Citation: Young, A.; Rogers, W.P. Modelling Large Heaped Fill Stockpiles Using FMS Data. *Minerals* **2021**, *11*, 636. https://doi.org/ 10.3390/min11060636

Academic Editor: Simon Dominy

Received: 21 April 2021
Accepted: 8 June 2021
Published: 15 June 2021

Publisher's Note: MDPI stays neutral with regard to jurisdictional claims in published maps and institutional affiliations.

1. Introduction

1.1. Objective

As mines are depleted and average global ore grades decline, the desire for an improved working model for stockpile management is likely to grow [1]. Fortunately, digitization efforts have increased throughout the mining industry in recent years, and there is already pressure for mining companies to make use of their new data [2]. Stockpiles are composed of not just valuable material, but also valuable and often misunderstood data, which makes them prime targets for the benefits of digital innovation [3]. This paper provides a review of base and precious metal run-of-mine (ROM) pre-crusher stockpiles in the mining industry, and demonstrates how to build a spatial model of a large long-term stockpile using fleet management system (FMS) data and geostatistical code in Python and R Studio.

1.2. Background

Early base and precious metal mining endeavored to extract the richest ores available. In this scenario, miners would send rich ore directly to milling facilities. With the advent of industrial mining during the 19th century, partly due to Watt and the steam engine [4], mines scaled up into large tonnage operations capable of processing large tonnages of ore. For the first time, low-grade ores could be mined profitably [5]. As a result of these larger mines, the need for stockpiling gradually grew until they were justified to be economically advantageous over the course of a mine life [6–8].

According to Yi [7], the initial work around cut-off grades began in the 1960s and incorporated stockpiles in the 1980s [9]. These works determined that a stockpile will be required for any case in which a mine plans to change cut-off grade over the course of its life. Presently, the stockpiling of ore has become an integral part of mine planning [10,11]. Effective mine planning requires a detailed understanding of the ore deposit. This understanding is then leveraged into a production strategy that considers both the current mining environment as well as predictions of future demand, commodity prices, and other conditions.

Much work has been carried out to determine when base and precious metal mines should stockpile (cut-off grades, net present value (NPV) calculations, rehandle costs, milling costs, production contracts, mine planning, etc.) [1,11–17]. The discipline of managing stockpiles throughout mining operations to maximize the profit gained from their eventual processing is still emerging [18]. Often stockpiles are created as an afterthought or side effect to increased production. Occasionally, they are inherited from previous operators upon the purchase of a mine [19]. In some operations, stockpiles are required to fit into a space confined to a permit area they were not originally intended for.

The use of stockpiles has become paramount to the economic viability of production strategies for several reasons, which include the following:

1. For strategic leverage in labor negotiations [20,21],
2. As a hedge to price changes and market conditions [22],
3. To buffer against production variability [1],
4. As a way to gain additional income after the mining of high-grade ores or at the end of the mine life [11,12],
5. For ore blending purposes to ensure product quality [1,11,13,14], and
6. To counteract geological uncertainty [23].

1.3. Classification of Stockpiles

As is the case with many processes, blending or mixing is a necessary step to ensure optimal product quality. Typically, these blending processes are performed after the primary crusher step. Within mining, pre-crusher stockpiling is often used for its operational simplicity, but it typically lowers the confidence of the ore grade and reduces certainty in feed quality [15]. Pre-crusher stockpiling takes on five forms which are illustrated in Figure 1 [15–17]:

Just as there are operational differences between iron, coal and base and precious metal mining, there are also differences in how these operations stockpile. For example, iron ore is often shipped directly from the mine to the customer. Since the tonnages involved are immense, extensive work has already been carried out to model the way that iron ore should be stockpiled for re-handle [16,17,24,25]. Likewise, due to hazards around coal (loss of energy through oxidation, spontaneous combustion, etc.) and since it is typically directly shipped, there has been much work dedicated to understanding how it behaves during stockpiling [26].

For large-scale base and precious metal mine stockpiles, however, little scientific literature exists to document how to spatially model multivariate distribution of characteristics such as grade, hardness, rock type, and other geochemical properties. Kasmaee et al. [27] demonstrated a method using sampling from the base of stockpiles in conjunction with monthly surveying to create a kriging of an iron ore stockpile in Iran. Morley and Arvidson [28] mention that stockpile modelling typically involves arithmetic weighted averages of grade and tonnage values determined from blast hole samples and survey volume records. They also state that spatial modelling of stockpiles is performed manually using engineering software tools. The disadvantage of these manual models is that they require constant engineering resources which could be spent elsewhere. They are also subject to human error, risk of employee turnover and inconsistencies in their design method. Furthermore, base and precious metal stockpiles are classified as temporary storage [29] and their contents shift over time. Dozers and other heavy equipment frequently displace stockpiled material from its original dump location, making it hard to know where material is located within the stockpile.

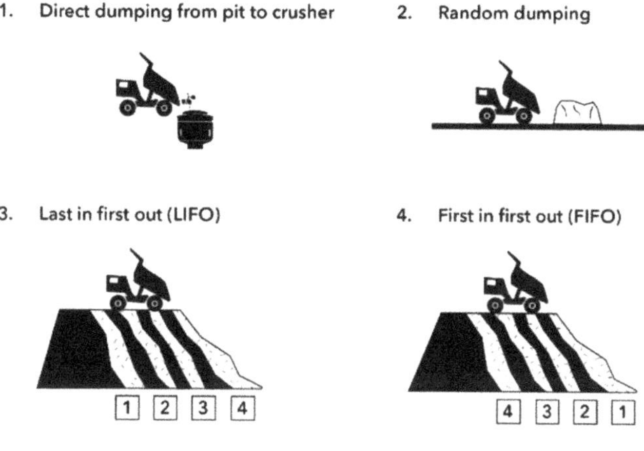

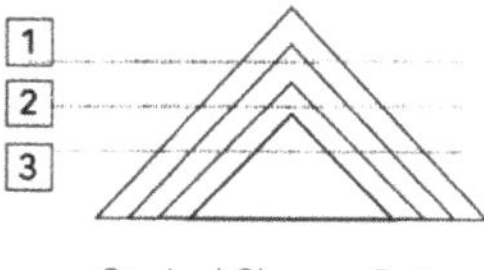

Figure 1. Five Types of Pre-Crusher Stockpiles (adapted from [15–17]).

Another way of classifying stockpiles is by their construction design. Figure 2 shows the different types of stockpile constructions which are common in mining operations.

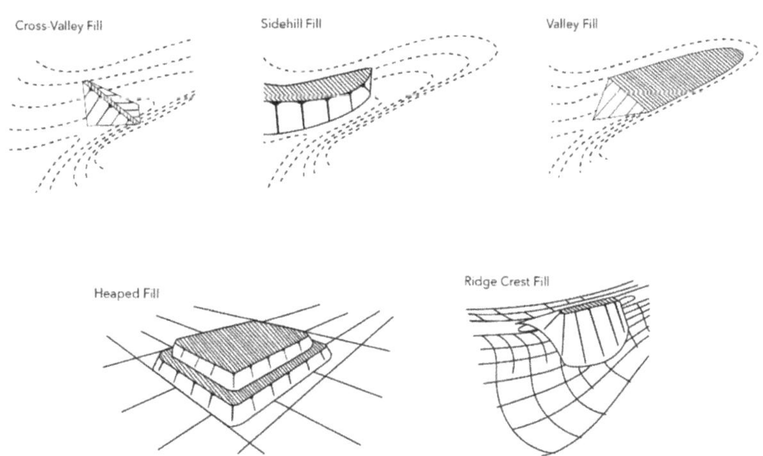

Figure 2. Types of Stockpile Fill (adapted from [29]).

According to Carter, solutions for stockpile management are built into roughly a dozen comprehensive mine scheduling software platforms [18]. These software systems integrate with short-term and small-scale stockpiles, but they do not serve well for large, long-term, truck-dumped stockpiles. Generally, the stockpile modelling software that exists is tailored to conveyor systems for intermediate stockpiles [24,25], but it may be possible to apply some of their concepts to larger stockpiles. Discrete element modelling (DEM) [30] has demonstrated the ability to model material flow of intermediate stockpiles, but remains unproven for large stockpiles where particles are exceedingly numerous and variable.

In addition to modelling, documented technical methods for tracking ROM ore from haul trucks into large stockpiles and onward through downstream processes are perfunctory. Surface base and precious metal mines typically use global positioning systems (GPS) to track haul truck movement and fleet management systems (FMS) to optimize productivity. Afrapoli and Askari-Nasab [31] provide a detailed and current review of FMS technology. FMS systems perform best for tracking material directly dumped from pit to crusher, but lack the capability to track ore once it has been dumped into a stockpile.

Material tracking technology exists, but is rarely used in large stockpiles due to the long duration of time from the placement of the ore to processing. Jansen et al. [32] demonstrated proof of concept case studies involving SmartTag™ radio-frequency identification (RFID) tracer technology for both source-to-product as well as process hold-up ore tracking for the Northparkes mines in Australia. Jurdziak et al. [33] discuss the challenges, propose a workflow and estimate savings of 2.5 million euros for implementing RFID ore tracking at the KGHM Lubin mine in Poland. In addition to RFID, unmanned aerial vehicle (UAV) technology, while promising for its ability to determine stockpile volumes [34,35], is still incipient for the purposes of predicting grade distribution within the stockpile or tracking material movement.

Failing to adequately track and model ore stockpiles results in a loss of data already gained from geological exploration and mine production as illustrated in Figure 3. Such a failure represents not only data and money wasted, but also an opportunity lost in saving additional money during the processing of the stockpiled materials. This is a double lost opportunity for the operator who is probably already operating on relatively thin margins.

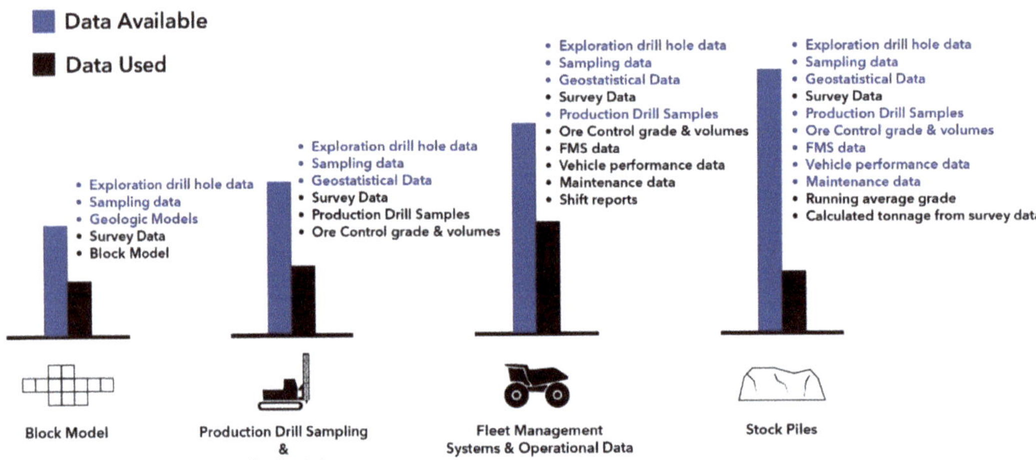

Figure 3. Digital Gaps in Different Phases of Mining Operation.

Figure 3 illustrates a conceptual loss of data associated with stockpiled material that was previously available from exploration models and operational logs. Moreover, the figure illustrates how information is added about the material at each processing step, including production drill sample analyses, ore control modelling and fleet management system information, until the time of stockpiling. Data from each of these operational steps potentially add value to the material if appropriately used, but it is also wasted if the material is not tracked into the stockpile.

If a modern mine can be considered as a digital system of production, stockpiles may be thought of as digital bottlenecks, since they output less information into downstream processes than was input into them [36]. According to Kahraman et al. [37], identifying, tracking, and managing bottlenecks will enable significant improvement in mining operations. However, data gaps, such as stockpiles, short circuit the bottleneck tracking capabilities of sites not to mention, automation, mine-to-mill, and other operational improvement initiatives. While missing data can be thrown out to little detriment for estimation purposes (grade, tonnage, amount of explosive needed, etc.), they need to be included for predictive purposes (capital asset performance, maintenance, throughput, mine-to-mill, etc.) [38].

1.4. Reconciliation and Metallurgical Accounting

Despite little documentation on technical modelling methods, literature contains robust discussion on stockpiles as they pertain to reconciliation and metal accounting [32]. After the Sarbanes-Oxley Act of 2002 [39], the requirement for real-time disclosure of financial conditions to stakeholders in the mining business has inspired more systematic approaches to reconciliation and therefore more scrutiny of stockpiles. Macfarlane [40] contends that a full understanding of metal flow is necessary in order for there to be a systematic approach to reconciliation. Random stockpiling without a clearly defined process map can make accounting for tonnage discrepancies futile. Misunderstanding of stockpile characteristics reduces conversion of mineral resource inventory into saleable product. Motivations for stockpile misunderstandings, underappreciation and neglect are numerous, but generally include issues from the following categories [40,41]:

- Sampling accuracy,
- Data recording and material tracking,
- Geological modelling and estimation errors,
- Short-term and intermediary stockpile accounting,
- Mine design and mine planning,
- Grade control,
- Dispatch,
- Survey inconsistencies,
- Interdepartmental communication,
- Mine operations,
- Management,
- Training and turnover of employees, and
- Dilution, natural leaching and other factors.

Ghorbani and Nwalia [41] affirm that mass flow measurement, stream sampling, mass balancing, and data handling and reporting are the four components of metallurgical accounting. Large, long-term stockpiles pose challenges to each of these components. Stockpiles are difficult to measure and sample [28]. They also risk being manipulated for the purposes of balancing the overall deposit metal value, may become orphaned by multiple operational departments and offer low incentives to be accurately recorded or reported on a frequent basis [41,42]. Improvements in technical capacity to model and manage stockpiles will therefore benefit reconciliation and metallurgical accounting. However, these technical improvements must be incorporated into mining workflows for such benefits to be reaped. Fortunately, existing reserve and operational models provide a foundation for the incorporation of a stockpile working model.

1.5. Reserve and Operational Models

The primary method currently used to model stockpile characteristics in mining is averaging the materials stockpiled and reconciling the values against reserve and operational models. Figure 4 illustrates key aspects of reserve and operational models. A reserve model accounts for the entire life of mining operations and is based primarily on exploration drill holes that are widely spaced apart compared to production drilling. This model is updated biannually and contains more risk and uncertainty than the other models. Short term operational models are a type of hybrid between ore control models and the reserve model and are used for the purposes of planning on a one-to-three-year time horizon. Ore control models are made from additional data obtained by production drilling and contain only the information pertinent to areas of immediate mining within the next week to month for operational purposes. While some details vary in the drillhole spacing and the amount of hybridization between the short-term model, the reserve model and the ore control model, most mining operations follow some version of this established working model.

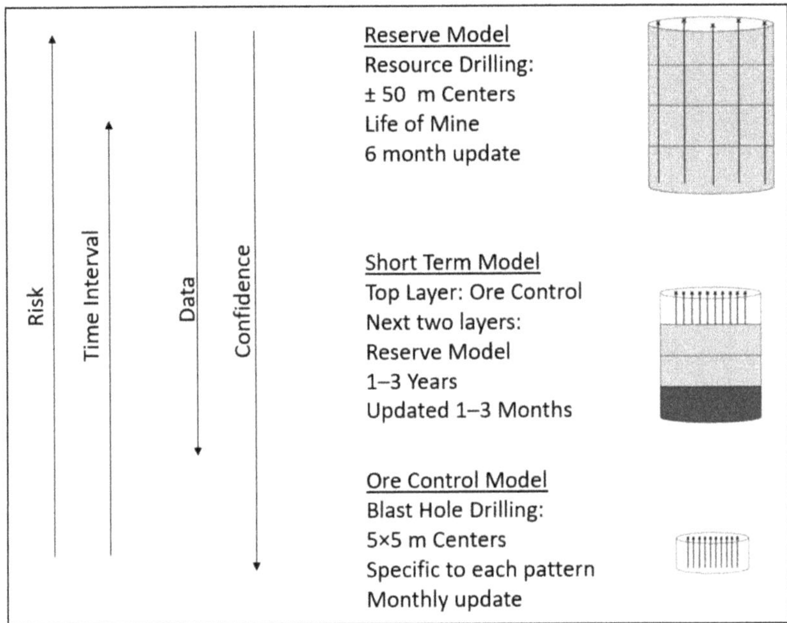

Figure 4. Reserve and Operational Models.

Unlike the reserve and operational models described previously, to the knowledge of the authors, there is no documented working model for large long-term stockpile characteristic distribution, which poses some issues. First, since stockpiles are not income-generating assets, without a working model that is easily implemented, they typically suffer neglect. Moreover, since many short-term discrepancies in ore reserve accounting can be written-off over time under the guise of environmental degradation [42] of stockpiles at the end of the mine life, the absence of a working model may be a disincentive to dedicate engineering resources to their concurrent management. Ultimately, these practices reduce confidence in the characteristics of stockpiled material and lead to a wasteful duplication of time and energy in drilling, and sampling, when the stockpile is finally set to be processed. Many stockpile working models have been developed manually by mining engineers to meet industry needs. However, there is a knowledge gap between the practical methods developed within industry and scientific documentation.

1.6. Sampling

Many sampling issues have been identified to exist for both stockpiles as well as ore control operational models. In the case of ore control, blast hole sampling from drill collars has undergone intense scrutiny from the sampling community. Engström [43] provides a recent literature review on this scrutiny and states that blast hole sampling problems include loss of fines (or inaccurate particle size distribution representation), upward/downward contamination, influx of sub-drill material, pile segregation, pile shape irregularities, operator-dependent sampling, too small sample size, frozen (or weathered) blast hole cones and non-equiprobabilistic sampling equipment. Further complications to ore control understanding exist due to material movement after blasting. Thornton [44] states that material moves more than 4 m on average and movements of 10 m are common. Depending on the drill pattern, this movement could equate to a displacement from the initial sample location to up to three sample locations away.

The probability of a misclassification of ore to waste or waste to ore from sampling errors alone is commonly between 5% and 20% for base and precious metal mines [45,46]. In addition, ore loss of 9% to 19% due to blast movement and dilution can be expected [44]. The negative impacts of these issues may become more critical when handling precious metals and FMS data do not typically contain adjustments for them. However, the problem of data accuracy is separate from that of data utilization. If FMS data are used more frequently for modelling, then improvements in data accuracy will be encouraged through data feedback loops. Improvements in sensor technology could lead to increased sampling of mining operations and eventually better account for blast movement, dilution and sampling errors. It is probable that incorporating these more accurate measuring methods may help to improve the precision of the entire approach.

1.7. Current Practice and Scope

Industry sources have revealed that current practice is to take a running average of the material characteristics that are fed into the stockpile over time. These running averages are based on the reconciled survey volumes and grade values of materials and are updated concurrent with reserve and operating models. This process is designed to fulfill legal reporting requirements. Despite there being a large amount of data that are tracked in the FMS, the resulting stockpile block model currently in place is merely one large, homogenized block value containing the rolling average grade.

At this time, it is the understanding of the authors that no documented methodology exists in academic literature for determining the grade distribution of large, low-grade, randomly truck-dumped, pre-crusher stockpiles via FMS data. In such cases, drilling and grab sampling of the stockpile are often performed in order to determine the grade distribution and model the stockpile. These techniques have proven to be erroneous and biased [28]. We therefore present and explore an initial approach believed to be readily workable for most modern mines through use of established geostatistical modelling techniques applied to the type of data generated by FMS. This method is part of an ongoing study into developing an engineering methodology for greater understanding of large ROM truck-dumped stockpiles.

The working model for stockpile management conceived in this paper can attenuate the negative impacts of the present stockpiling scenario. It requires minimal engineering resources, is easy to setup and maintain over a long period of time, makes use of readily available data already existing in most modern mines and yields a high amount of detail. Our method could be used as a stand-alone model, or it could be used to enhance or verify other models. This method is software agnostic and can be integrated with mining software packages, as we will demonstrate.

2. Materials and Methods

2.1. Approach

A data-driven approach was used to develop a working model of the characteristics of the stockpile. This approach combined the knowledge of the mine planners with that of the researchers in a process similar to the one demonstrated by Subramaniyan et al. [47]. For the purposes of this paper, as it is an example of the modelling technique, the exact data presented are not real data but exemplify real data seen by researchers and demonstrate a hypothetical stockpile created to show the design method. Figure 5 outlines the methodology used.

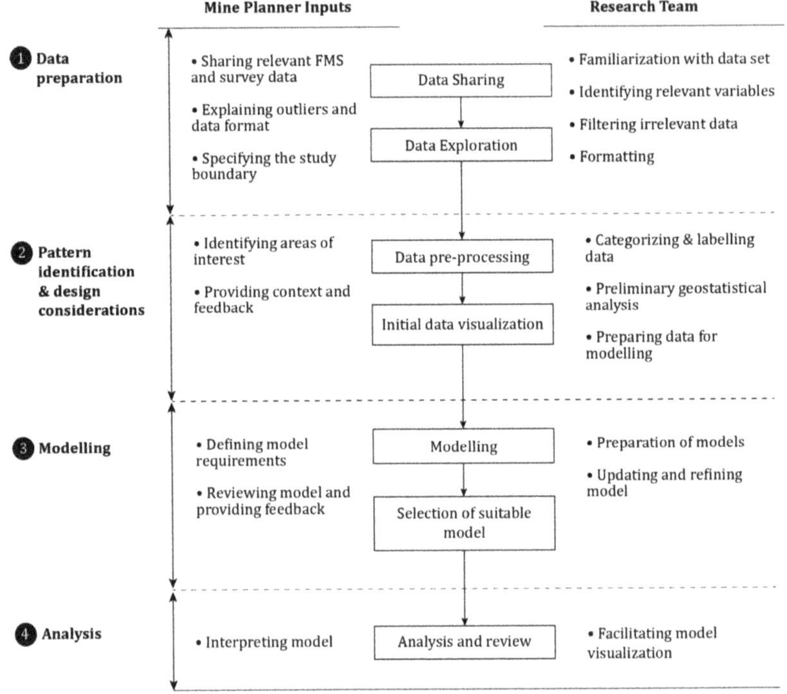

Figure 5. Research Approach Following a Data-Driven [47] Technique.

As demonstrated in Figure 5, the inputs of the mine planners were used during each step of model development. In Step 1 mine planners shared relevant data with researchers, helped clarify questions about the data itself along with explaining any outliers and also defined the project boundary while researchers explored and cleaned the data. During Step 2 researchers presented some early visualizations of the raw data and prepared the modelling concepts. Mine planners provided feedback and context for the data visualization. Step 3 involved preparing and refining the model in accordance with mine planner requirements. Step 4 consisted of a final analysis and review. Each step provides the opportunity for feedback to prior and subsequent steps in the process for the purposes of tuning the final model.

2.2. Data

For the actual case study, a variety of survey, FMS, geological look-up tables, design files, and drone photogrammetry data were made available to the research team. These data have been kept proprietary and this paper contains only a hypothetical stockpile created in likeness of the real data. The flow of data used for creating the model for the actual case study is shown in Figure 6.

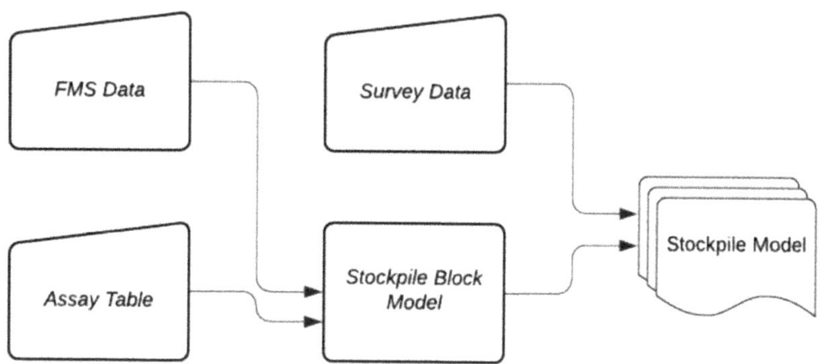

Figure 6. Data Flow Overview Figure.

FMS data came from the mines dispatching system, which contained information about trucks, grade IDs (ore control patterns), dumping locations, dump tonnage, and dumping coordinates. Assay table data contained grade values along with other details about the material such as hardness, rock type, concentration of deleterious elements, etc. Assay table data were matched to corresponding grade IDs in the FMS data. These values were interpolated into a block model by the same method demonstrated subsequently in this paper. Survey data consisted of .dxf files from survey pickups. Survey data were used to create the topographical extents of the block model and ensure that blocks were sequenced correctly. Historical surveys were used to verify that interpolated block values existed matched the real time frame.

A hypothetical dataset was made to illustrate the design method in this paper without revealing the proprietary data of the site. Tables 1 and 2 describe this hypothetical dataset, which represents the results of the combined FMS and assay table data shown in Figure 6. These data are statistically similar to the data used at the case study location and are similar to operational data available to most modern mines. The hypothetical example is that of an open pit gold mine, but the methodology could easily be applied to any other base or precious metal mine.

Table 1. Example FMS and Assay Data Categories.

Timestamp	Grade ID	Au g/t	Actual Dumping Location	Dump Coordinate Easting	Dump Coordinate Northing	Dump Tonnage
Date and Time of Dump	Unique Grade Pattern of Material Dumped	Au Grade value in g/t	Name and bench height of dump location polygon	Easting coordinate of dump location	Northing coordinate of dump location	Tonnage of material dumped

Table 2. Statistical Characteristics of the Modelling Dataset.

Spatial Data			
Number of Points		963	
X min (m)	X Max (m)	Y Min (m)	Y Max (m)
5	145	5	150
Grade Data			
Grade Unit		Grams per ton (g/t)	
Minimum	1st Quartile	Mean	Median	3rd Quartile	Maximum
0.3	0.33	0.3952	0.37	0.47	0.57

As this model is a hypothetical example, survey volumes were designed in Maptek Vulcan for the creation of the stockpile survey data. Maptek Vulcan was chosen only because it was familiar and accessible to both the research team and the mine involved in the study. The model is software agnostic and compatible with any engineering software which uses .dxf file format. Other file formats may also prove compatible with the model in the future. The dimensions of the hypothetical heaped fill stockpile are 150 m × 150 m × 5 m. Figure 7 shows a screenshot of the designed hypothetical stockpile.

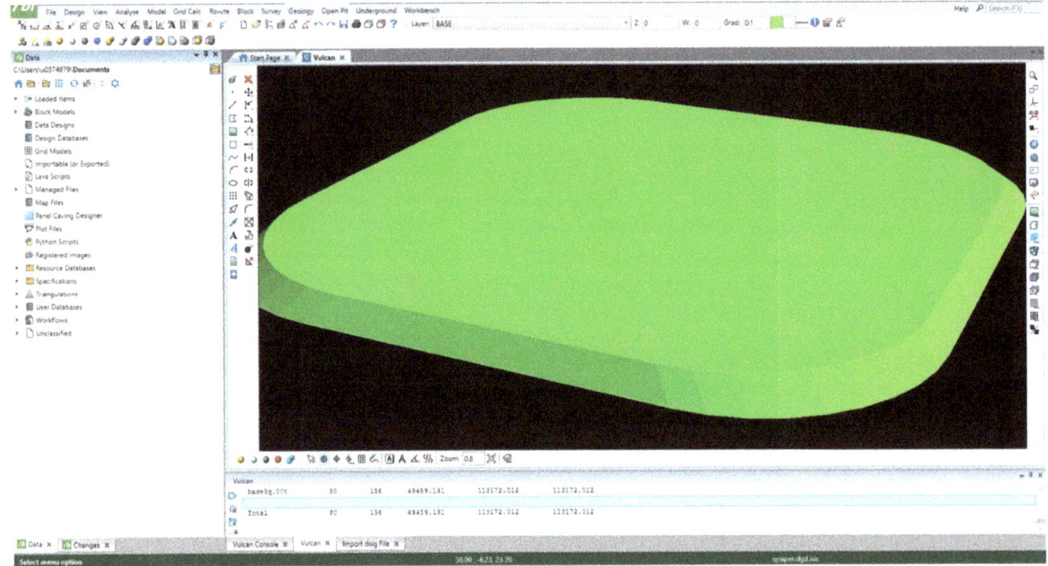

Figure 7. Screenshot of Hypothetical Heaped Fill Stockpile Designed in Maptek Vulcan.

Since the accuracy of each elevation value was limited to the value of the bench level, each bench was modeled as a two-dimensional plane. In the actual case study, each bench was aggregated to create the final block model. For the example shown in this paper, only one bench is demonstrated. While there is some difference in the tonnage values of each truck load in the data, for the purposes of this model, all dump locations were assumed to contain the same amount of material. Only one bench is demonstrated in this paper, but by this method multiple benches can be modelled as independent domains and combined into a block model for an entire stockpile.

2.3. Model

The block model was computed using R scripts running on a Python Jupyter Notebook Kernel. This notebook was developed using concepts from Pyrcz [48]. First, the combined values given by the FMS and assay data were imported into a dataframe. Then, the bench height given by the dump location was converted into an elevation or Z value for each of the dump coordinates. A dataframe consisting of the centroid locations for each block in the desired block model extents was then created. Finally, values for the block centroids were interpolated from the initial dataframe and assigned based on the inverse distance weighting (IDW) function from the gstats package in the R programming language to the centroids of the dataframe representing the block model. These centroids were then imported into Maptek Vulcan to demonstrate the capability of the model to be made useable by an engineering team.

The gstats package, developed by Edzer Pebesma [49], in the R programming language performs a number of common geostatistical functions. Pebesma outlines the modelling approach which should be used with gstats which is shown as Figure 8.

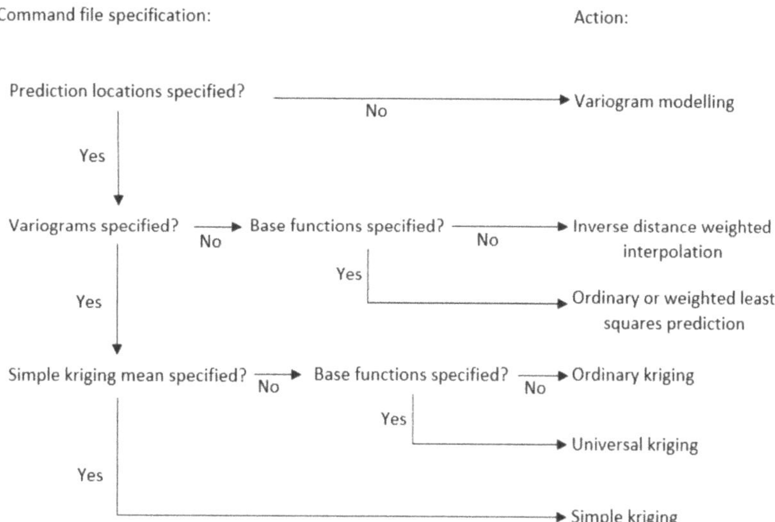

Figure 8. Decision Tree for Default gstats Program Action (from Pebesma [49]).

Inverse distance weighting (IDW) was selected as the modelling method in accordance with Figure 8, in that prediction locations were specified while variograms and base functions were not. Within gstats, the IDW function works the same way as the ordinary kriging function only without a model being passed and instead the inverse distance weighting power is directly specified by the user ($\beta = 2$) (see Equation (1)). A global search neighborhood (default parameter) was used for the model, meaning that all data points were used for estimating the value at each location.

IDW is a form of interpolation. Interpolation means to predict an attribute value $\hat{z}$ from sampled locations xi at unsampled sites (x0) of a given neighborhood [50]. In this case, each of the dump locations along with corresponding grade values ($\hat{z}$) were considered as the "sampled locations" (xi) and the unsampled sites (x0) were an array of block model centroid values (x and y coordinates within the model space at 15 m × 15 m spacings). While the general IDW equation varies slightly [51], for the purposes of the model shown in this paper, the equation used for interpolation is

$$Z_{x,y} = \frac{\sum_{i=1}^{n} Z_i d_{x,y,i}^{-\beta}}{\sum_{i=1}^{n} d_{x,y,i}^{-\beta}} \tag{1}$$

where $Z_{x,y}$ is the centroid point to be estimated, n is the number of samples, z_i represents the value of the ith sample, $d_{x,y,i}$ is the distance between $Z_{x,y}$ and z_i, and β is the user specified weight value.

This paper does not address which geostatistical method is best suited for modelling stockpiles. It is meant to demonstrate that FMS data are capable of being used along with geostatistical methods for greater operational understanding of stockpile characteristics. IDW was selected to show that FMS data are amicable to such methods of interpolation commonly employed in geostatistics. It was also chosen because it runs quickly without requiring additional model parameters as an input, whereas variogram modelling is more time consuming and does not create a block model as a final product. The discussion section of this paper includes more information on how to further refine the initial model.

3. Results

In a heaped fill stockpiling scenario, dumping occurs in two phases. The first phase is a series of paddock dumps to form the base layer of the stockpile. The second phase involves building an upper layer above an area of the paddock dumps from which a campaign of edge dumps occurs until the bench is completed. Our results are broken down into the scenarios described, being first paddock dumping, second edge dumping and third a look at both in combination.

3.1. Data Visualization

Figure 9 shows an initial visualization of the hypothetical dataset. In Figure 9 each dot represents a single dump of similar volume. The positions of the dots represent the FMS tracked dumping position of the truck at the time of dumping. The dot colors represent the interval of their grade values in accordance with the legend shown.

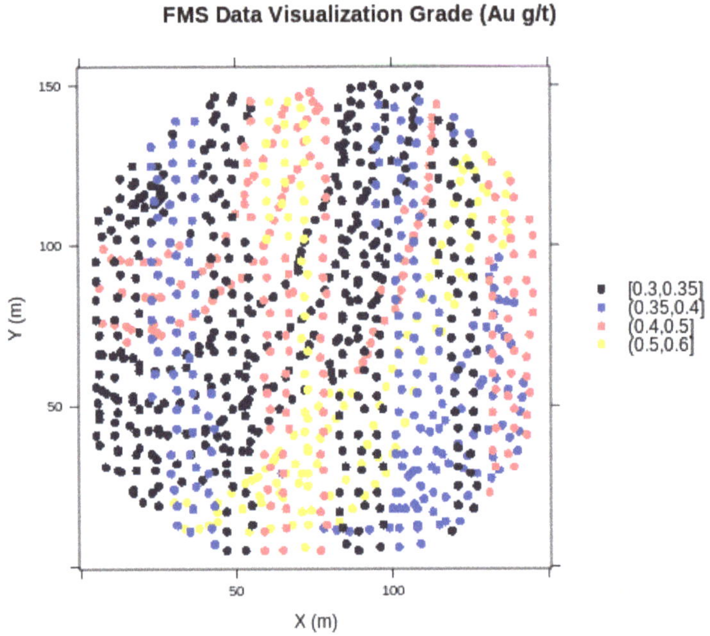

Figure 9. Visualization of FMS Data.

Visualizing FMS data in this manner allows for the identification of some areas of homogeneity within the stockpile. One area would be the bottom left part of the figure, where many black dots are near each other. It also reveals that most of the stockpile contains areas of mixed values and it is not easily visually interpreted into a workable model.

The data from Figure 9 can be broken down into two types of data for greater understanding and better modelling. Each of the data types refer directly to the dumping process used to create the data. The first type of data is termed base layer or paddock dumping data and refers to instances in the data where the material was dumped as a heap on a roughly flat surface. The second type of data is called upper layer or edge dumping data and refers to situations where material was dumped down an existing stockpile face. Each of these data types were also explored and modeled.

3.2. Base Layer/Paddock Dumping Model

Via preliminary data analysis, FMS data which corresponds to paddock dumps and forms the base layer of a stockpile in a hypothetical scenario are shown in Figure 10. In Figure 10 each dot represents a single dump of similar volume. The positions of the dots represent the FMS tracked dumping position of the truck at the time of dumping. The dot colors represent the interval of their grade values in accordance with the legend shown.

Base Layer/Paddock Dumps Grade (Au g/t)

Figure 10. Base Layer Grade Dumps of Example Stockpile Colored by Grade.

As is the case with paddock dumping for the base layer of a stockpile, the rows space evenly and maintain homogeneity along their respective row or column as the material originated from various low-grade areas in the pit of different grade amounts. This scenario occurs at the beginning of each bench of the stockpile or from paddock dumping areas where no upper layer is added. Material dumped this way is only handled by dozers or other heavy equipment in areas around the perimeter of the stockpile. Areas where dumps occurred too close to one another tend to settle into more vacant areas over time.

In the paddock dumping scenario, the area of influence of each truck load is closely related to the dimensions of the haul truck used. Figure 11 illustrates how truck dimensions influence the size and shape of the dumped material.

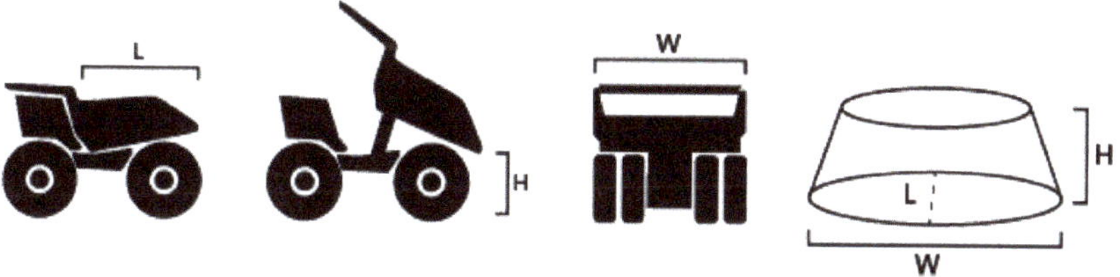

Figure 11. Volume of Influence of Haul Truck in Paddock Dumping Case.

From Figure 11, the resulting heap of material takes on geometric form similar to that of an elliptical frustum. The height, length and width (H, W, L in Figure 11) of the heap are dependent on the respective height, width and length of the haul truck used. These dimensions may be extended in horizontal directions if the haul truck moves excessively during the dumping of its material, which will also coincide with a reduction in the height of the corresponding heap. The width may also expand beyond the original width of the truck if the truck does not move forward. A movement of less than two meters during dumping is a common occurrence and this movement typically only affects the length of the heap, not the width. Movements occur most frequently when a truck has backed up too close to the neighboring heap before beginning its dumping phase, which acts to smooth out the overall average area of influence of multiple paddock dumps for a given area. At the time of dumping, heaps maintain an approximate 2:1 slope, consistent with that of heaped loaded material. Over time the slope decreases to that of the natural angle of repose of the material.

Figure 10 alone is sufficient to create a working model for a paddock-dumped stockpile without the need of additional modelling. If operators have an approximate understanding of where heaps of given ore values are located, they can easily match those locations to the corresponding heaps during operation. Blending the example shown in Figure 10 can be intuitively performed by processing the stockpile in parallel vertical approaches as needed.

3.3. Upper Layer/Edge Dumping Model

Figure 12 shows hypothetical FMS data for the second phase of heaped fill stockpile construction. Like Figure 10, each dot in Figure 12 represents a single dump of similar volume. The positions of the dots represent their FMS tracked position. The dot colors represent the interval of their grade values in accordance with the legend shown. During this phase, material is dumped on top of the material of the base layer. However, data from the previous layer have been removed to facilitate visualization.

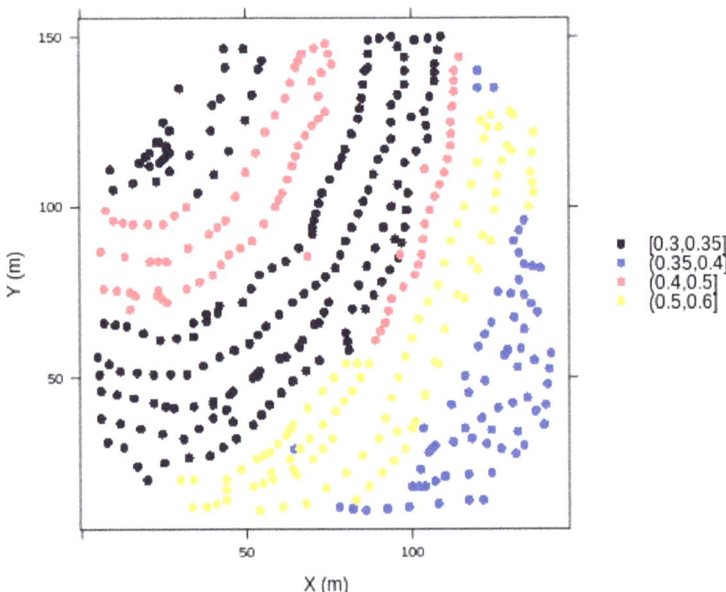

Figure 12. Upper Layer Dumps of an Example Stockpile Colored by Grade.

Figure 12 demonstrates that many dumping locations are clustered together in the initial area where the upper layer of the stockpile is first made (top left of Figure 12). Afterwards, the stockpile is built by edge dumping material from the initial dumping area until it fills the designed volume for the given bench. Dozers and other heavy equipment help to ensure that material is cascaded without clumping, ensure compaction and maintain safety berms during operation. These actions mix the material from its initial dumping location in intractable ways.

While the FMS data in the previous phase could be defined by orderly row and column behavior, within this upper layer, the dumping process is defined by radial progression from an initial cluster point. Variation in the grade distribution is therefore defined by the tendency of the material to be added in sweeping radial movements which correlate with low-grade patterns of various grades in the pit.

Generally, its assumed that the volume of influence for of the haul truck in the upper layer/edge dump case is that of dimensions in width equal to the width of the haul truck, length equal to the horizontal component of the bench slope and variable height which is influenced by width, length and material characteristics. This volume runs perpendicular to the tangent of the dump location and is illustrated in Figure 13.

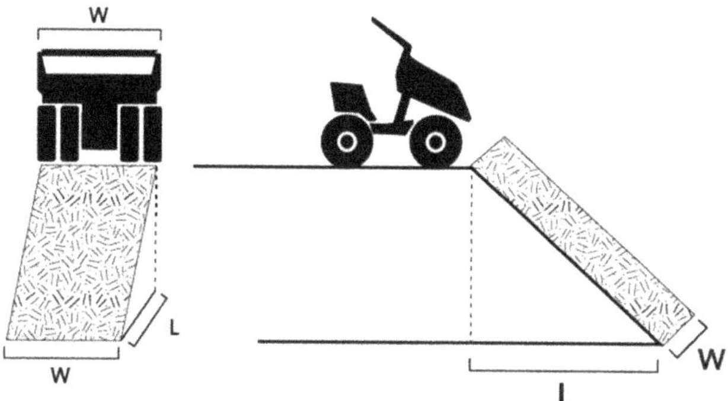

Figure 13. Assumed Area of Influence of Haul Truck in Edge Dumping Scenarios.

While the total volume of material is influenced by the capacity of the haul truck, during this phase the shape of the material dumped is mostly influenced by bench height and material characteristics. When edge dumping occurs normally, that is without rehandling from dozers or other equipment, the resulting shape is a streak of material cascaded along the entire face of the stockpile. This cascade of material typically aggregates more at the bottom of the dumping area under normal conditions and less near the top crest of the dump. Depending on the face of the stockpile larger amounts of material may clump at various places along the face. Round and large material may also roll beyond the floor of the bench, especially at higher bench heights. The authors recognize that these are initial assumptions and more discussion on improvement of the model may be found in the discussion section of this paper.

It should be noted that paddock dumping still occurs on top of the dump area during the edge dumping phase. These dumps are usually to patch and level the floor of the dump area, create safety berms or protect light stations. Occasionally, paddock dumping occurs near the edge before subsequentially being pushed over by a dozer. Operator error may cause trucks to misalign with the edge of the dump creating a more turbulent cascade. Such dumping scenarios are extremely difficult to identify from the data alone.

While it does not offer an operational model on its own, Figure 12 still serves as a starting point. Most homogeneity occurs in radial/diagonal left-to-right directions and at the corners. As with Figure 10, optimizing stockpile processing can intuitively follow the observable pattern of homogeneity (bottom left to top right or vice versa). Unlike Figure 10, operators will have a harder time identifying which area corresponds to which heap shown in the data. This scenario also contains more natural mixing of material due to settling and dozer handling than does the previous scenario.

3.4. Interpolation Model

Unlike Figures 10 and 12, Figure 9 offers no visual foundation for an operational model. The approach considered for such a situation is that of interpolation. Figure 14 shows an example of an interpolated model for the hypothetical stockpile using the inverse distance weighting method described in the method section. In this scenario, the area of influence of each truck load is difficult to determine since material that was dumped first affects the form that material dumped later takes.

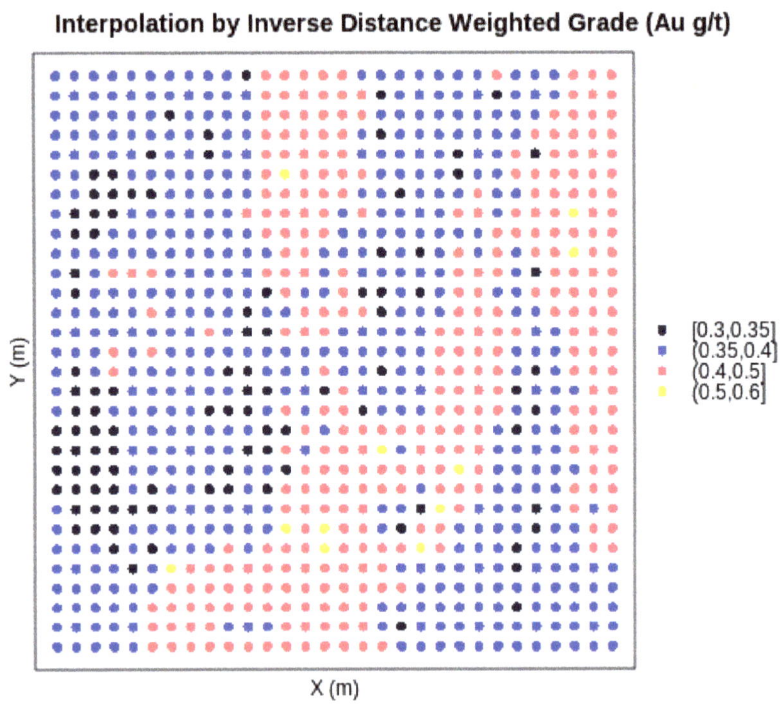

Figure 14. Grade Distribution Map of Combination of Layers.

From the map shown in Figure 14, a block model may be created, since each coordinate is located on the same elevation, the block centroid Z-value is given by the elevation level midway through the bench. The block model shown in Figure 15 is the resulting block model from the interpolation shown in Figure 15 in Maptek Vulcan and bounded by the dimensions of the stockpile shown in Figure 7. To be clear, this block model contains only one block level. In the event that a stockpile contains multiple benches or levels, this same modelling process may be repeated for each level. Because the model was an interpolation, it populated values for every position within the sample space, including the corners where no data were shown. This interpolation is trimmed by the survey volume to create the block model shown in Figure 15.

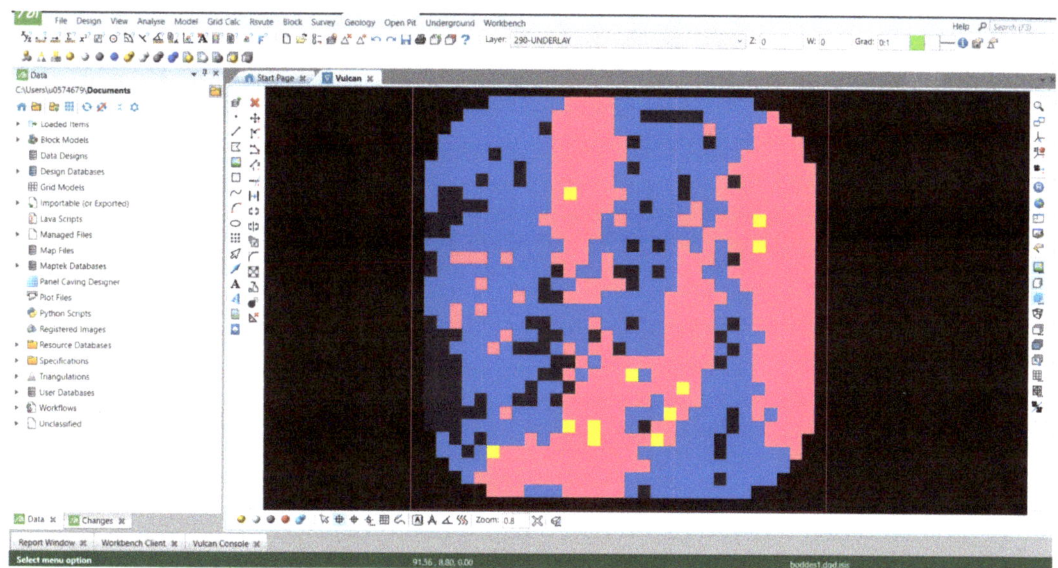

Figure 15. Screenshot in Maptek Vulcan of Resulting Block Model from FMS Data Interpolation.

Figure 15 shows grade values for each 5 m × 5 m × 5 m block in the example stockpile volume. Grade values are colored in accordance with the legend shown in the top left of the black viewing window of the screenshot. The red box represents the model extent area.

3.5. Analysis

To showcase the variance of the model by region and see how it compares with the data, confidence intervals on the mean values were used on each quadrant of the model area. A summary of the confidence intervals is shown in Table 3. Confidence intervals measure the degree of uncertainty in a dataset. They can take any number of probability limits, but Table 3 shows only 90%, 95% and 99% confidence levels. The greater the confidence interval value, the farther the mean value is from the remainder of the data.

Table 3. Confidence Intervals of Data and Model Areas.

Example Data	All Area	Bottom Left	Bottom Right	Top Left	Top Right
Confidence Level (90.0%)	0.004606	0.005712376	0.005165079	0.005724806	0.005109
Confidence Level (95.0%)	0.00549	0.006809497	0.006156908	0.006824275	0.006090
Confidence Level (99.0%)	0.007221	0.008958197	0.008099109	0.008977509	0.008011
IDW Model	**All Area**	**Bottom Left**	**Bottom Right**	**Top Left**	**Top Right**
Confidence Level (90.0%)	0.001927	0.003029155	0.003031626	0.002473611	0.002453
Confidence Level (95.0%)	0.002297	0.003611474	0.003614413	0.002948945	0.002924
Confidence Level (99.0%)	0.003021	0.004752802	0.004756644	0.003880283	0.003847

Table 3 shows that for each quadrant and confidence level, the model has a smaller confidence interval value than that of the example data. Confidence interval values for the entire data area are only lower than two of the confidence intervals for the IDW model. These areas are in the bottom left and bottom right areas of the model and only when compared between a 90% confidence level with the data to a 99% confidence level with the model. Overall, these lower values demonstrate that the model has less variance than the dataset. The top half of the model also has less variance than the bottom half.

4. Discussion

4.1. Workflow

The controversy in using a method which incorporates FMS data is that it frequently disagrees with the reconciled grade values which are required for financial reporting. FMS data also contain errors and need to be cleaned before modelling can occur smoothly. However, FMS data can be integrated into an automatic modelling process which frees up engineering resources. Using a data-driven approach also acts as a step towards digitization, which will improve after each iteration.

Figure 15 demonstrates that this modelling method may be passed into conventional engineering software and used by mine operations to optimize the processing of the stockpile. Effectively, this makes stockpile processing similar to mining of a large muck pile and subject to the same methods of ore control and mine planning previously established. Engineering software can create an optimized sequence for processing the stockpile which reduces the variability of the feed entering the mill. Block information can be uploaded into modern shovels with tracking technology so that operators can more tightly control processing along expected grade boundaries.

Having knowledge of zones within the stockpile that are trending to higher or lower values could potentially lead the operation to plan the dumping locations of each blast pattern more thoroughly. The added cost of organizing the stockpile in a real-time optimization would need to be weighed against the expected savings in rehandle cost at the time of the processing of the stockpile years in the future. This cost analysis would be tricky to perform since there are many variables to consider and most mines do not know the exact values and tonnages of their low-grade ore to be mined over the course of the entire mine life.

4.2. Data

Some dump coordinates exist outside of survey areas. This may be due to errors in the FMS tracking capability, or some material may be initially dumped outside of the boundary and later moved into it via dozers or other heavy equipment. Since FMS data alone makes it hard to determine the extents of the topography, survey data will always need to be used to create a final trim. There are also issues in variability in the physical location of the GPS coordinate of the haul truck and how that exact position best relates to the centroid of the corresponding dump location. This difference makes an even greater impact in the upper layer or edge dumping case, where the area of influence of the truck is more one-dimensional.

Future data inputs from real-time surveying of stockpiles via UAVs can improve the z values substantially. Drone data can also be used to model the dumping by truck and the movement of material by dozers. Furthermore, drones and additional sensors, along with improved modeling of the spreading behavior can improve the basic assumptions of the model, thus improving understanding the relationships between the data and the model.

Despite including drone data, the ability to track material flow over the entirety of its life in a heap fill stockpile and through processing will likely remain out of reach. However, a generalized model of key components of truck-dumped stockpile material behavior (material flow during dumping and movement by dozers for example) will improve the ability to estimate characteristic distribution within stockpiles. This modelling could further reduce processing downtime and increase throughput during the eventual processing of the stockpile.

4.3. Modelling

Sampling can be used to verify or calibrate the model. If each truck is sampled individually, the average sampling density for the stockpile modelled would be equal to the haulage capacity of the truck used (100–400 t), which is an improvement in sampling density orders of magnitude above conventional sampling campaigns for heaped filled stockpiles and leach pads (175,000 tons in the use case of Winterton) [52]. While modelling will never replace sampling, both problems can be worked interdependently.

In the event that the grade and tonnage values differ substantially from official values given for financial reporting, the block model may be adjusted to represent the grade distribution of the stockpile as weighted by each block in comparison to all others and not strictly on block values given by interpolation of the FMS data. This adjustment to the model could also be completed in an autonomous manner and would involve data which are readily available at the mine.

The demonstration of FMS data incorporated into a stockpile block model in this paper opens the door to additional discussion around optimal stockpile modelling from FMS data. While the characteristic distribution of the material may not follow the inverse distance weighting method presented in this paper, it may be adequately modelled via other geostatistical methods, such as triangulation or nearest neighbors. Variograms, ordinary least squares (OLS) prediction methods, as well as kriging may also prove useful for stockpile modelling. These methods follow the modelling flowchart created by Pebesma [49] and shown as Figure 8.

Further studies and sampling could fine tune the model, such as using discrete element modeling for tracking flow of material for various settling scenarios. Once a model is deemed precise, which accurately accounts for the stockpile characteristics "as is", more advanced models may account for metal degradation over time, natural leaching, and other environmental factors. These models may apply directly to heap leach operations by indicating target areas for hydrological work that could increase leaching capacity.

4.4. Other Factors

While the stockpile used in this study acts as a strategic form of storage, many of the stockpiles used in mining act as a buffer to ensure continuity of the processing stage. Stockpiles dedicated to maintaining a steady flow of material into the plant are generally beneficial. However, bottlenecks and data loss may affect the intention of keeping control of the flow of material from mine to plant. The short timeframe between stockpiling and processing these types of stockpiles means a different approach for modelling them should be used when compared with long-term stockpiles.

Exposure of stockpiles to dilution, weathering and other elements causes significant changes to their characteristics. These changes amount to degradation, which is more prevalent in long-term stockpiles. Rezakhah and Newman [42] quantify degradation to be between 5 and 10% annually and recognize that literature on the problem of degradation is often seen as isolated from stockpiling or mine planning problems. More robust models of stockpiles from historical FMS data could provide a more granular look into the distribution and rate of degradation which occurs within stockpiles over time.

Base and precious metal stockpiles have much in common, especially if they are made using haul trucks and dozers of the same size. Where base and precious metal stockpiles differ is due to geochemical characteristics of the ore as well as grade distribution differences (nugget effect). Base metals also oxidize quickly, which can affect the geotechnical stability of their stockpiles. These differences should be considered when developing a modelling approach for each of these types of stockpiled materials.

The model demonstrated in this paper takes a simplistic approach to the material in each truck load. In reality each truck load will likely have undergone some degree of pre-crusher blending. Shovels and loaders constantly mix the material and blast movement and dilution cause it to not reflect what is in the FMS data. Stockpiles that act as process buffers may also undergo additional blending both before and after crushing. In either type of stockpile, the effects of blending must be considered when modelling.

5. Conclusions

In conclusion, the type of working method demonstrated in this paper shows how leveraging FMS data and existing interpolation techniques can lead to increased understanding of the grade distribution within stockpiles. Figures 14 and 15 indicate interpolated grade values at each location within the stockpile in a way which is directly incorporable into mining operations. Knowledge of the types of zones illustrated by Figures 14 and 15 could enable miners to optimize the processing of their stockpiles. While this method is merely illustrative, it shows that through a systematic process of validation and modelling improvements, a given mine can use this method to come to a better understanding of the grade distribution within its long-term, low-grade stockpile. The implications of better modelling ROM stockpiles will enhance the overall mine optimization such as mine-to-mill and other continuous improvement initiatives at mines.

Author Contributions: Conceptualization, A.Y.; Data curation, A.Y.; Formal analysis, A.Y.; Funding acquisition, W.P.R.; Investigation, A.Y.; Methodology, A.Y.; Project administration, W.P.R.; Resources, W.P.R.; Software, A.Y.; Supervision, W.P.R. All authors have read and agreed to the published version of the manuscript.

Funding: This research received no external funding.

Data Availability Statement: The dataset used was made by the authors and is available upon request.

Acknowledgments: The authors would like to thank Newmont Mining Corporation for the collaboration and support of this project.

Conflicts of Interest: The authors declare no conflict of interest.

References

1. Hustrulid, W.A. *Open Pit Mine Planning & Design*, 3rd ed.; Kuchta, M., Martin, R.K., Eds.; CRC Press: Boca Raton, FL, USA; Taylor & Francis Distributor: London, UK, 2013.
2. Rogers, W.P.; Kahraman, M.M.; Dessureault, S. Exploring the value of using data: A case study of continuous improve-ment through data warehousing. *Int. J. Min. Reclam. Environ.* **2019**, *33*, 286–296. [CrossRef]
3. Young, A.; Rogers, P. A Review of Digital Transformation in Mining. *Min. Met. Explor.* **2019**, *36*, 683–699. [CrossRef]
4. Kanefsky, J.; Robey, J. Steam Engines in 18th-Century Britain: A Quantitative Assessment. *Technol. Cult.* **1980**, *21*, 161. [CrossRef]
5. Mudd, G.M. Gold mining in Australia: Linking historical trends and environmental and resource sustainability. *Environ. Sci. Policy* **2007**, *10*, 629–644. [CrossRef]
6. Zhang, K.; Kleit, A.N. Mining rate optimization considering the stockpiling: A theoretical economics and real option model. *Resour. Policy* **2016**, *47*, 87–94. [CrossRef]
7. *Optimum Determination of Sub-Grade Stockpiles in Open Pit Mines*; The Australasian Institute of Mining and Metallurgy: Carlton, Australia, 1988.
8. Brennan, M.J.; Schwartz, E.S. Evaluating Natural Resource Investments. *J. Bus.* **1985**, *58*, 135. [CrossRef]
9. Lane, K.F. *The Economic Definition of Ore: Cut-Off Grades in Theory and Practice*; Mining Journal Books: London, UK, 1988.
10. Menabde, M. Mining Schedule Optimisation for Conditionally Simulated Orebodies. In *Advances in Applied Strategic Mine Planning*; Dimitrakopoulos, R., Ed.; Springer International Publishing: Cham, Switzerland, 2018; pp. 91–100.
11. Asad, M.W.A. Cutoff grade optimization algorithm with stockpiling option for open pit mining operations of two economic minerals. *Int. J. Surf. Min. Reclam. Environ.* **2005**, *19*, 176–187. [CrossRef]
12. Asad, M.W.A.; Dimitrakopoulos, R. Optimal production scale of open pit mining operations with uncertain metal supply and long-term stockpiles. *Resour. Policy* **2012**, *37*, 81–89. [CrossRef]
13. Chanda, E.; Dagdelen, K. Optimal blending of mine production using goal programming and interactive graphics systems. *Int. J. Surf. Min. Reclam. Environ.* **1995**, *9*, 203–208. [CrossRef]
14. Gu, Q.; Lu, C.; Guo, J.; Jing, S. Dynamic management system of ore blending in an open pit mine based on GIS/GPS/GPRS. *Min. Sci. Technol. (China)* **2010**, *20*, 132–137. [CrossRef]
15. Jupp, K.; Howard, T.J.; Everett, J.E. Role of pre-crusher stockpiling for grade control in iron ore mining. *Appl. Earth Sci.* **2013**, *122*, 242–255. [CrossRef]
16. Everett, J. Simulation Modeling of an Iron Ore Operation to Enable Informed Planning. *Interdiscip. J. Inf. Knowl. Manag.* **2010**, *5*, 101–114. [CrossRef]
17. Everett, J.E.; Howard, T.J.; Jupp, K. Simulation modelling of grade variability for iron ore mining, crushing, stockpiling and ship loading operations. *Min. Technol.* **2010**, *119*, 22–30. [CrossRef]
18. Carter, R.A. Staying on Top of Stockpile Management. *Eng. Min. J.* **2018**, *219*, 58.

19. Vangold Resources Completes Acquisition of Historic El Pinguico Mine, Guanajuato, Mexico. Accesswire, Junior Mining Network. 2017. Available online: https://www.juniorminingnetwork.com/junior-miner-news/press-releases/1961-tsx-venture/gsvr/31757-vangold-resourcescompletes-acquisition-of-historic-el-pinguico-mine-guanajuato-mexico.html (accessed on 10 June 2021).
20. Preece, H. Gencor is Building up Stockpile of Gold Ore, in American Metal Market. 1986. Gale General OneFile. Available online: www.gale.com/apps/doc/A4256121/ITOF?u=marriottlibrary&sid=bookmark-ITOF&xid=edcc5334 (accessed on 10 June 2021).
21. Bulbeck, C. The Iron Ore Stockpile and Dispute Activity in the Pilbara. *J. Ind. Relat.* **1983**, *25*, 431–444. [CrossRef]
22. Ozols, V. Anvil Range to Shutter Mine: Company Will Process Ore from Existing Stockpile, in American Metal Market. 1996. Gale General OneFile. Available online: www.gale.com/apps/doc/A18883408/ITOF?u=marriottlibrary&sid=bookmark-ITOF&xid=c27bb2d5 (accessed on 10 June 2021).
23. Koushavand, B.; Askari-Nasab, H.; Deutsch, C.V. A linear programming model for long-term mine planning in the presence of grade uncertainty and a stockpile. *Int. J. Min. Sci. Technol.* **2014**, *24*, 451–459. [CrossRef]
24. Zhao, S.; Lu, T.-F.; Koch, B.; Hurdsman, A. 3D stockpile modelling and quality calculation for continuous stockpile management. *Int. J. Min. Process.* **2015**, *140*, 32–42. [CrossRef]
25. Zhao, S.; Lu, T.-F.; Koch, B.; Hurdsman, A. Stockpile Modelling Using Mobile Laser Scanner for Quality Grade Control in Stockpile Management. In Proceedings of the 2012 12th International Conference on Control Automation Robotics & Vision (ICARCV), Guangzhou, China, 5–7 December 2012; pp. 811–816.
26. Sensogut, C.; Ozdeniz, A.H. Statistical modelling of stockpile behaviour under different atmospheric condi-tions—Western Lignite Corporation (WLC) case. *Fuel* **2005**, *84*, 1858–1863. [CrossRef]
27. Kasmaee, S. Reserve estimation of the high phosphorous stockpile at the Choghart iron mine of Iran using geostatisti-cal modeling. *Min. Sci. Technol. (China)* **2010**, *20*, 855–860. [CrossRef]
28. Morley, C.; Arvidson, H. Mine value chain reconciliation-demonstrating value through best practice. In Proceedings of the Tenth International Mining Geology Conference, Hobart, Tasmania, 20–22 September 2017; pp. 20–22.
29. Hawley, M.; Cunning, J. *Guidelines for Mine Waste Dump and Stockpile Design*; CSIRO Publishing: Clayton, Australia, 2017.
30. Parker, B.M. The Simulation and Analysis of Particle Flow Through an Aggregate Stockpile. Ph.D. Thesis, Virginia Tech, Blacksburg, VA, USA, 2009.
31. Afrapoli, A.M.; Askari-Nasab, H. Mining fleet management systems: A review of models and algorithms. *Int. J. Min. Reclam. Environ.* **2017**, *33*, 42–60. [CrossRef]
32. Jansen, W. Tracer-based mine-mill ore tracking via process hold-ups at Northparkes mine. In Proceedings of the Tenth Mill Operators' Conference, Adelaide, Australia, 12 October 2009; pp. 12–14.
33. Jurdziak, L.; Kawalec, W.; Król, R. Study on Tracking the Mined Ore Compound with the Use of Process Analytic Tech-nology Tags. In *Intelligent Systems in Production Engineering and Maintenance—ISPEM 2017*; Springer International Publishing: Cham, Switzerland, 2018.
34. Ge, L.; Li, X.; Ng, A.H.-M. UAV for mining applications: A case study at an open-cut mine and a longwall mine in New South Wales, Australia. In Proceedings of the 2016 IEEE International Geoscience and Remote Sensing Symposium (IGARSS), Beijing, China, 10–15 July 2016; pp. 5422–5425. [CrossRef]
35. Rathore, I.; Kumar, N.P. Unlocking the potentiality of uavs in mining industry and its implications. International Journal of Innovative Research in Science. *Eng. Technol.* **2015**, *4*, 852–855.
36. Goldratt, E.M.; Cox, J. *The Goal: A Process of Ongoing Improvement*; North River Press: Great Barrington, MA, USA, 2012.
37. Kahraman, M.M.; Rogers, W.P.; Dessureault, S. Bottleneck identification and ranking model for mine operations. *Prod. Plan. Control* **2020**, *31*, 1178–1194. [CrossRef]
38. Sarle, W.S. Prediction with missing inputs. *Jcis 98 Proc.* **1998**, *2*, 399–402.
39. Sarbanes, P. Sarbanes-Oxley act of 2002. In *The Public Company Accounting Reform and Investor Protection Act*; US Congress: Washington, DC, USA, 20 July 2002.
40. Macfarlane, A. Reconciliation along the mining value chain. *J. S. Afr. Inst. Min. Met.* **2015**, *115*, 679–685. [CrossRef]
41. Ghorbani, Y.; Nwaila, G.T.; Chirisa, M. Systematic Framework toward a Highly Reliable Approach in Metal Accounting. *Min. Process. Extr. Met. Rev.* **2020**, 1–15. [CrossRef]
42. Rezakhah, M.; Newman, A. Open pit mine planning with degradation due to stockpiling. *Comput. Oper. Res.* **2020**, *115*, 104589. [CrossRef]
43. Engström, K. A comprehensive literature review reflecting fifteen years of debate regarding the representativity of reverse circulation vs blast hole drill sampling. *TOS Forum* **2013**, *2013*, 36. [CrossRef]
44. Thornton, D. The implications of blast-induced movement to grade control. In Proceedings of the Seventh International Mining Geology Conference, Perth, Australia, 17 August 2009; pp. 287–300.
45. Abzalov, M.Z.; Menzel, B.; Wlasenko, M.; Phillips, J. Optimisation of the grade control procedures at the Yandi iron-ore mine, Western Australia: Geostatistical approach. *Appl. Earth Sci.* **2010**, *119*, 132–142. [CrossRef]
46. Ortiz, M.J.; Magri, E.J. Designing an advanced RC drilling grid for short-term planning in open pit mines: Three case studies. *J. S. Afr. Inst. Min. Met.* **2014**, *114*, 631–639.
47. Subramaniyan, M.; Skoogh, A.; Muhammad, A.S.; Bokrantz, J.; Johansson, B.; Roser, C. A data-driven approach to diagnosing throughput bottlenecks from a maintenance perspective. *Comput. Ind. Eng.* **2020**, *150*, 106851. [CrossRef]
48. Pyrcz, M.J.; Deutsch, C.V. *Geostatistical Reservoir Modeling*; Oxford University Press: Hong Kong, China, 2014.

49. Pebesma, E.; Wesseling, C.G. Gstat: A program for geostatistical modelling, prediction and simulation. *Comput. Geosci.* **1998**, *24*, 17–31. [CrossRef]
50. Burrough, P. Principles of geographical information systems for land resources assessment. *Geocarto Int.* **1986**, *1*, 54. [CrossRef]
51. Bartier, P.M.; Keller, C. Multivariate interpolation to incorporate thematic surface data using inverse distance weighting (IDW). *Comput. Geosci.* **1996**, *22*, 795–799. [CrossRef]
52. Winterton, J. *Estimating the Residual Inventory of A Large Gold Heap Leach Pad*; Society for Mining, Metallurgy & Exploration: Englewood, CO, USA, 2013.

Article

Use of Decision Trees for the Development of Decision Support Systems for the Control of Grinding Circuits

Jacques Olivier [1] and Chris Aldrich [1,2,*]

1 Western Australian School of Mines: Minerals, Energy and Chemical Engineering, Curtin University, GPO Box 1987, Perth, WA 6845, Australia; jacques.olivier@postgrad.curtin.edu.au
2 Department of Process Engineering, Stellenbosch University, Private Bag X1, Matieland, Stellenbosch 7602, South Africa
* Correspondence: chris.aldrich@curtin.edu.au

Abstract: Grinding circuits can exhibit strong nonlinear behaviour, which may make automatic supervisory control difficult and, as a result, operators still play an important role in the control of many of these circuits. Since the experience among operators may be highly variable, control of grinding circuits may not be optimal and could benefit from automated decision support. This could be based on heuristics from process experts, but increasingly could also be derived from plant data. In this paper, the latter approach, based on the use of decision trees to develop rule-based decision support systems, is considered. The focus is on compact, easy to understand rules that are well supported by the data. The approach is demonstrated by means of an industrial case study. In the case study, the decision trees were not only able to capture operational heuristics in a compact intelligible format, but were also able to identify the most influential variables as reliably as more sophisticated models, such as random forests.

Keywords: grinding circuits; minerals processing; random forest; decision trees; machine learning; knowledge discovery; variable importance

Citation: Olivier, J.; Aldrich, C. Use of Decision Trees for the Development of Decision Support Systems for the Control of Grinding Circuits. *Minerals* **2021**, *11*, 595. https://doi.org/10.3390/min11060595

Academic Editor: Yosoon Choi

Received: 12 April 2021
Accepted: 28 May 2021
Published: 31 May 2021

Publisher's Note: MDPI stays neutral with regard to jurisdictional claims in published maps and institutional affiliations.

1. Introduction

Grinding circuits are well-known for their inefficiency and disproportionate contributions to the energy expenditure of mineral processing plants [1]. Studies have estimated that comminution circuits account for between 2–3% of global energy consumption [1,2], and up to 50% of the energy consumption on mineral processing plants [3], with the cost rising exponentially, the finer the grind. Moreover, the pronounced effect of grinding on downstream separation processes has long been a driving force for more efficient operation of these circuits through process control. However, advanced control is often hindered by the complexity of grinding operations, characterised by strongly nonlinear relationships between coupled circuit variables and long time delays. In addition, frequent changes in feed ore characteristics and other disturbances affect the operational state of the circuit, requiring frequent adjustment of set points.

This is partly the reason why in the mineral processing industry, the majority of regulatory grinding circuit control systems is realised through PID control loops [4,5]. In contrast, supervisory control functions designed to maintain set points for regulatory control and adherence to process constraints are entrusted to either process operators or advanced process control (APC) systems. The latter includes expert control systems (ECS) [6,7], fuzzy controllers [8,9], and model predictive control systems [10,11].

Despite the advantages of APC, a recent survey [12] has indicated that it is still not well-established on most mineral processing plants. As a consequence, grinding circuit performance is still largely dependent on operator decision-making processes.

Through trial and error, operators accumulate experience and heuristic knowledge to perform these supervisory functions. However, application of this knowledge is dependent on a subjective assessment of the state of the circuit and appropriate corrective action. Naturally, this can lead to inconsistent operator decision-making during similar operational states, as well as inconsistent operation between different individuals.

This work explores a methodology to support operator decision-making for the control of grinding circuits by extracting knowledge from plant data. Generally speaking, this may require feature extraction from raw process signals to be interpretable, which could be visualised as sets of univariate time series signals or structured data by the operator, or via more sophisticated infographic plots or displays, as indicated in Figure 1. This information may also captured by diagnostic process models designed for anomaly or change point detection or fault diagnosis, as well as models that could be used in automatic control systems, as indicated in Figure 1.

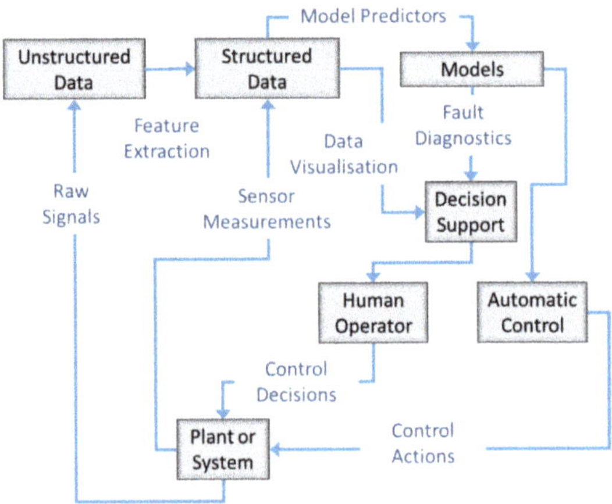

Figure 1. Data-driven modelling and decision support framework in support of human plant operator and automatic control systems.

It may also be possible to build diagnostic models that could include *if–then* rules that could be used for higher level interpretation of the data, and it is this aspect that is considered in this paper.

More specifically, operator behaviour embedded in process data are extracted using decision trees to provide explicit and actionable rules, such as those found in ECS, presented in an *if–then* format. These rules are used to analyse current operator behaviour and guide future operator decisions. Application of the methodology is demonstrated in a case study using data from an industrial grinding circuit.

Section 2 provides a brief overview of knowledge discovery from data relating to grinding circuits, while Section 3 describes the specific methodology followed in the investigation, based on the use of decision trees. Section 4 demonstrates the methodology on an industrial grinding circuit, with a discussion of the results and general conclusions presented in Section 5.

2. Knowledge Discovery for Grinding Circuit Control

Control of grinding circuits requires knowledge of the fundamental principles governing circuit operation. This knowledge allows the transformation of observed data and information into sets of instructions [13]. The knowledge acquisition process is often referred to as knowledge discovery, or knowledge engineering in ECS literature [14,15].

During traditional knowledge engineering, such as used during the crafting of ECS, rules are extracted directly from the heuristic knowledge of operators or circuit experts. This is generally facilitated through interviews, questionnaires, or observation of circuit operation by the knowledge engineer [15,16]. However, situations often arise where human experts have difficulties articulating their own decision-making rationale, or the human expert knowledge is inadequate for a comprehensive knowledge-base. Alternatively, data mining tools can be applied to extract knowledge from process data through inductive learning.

Data mining usually involves fitting models to or extracting patterns from systems by learning from examples. To facilitate decision support, the specific representation of knowledge in these models are of great importance. For knowledge discovery purposes, data mining models can be categorised according to the manner in which this knowledge is represented, being either implicit or explicit [17]. Implicit representations lack the formal mathematical notations of explicit representations, thus requiring experience to understand and possibly enabling subjective interpretations.

Black box models with implicit knowledge representations have been applied successfully for grinding circuit control using neural networks [18–20], support vector machines [21,22], reinforcement learning [23,24], or hybrid model systems [25], among others. However, the difficulty associated with interpreting and transferring such implicit knowledge make these models undesirable for operator decision support.

In contrast, rule-based classifiers are more suitable for the development of decision support systems, as they generate explicit *if–then* rules that can in principle be interpreted easily by human operators. In mineral processing, rule-based classifiers include evolutionary algorithms, such as genetic algorithms [26], genetic programming [27,28], rough sets [29], as well as decision trees [30]. In addition to this, some efforts have also been made to extract rules indirectly from the data, via from black box models, such as neural networks [31–34].

Of these methods, decision trees are by far the most popular, as the rules generated by these models are easily interpretable by operators and can provide actionable steps to move from one operational state to another. These rules consist of a sequence of conditional statements combined by logical operators; an example is given below.

$$IF\ (X_1 < C_1)\ and\ (X_2 > C_2)\ and\ (X_3 = C_3)\ and \ldots THEN\ (Class = k) \tag{1}$$

While this method for rule induction has been applied to numerous problems, it has found sparse application in the chemical and mineral processing industries. Saraiva and Stephanopoulos [35] demonstrated the use of decision rules extracted from decision trees to developing operating strategies for process improvement in case studies from the chemical processing industry. Leech [36] developed a knowledge base from rules induced from decision trees to predict pellet quality of uranium oxide powders for nuclear fuels. Reuter et al. [37] generated a rule base to predict the manganese grade in an alloy from slag characteristics in a ferromanganese submerged-arc furnace.

The applications of rule-based classifiers mentioned above mostly focus on the generation of rule sets for automatic control systems. Accordingly, these systems are often not suitable for interpretation by humans. In this study, the use of decision trees to extract rules from grinding circuit data that can be used to support control operators is considered. Decision trees have received some focus in the development of decision support systems, but applications to the processing industries were rarely encountered [38,39]

3. Methodology

3.1. Classification and Regression Trees

Decision trees are a class of machine learning algorithms that split the data space into hyper-rectangular subspaces. Each subspace is associated with a single class label, for categorical data, or numerical value for continuous data. These subspaces are identified by recursively searching for partitions based on a single input variable that cause the largest reduction in impurity of the output variable in the associated hyperrectangle.

Tree induction algorithms, such as CART (Breiman, et al., 1984) and C4.5 (Quinlan, 1993), utilise different concepts for this notion of impurity. Different impurity measures are also used depending on whether the tree is used for classification or regression. For classification purposes, CART, the algorithm used during this investigation, calculates the Gini Index at each split point. Consider a classification task where the proportion of class k, at node η, is given by $p(k|\eta)$. The Gini Index, $i(\eta)$, at node η is given by:

$$i(\eta) = 1 - \sum_{k=1}^{C} p(k|\eta)^2 \tag{2}$$

The Gini Index increases as the impurity or mixture of classes increases at the node. For example, in a binary classification problem the Gini Index reaches a maximum value of 0.5 when both classes are of equal proportion in the node. A node containing examples from a single class will have a Gini Index of 0. The reduction in impurity for a proposed split position, ξ, depends on the impurity of the current node, the impurity of proposed left and right child nodes (η_L and η_R), as well as the proportion of samples reporting to each child node (p_L and p_R):

$$\Delta i(\xi, \eta) = i(\eta) - p_R \times i(\eta_R) - p_L \times i(\eta_L) \tag{3}$$

The split position resulting in the largest decrease in impurity is selected. In regression trees, splits are selected to minimise the mean squared error from predictions of the child nodes.

Without a stopping criterion specified, this procedure is repeated until all examples in a node belong to the same class, have the same response value or the node contains only a single training example. In classification models, these terminal (leaf) nodes will predict the label of the class present in the largest proportion. For regression models, leaf nodes predict the average value of all samples belonging to the node.

The recursive nature of tree induction algorithms allow the decision trees to be represented as tree diagrams, as shown for the classification tree in Figure 2.

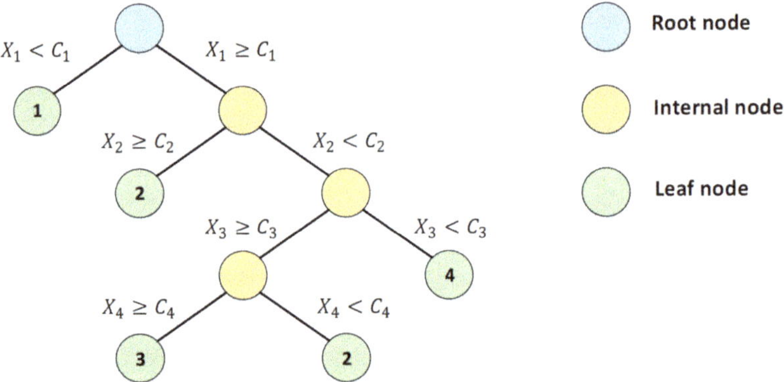

Figure 2. General classification tree diagram.

The trees are readily converted to discrete normal form (DNF) [40] as sets of *"if–then"* decision rules by following the decision paths from the root node to each leaf node. Table 1 presents the results of this procedure illustrated for the tree in Figure 2.

Table 1. Decision rules extracted from tree in Figure 2.

Rule 1	Rule 2	Rule 3	Rule 4	Rule 5
If ($X_1 < C_1$), then (Class = 1)	If ($X_1 >= C_1$), and ($X_2 >= C_2$), then (Class = 2)	If ($X_1 >= C_1$), and ($X_2 < C_2$), and ($X_3 >= C_3$), and ($X_4 >= C_4$), then (Class = 3)	If ($X_1 >= C_1$), and ($X_2 < C_2$), and ($X_3 >= C_3$), and ($X_4 < C_4$), then (Class = 2)	If ($X_1 >= C_1$), and ($X_2 < C_2$), and ($X_3 < C_3$), then (Class = 4)

Without any restrictions, decision tree models can grow to fit almost any data distribution. However, this generally results in the tree overfitting training data with reduced generalization performance. Stoppage criterions can be imposed to reduce the complexity of the tree and reduce this probability of overfitting the dataset.

These are enforced by requiring a minimum amount of samples belonging to a proposed node for the node to be formed, or placing a hard limit on the amount of allowable splits or overall tree depth.

3.2. Evaluating the Utility of Decision Rules

In this investigation, the focus was on using decision tree algorithms to extract decision rules which have significant support in the dataset and are sufficiently accurate, while remaining compact and interpretable for decision support. The utility of a rule could be considered a function of these factors, as indicated in Equation (4) below.

$$utility(rule) = f(support, \; accuracy, \; complexity) \tag{4}$$

Conceptually, there should exist some optimal configuration of the parameters wherein the utility of a rule is maximised. This concept is explored qualitatively in the case study. Each of the three requirements are briefly discussed below.

3.2.1. Supporting Samples in the Dataset

For a rule to be of any utility, it needs to be applicable in the modelled system for significant periods of time. The higher the number of samples in the dataset belonging to a specific rule, the higher the support is for the rule, and the larger the fraction of time for which the rule is valid for decision support. This metric is analogous to the support metric used to quantify the strength of association rules [41]. An acceptable level of support for a rule is dependent on the modelling problem. When modelling common operational practices, a high number of supporting samples would be required. However, if fault or rare events are investigated, the level of support could be considerably less.

3.2.2. Rule Accuracy

The accuracy of the rule refers to the dispersion of target values of the data samples belonging to the rule. For classification trees, the accuracy of a rule is represented by the proportion of training samples belonging to the class predicted by the node. For regression trees, the accuracy is well represented by the standard deviation of the target values of samples belonging to the rule. Low standard deviation of the target values of samples indicates a relatively accurate prediction by the regression tree, with sample target values close to the predicted mean.

Both accuracy measures are closely related to the impurity measures used during construction of the trees. Ideally, emphasis is placed upon rules with high accuracy.

3.2.3. Complexity and Rule Interpretability

For a rule to be of utility for decision support, the rule must remain interpretable by humans. While this notion of interpretability is naturally subjective, longer rules with a large amount of splits are difficult to interpret and tie to physical phenomena. The shorter the rule, the easier it is to interpret and possibly act upon.

Here, the decision tree algorithm is forced to generate shorter rules, and a fewer number of rules, by specifying the maximum amount of splits allowed in the tree. However, these restrictions will come at a cost, possibly decreasing the accuracy of the rules, since the capacity of the model has been decreased. However, as will be shown, restricting the number of splits does always not have a significant effect on the model generalisation ability, while it significantly increases the interpretability of the rules.

3.3. Decision Rule Extraction Procedure

This section describes the methodology used to extract useful rules from decision trees to support operator decision-making. The overall procedure is displayed in Figure 3. The process is naturally iterative, and in practice, a practitioner would repeat the process until a rule set of sufficient utility is discovered. A short discussion of each step is presented below.

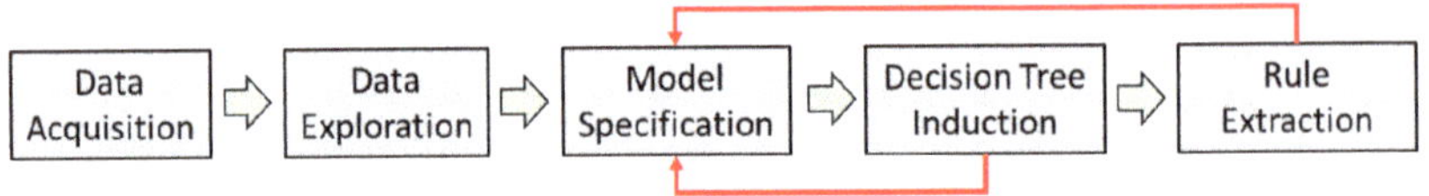

Figure 3. Procedure for extracting rules from decision trees.

3.3.1. Data Acquisition and Exploration

Operational data is collected from a mineral processing plant. Ideally, this dataset would span a period of operation capturing some variation or drift in the process. To successfully evaluate the utility of identified rules, the practitioner has to be very familiar with the intricacies of the operation, or a circuit expert needs to be consulted. Next, an exploratory analysis of the collected data can be conducted. The presence of frequently recurring operating states are identified and the conditions around these states inspected. Tying decision rules to specific operational states could provide guidance to move from less, to more favourable states.

3.3.2. Model Specification and Tree Induction

The modelling problem needs to be carefully formulated to ensure rules are extracted to address a specific variable that can solve an existing problem, or address specific operational patterns. This leads to the identification of candidate input, $\mathbf{X}$, and target variables, $\mathbf{y}$, for the decision tree algorithm.

Both controlled and manipulated variables are suitable targets for knowledge discovery. Decision tree models constructed with manipulated variables as the target leads to rules with a direct control action as its prediction. Modelling a controlled variable does not have this feature, but serves to discover common operational patterns leading to different operational states.

In addition to traditional operational variables, the role of the operator is embedded as a latent variable into the dataset. The operator's contribution will usually be revealed as a set point change in the manipulated variables. Thus, in many situations rules are actually describing common decision-making patterns by operators.

Once the input and output variables are designated, decision trees are induced on a training partition of the data, and a test set is used to measure the generalisation ability of the tree. If the accuracy of the tree proved to be too low, previous steps were repeated. Variable importance measures were used to quantify the relative contributions of different variables to the model. The results should be evaluated for consistency with heuristic circuit knowledge.

3.3.3. Rule Extraction and Evaluation

Decision trees are readily converted into a DNF rule set. In a decision tree, each path from the root node to a terminal node can be represented as a rule consisting of the conjunction of tests on the internal nodes on the path. The outcome of the rule is the class label or numerical value in the leaf node. Such a rule is extracted for each terminal node in the decision tree.

The utility of each rule was evaluated using the measures proposed above. Rules with high utility are considered valuable for operator decision support and were further analysed for the knowledge the rule contains and its practical usage.

4. Case Study

In this section, the methodology is applied to a dataset from an industrial semi-autogenous grinding (SAG) circuit. The circuit is operated under human-supervisory control, with set points primarily determined by process operators based on production targets from management. Classification trees were used to analyse the operational patterns surrounding periods wherein the mill overloaded, requiring drastic action from process operators. Identifying and addressing the circuit operating patterns during such events could reduce the frequency of similar events in future operation.

4.1. SAG Circuit Description

A schematic of the SAG circuit is shown in Figure 4. Crushed ore and water are fed to the SAG mill. Fine SAG mill product leaves the mill through a trommel screen and enters a sump before being pumped to a reverse configuration ball milling circuit. Pebbles, consisting of the mid-size fraction material building up, exit the mill through pebble ports and crushed in a cone crusher before being fed back to the mill.

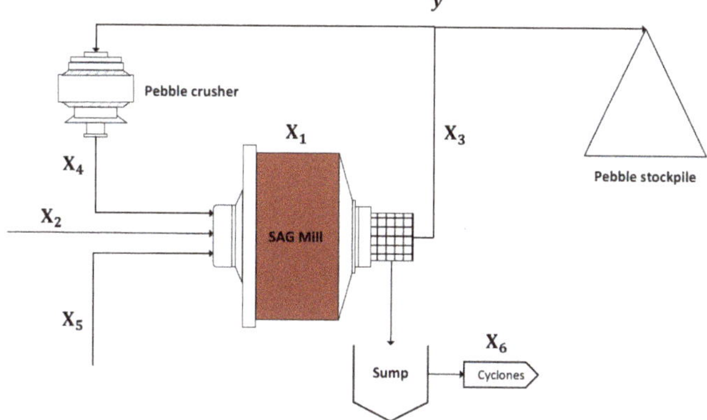

Figure 4. SAG circuit diagram, with variable descriptions in Table 2.

Table 2. Description of SAG circuit variables in Figure 4.

Name	Description	Unit
X_1	Mill power draw	kW
X_2	Dry feed rate	Tonnes/hour
X_3	Pebble discharge rate	Tonnes/hour
X_4	Pebble returns rate	Tonnes/hour
X_5	Water addition rate	m^3/hour
X_6	Cyclone Pressure	kPa
y	Pebble circuit bypass	Binary control variable

The SAG circuit is operated to achieve maximum throughput, by maximising dry feed rate, while operating within the power draw limits imposed by the SAG mill drive system. Operators continuously monitor the mill power draw and respond by changing the feed rate accordingly. Distinction is made between the pebble discharge, X_3, and pebble returns rate, X_4, since operators have the option to drop the whole pebble stream to a stockpile, allowing near instantaneous mill load control.

This action was observed most often when the mill power draw reached a very high level and risked tripping the mill. This action can be considered as another binary "on-or-off" manipulated variable. A description of measured circuit variables, as indicated in Figure 4, are given in Table 2.

4.2. Modelling Problem Description

In this case study, the focus was on gaining an understanding of the sequence of events leading to operators deciding to bypass the pebble circuit. It was generally understood that these events occurred in reaction to impending mill overloads, by removing the mid-size fraction from the mill charge. However, metallurgists wanted to discover common operating patterns leading to these overload and subsequent pebble circuit bypass events.

While dropping the pebbles to a stockpile can dramatically reduce the mill load, and subsequently power draw, this action essentially just postpones the problem. Generation of the pebble material has consumed significant amounts of energy, without resulting in product sent for downstream concentration. The pebbles contain a significant amount of valuable material that will require regrinding in the future. Additionally, the drastic change in mill load results in a coarser overall grind and forces the mill into subsequent cycles of instability.

Accordingly, the modelling problem was formulated to predict the status of the pebble circuit as a function of the remaining operational variables. The model specification is summarised in Table 3.

Table 3. Classification model specification to predict the status of the SAG pebble circuit.

Inputs	Output
X_1, X_2, X_3, X_5, X_6	y

Since the status of the pebble circuit can be represented as a binary "on/off" variable, the problem was suitable for a classification model. Alternatively, modelling the variable X_4 as a regression target should lead to similar results.

Notably missing from the inputs in Table 3 is X_4, the pebble returns rate. Pebble circuit bypass events correspond to normal tonnages on X_3, but no pebble returns to the feed conveyor (zero on X_4). Thus, the status of the pebble circuit can be perfectly predicted from knowledge of X_3 and X_4 alone. Combining these two variables in a model will result in perfect predictions, but no meaningful insights will be obtained from decision rules.

Since it is a manually triggered event, modelling the status of the pebble circuit essentially attempts to model the operators' decision-making processes. Decision rules induced during the analysis should identify the most common operational patterns leading to bypass events. Once these patterns are identified, the behaviours can be addressed in an attempt to decrease the frequency of these occurrences.

4.3. Raw SAG Circuit Data Exploration

Data samples were collected at a frequency of 5 min from the plant spanning a period of approximately six weeks of operation. The normalised data samples are shown in Figure 5. Regarding the binary variable y, a bypass of the pebble circuit is designated with the value 1, while normal operation of the circuit is denoted with the value 0. From the 9500 samples present in the dataset, 435 samples corresponded to periods of bypassing the pebble circuit, from 141 unique bypass events. The data were cleaned to remove downtime and any equipment maintenance periods.

Figure 6 shows the effect of the pebble circuit bypass on the overall variability in the circuit. The figure shows the principal component scores of the dataset on the first three principal axes, with the pebble returns superimposed as a colour map. Bypassing the pebble circuit corresponds to no pebbles returned to the mill. From the figure, it can be seen that periods of bypass lie outside the edges of the central cluster of normal operation. This suggests that reducing the frequency of these events would decrease the overall variability in the SAG circuit.

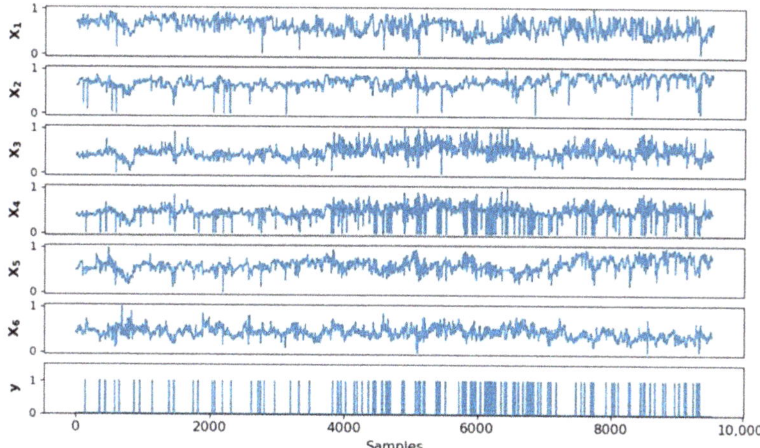

Figure 5. Normalised SAG circuit operational data spanning six weeks of operation.

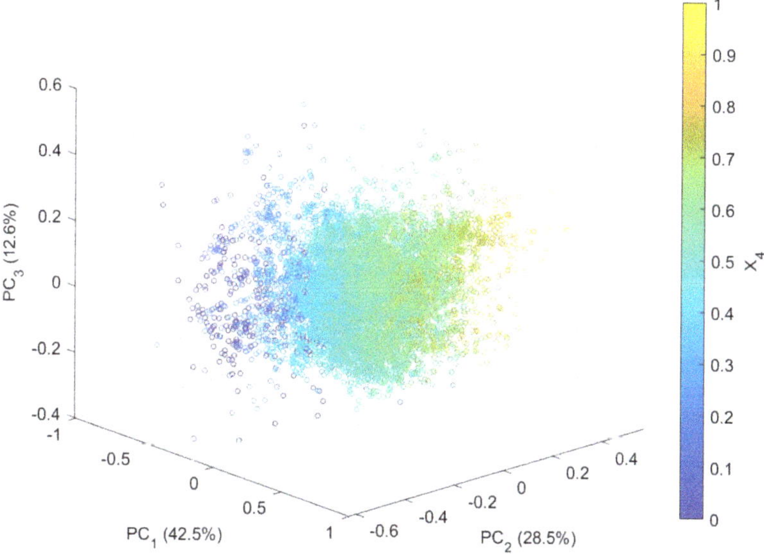

Figure 6. Principal component scores of SAG circuit operational data projected on first three principal axes. Percentage of variance explained by each principal component shown in brackets. Pebble return rates superimposed as a colour map.

4.4. Random Forest Classification Model

To gain a baseline indication of the predictability of the power draw from other circuit variables, a random forest model [42] was trained for the classification task. The random forest model, consisting of the bagged ensemble of trees and bootstrap samples used to train each tree in the forest, provides an upper limit for comparison to the predictive performance of individual tree models. Variable importance estimates from random forest models are also generally more reliable than decision trees because of the bootstrap aggregating procedure.

A random forest model was trained for the classification task specified in Table 3, using the parameters summarised in Table 4. The number of trees in the forest was selected to be large enough such that a further increase in the number of trees does not increase the model generalization. The number of predictors to sample at each split in the tree, from the total number of variables M, was maintained at the default value as suggested by Liaw and Weiner [43].

Table 4. Random forest model parameters for classification of SAG circuit data.

Parameter	Value
Maximum number of trees	50
Number of predictors sampled at each split	$floor\left(\sqrt{M}\right) = 2$
Minimum leaf size	1
Misclassification costs	Table 5

Table 5. Custom cost matrix for random forest models to reduce false negatives.

		Predicted Class	
		0	1
True Class	0	0	1
	1	20	0

The dataset was split into a training and test dataset in an 80/20 ratio. Since the number of bypass events are highly outnumbered by normal operation, the target dataset was highly imbalanced. The imbalance was negated by imposing a higher misclassification cost on bypass samples misclassified as normal pebble circuit operation. The higher misclassification cost was set equal to the proportion of normal samples to pebble circuit bypass samples, to reduce the amount of predictions resulting in false negatives. The custom cost matrix is shown in Table 5.

To quantify the model accuracy on the imbalanced dataset, the F1-score was used, as defined below:

$$F1\ Score = 2 \times \frac{Precision \times Recall}{Precision + Recall} \tag{5}$$

In this context, the precision designates the fraction of samples correctly classified as bypass events (true positives) against the total number of samples classified as bypass events (true positives and false positives). The recall designates the fraction of samples correctly classified as bypass events (true positives) against the actual number of bypass event samples (true positives and false negatives). Ideally, a model should obtain high precision and recall. Since the F1-score is simply the harmonic mean of these two measures, a high F1 score is also desired.

The F1 score as a function of the number of trees in the random forest model, as calculated on the held-out test set, is shown in Figure 7. The figure demonstrates that the F1-score improves sharply until ten trees are added to the model, after which the score plateaus around 0.7 and less significant increases to the generalization performance is observed. Notably, a single decision tree achieves a F1-score of only 0.47, indicating a significant number of misclassifications.

The misclassifications are shown in the confusion matrix in Table 6 for a random forest with 50 trees. The confusion matrix shows that the model predicts a low number of false positives, corresponding to a precision of 0.77. However, a significant number of false negatives, corresponding to a recall of 0.6, arises because of misclassification of actual bypass events.

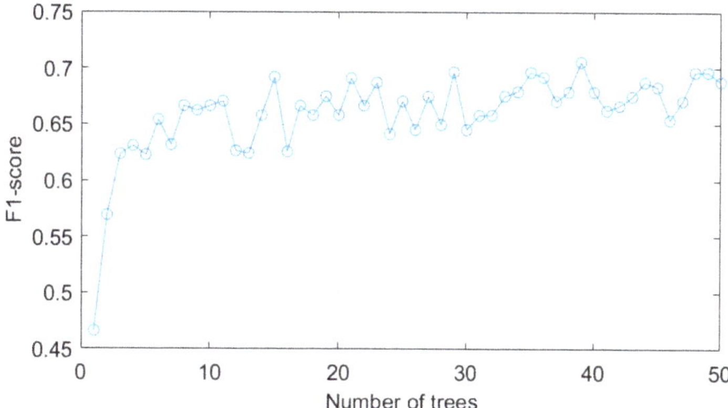

Figure 7. Random forest classification accuracy as a function of number of trees in the model.

Table 6. Confusion matrix of a random forest models with 50 trees on an independent test set.

		Predicted Class	
		0	**1**
True Class	0	1805	15
	1	35	52

The results demonstrate the difficulty of classifying the pebble circuit bypass events. This is likely a consequence of the fact that the model is attempting to describe operator decision-making. While it is thought that there is a general pattern leading to these events, the decisions made ultimately rely on subjective assessments of conditions and inconsistent choices between different individuals. The complexity of the modelling task is further increased by the general uncertainty present in the circuit, related to the disturbances of feed ore characteristics. However, the majority of the events are correctly classified, and the rule extraction procedure can be used to identify the most prominent behavioural patterns leading to these events.

The permutation importance and Gini importance measures [42,44] were calculated to quantify the importance of each variable in the random forest model. A random or dummy variable was added to the set of predictor variables, as was proposed by [45]. This random variable had no relationship with the target variable and serves as an absolute benchmark against which to measure the contributions of the variables. Both measures were calculated for 30 instances of the model, with each instance trained on a different subset of the data. The distributions of the importance measures calculated based on the permutation importance and Gini importance criteria are shown in Figures 8 and 9, respectively. In these figures, the red horizontal bars in the centres of the boxes show the median values of the importance measures, while the upper and lower edges of the boxes correspond with the 25th and 75th percentiles of the measures. The whiskers extend to the most extreme points not considered outliers, which are indicated by '+' markers. The notches in the boxes can be used to compare the median values of the importance measures, i.e., non-overlapping notches indicate a difference between the median values of the importance measures with 95% certainty

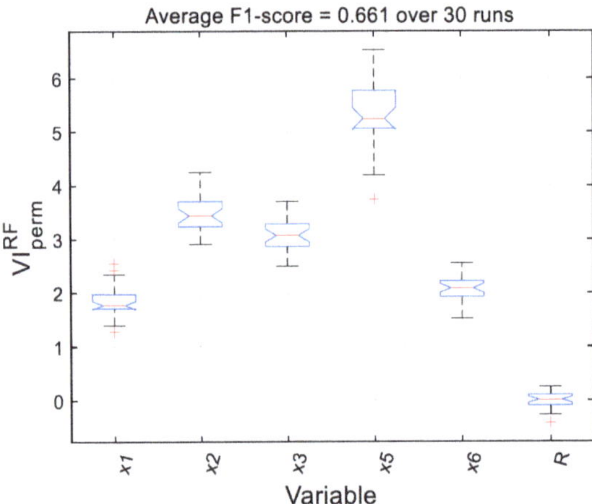

Figure 8. Box plots of permutation variable importance measures of a random forest model with 50 trees for the predictor set and a dummy variable (R), showing the median values (red bar), 25% and 75% percentiles (upper and lower box edges), extreme points (whiskers), as well as outliers (red '+' markers).

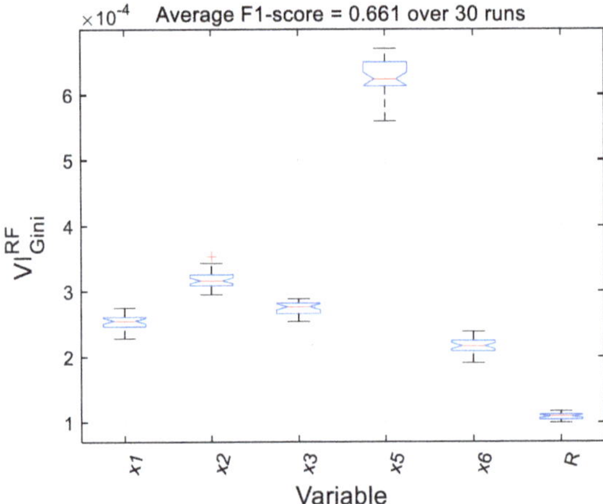

Figure 9. Box plots of Gini variable importance measures of random forest model with 50 trees for the predictor set and a dummy variable (R), showing the median values (red bar), 25% and 75% percentiles (upper and lower box edges), extreme points (whiskers), as well as outliers (red '+' markers).

In both figures, all variables contributed significantly more to the target variable than the random variable. Both measures shows markedly similar variable importance distributions. Both measures identify significant model contributions from X_5, the water addition rate, and X_2, the dry feed rate. A lesser contribution from X_3, the pebble discharge rate, is noted. Although the pebble circuit bypass is generally thought to be a response to rapid increases in the power draw, this variable was deemed less significant.

Ideally, a decision tree analysed for decision support should prioritise the same variables as the random forest model. Apart from the F1-score, a similarity in the variable importance distributions serve as additional indication that the structure of a decision tree is sufficiently representative of the more accurate and robust random forest model. Accordingly, these measures were compared with the variable importance measures obtained from a single decision tree, as demonstrated in the next section.

4.5. Decision Tree Induction and Simplification

The previous section demonstrated that the status of the pebble circuit is to an extent predictable from the set of input variables. The RF model demonstrated that a F1-score of 0.47 could be obtained using a single decision tree. While this constitutes a considerable drop in accuracy from the unrestricted random forest model, simpler decision tree models should still be able to extract simple rules describing the most common patterns leading to bypass events.

The trees in the random forest model are constructed without any restrictions on tree or branch growth. This impedes the extraction of short, interpretable decision rules from the tree.

A decision tree model was trained for the classification problem using the parameters in Table 7 below. A restriction on the minimum parent (branch) node size is usually imposed as a default setting in software packages to prevent the tree from growing separate branches for each training example. However, the minimum leaf size of one member still allows the tree to overfit the training data.

Table 7. CART model parameters for decision tree induction on the SAG circuit data.

Parameter	Value
Minimum parent node size	10
Minimum leaf size	1
Number of predictors sampled at each split	All (M)
Misclassification costs	Table 5

A decision tree trained using the parameters in Table 7 is shown in Figure 10. With no restrictions placed on the branch growth, the tree contains 173 branch nodes and 174 leaf nodes. The tree achieved a classification accuracy, in terms of the F1 score, of 0.422.

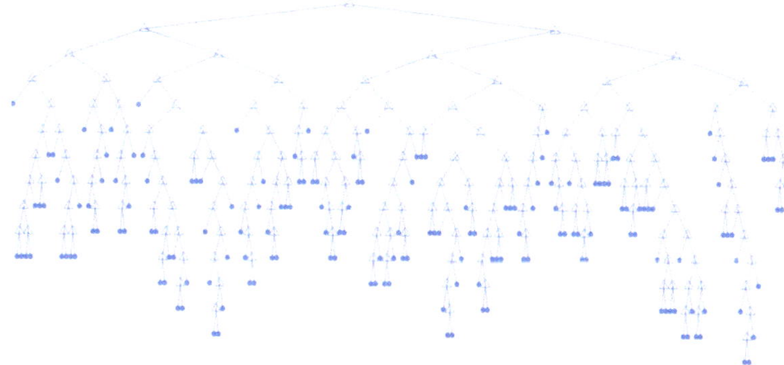

Figure 10. Decision tree predicting SAG pebble circuit status from operational data, with no restrictions on tree growth capabilities.

While the large number of splits and leaf nodes allow the tree to more closely fit the training data, the interpretability of the decision tree and individual tree branches is lost. The absence of restrictions on tree splitting parameters, such as the number of splits or tree depth, also reduces the generalisation ability of the tree by overfitting to the training dataset. This is demonstrated in Figure 11, where the training and test set accuracy of a decision tree is plotted as a function of the maximum number of splits allowed.

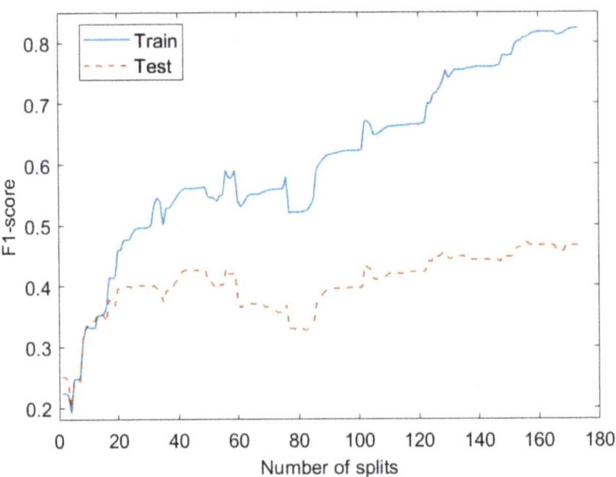

Figure 11. Classification tree prediction accuracy as a function of the maximum number of split nodes allowed.

Figure 11 shows a slight increase in the F1-score on the test set with increasing number of splits. There is a sharp increase in the F1-score up until 20 splits, after which the increases become less significant. However, even at 20 splits, the F1-score is only 0.4, indicating a significant number of misclassifications. This stresses the importance of the accuracy of rules extracted from such a tree.

Further, Figure 11 shows that above 20 split nodes the tree is starting to overfit the training data with only marginal increases to the generalization ability. This result indicates that we can significantly reduce the number of splits for construction of the decision trees, while maintaining acceptable accuracy and generalisation ability. Restricting the number of splits and leaf nodes will somewhat decrease the reliability of the tree, but will also simplify the tree branches greatly to allow the extraction of interpretable decision rules. This simplification is demonstrated in Figure 12.

The trees in Figure 12 were constructed with the parameters indicated in Table 7, as well as an additional parameter restricting the maximum number of splits allowed in the tree. The figure demonstrates that small numbers of short, interpretable rule sets can be generated with 20 or less splits in the tree, while maintaining acceptable accuracy on the test dataset.

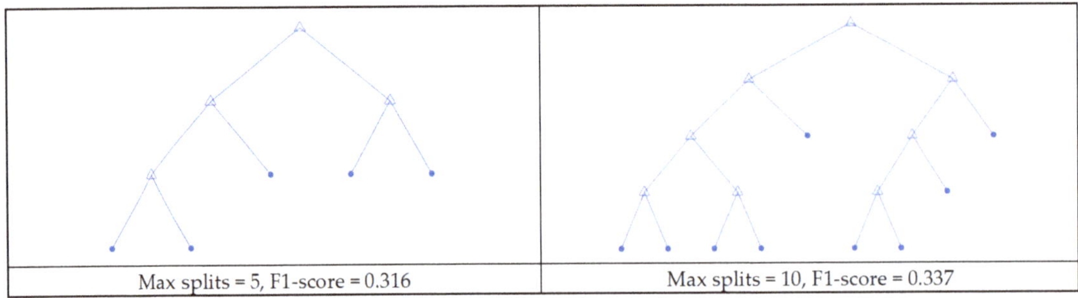

Figure 12. *Cont.*

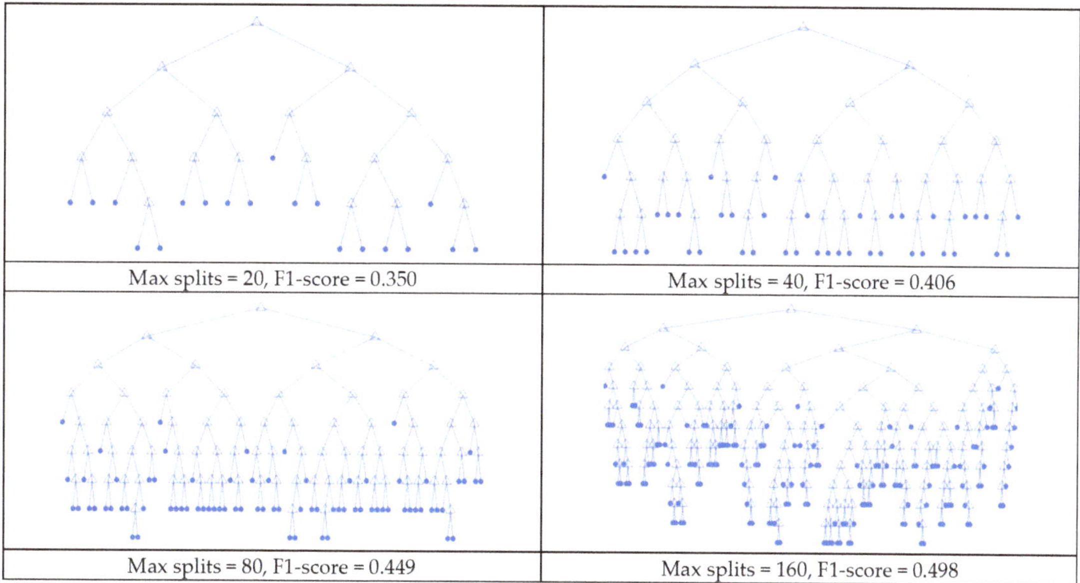

| Max splits = 20, F1-score = 0.350 | Max splits = 40, F1-score = 0.406 |
| Max splits = 80, F1-score = 0.449 | Max splits = 160, F1-score = 0.498 |

Figure 12. Decision trees generated on the SAG circuit data set with maximum number of splits imposed. F1-score of each tree on an independent test set is indicated.

For illustrative purposes, the rest of the analysis considers a tree with ten split nodes. Table 8 shows the confusion matrix for such a tree, which achieved an F1-score of 0.331. Because of the increased misclassification cost, the majority of bypass events are correctly classified. However, this also leads to an increased number of false positives.

Table 8. Confusion matrix of a decision tree with a maximum of ten splits on an independent test set.

Confusion Matrix		Predicted Class	
		0	1
True class	0	1616	204
	1	29	58

The above tree induction was simulated 30 times and the variable importance measures were calculated at each iteration. The permutation and Gini variable importance measures are shown in Figures 13 and 14, respectively. Both measures rank the importance of the input variables similarly to that of the random forest model in Figures 8 and 9. However, the importance of the hydrocyclone has diminished in the underfit decision trees. All inputs are again at least as important as the random variable.

The average F1-score over the 30 model instances is notably lower than the above result. This demonstrates the sensitivity of the generated models to the specific partition of data used during training.

The variable importance measures were calculated to directly compare with the results in Figures 8 and 9. This analysis is required to analyse the importance of variables in a random forest, since the forest of trees obstructs simple interpretation of the model. However, a single decision tree is more interpretable, and the most significant variables should be recognisable from the top branches in the tree.

The tree corresponding to the results in Table 8 is presented in Figure 15. The variables close to the root node in the tree correspond to those identified as most significant by the variable importance measures. In the following section, decision rules are extracted from this tree and analysed for their utility in decision support.

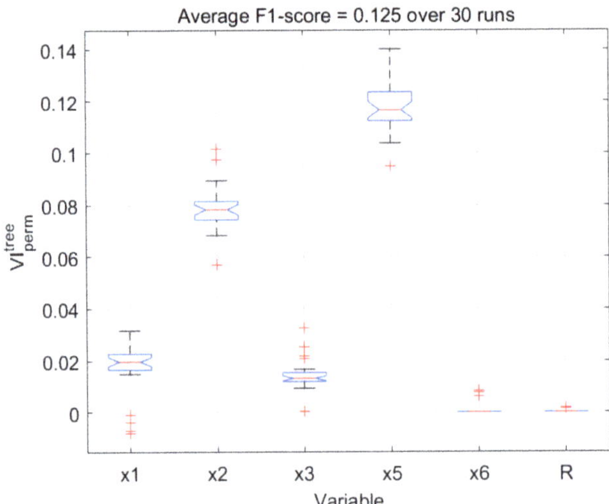

Figure 13. Box plots of permutation variable importance measures of a single decision tree with a maximum of ten splits, showing the median values (red bar), 25% and 75% percentiles (upper and lower box edges), extreme points (whiskers), as well as outliers (red '+' markers). Distributions were calculated over 30 model realisations.

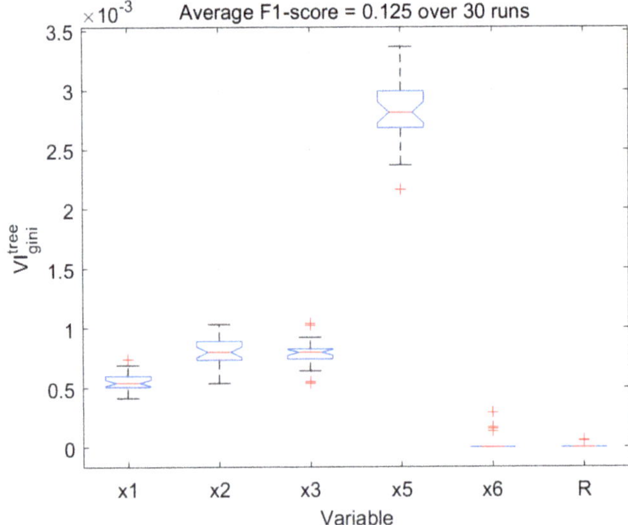

Figure 14. Box plots of Gini variable importance measures of a single decision tree with a maximum of ten splits, showing the median values (red bar), 25% and 75% percentiles (upper and lower box edges), extreme points (whiskers), as well as outliers (red '+' markers). Distributions were calculated over 30 model realisations.

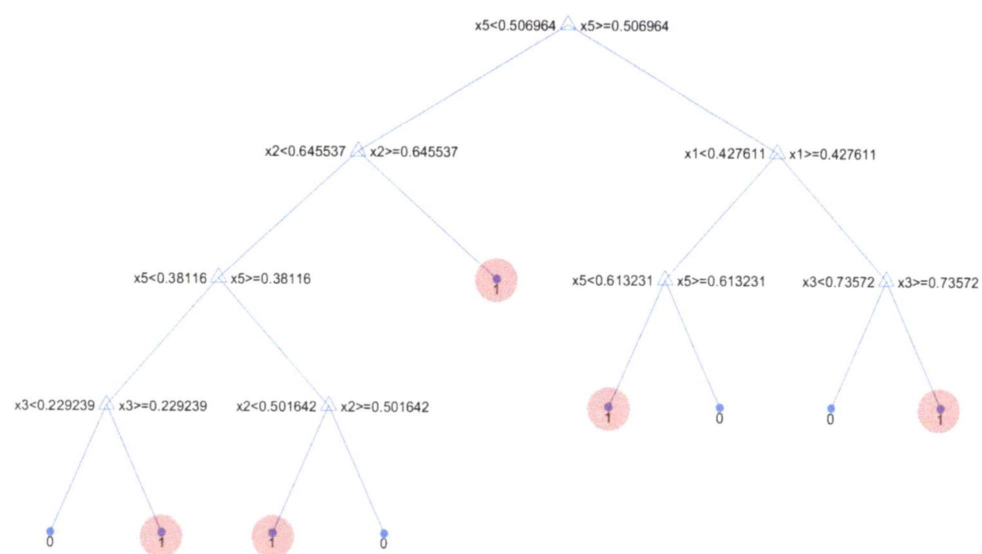

Figure 15. Classification decision tree with a maximum of 10 node splits. Leaf nodes resulting in pebble circuit bypass events are circled in red.

4.6. Extracting and Evaluating Decision Rules

The pathways from the root node to leaf nodes in the tree in Figure 15 are represented as decision rules in Table 9. Where the pathway contains multiple partitions on the same variable, the rule was simplified to contain a single expression for each unique variable. The support, accuracy and number of splits for each rule is summarised in Table 10, with which the utility of each rule could be evaluated.

Table 9. Simplified decision rules extracted from the tree in Figure 15.

Number	Rule
1	IF $X_5 < 0.381$ AND $X_2 < 0.646$ AND $X_3 < 0.229$; THEN $y = 0$
2	IF $X_5 < 0.381$ AND $X_2 < 0.646$ AND $X_3 \geq 0.229$; THEN $y = 1$
3	IF $0.381 \leq X_5 < 0.507$ AND $X_2 < 0.502$; THEN $y = 1$
4	IF $0.381 \leq X_5 < 0.507$ AND $0.502 \leq X_2 < 0.646$, THEN $y = 0$
5	IF $X_5 < 0.507$ AND $X_2 \geq 0.646$; THEN $y = 1$
6	IF $0.507 \leq X_5 < 0.613$ AND $X_1 < 0.428$; THEN $y = 1$
7	IF $X_5 \geq 0.613$ AND $X_1 < 0.428$; THEN $y = 0$
8	IF $X_5 \geq 0.507$ AND $X_1 \geq 0.428$ AND $X_3 < 0.736$; THEN $y = 0$
9	IF $X_5 \geq 0.507$ AND $X_1 \geq 0.428$ AND $X_3 \geq 0.736$; THEN $y = 1$

Table 10. Support, accuracy, and number of splits per rule in Table 8.

Number	Supporting Samples (% of Dataset)	Accuracy (Probability of Predicted Class)	Number of Splits
1	1.45	1.000	3
2	2.24	0.246	3
3	0.48	0.189	2
4	9.82	0.988	4
5	5.79	0.360	2
6	2.35	0.196	2
7	6.21	0.970	2
8	68.17	0.989	3
9	3.49	0.102	3

The rules in Table 9 leading to bypass events naturally have relatively small numbers of supporting samples because of the prevalence of these events, and this state essentially representing fault conditions. In Table 10, the rules predicting bypass events all have an accuracy below 0.5. If the prediction was a simple majority vote of all the samples belonging to the rule, the rules would naturally predict normal operation of the pebble circuit. However, the higher cost imposed on misclassifying actual bypass events outweighs the cost of misclassifying normal operation samples. Thus, the higher misclassification cost allows for the identification of the operational states wherein these bypass events are most likely to occur. Intuitively, the accuracy is thus better interpreted as an indication of the probability of a bypass event occurring in the operational state specified by the rule. The number of splits are low and interpretable because of the maximum number of splits restriction imposed on the decision tree.

Three of the rules extracted are critically analysed below. Consider a closer inspection of rule 5:

$$\text{IF Water addition rate} < 0.507$$
$$\text{AND Dry feed rate} \geq 0.646; \tag{6}$$
$$\text{THEN Bypass pebble circuit (Probability} = 0.36)$$

The rule states that at a higher dry feed rate coupled with a lower water addition rate, corresponding to an increased solids density in the SAG mill, there is a 36% chance the circuit would be bypassed. Depending on ore characteristics at the time, the inadequate water addition is causing the mill to retain more fines than usual, leading to an increase in the mill load and power draw. To decrease the probability of this event in the future, metallurgists could reconsider the SAG discharge density targets given to operators based on different ore sources.

Rule 9 states the following:

$$\text{IF Water addition rate} \geq 0.507$$
$$\text{AND Mill power draw} \geq 0.428$$
$$\text{AND Pebble discharge rate} \geq 0.736; \tag{7}$$
$$\text{THEN Bypass pebble circuit (Probability} = 0.102)$$

Rule 9 states that when the water addition rate and power draw are at medium levels or higher, while the pebble discharge rate is high, there is a 10% chance that the pebble circuit would be bypassed. This situation might arise when the mill feed suddenly changes to a more competent ore source, or a larger portion of mid-size fraction material is being fed. The mill load and power draw are not necessarily high, but the fraction of mid-size material being discharged from the mill is increasing, possibly to a level where the pebble crusher and circuit conveyors are unable to deal with the increased load. Depending on the mill fill level and the amount of power available, operators might choose to draw a higher portion of large rocks from the stockpile to attempt to break down some of the mid-size material. Alternatively, operators may choose to draw an increased fraction of finer material from the stockpile to maintain the mill throughput while not further contributing to the generation of pebbles. Metallurgists could further investigate the particle size distributions received from the preceding crusher section to deal with these occurrences.

Rule number 9 is directly contrasted by rule 8, which received the highest amount of support in the dataset:

$$\text{IF Water addition rate} \geq 0.507$$
$$\text{AND Mill power draw} \geq 0.428$$
$$\text{AND Pebble discharge rate} < 0.736; \tag{8}$$
$$\text{THEN Normal pebble circuit operation (Probability} = 0.989)$$

As seen in the tree in Figure 15, rule 8 and rule 9 split the data space according to the specific value of the pebble discharge rate. In contrast to rule number 9, rule 8 predicts that at lower pebble discharge rates, the circuit was only bypassed 1.1% of the time. Thus, the combination of the two rules discover the explicit value of the pebble discharge rate, such that when this value is exceeded, the operator is ten times more likely to bypass the pebble circuit. The rule can alert an operator when approaching this specific operational state, hopefully triggering faster control action and avoiding the bypass event.

5. Discussion and Conclusions

In the case study, classification trees were used to model operator decision-making, when deciding whether to remove the critical size material from the circuit to prevent the mill from overloading. It was demonstrated how the model specification can be exploited to identify the causes of rare events. It was demonstrated that rules can be extracted to understand why and when operators were making this particular decision.

This type of knowledge can be utilised by metallurgists to aid in determining circuit operational parameters, or provided to the operator as decision support on a human-machine interface (HMI). Decision support on a HMI could nudge the more cautious operator to increase throughput, or restrain aggressive operators when their ambitions might push equipment towards its limit and require drastic action. This decision support could take the form of explicitly displaying rules extracted on a HMI, or process alarms alerting the operator when entering a state governed by a specific rule. Depending on the specific problem investigated, such decision support systems could either increase overall throughput, increase the energy efficiency of the grinding task, or reduce the wear to mill consumables and liners.

The merits of this type of rule induction is based on its simplicity. Site experts or metallurgists can identify a problem and formulate a model to answer questions regarding the problem. Rules are then easily induced using pre-packed CART implementations. There is no guarantee that the rules will contain valid or insightful knowledge, so the expert is required to critically analyse to ensure they are reasonable. The greatest inhibitor of extracting rules for successful decision support would be the unavailability of quality data sets, or a lack of site-specific knowledge to interpret and critically evaluate the patterns such rules discover. Neither of these should be of any concern to a plant metallurgist.

While it is unlikely that this induction is used for the generation of a complete ECS, it can certainly augment heuristic knowledge from experts in such systems. Experts often have difficulty explaining the procedures they follow to arrive at decisions [15]. Rule induction could aid in formalizing some of the procedures.

As noted by Li et al. [46], the integration of human operators and technology in the control room is lacking in the minerals processing industry. The successful implementation of any process control or decision support system is reliant on effective HMI visualisations, and training operators to effectively utilise such tools.

In summary, decision trees can be used as an effective approach to extract intelligible rules from data that can be used to support operators controlling grinding circuits. Three criteria are considered in the process, i.e., the accuracy of the rule, the support of the rule, and the complexity of the rule. While the case study described the application of classification trees, the methodology is easily extended to regression problems.

In addition, in some instances, as was the case in this investigation, they can be used to identify the most influential variables as reliable as more complex models, such as random forests.

Future work will focus on the industrial operationalization of the approach, making use of various online sensors in grinding circuits.

Author Contributions: Conceptualization, C.A. and J.O.; methodology, C.A. and J.O.; software, J.O.; validation, J.O.; formal analysis, J.O.; investigation, J.O.; resources, C.A.; data curation, C.A. and J.O.; writing—original draft preparation, J.O.; writing—review and editing, C.A. and J.O.; supervision, C.A.; funding acquisition, C.A. All authors have read and agreed to the published version of the manuscript.

Funding: Financial support for the project by Molycop (https://molycop.com) is gratefully acknowledged.

Data Availability Statement: The data are not publicly available for proprietary reasons.

Conflicts of Interest: The authors declare no conflict of interest.

References

1. Fuerstenau, D.W.; Abouzeid, A.Z. The energy efficiency of ball milling in comminution. *Int. J. Miner. Process.* **2002**, *67*, 161–185. [CrossRef]
2. Napier-Munn, T. Is progress in energy-efficient comminution doomed? *Miner. Eng.* **2015**, *73*, 1–6. [CrossRef]
3. Napier-Munn, T.J.; Morrell, S.; Robert, D.; Kojovic, T. *Mineral Comminution Circuits: Their Operation and Optimisation*; Julius Kruttschnitt Mineral Research Centre; University of Queensland; Indooroopilly, Australia, 1996.
4. Wei, D.; Craig, I.K. Grinding mill circuits—A survey of control and economic concerns. *Int. J. Miner. Process.* **2009**, *90*, 56–66. [CrossRef]
5. Zhou, P.; Lu, S.; Yuan, M.; Chai, T. Survey on higher-level advanced control for grinding circuits operation. *Powder Technol.* **2016**, *288*, 324–338. [CrossRef]
6. Chen, X.S.; Li, Q.; Fei, S.M. Supervisory expert control for ball mill grinding circuits. *Expert Syst. Appl.* **2008**, *34*, 1877–1885. [CrossRef]
7. Chen, X.; Li, S.; Zhai, J.; Li, Q. Expert system based adaptive dynamic matrix control for ball mill grinding circuit. *Expert Syst. Appl.* **2009**, *36*, 716–723. [CrossRef]
8. Van Drunick, W.I.; Penny, B. Expert mill control at AngloGold Ashanti. *J. S. Afr. I. Min. Metall.* **2005**, *105*, 497–506.
9. Hadizadeh, M.; Farzanegan, A.; Noaparast, M. A plant-scale validated MATLAB-based fuzzy expert system to control SAG mill circuits. *J. Process Control.* **2018**, *70*, 1–11. [CrossRef]
10. Le Roux, J.D.; Padhi, R.; Craig, I.K. Optimal control of grinding mill circuit using model predictive static programming: A new nonlinear MPC paradigm. *J. Process Control.* **2014**, *24*, 29–40. [CrossRef]
11. Botha, S.; le Roux, J.D.; Craig, I.K. Hybrid non-linear model predictive control of a run-of-mine ore grinding mill circuit. *Miner. Eng.* **2018**, *123*, 49–62. [CrossRef]
12. Olivier, L.E.; Craig, I.K. A survey on the degree of automation in the mineral processing industry. In Proceedings of the 2017 IEEE AFRICON, Cape Town, South Africa, 18–20 September 2017; pp. 404–409. [CrossRef]
13. Ackoff, R.L. From Data to Wisdom. *J. Appl. Syst. Anal.* **1989**, *16*, 3–9. [CrossRef]
14. Fayyad, U.; Piatetsky-Shapiro, G.; Smyth, P. The KDD Process for Extracting Useful Knowledge from Volumes of Data. *Commun. ACM.* **1996**, *39*, 27–34. [CrossRef]
15. Johannsen, G.; Alty, J.L. Knowledge engineering for industrial expert systems. *Automatica* **1991**, *27*, 97–114. [CrossRef]
16. Sloan, R.; Parker, S.; Craven, J.; Schaffer, M. Expert systems on SAG circuits: Three comparative case studies. In Proceedings of the Third International Conference on Autogenous and Semiautogenous Grinding Technology, Vancouver, BC, Canada, 30 September–3 October 2001. [CrossRef]
17. Bandaru, S.; Ng, A.H.C.; Deb, K. Data mining methods for knowledge discovery in multi-objective optimization: Part A—Survey. *Expert Syst. Appl.* **2017**, *70*, 139–159. [CrossRef]
18. Stange, W. Using artificial neural networks for the control of grinding circuits. *Miner. Eng.* **1993**, *6*, 479–489. [CrossRef]
19. Chai, T.; Zhai, L.; Yue, H. Multiple models and neural networks based decoupling control of ball mill coal-pulverizing systems. *J. Process Control.* **2011**, *21*, 351–366. [CrossRef]
20. Aldrich, C.; Burchell, J.J.; De, J.W.; Yzelle, C. Visualization of the controller states of an autogenous mill from time series data. *Miner. Eng.* **2014**, *56*, 1–9. [CrossRef]
21. Zhao, L.; Tang, J.; Yu, W.; Yue, H.; Chai, T. Modelling of mill load for wet ball mill via GA and SVM based on spectral feature. In Proceedings of the IEEE Fifth International Conference on Bio-Inspired Computing: Theories and Applications (BIC-TA), Changsha, China, 23–26 September 2010. [CrossRef]
22. Jemwa, G.T.; Aldrich, C. Kernel-based fault diagnosis on mineral processing plants. *Miner. Eng.* **2006**, *19*, 1149–1162. [CrossRef]
23. Valenzuela, J.; Najim, K.; del Villar, R.; Bourassa, M. Learning control of an autogenous grinding circuit. *Int. J. Miner. Process.* **1993**, *40*, 45–56. [CrossRef]
24. Conradie, A.V.E.; Aldrich, C. Neurocontrol of a ball mill grinding circuit using evolutionary reinforcement learning. *Miner. Eng.* **2001**, *14*, 1277–1294. [CrossRef]
25. Zhou, P.; Chai, T.; Sun, J. Intelligence-based supervisory control for optimal operation of a DCS-controlled grinding system. *IEEE Trans. Control Syst. Technol.* **2013**, *21*, 162–175. [CrossRef]
26. Gouws, F.S.; Aldrich, C. Rule-based characterization of industrial flotation processes with inductive techniques and genetic algorithms. *Ind. Eng. Chem. Res.* **1996**, *35*, 4119–4127. [CrossRef]
27. Greeff, D.J.; Aldrich, C. Development of an empirical model of a nickeliferous chromite leaching system by means of genetic programming. *J. S. Afr. I. Min. Metall.* **1998**, *98*, 193–199.
28. Chemaly, T.P.; Aldrich, C. Visualization of process data by use of evolutionary computation. *Comput. Chem. Eng.* **2001**, *25*, 1341–1349. [CrossRef]
29. Zhang, Y.; Fang, C.-G.; Wang, F.; Wang, W. Rough sets and its application in cation anti-flotation control. *IFAC Proc. Vol.* **2003**, *36*, 293–297. [CrossRef]
30. Aldrich, C.; Moolman, D.W.; Gouws, F.S.; Schmitz, G.P.J. Machine learning strategies for control of flotation plants. *Control Eng. Pract.* **1997**, *5*, 263–269. [CrossRef]

31. Schmitz, G.P.J.; Aldrich, C.; Gouws, F.S. ANN-DT: An algorithm for extraction of decision trees from artificial neural networks. *IEEE Trans. Neural Netw.* **1999**, *10*, 1392–1401. [CrossRef]
32. Hayashi, Y.; Setiono, R.; Azcarraga, A. Neural network training and rule extraction with augmented discretized input. *Neurocomputing* **2016**, *207*, 610–622. [CrossRef]
33. Zeng, Q.; Huang, H.; Pei, X.; Wong, S.C.; Gao, M. Rule extraction from an optimized neural network for traffic crash frequency modeling. *Accid. Anal. Prev.* **2016**, *97*, 87–95. [CrossRef]
34. Bondarenko, A.; Aleksejeva, L.; Jumutc, V.; Borisov, A. Classification Tree Extraction from Trained Artificial Neural Networks. *Procedia Comput. Sci.* **2016**, *104*, 556–563. [CrossRef]
35. Saraiva, P.M.; Stephanopoulos, G. Continuous process improvement through inductive and analogical learning. *AIChE J.* **1992**, *38*, 161–183. [CrossRef]
36. Leech, W.J. A rule-based process control method with feedback. *Adv. Instrum.* **1986**, *41*, 169–175.
37. Reuter, M.A.; Moolman, D.W.; Van Zyl, F.; Rennie, M.S. Generic Metallurgical Quality Control Methodology for Furnaces on the Basis of Thermodynamics and Dynamic System Identification Techniques. *IFAC Proc. Vol.* **1998**, *31*, 363–368. [CrossRef]
38. Eom, S.B.; Lee, S.M.; Kim, E.B.; Somarajan, C. A survey of decision support system applications (1988–1994). *J. Oper. Res. Soc.* **1998**, *49*, 109–120. [CrossRef]
39. Eom, S.; Kim, E. A survey of decision support system applications (1995–2001). *J. Oper. Res. Soc.* **2006**, *57*, 1264–1278. [CrossRef]
40. Weiss, S.M.; Indurkhya, N. Optimized Rule Induction. *IEEE Expert* **1993**, *8*, 61–69. [CrossRef]
41. Agrawal, R.; Imielinski, T.; Swami, A. Mining association rules between sets of items in large databases. In Proceedings of the 1993 ACM SIGMOD International Conference on Management of Data—SIGMOD '93, Washington, DC, USA, 26–28 May 1993; pp. 207–216.
42. Breiman, L. Random forests. *Mach. Learn.* **2001**, *45*, 5–32. [CrossRef]
43. Liaw, A.; Wiener, M. Classification and Regression by randomForest. *R News.* **2002**, *2*, 18–22.
44. Breiman, L.; Friedman, J.H.; Olshen, R.A.; Stone, C.J. *Classification and Regression Trees*; Wadsworth Int. Group: Monterey, CA, USA, 1984.
45. Auret, L.; Aldrich, C. Interpretation of nonlinear relationships between process variables by use of random forests. *Miner. Eng.* **2012**, *35*, 27–42. [CrossRef]
46. Li, X.; McKee, D.J.; Horberry, T.; Powell, M.S. The control room operator: The forgotten element in mineral process control. *Miner. Eng.* **2011**, *24*, 894–902. [CrossRef]

minerals

Article

A Case Study of Rock Type Prediction Using Random Forests: Erdenet Copper Mine, Mongolia

Narmandakh Sarantsatsral [1], Rajive Ganguli [1,*], Rambabu Pothina [1] and Batmunkh Tumen-Ayush [2]

[1] Department of Mining Engineering, University of Utah, Salt Lake City, UT 84112, USA; n.sarantsatsral@utah.edu (N.S.); rambabu.pothina@utah.edu (R.P.)

[2] Erdenet Mining Corporation, Erdenet 61027, Mongolia; tbatmunkh@erdenetmc.mn

* Correspondence: rajive.ganguli@utah.edu

Abstract: In a mine, knowledge of rock types is often desired as they are important indicators of grade, mineral processing complications, or geotechnical attributes. It is common to model the rock types with visual graphics tools using geologist-generated rock type information in exploration drillhole databases. Instead of this manual approach, this paper used random forest (RF), a machine learning (ML) algorithm, to model the rock type at Erdenet Copper Mine, Mongolia. Exploration drillhole data was used to develop the RF models and predict the rock type based on the coordinates of locations. Data selection and model evaluation methods were designed to ensure applicability for real life scenarios. In the scenario where rock type is predicted close to locations where information is available (such as in blocks being blasted), RF did very well with an overall success rate (OSR) of 89%. In the scenario where rock type was predicted for two future benches (i.e., 30 m below known locations), the best OSR was 86%. When an exploration program was simulated, performance was poor with a OSR of 59%. The results indicate that EMC can leverage RF models for short-term and long-term planning by predicting rock types within drilling blocks or future blocks quite accurately.

Keywords: machine learning; random forest; rock type; mining geology

Citation: Sarantsatsral, N.; Ganguli, R.; Pothina, R.; Tumen-Ayush, B. A Case Study of Rock Type Prediction Using Random Forests: Erdenet Copper Mine, Mongolia. *Minerals* **2021**, *11*, 1059. https://doi.org/10.3390/min11101059

Academic Editor: Yosoon Choi

Received: 30 August 2021
Accepted: 24 September 2021
Published: 28 September 2021

Publisher's Note: MDPI stays neutral with regard to jurisdictional claims in published maps and institutional affiliations.

1. Introduction

Machine learning (ML) has been applied to mining and geology problems for at least two decades now [1–6]. On the mining geology side, grade estimation has been a major area of focus [7–11]. Machine learning techniques that were commonly applied were neural networks (NN) and support vector machines. Many also tried hybrid approaches [12]. In order to estimate iron ore grades at a mine, researchers [6] used an "extreme learning machine" (a feed forward NN) algorithm in combination with a "particle swarm optimization" approach. To fill the data gaps for geochemical element grades in a porphyry copper deposit, a multi-layer NN was used [13] along with a Gustafson-Kessel clustering algorithm. In a case study to generalize assay values for known and unknown sampled locations of a mineral sand deposit a hybrid NN was deployed. The combination included a trained, tested, and validated feed forward NN along with a geostatistics model [14]. In another instance, a genetic algorithm (GA) was used to train a NN [11] for predicting iron grades.

Researchers investigated methods for generalization, considering the complications typical in earth science data [2,15,16]. Addressing these issues, some researchers have used GA to split datasets properly into training and testing subsets [17,18]. To be method agnostic, recommendations were made on how data should be split to ensure proper evaluation of artificial intelligence models [19].

Some recent examples used ML to identify rock types based on machine operation data from drills (such as drill penetration rate) or other sensor data. Logistic regression, neural networks and gradient boosting were used by [20] to identify rock types based on sensor data in oil well directional drilling. Clustering and other techniques were applied to "measurement-while-drilling" data to identify rock types in an iron ore mine [21]. Though the nature of the application is different, it is worth mentioning that some have also used machine learning to identify rock types from images [22,23].

Detecting rock types is also a focus of this paper. The large exploration drillhole database of Erdenet Copper Mine (EMC), Mongolia, is utilized in this paper to identify rock types. Traditionally, exploration databases are used primarily for grade estimation. However, rock type modeling is also undertaken in support of grade estimation, geotechnical modeling or mineral processing operations. For example, if rock type is known along with grade, it may be processed a particular way. If rock type can be estimated at depths below current operational depths, it can be used in developing future plans. Currently, rock type modeling is performed manually using visual tools.

Manual modeling performed using 3D visual tools can be difficult and time consuming. Making changes to manual models because of new data is also difficult. ML, on the other hand, not only makes the job easier but also allows incorporation of data from other sources. Therefore, the objective of this paper is to evaluate the effectiveness of using ML in modeling the rock type.

EMC, about 350 km northwest of capital city Ulaanbaatar, mines the Erdenetiin Ovoo copper porphyry deposit, one of the largest copper-molybdenum deposits in Mongolia. The deposit is hosted by an intrusive complex in the Orkhon-Selenge trough [24]. The mine, which started operations in 1978, splits the mining area primarily into four deposits, Central, Northwest, Shand and Oyut. This paper focuses only on the Northwest and Central deposits, as they are the only two deposits being mined currently.

Though the exploration holes were drilled from 1963 to 2018, the drillhole information was only recently entered into a database as part of a relatively new digitization effort at the mine. Therefore, there were several issues with the database, all of which had to be dealt with prior to starting work on this paper. The issues primarily included duplicate holes, irrelevant columns (or fields), terminology issues, missing critical values, and spelling. After cleaning, the database consisted of 2823 exploration drillholes for the Northwest and Central deposits. The total number of lithological "segments" were 90033. Segments are explained later in the paper. Four fields (or columns in tables) in the database were used in this research, three for the coordinates, and one for the rock type. As is common in exploration databases, rock types in the database are geologist's interpretation of the rock.

Figure 1 shows two views of the drillholes. Some holes were drilled from the surface before the start of operations, while other holes were drilled inside the pit. Therefore, hole lengths ranged from 28 m to 1054 m, with a median length of 75 m. About 140 holes were above 485 m in depth (95th percentile). Hole bottom elevations range from 166 m to 1505 m, with the median bottom elevation being 1310 m.

EMC uses the drillhole database to classify the main domains by lithology and fault zones. These zones are then related to mining and mineral processing conditions. Rocks are grouped into five major zones: andesite, granodiorite (GDIR), biotite granodiorite porphyry, dyke and fault zones, and finally, unknown. About 43% of the copper comes from GDIR. Therefore, the goal in this paper is to predict if the rock type in a given location is GDIR or not.

ML, as with most modeling methods, requires data to be split into modeling (or training) subset and testing subset. Usually, data is split into training and testing subsets to ensure that both subsets are similar [15]. However, a model can be developed and evaluated using different strategies to reflect the various ways it can be used in real life. Therefore, a novelty of this paper is in how data is split for modeling and evaluation. This is explained in the next section.

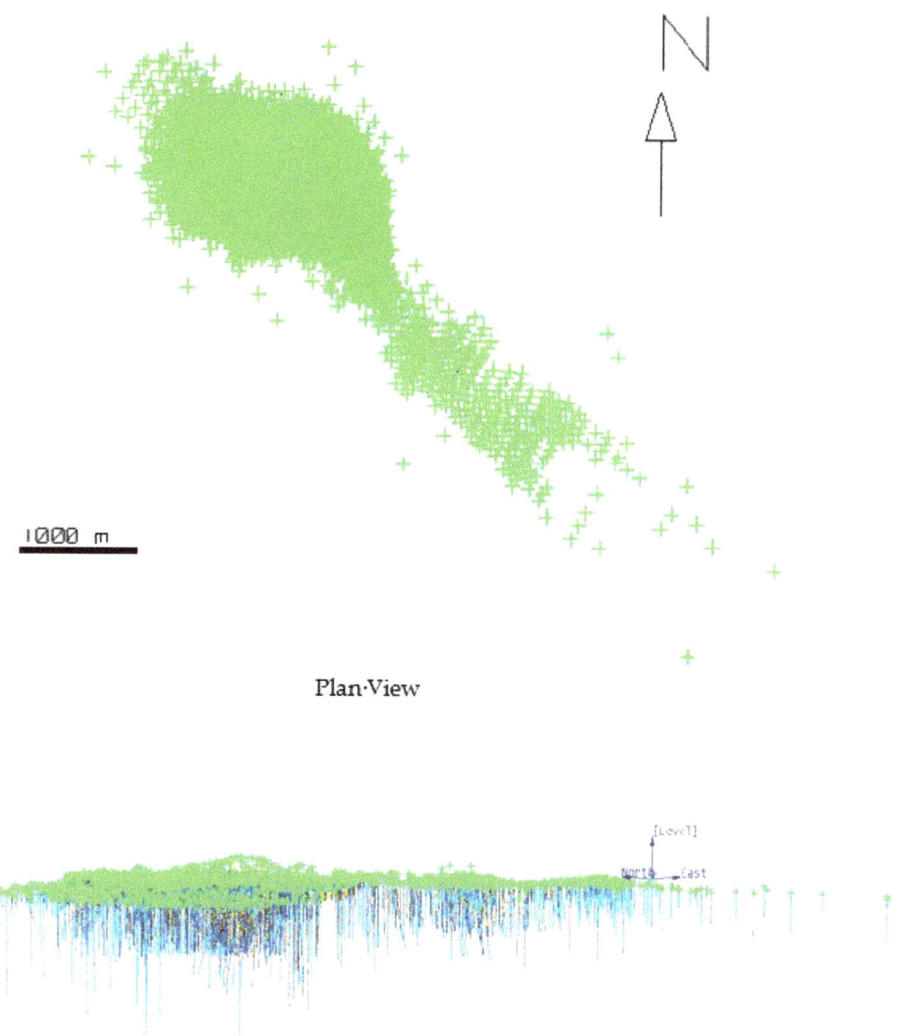

Figure 1. Plan view (top) and cross-sectional view of the 2823 drillholes. Arrow points north. Scale is shown in meters. A total of 90,033 drillhole segments are depicted.

2. Methodology

2.1. Data Selection Approaches

This paper uses two approaches for selecting data for training and testing subsets, segment-based (SB) and hole-based (HB). The reasoning for the two approaches is explained in a subsequent section.

SB and HB approaches are demonstrated using Figure 2. The figure shows a dataset consisting of four holes, H1–H4. Each hole contains several lithological segments. Segments are 5 m in thickness, except when the lithological segment is less than 5 m in thickness or not a perfect multiple of 5 m. For example, consider a granodiorite intersection of 23 m, followed by 3.5 m of diorite. The granodiorite intersection will be split into five segments of lengths 5 m, 5 m, 5 m, 5 m, 5 m and 3 m. The diorite will be on a separate segment of 3.5 m.

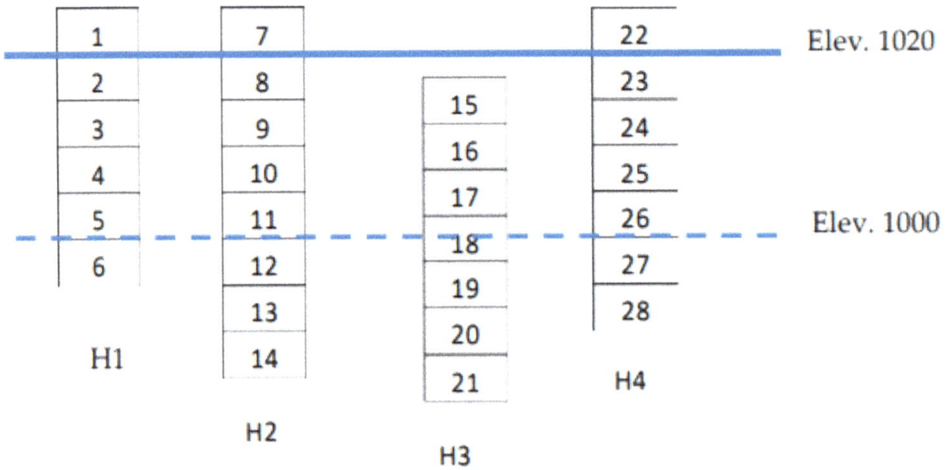

Figure 2. Cross section view of the example holes (four) containing a total of 28 lithological segments. Two lines show elevations ("Elev.") of 1020 and 1000.

In Figure 2, there are a total of 28 segments between the four holes. The figure also shows two lines that indicate two arbitrary elevations (1020 and 1000). These lines will be used later to explain additional concepts.

Assume that it is determined that 75% of the data will be selected for training. In the SB method, 21 segments are selected for training. Of course, segments are selected so that the training and testing subsets are similar in their distribution of rock types [19] or meet the real life considerations. In the SB method, each hole will likely contribute to both training and testing subsets. In the HB method, selection is made by holes and not by segments. Therefore, 75% of the holes are selected for the training subset. Each segment in the selected hole contributes only to the training subset. Segments in the other holes are all in the testing subset.

Note that regardless of method, there would be exactly 28 rows of total data in the data set. However, while the number of rows in the training subset will be 21 in the SB approach, this will be different for the HB approach. It depends on which holes are selected for training and testing subsets. For example, if H1 is sent to the testing set, the training subset would have 22 rows.

2.2. Operational Situations and Their Relationship to Evaluation Methods

In a mine, there is information about rock type in areas that are drilled. However, information is often preferred at a more granular level for operational reasons. Many times, in this scenario, there is information available close to and surrounding the non-drilled location. This operational situation is reflected in the SB strategy, where rock types are predicted at locations close to where information is available. For example, if segments 3 and 5 in Figure 2 are in the test set, they are locations close to where information is available (segments 1, 2, 4, 6). Segments are about 5 m apart. Therefore, this is similar to desiring to know the rock type in a particular production blast, since drillhole spacing in a typical blast is 5-by-5 m at EMC. Knowing the rock type has immediate operational value as it can help predict grades or mineral processing complexities.

Another situation that occurs at a mine is when information is needed for areas where hole density is sparse. This scenario is captured by the HB method. Since the holes in the test set are not known to the model, this method simulates predicting an entire drillhole between known drillholes. The difference with SB is that the distance of testing segments from training segments is much larger in HB. In HB, when a prediction is made for a test segment, it is made based on segments (training data) that are in other holes. Since holes are 50 m or more apart, predictions are essentially for locations 50 m or more away from known data. In SB, however, predictions are made based on segments, some of which are in the same hole (perhaps as close as 5 m away). SB is thus a scenario where predictions are for locations that are near to locations with known data. Hence, SB-versus-HB is also a near-versus-far comparison.

A variant of the above scenario is when information is required at depths beyond the current drilling depth. In this situation, named "SB specific to elevation" (SBE), information is available up to a given elevation, while there is interest in knowing the rock types below this elevation. Therefore, using information up to this elevation, rock type has to be predicted for deeper locations (future benches) for short-term or long-term planning purposes. In this method, all segments above the specific elevation are in training subset, while locations deeper than that are in the test subset. To define terminology, SBE-1600-1300-30 indicates the SB evaluation method where segments between 1600 m and 1300 m elevations are part of the training subset. The "30" refers to the segments in the next 30 m of depth (1270–1300 m elevation). This 30 m forms the test set. Thus, the evaluation is occurring at 1300 m elevation, with 1600–1300 m being the training set and 1270–1300 being the test set.

In the label SBE-1600-1300-30, 1600–1300 is referred to as the training interval (TI) with a training width (TW) of 300 (1600–1300 = 300), while 30 is the evaluation width. Incidentally, the highest collar elevation is 1600 m and, therefore, when the training interval starts at 1600 m, it implies all segments up to a certain depth are included in the training subset.

One may also use Figure 2 to understand this method. When applied to Figure 2, SBE-1020-1000-5 would imply that all segments of the dataset between the thick blue line and the dashed blue line would be used in the training set. Predictions will be made for 5 m below this line, i.e., one segment below the dashed line. Note that in the dataset a segment is represented by the coordinates of its centroid. Therefore, unlike Figure 2, it is always clear whether a segment is above or below a line.

In the SB and HB strategies, training and testing subsets are selected by randomly splitting the datasets [25]. In the results section, it is shown that despite the random shuffling, the characterization of the subsets is almost identical in both strategies. In the SBE strategy, training data is everything within a particular training interval, while testing data is everything within a particular evaluation width that is just outside the training interval. Since the two subsets represent different 3D spaces, there is no reason for them to be similarly characterized. Normally, this would be an improper modeling approach. However, that concern does not apply here as the intention is to test if ML can predict just outside its training area.

The ML method used in the paper is random forest (RF). RF were used for two major reasons [26]. One, unlike geostatistics, RF do not require any assumptions on the distribution of data. Two, as explained in the section below, RF tend to generalize well. RF are not new to mining geology [27,28], but since they are not a common technique in mining they are briefly presented next.

2.3. Random Forest: Background

This paper is not intended to be a manual on random forest (RF). Those seeking a deeper understanding are referred to [29], the source for this introduction. First, a note on terminology. In machine learning terminology, 'feature' refers to a database field. A drillhole database that contains the coordinates (northing, easting, elevation) and the rock type code has four features. A RF developed to determine the rock type will then be based on three features (northing, easting, elevation).

To understand random forests, one must first understand decision trees. A decision tree is a series of yes/no questions that are used to sub-divide the samples in the training set. A question applied to a group of data acts like a boundary, as it splits the parent group into two. The child groups can then be further split using boundaries of their own. The application of decision trees is explained through an example.

Consider the training set in Figure 3 where each sample consists of x-coordinates, y-coordinates, and a binary class indicator (1 or 0). In this example, the goal of the decision tree is to determine the class for a given (x, y) location.

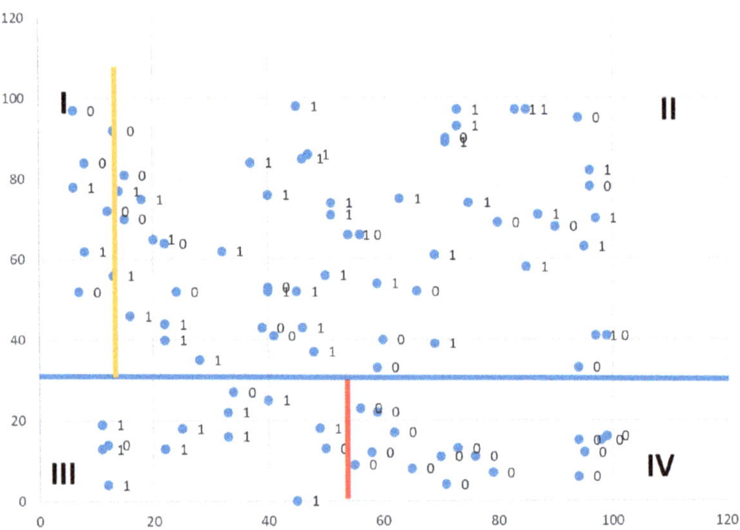

Figure 3. Example training data set showing the two classes (1 and 0) and their coordinates, x (horizontal axis) and y. The three lines shows three boundaries.

Assume that the tree starts with the blue boundary (Y > 36), splitting the data into two. The two resultant groups are further split using the red (bottom group) and yellow (top group) boundaries. The four subgroups are numbered I-IV to assist in the description. Assume that the above was the extent of the tree, and the modeler wishes to know the class for the test point (20,5). When the decision tree is applied to the point, it lands in Group III. Therefore, the class assigned to (20,5) is the class implied by the samples in Group III. Since 1's form the majority in Group III, the class assigned to (20,5) is 1. In a regression decision tree, the assigned value can be the mean or median (or any other appropriate statistic) of the group into which the point lands. In this example, any point being evaluated will face at most two boundaries. Therefore, the depth of the tree is 2. Figure 4 shows a representation of the decision tree, with the "yes" branch progressing to the left. The location at which a boundary exists is called a node, i.e., a group of data points is a node. The final nodes are also shown (I, II, III, and IV).

When a node is to be divided, one must first decide which feature to use for the boundary. In this example, two features are available to be used as a basis for dividing the boundary. The first boundary in the above example could have been on the X-axis instead of the Y-axis. The next design choice is to identify where to locate the boundary on the selected feature. In this example, the choice was to locate the first boundary at 36 (i.e., Y > 36). Most decision tree algorithms make both choices at once. If the number of features is low, one could systematically apply boundaries in all the features, and then pick the one where the resultant child groups have the least error (i.e., each node is homogenous and contains only or mostly samples from the same category). Notice how group IV contains only 0. This node can no longer be divided as it is fully homogeneous. The process of dividing nodes can continue till the final nodes are all homogenous or have at least one sample. One may also choose to limit the depth of the tree. Usually, a tree that is too deep may not be generalized. When the number of features is large, to reduce computations, the algorithm may randomly choose a set of features to be used a basis for the boundary. Different features are then considered for different boundaries.

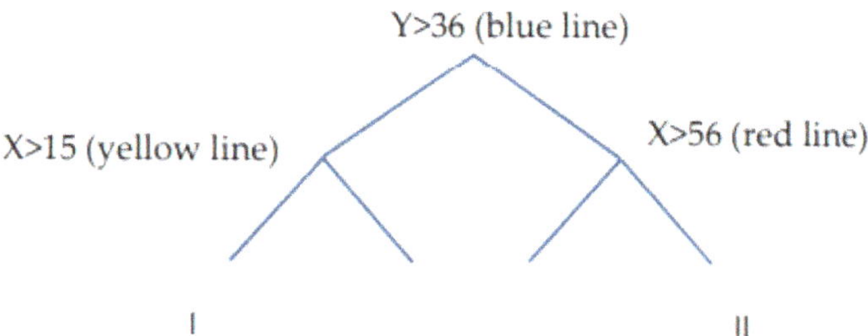

Figure 4. The example decision tree showing a tree depth of two. The labels (such as Y > 36) describe the decision boundary at a node.

In a decision tree, algorithms will generally yield the same set of boundaries for a given training set if all the features are considered for every boundary. In a random forest with N training data points, decision trees are formed by randomly selecting (with replacement) N of the training data points. Thus, the same data point may be selected many times for modeling a tree, at the cost of other data points that are not selected. Multiple trees are formed this way to make the forest. When the forest is applied to determine the category for a given test point, the decisions of the various trees in the forest are combined to form the final decision. One may use different strategies to combine the decisions. Random forests have been found to be superior to a single decision tree, with generalization not being an issue [26].

3. RF Modeling and Results

RF models were developed using the RandomClassifier() tool in scikit [30]. Only one hyper parameter was set: maximum tree depth (MTD). It was set using trial and error runs. Tree depth was increased until performance did not increase. In other words, the shortest tree depth for the highest performance was used as the setting. The task of the RF was to predict the rock class, GDIR (1) or not (0). Table 1 shows the distribution of GDIR rock type in the training and testing subsets for the various strategies. Table 2 shows the performance of the RF models for the various strategies.

Table 1. Data characterization for various evaluation strategies.

Strategy	MTD	NTrain	GDIR_Train	GDIR_Train_Prop	NTest	GDIR_Test	GDIR_Test_Prop	nonGDIR_Test
SB	20	45,016	18,696	42%	45,017	18,404	41%	26,613
SBE-1600-1300-30	25	45,603	20,872	46%	5473	2198	40%	3275
SBE-1600-1300-45	25	45,603	20,872	46%	7995	3216	40%	4779
SBE-1600-1300-60	25	45,603	20,872	46%	10,468	4230	40%	6238
SBE-1400-1300-30	25	28,531	12,744	45%	5473	2198	40%	3275
SBE-1400-1300-45	25	28,531	12,744	45%	7995	3216	40%	4779
SBE-1400-1300-60	25	28,531	12,744	45%	10,468	4230	40%	6238
SBE-1500-1200-30	25	61,589	27,411	45%	4093	1490	36%	2603

Table 1. *Cont.*

Strategy	MTD	NTrain	GDIR_Train	GDIR_Train_Prop	NTest	GDIR_Test	GDIR_Test_Prop	nonGDIR_Test
SBE-1500-1200-45	25	61,589	27,411	45%	6008	2171	36%	3837
SBE-1500-1200-60	25	61,589	27,411	45%	7786	2804	36%	4982
SBE-1300-1200-30	25	16,632	6590	40%	4093	1490	36%	2603
SBE-1300-1200-45	25	16,632	6590	40%	6008	2171	36%	3837
SBE-1300-1200-60	25	16,632	6590	40%	7786	2804	36%	4982
HB	25	45,154	18,467	41%	44,879	18,632	42%	26,247

MTD = Maximum Tree Depth; NTrain = Total rows in training subset; GDIR_Train = Number of rows in training set with GDIR; GDIR_Train_Prop = Proportion of GDIR in training subset; NTest = Total rows in testing subset; GDIR_Test = Number of rows in testing set with GDIR; GDIR_Test_Prop = Proportion of GDIR in testing subset; nonGDIR_Test = Number of rows in testing set with rocks other than GDIR

Table 2. Performance of RF models for various evaluation strategies.

Strategy	GDIR_success_num	GDIR_success_prop	GDIR False Positive	nonGDIR_success_num	nonGDIR_success_prop	OSR
SB	15,760	86%	9%	24,246	91%	89%
SBE-1600-1300-30	1584	72%	13%	2865	87%	81%
SBE-1600-1300-45	2179	68%	14%	4115	86%	79%
SBE-1600-1300-60	2758	65%	15%	5301	85%	77%
SBE-1400-1300-30	1414	64%	11%	2909	89%	79%
SBE-1400-1300-45	1939	60%	13%	4175	87%	76%
SBE-1400-1300-60	2444	58%	14%	5376	86%	75%
SBE-1500-1200-30	1209	81%	12%	2302	88%	86%
SBE-1500-1200-45	1704	78%	13%	3353	87%	84%
SBE-1500-1200-60	2146	77%	14%	4304	86%	83%
SBE-1300-1200-30	1415	95%	71%	763	30%	53%
SBE-1300-1200-45	2053	95%	71%	1100	29%	52%
SBE-1300-1200-60	2656	95%	71%	1424	29%	52%
HB	7756	42%	29%	18727	71%	59%

GDIR_success_num = Number of GDIR test rows successfully classified; GDIR_success_prop = Proportion of GDIR test rows successfully classified (100 × GDIR_success_num/ GDIR_Test); nonGDIR_success_num = Number of non-GDIR test rows successfully classified; nonGDIR_success_prop = Proportion of non-GDIR test rows successfully classified (100 × nonGDIR_success_num/non-GDIR_Test);

The results demonstrate the following:

- The proportion of GDIR in the training and testing subsets depend on the evaluation strategy.
  - In SB and HB, despite random shuffling, GDIR is split about evenly between training and testing subsets. This similarity between training and testing subsets is appropriate as both represent the same 3D space.
  - In the SBE strategies, the training subsets are much larger than the testing subsets, since the training interval (e.g. 1600–1300 implies a 300 m training interval) is much larger than the evaluation widths (e.g. 30 m). Since the two

subsets represent completely different 3D spaces, the proportion of GDIR and non-GDIR in the two subsets can be quite different.

- SBE models were developed for elevations of 1300 and 1200 m, as the mine is currently operating approximately between those levels.
- RF performs quite well in the SB strategy. 81% of GDIR in the test subset is detected, while 90% of non-GDIR is detected. The overall success rate (OSR) was 87%, i.e., 87% of the rocks are recognized correctly as GDIR or non-GDIR.
- In the SBE strategy (also see Figure 5):
  - ○ Notice how the performance lines in Figure 5 are inclined downwards to the right. In each scenario, the performance falls as the evaluation width increases from 30 m to 60 m. This is not surprising, as a larger evaluation width tests space farther away from the modeling space.
  - ○ The overall accuracy is higher for higher training intervals (Figure 6). Thus, at 1300 m, 1600–1300 (training interval = 300) outperforms 1400–1300 (training interval = 100). Similarly, at 1200 m, 1500–1200 outperforms 1300–1200. The effect is more pronounced at 1200 m elevation.
  - ○ The seemingly flawless performance for SBE-1300-1200 is misleading (Table 2, column GDIR_success_prop). The ability to classify 95% of the GDIR rock type as GDIR is paired with a 71% false positive rate. In other words, the classification of rock as GDIR is unreliable. This strategy classifies most segments as GDIR. Though that results in capturing all the GDIR, it also ends up classifying non-GDIR as GDIR. This is seen in the low success rate for classifying non-GDIR.
- The false positive rate of 9–15% (for most cases) is decent. This means that when a rock is classified as GDIR, it is most likely to be GDIR.
- HB strategy showed that predicting entire holes is difficult. When a hole is hidden in its entirety, only 42% of the GDIR rock segments in the hole are classified accurately. This is accompanied by a 29% false positive rate, which is not good.

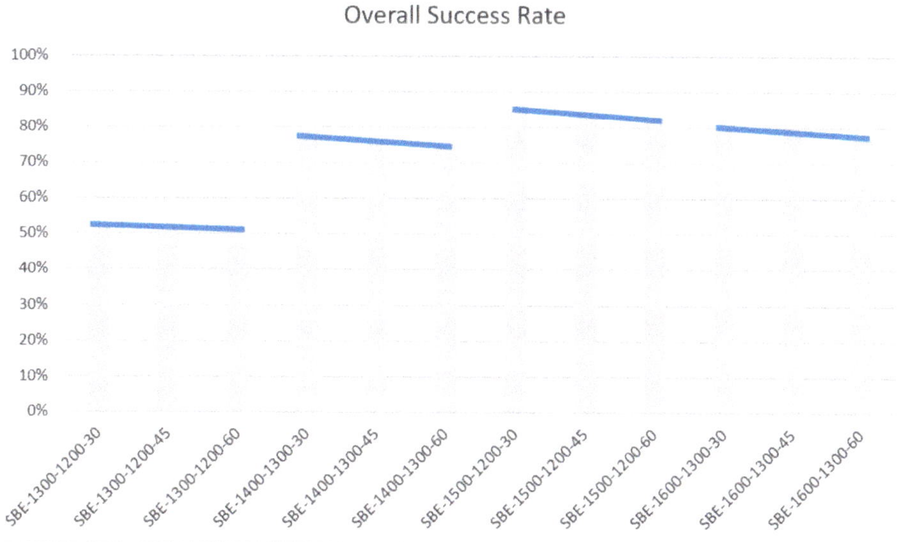

Figure 5. The overall success rate for the SBE strategy at each of the elevations, for different evaluation widths. Each blue line represents performance at a particular elevation. At each depth, performance falls as evaluation width increases. OSR = Overall success rate =100 × (GDIR_success_num + nonGDIR_success_num)/NTest.

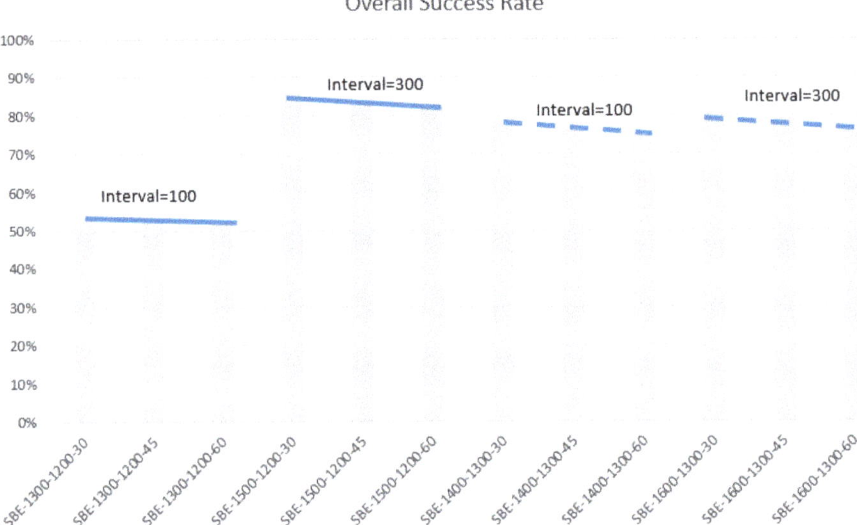

Figure 6. The overall success rate for the SBE strategy arranged by training width. Each line represents a particular elevation, with dashed lines representing 1300 m, while thick line representing 1200 m elevation.

4. Discussion

Most mining operations either use the manually developed rock type models or sensor technologies to make assumptions on the rock types contained within a drill block, or in future benches/blocks. This paper tested ML algorithms as an alternative to both approaches.

The SB strategy demonstrated that given a good density of information, the gaps can be predicted with high accuracy. This would suggest that ML of existing information may be a good substitute for using technologies to detect rock types, when information is available for nearby locations.

The SBE strategies demonstrated that mine planning can benefit from ML. Erdenet Copper Mine, with a bench height of 15 m, can predict rock type two to three benches below the current depth with significant reliability.

The HB strategy demonstrated that RF machine learning cannot yet replace a drilling campaign. The HB strategy simulated data sparsity. Without data density, ML can have problems. A research team [31] cited inadequate data as the reason for overfitting when applying neural networks to estimate grades based on sample locations, lithological features and alteration levels. Another team [28] cited data density as a concern when applying RF for mineral prospectivity mapping.

Despite the mixed results, there are advantages to using RF. Unlike geostatistics, no assumptions are made about the statistical characterization of drillhole data. However, RF performs about as well as geostatistics [32]. Performance aside, geostatistical methods take advantage of spatial relationships as defined by variograms. RF does not explicitly take advantage of spatial relationships. The K-nearest neighbor machine learning technique [33], which is a version of the common inverse distance squared technique in geostatistics, does take distances into consideration. However, it is not a sophisticated algorithm. It is possible that by incorporating spatial relationships such as variograms, RF or other machine learning techniques may perform better. This would be an excellent topic for future research, would be along approaches being attempted in recent times [18].

5. Conclusions

The machine learning technique random forest was applied to the exploration drillhole database at Erdenet Copper Mine in Mongolia to predict the presence of rock type granodiorite. Granodiorite is an important rock type at the mine as it contains 43% of the copper. The data consisted of 90,033 drillhole segments from 2823 drillholes. Most segments were 5 m in thickness. Two data selection approaches, segment-based and hole-based, were utilized to ensure that models could be tested to align with real life needs. Models were developed to test for three operational scenarios. The base SB method tested for the scenario when rock type is predicted at locations close to where rock types are known. This simulates the typical block that is blasted as part of day-to-day operation, where rock type is known in a relatively dense grid. The base HB method tested for the scenario where rock type is unknown for the entire length of a drillhole in between other drill holes. The SBE method tested for the scenario where rock type is known up to a given elevation but is unknown beyond that elevation. In the SBE method, rock types were predicted for 30, 45 and 60 m (evaluation width) beyond a specific elevation. The information made available to the models in the SBE method, or the training interval, varied from 100 m to 300 m. Given the 15 m benches at the mine, the 30, 45 and 60 m evaluation widths implied predictions to 2, 3 and 4 benches below where rock types were known.

The models performed very well in the SB scenario, with 86% of granodiorite being predicted accurately, with a false positive rate of 9%, resulting in an overall accuracy level of 89%. In the SBE method, the overall accuracy varied from 52% to 86%. Performance was better for higher training intervals, and for shorter evaluation widths. Performance was best in the SBE method at 1200 m, i.e., rock type was predicted better at 1200 m than at other elevations. The highest performance was achieved at 1200 m elevation with a training interval of 300 m and evaluation width of 30 m. The performance in the HB method was not encouraging, with an overall success rate of 59%.

This paper demonstrated that random forest-based machine learning can be very effective for predicting rock types in near distances. Predicting the entire length of a missing drillhole is, however, another story. The good performance of near-distance predictions should prompt mines to perhaps switch to machine learning over traditional manual modeling (or imperfect sensor technologies) to predict rock types in ore blocks blasted for production.

Author Contributions: Conceptualization, R.G.; data curation, N.S. and B.T.-A.; formal analysis, N.S, R.G.; funding acquisition, R.G.; investigation, N.S, R.G.; methodology, N.S, R.G, R.P; validation, N.S., R.G., R.P., B.T.-A.; visualization, N.S., R.G., R.P. and B.T.-A.; writing—original draft, N.S., R.G., R.P., and B.T.-A.; writing—review & editing, N.S., R.G., R.P. and B.T.-A. All authors have read and agreed to the published version of the manuscript.

Funding: This work was performed as part of an agreement between the University of Utah and Erdenet Mining Corporation.

Data Availability Statement: Not applicable.

Acknowledgments: The help of the geologists at EMC is gratefully acknowledged.

Conflicts of Interest: The authors declare no conflict of interest. The funders had no role in the design of the study; in the collection, analyses, or interpretation of data; in the writing of the manuscript, or in the decision to publish the results.

References

1. Dutta, S.; Ganguli, R.; Samanta, B. Investigation of Two Neural Network Methods in an Automatic Mapping Exercise. *Appl. GIS* **2005**, *1*, 1–19. [CrossRef]
2. Dutta, S.; Bandopadhyay, S.; Ganguli, R.; Misra, D. Machine Learning Algorithms and Their Application to Ore Reserve Estimation of Sparse and Imprecise Data. *J. Intell. Learn. Syst. Appl.* **2010**, *2*, 86–96. [CrossRef]
3. Yu, S.; Ganguli, R.; Bandopadhyay, S.; Patil, S.L.; Walsh, D.E. Calibration of online ash analyzers using neural networks. *Min. Eng.* **2004**, *56*, 99–102.
4. LaBelle, D. *Lithological Classification by Drilling*; Carnegie Mellon University: Pittsburgh, PA, USA, 2001.
5. Wu, X.; Zhou, Y. Reserve estimation using neural network techniques. *Comput. Geosci.* **1993**, *19*, 567–575. [CrossRef]
6. Fathi, M.; Alimoradi, A.; Hemati Ahooi, H.R. Optimizing Extreme Learning Machine Algorithm using Particle Swarm Optimization to Estimate Iron Ore Grade. *J. Min. Environ.* **2021**, *12*, 397–411. [CrossRef]

7. Samanta, B.; Bandopadhyay, S.; Ganguli, R.; Dutta, S. A comparative study of the performance of single neural network vs. Adaboost algorithm based combination of multiple neural networks for mineral resource estimation. *J. S. Afr. Inst. Min. Metall.* **2005**, *105*, 237–246.

8. Samanta, B.; Bandopadhyay, S.; Ganguli, R.; Dutta, S. An Application of Neural Networks to Gold Grade Estimation in Nome Placer Deposit. *J. S. Afr. Inst. Min. Met.* **2005**, *105*, 237–246.

9. Chatterjee, S.; Bandopadhyay, S.; Ganguli, R.; Bhattacherjee, A.; Samanta, B.; Pal, S.K. General regression neural network residual estimation for ore grade prediction of limestone deposit. *Min. Technol.* **2007**, *116*, 89–99. [CrossRef]

10. Tahmasebi, P.; Hezarkhani, A. A hybrid neural networks-fuzzy logic-genetic algorithm for grade estimation. *Comput. Geosci.* **2012**, *42*, 18–27. [CrossRef]

11. Mahmoudabadi, H.; Izadi, M.; Menhaj, M.B. A hybrid method for grade estimation using genetic algorithm and neural networks. *Comput. Geosci.* **2009**, *13*, 91–101. [CrossRef]

12. Jafrasteh, B.; Fathianpour, N. A hybrid simultaneous perturbation artificial bee colony and back-propagation algorithm for training a local linear radial basis neural network on ore grade estimation. *Neurocomputing* **2017**, *235*, 217–227. [CrossRef]

13. Jahangiri, M.; Ghavami Riabi, S.R.; Tokhmechi, B. Estimation of geochemical elements using a hybrid neural network-Gustafson-Kessel algorithm. *J. Min. Environ.* **2018**, *9*, 499–511. [CrossRef]

14. Jalloh, A.B.; Kyuro, S.; Jalloh, Y.; Barrie, A.K. Integrating artificial neural networks and geostatistics for optimum 3D geological block modeling in mineral reserve estimation: A case study. *Int. J. Min. Sci. Technol.* **2016**, *26*, 581–585. [CrossRef]

15. Dutta, S.; Misra, D.; Ganguli, R.; Samanta, B.; Bandopadhyay, S. A hybrid ensemble model of kriging and neural network for ore grade estimation. *Int. J. Min. Reclam. Environ.* **2006**, *20*, 33–45. [CrossRef]

16. Ganguli, R. A critical review of on-line quality analyzers. *Miner. Resour. Eng.* **2001**, *10*, 435–444. [CrossRef]

17. Samanta, B.; Ganguli, R.; Bandopadhyay, S. Comparing the predictive performance of neural networks with ordinary kriging in a bauxite deposit. *Min. Technol.* **2005**, *114*, 129–139. [CrossRef]

18. Samanta, B.; Bhattacherjee, A.; Ganguli, R. A genetic algorithms approach for grade control planning in a bauxite deposit. In *Proceedings of the 32nd International Symposium on the Application of Computers and Operations Research in the Mineral Industry*; APCOM: Tucson, AZ, USA, 2005.

19. Ganguli, R.; Dagdelen, K.; Grygiel, E. Systems engineering. In *Mining Engineering Handbook*; Darling, P., Ed.; Society for Mining, Metallurgy and Exploration, Inc.: Littleton, CO, USA, 2011.

20. Klyuchnikov, N.; Zaytsev, A.; Gruzdev, A.; Ovchinnikov, G.; Antipova, K.; Ismailova, L.; Muravleva, E.; Burnaev, E.; Semenikhin, A.; Cherepanov, A.; et al. Data-driven model for the identification of the rock type at a drilling bit. *J. Pet. Sci. Eng.* **2019**, *178*, 506–516. [CrossRef]

21. Zhou, H.; Hatherly, P.; Monteiro, S.T.; Ramos, F.; Oppolzer, F.; Nettleton, E.; Scheding, S. Automatic rock recognition from drilling performance data. In Proceedings of the 2012 IEEE International Conference on Robotics and Automation, Saint Paul, MN, USA; 2012; pp. 3407–3412.

22. Koch, P.-H.; Lund, C.; Rosenkranz, J. Automated drill core mineralogical characterization method for texture classification and modal mineralogy estimation for geometallurgy. *Miner. Eng.* **2019**, *136*, 99–109. [CrossRef]

23. Sinaice, B.; Owada, N.; Saadat, M.; Toriya, H.; Inagaki, F.; Bagai, Z.; Kawamura, Y. Coupling NCA Dimensionality Reduction with Machine Learning in Multispectral Rock Classification Problems. *Minerals* **2021**, *11*, 846. [CrossRef]

24. Gerel, O.; Munkhtsengel, B. Erdenetiin Ovoo Porphyry Copper-Molybdenum Deposit in Central Mongolia. In *Super Porphyry Copper & Gold Deposits: A Global Perspective*; Porter, T.M., Ed.; PGC Publishing: Adelaide, Australia, 2004.

25. Humphries, G.W.; Magness, D.R.; Huettmann, F. *Machine Learning for Ecology and Sustainable Natural Resource Management*; Springer: New York, NY, USA, 2018; ISBN 9783319969763.

26. Breiman, L. Random Forests. *Mach. Learn.* **2001**, *45*, 5–32. [CrossRef]

27. Jafrasteh, B.; Fathianpour, N.; Suárez, A. Comparison of machine learning methods for copper ore grade estimation. *Comput. Geosci.* **2018**, *22*, 1371–1388. [CrossRef]

28. McKay, G.; Harris, J.R. Comparison of the Data-Driven Random Forests Model and a Knowledge-Driven Method for Mineral Prospectivity Mapping: A Case Study for Gold Deposits Around the Huritz Group and Nueltin Suite, Nunavut, Canada. *Nat. Resour. Res.* **2016**, *25*, 125–143. [CrossRef]

29. Mitchell, T.M. *Machine Learning*; McGraw-Hill: New York, NY, USA, 1997; Volume 45, ISBN 0070428077.

30. Scikit-Learn. Scikit-Learn: Machine Learning in Python. *Scikit-Learn.* 2020. Available online: https://scikit-learn.org/stable/ (accessed on 15 July 2021).

31. Kaplan, U.; Topal, E. A New Ore Grade Estimation Using Combine Machine Learning Algorithms. *Minerals* **2020**, *10*, 847. [CrossRef]

32. Hengl, T.; Nussbaum, M.; Wright, M.; Heuvelink, G.; Gräler, B. Random forest as a generic framework for predictive modeling of spatial and spatio-temporal variables. *PeerJ* **2018**, *6*, e5518. [CrossRef] [PubMed]

33. Scikit-Learn. Nearest Neighbors. Available online: https://scikit-learn.org/stable/modules/neighbors.html#neighbors (accessed on 15 September 2021).

 *minerals*

Article

Integrating Production Planning with Truck-Dispatching Decisions through Reinforcement Learning While Managing Uncertainty

Joao Pedro de Carvalho * and Roussos Dimitrakopoulos *

COSMO—Stochastic Mine Planning Laboratory, Department of Mining and Materials Engineering, McGill University, 3450 University Street, Montreal, QC H3A 0E8, Canada
* Correspondence: joao.decarvalho@mail.mcgill.ca (J.P.d.C.); roussos.dimitrakopoulos@mcgill.ca (R.D.)

Abstract: This paper presents a new truck dispatching policy approach that is adaptive given different mining complex configurations in order to deliver supply material extracted by the shovels to the processors. The method aims to improve adherence to the operational plan and fleet utilization in a mining complex context. Several sources of operational uncertainty arising from the loading, hauling and dumping activities can influence the dispatching strategy. Given a fixed sequence of extraction of the mining blocks provided by the short-term plan, a discrete event simulator model emulates the interaction arising from these mining operations. The continuous repetition of this simulator and a reward function, associating a score value to each dispatching decision, generate sample experiences to train a deep Q-learning reinforcement learning model. The model learns from past dispatching experience, such that when a new task is required, a well-informed decision can be quickly taken. The approach is tested at a copper–gold mining complex, characterized by uncertainties in equipment performance and geological attributes, and the results show improvements in terms of production targets, metal production, and fleet management.

Keywords: truck dispatching; mining equipment uncertainties; orebody uncertainty; discrete event simulation; Q-learning

Citation: de Carvalho, J.P.; Dimitrakopoulos, R. Integrating Production Planning with Truck-Dispatching Decisions through Reinforcement Learning While Managing Uncertainty. *Minerals* **2021**, *11*, 587. https://doi.org/10.3390/min11060587

Academic Editors: Rajive Ganguli, Sean Dessureault and Pratt Rogers

Received: 16 April 2021
Accepted: 26 May 2021
Published: 31 May 2021

Publisher's Note: MDPI stays neutral with regard to jurisdictional claims in published maps and institutional affiliations.

1. Introduction

In short-term mine production planning, the truck dispatching activities aim to deliver the supply material, in terms of quantity and quality, being extracted from the mining fronts by the shovels to a destination (e.g., processing facility, stockpile, waste dump). The dispatching decisions considerably impact the efficiency of the operation and are of extreme importance as a large portion of the mining costs are associated with truck-shovel activities [1–4]. Truck dispatching is included under fleet optimization, which also comprises equipment allocation, positioning shovels at mining facies and defining the number of trucks required [2,5,6]. Typically, the truck dispatching and allocation tasks are formulated as a mathematical programming approach whose objective function aims to minimize equipment waiting times and maximize production [7–11]. Some methods also use heuristic rules to simplify the decision-making strategy [12–14]. In general, a limiting aspect of the structure of these conventional optimization methods is related to the need to reoptimize the model if the configuration of the mining complex is modified, for example, if a piece of fleet equipment breaks. Alternatively, reinforcement learning (RL) methods [15] provide means to make informed decisions under a variety of situations without retraining, as these methods learn from interacting with an environment and adapt to maximize a specific reward function. The ability to offer fast solutions given multiple conditions of the mining complex is a step towards generating real-time truck dispatching responses. Additionally, most methods dealing with fleet optimization are applied to single mines, whereas an industrial mining complex is a set of integrated operations and facilities transforming geological resource supply into sellable products. A mining complex can include multiple mines, stockpiles, tailing dams, processing routes, transportation systems, equipment types and sources of uncertainty [16–27].

The truck-dispatching model described herein can be viewed as a particular application belonging to the field of material delivery and logistics in supply chains, commonly modelled as vehicle routing problems and variants [28–30]. Dynamic vehicle routing problems [31,32] are an interesting field which allows for the inclusion of stochastic demands [33] and situations where the customer's requests are revealed dynamically [34]. These elements can also be observed in truck-dispatching activities in mining complexes, given that different processors have uncertain performances and that production targets may change, given the characteristics of the feeding materials. One particularity of the truck-dispatching model herein is that the trips performed between shovels and destinations usually have short lengths and are repeated multiple times. Another important aspect is that uncertainty arises from the geological properties of the transported materials and the performance of different equipment. Over the last two decades, there is an effort to develop frameworks accommodating uncertainties in relevant parameters within the mining complex operations to support more informed fleet management decisions. Not accounting for the complexity and uncertainties inherent to operational aspects misrepresent queue times, cycling times and other elements, which inevitably translates to deviation from production targets [6,35]. Ta et al. [9] allocate the shovels by a goal programming approach, including uncertainties in truckload and cycle times. Few other approaches optimize fleet management and production scheduling in mining complexes under both geological and equipment uncertainty [22,36,37].

A common strategy to model the stochastic interactions between equipment and processors in an operating mining environment is through the use of discrete event simulation (DES) approaches [35,38–43]. The DES allows for replacing an extensive mathematical description or rule concerning stochastic events by introducing randomness and probabilistic parameters related to a sequence of activities. The environment is characterized numerically by a set of observable variables of interest, such that each event modifies the state of the environment [44]. This simulation strategy can be combined with ideas from optimization approaches. Jaoua et al. [45] describe a detailed truck-dispatching control simulation, emulating real-time decisions, coupled with a simulated annealing-based optimization that minimizes the difference between tonnage delivered and associated targets. Torkamani and Askari-Nasab [35] propose a mixed integer programming model to allocate shovels to mining facies and establish the number of required truck trips. The solution's performance is assessed by a DES model that includes stochastic parameters such as truck speed, loading and dumping times, and equipment failure behavior. Chaowasakoo et al. [46] study the impact of the match factor to determine the overall efficiency of truck-shovel operations, combining a DES and selected heuristics maximizing production. Afrapoli et al. [47] propose a mixed integer goal programming to reduce shovel and truck idle times and deviations from production targets. A simulator of the mining operations triggers the model to be reoptimized every time a truck requires a new allocation. Afrapoli et al. [11] combine a DES with a stochastic integer programming framework to minimize equipment waiting times.

It is challenging to formulate all the dynamic and uncertain nature of the truck-shovel operation into a mathematical formulation. The daily operations in a mining complex are highly uncertain; for example, equipment failure, lack of staff or weather conditions can cause deviations in production targets and cause modifications in the dispatching policy. These events change the performance of the mining complex; thus, the related mathematical programming model needs to be reoptimized. The use of DES of the mining complex facilitates the modelling of such events. Note that some of the above mentioned approaches simulate the mining operations to assess the dispatching performance or improve it, using heuristic approaches. This strategy can generate good solutions, but the models do not learn from previous configurations of the mining complex.

Unlike the in the mentioned heuristic approaches, RL-based methods can take advantage of a mining complex simulator to define agents (decision-makers) that interact with this environment based on actions and rewards. The repetition of such interaction provides these agents with high learning abilities, which enables fast responses when a new assignment is required. Recent approaches have achieved high-level performances over multiple environments that require complex and dynamic tasks [48–55]. They have also been applied to some short-term mine planning aspects showing interesting results [56–58].

This paper presents a truck-dispatching policy based on deep Q-learning, one of the most popular RL approaches, in order to improve daily production and overall fleet performance, based on the work in Hasselt et al. [50]. A DES is used to model daily operational aspects, such as loading, hauling and dumping activities, generating samples, to improve the proposed truck dispatching policy. A case study applies the method to a copper–gold mining complex, which considers equipment uncertainty, modelled from historical data, and orebody simulations [59–63] that assess the uncertainty and variability related to metal content within the resource model. Conclusion and future work follow.

2. Method

The proposed method adapts the deep double Q-learning neural network (DDQN) method [50] for dispatching trucks in a mining environment. The RL agents continually take actions over the environment and receive rewards associated with their performances [15]. Herein, each mining truck is considered an agent; therefore, these terms are used interchangeably throughout this paper. The DES, described in Section 2.1, receives the decisions made by the agents, simulates the related material flow and a reward value evaluating each action. Section 2.2 defines the reward function and how the agents interact with the RL environment; where the observed states and rewards compose the samples used to train the DDQN. Subsequently, Section 2.3 presents the training algorithm based on Hasselt et al. [50], updating the agent's parameters (neural network weights). Figure 1 illustrates the workflow showing the interaction between the DES and the DDQN policy.

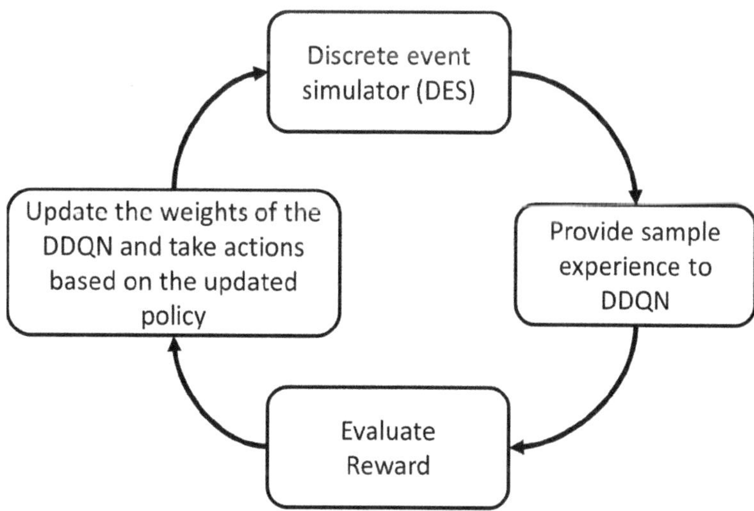

Figure 1. Workflow of the interaction between the DES and the DDQN method.

2.1. Discrete Event Simulator

The discrete event simulator presented in this work assumes a predefined sequence of extraction, the destination policy of each mining block and the shovel allocation. It also presumes that the shortest paths between shovels and destinations have been defined.

Figure 2 illustrates this relationship where the black arrow is the predefined destination path for the block being extracted by the shovel. After the truck delivers the material to the dumping point (waste dump, processing plant or leaching pad, for example), a dispatching policy must define the next shovel assignment. The red arrow illustrates the path options for dispatching.

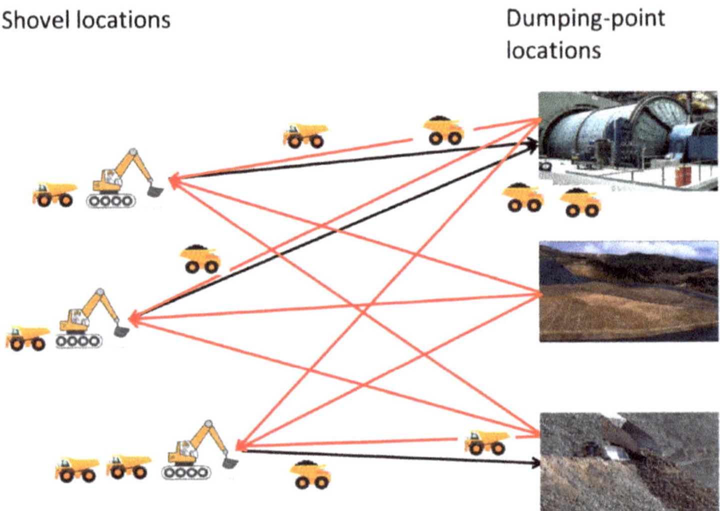

Figure 2. Representation of the possible paths a truck can follow: the pre-defined destination of each block (black arrow); the possible next dumping point (red arrows).

To simulate the operational interactions between shovels, trucks and dumping locations present in the mining complex, the DES considers the following major events:

Shovel Loading Event: The shovel loads the truck with an adequate number of loads. The total time required for this operation is stochastic, and once the truck is loaded, it leaves the shovel as the destination, triggering the "Truck Moving Event." If the shovel must move to a new extraction point, it incurs a delay, representing the time taken to reposition the equipment. After the truck leaves the loading point, this event can trigger itself if there is another truck waiting in the queue.

Truck Moving Event: This event represents the truck going from a shovel to a dumping location, or vice versa. Each travelling time is sampled from a distribution approximated from historical data. Travelling empty or loaded impacts on the truck speed, meaning that time values are sampled from different distributions in these situations. When the truck arrives at the loading point and the shovel is available, this event triggers a "Shovel Loading Event"; otherwise, it joins the queue of trucks. If the truck arrives at the dumping location, the event performs similarly; if the destination is empty, this event triggers a "Truck Dumping Event," otherwise, the truck joins the queue of trucks.

Truck Dumping Event: This event represents the truck delivering the material to its destination, to a waste dump or a processing plant, for example. The time to dump is stochastic, and after the event is resolved, a "Truck Moving Event" is triggered to send the truck back to be loaded. Here, a new decision can be made, sending the truck to a different shovel. Similar to the "Shovel Loading Event," once this event is finished, it can trigger itself if another truck is in the queue waiting for dumping.

Truck Breaking Event: Represents a truck stopping its activities due to maintenance or small failures. In this event, a truck is removed from the DES regardless of its current assignment. No action can be performed until it is fixed and can be returned to the operation.

Shovel Breaking Event: Represents the shovel becoming inaccessible for a certain period due to small failures or maintenance. No material is extracted during this period, and no trucks are sent to this location, being re-routed until the equipment is ready to be operational again.

Figure 3a shows a diagram illustrating a possible sequence of events that can be triggered. In the figure, the solid lines represent the events triggered immediately after the end of a particular event. The dashed lines are related to events that can be triggered if trucks are waiting in the queue. To ensure the sequence respects a chronological ordering, a priority queue is maintained, where each event is ranked by its starting time, as illustrated in Figure 3b.

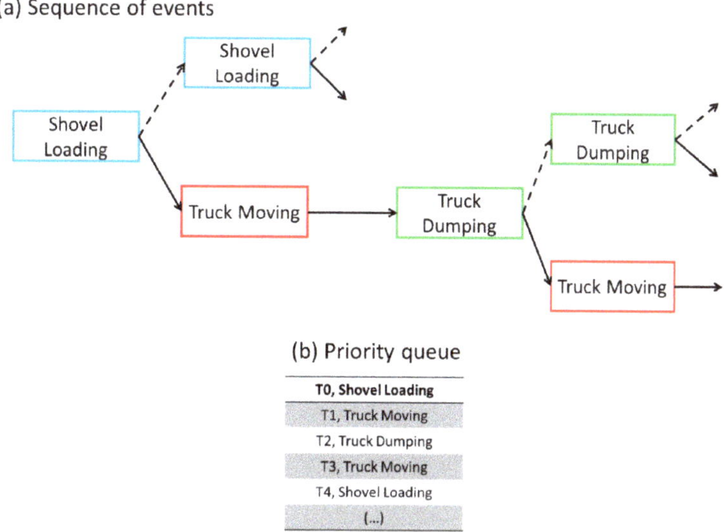

Figure 3. Discrete event simulation represented in terms of: (**a**) an initial event and possible next events that can be triggered; (**b**) a priority queue that ranks each event by its starting time.

The DES starts with all the trucks being positioned at their respective shovel. This configuration triggers a "Shovel Loading Event," and the DES simulates the subsequent events and how much material flows from the extraction point to their destinations by the trucks. Once the truck dumps, a new decision is taken according to the DDQN policy. The DES proceeds by simulating the subsequent operations triggered by this assignment. This is repeated until the predefined time horizon, which represents N_{days} of simulated activities, is reached by the DES. All events that occur between the beginning and the end of the DES constitute an episode. Subsequent episodes start by re-positioning the trucks at their initial shovel allocation.

2.2. Agent–Environment Interaction

2.2.1. Definitions

The framework considers N_{trucks} trucks interacting with the DES. At every time step $t \in T$, after dumping the material into the adequate location, a new assignment for truck $i \in N_{trucks}$ is requested. The truck-agent i observes the current state $S_t^i \in S$, where S_t^i represents the perception of truck i on how the mining complex is performing at step t and takes an action $A_t^i \in A$, defining the next shovel to which the truck will be linked. The state S_t^i is a vector encoding all attributes relevant to characterize the current status of the mining complex. Figure 4 illustrates these attributes describing the state space, such as current queue sizes, current GPS location of trucks and shovels, and processing plant requirements.

This state information is encoded in a vector and inputted into the DDQN neural network, which outputs action-values, one for each shovel, representing the probability that the truck be dispatched to a shovel-dumping point path. A more detailed characterization of the state S_t^i is given in Appendix A.

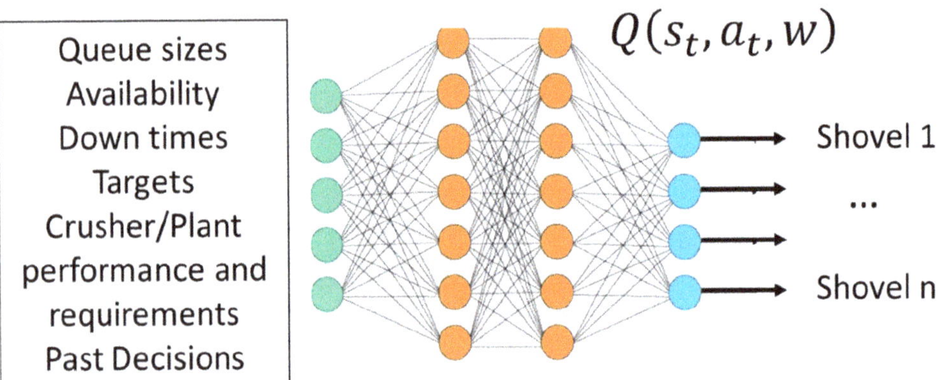

Figure 4. Illustration of the DDQN agent, which receives as input the state of the environment as input and outputs the desirability probability of choosing an action.

2.2.2. Reward Function

Once the agent outputs the action A_t^i, the DES emulates how the mining complex environment evolves by simulating, for example, new cycle times, the formation of queues, taking into consideration all other trucks in operation. The environment, then, replies to this agent's decision with a reward function, represented by Equation (1):

$$R_t^i = perc_t^i - pq_t^i \tag{1}$$

where $perc_t^i$ is the reward associated with delivering material to the mill and accomplishing a percentage of the destination's requirement (e.g., mill's daily target in tons/day). pq_t^i is the penalty associated with spending time in queues at both shovels and destinations. This term guides solutions with smaller queue formation while ensuring higher productivity.

In this multi-agent setting, each truck receives a reward R_t, which is the sum of each truck R_t^i, as shown in Equation (2), to ensure that all agents aim to maximize the same reward function.

$$R_t = \sum_i^{N_{trucks}} R_t^i \tag{2}$$

During each episode, the agent performs N_{steps} actions, the discounted sum of rewards is the called return presented by Equation (3):

$$G_t = R_{t+1} + \gamma R_{t+2} + \gamma^2 R_{t+3} + \ldots + \gamma^{N_{steps}-t-1} R_{N_{steps}} = \sum_{k=t+1}^{N_{steps}} \gamma^{k-t-1} R_k \tag{3}$$

where γ is a discounting factor parameter, which defines how much actions taken far in the future impact the objective function [15]. Equation (4) defines the objective, which is to obtain high-level control by training the agent to take improved actions so that the trucks can fulfil the production planning targets and minimize queue formation.

$$\max_{a \in A} E\left[G_t \middle| S = S_t^i, A = A_t^i\right] \tag{4}$$

The environment is characterized by uncertainties related to loading, moving, dumping times of the equipment, breakdowns of both trucks and shovels. This makes it very difficult to define all possible transition probabilities between states $\left(p\left(S_{t+1}^i \mid S = S_t^i, A = A_t^i\right)\right)$ to obtain the expected value defined in Equation (4). Therefore, these transition probabilities are replaced by the Monte Carlo approach used in the form of the DES.

The framework allows for future actions to be rapidly taken since providing the input vector S_t to the neural network and outputting the corresponding action is a fast operation. This means that the speed at which the decisions can be made depends more on how quickly the attributes related to the state of the mining complex can be collected, which has been recently substantially improved with the new sensors installed throughout the operation.

2.3. Deep Double Q-Learning (DDQN)

The approach used in the current study is the double deep Q-learning (DDQN) approach based on the work of Hasselt et al. [50]. Q-function $Q_i\left(S_t^i, A_t^i, w_t^i\right)$ is the action-value function, shown in Equation (5), which outputs values representing the likelihood of truck i choosing action A_t^i, given the encoded state S_t^i and the set of neural-network weights w_t^i, illustrated by Figure 4.

$$Q_i\left(S_t^i, A_t^i, w_t^i\right) = E\left[G_t \mid S = S_t^i, A = A_t^i\right] \tag{5}$$

Denote $Q_i^*\left(s_t^i, a_t^i, w^i\right)$ to be the theoretical optimal action-value function. Equation (6) presents the optimal policy $\pi^*\left(S_t^i\right)$ for the state S_t^i, which is obtained by using the action-function greedily:

$$\pi^*\left(S_t^i\right) = \underset{a' \in A}{\mathrm{argmax}} Q_i^*\left(S_t^i, a', w^i\right) \tag{6}$$

Note that, using Equation (6), the approach directly maximizes the reward function described in Equation (4). This is accomplished by updating the $Q_i\left(S_t^i, A_t^i, w_t^i\right)$ function to approximate the optimal action-value function $\left(Q_i\left(S_t^i, A_t^i, w_t^i\right) \rightarrow Q_i^*\left(S_t^i, A_t^i, w^i\right)\right)$.

By letting agent i interact with the environment, given the state S_t^i, the agent chooses A_t^i, following a current dispatching policy $\pi_i\left(S_t^i\right) = \underset{a' \in A}{\mathrm{argmax}} Q_i\left(S_t^i, a', w_t^i\right)$, the environment then returns the reward R_t and a next state S_{t+1}^i. The sample experience $e_k^i = \left(S_t^i, A_t^i, R_t, S_{t+1}^i\right)$ is stored in a memory buffer, $D_K^i = \left\{e_1^i, e_2^i, \ldots, e_K^i\right\}$, which is increased as the agent interacts with the environment for additional episodes. A maximum size limits this buffer, and once it is reached, the new sample e_k^i replaces the oldest one. This is a known strategy called experience replay, which helps stabilize the learning process [48,50,64].

In the beginning, $Q_i\left(S_t^i, A_t^i, w_t^i\right)$ is randomly initialized, then a memory tuple e_k^i is repeatedly uniformly sampled from the memory buffer D_K^i, and the related $e_k^i = \left(S_t^i, A_t^i, R_t, S_{t+1}^i\right)$ values are used to estimate the expected future return $\overline{G}_t$, as shown in Equation (7):

$$\overline{G}_t = \begin{cases} R_t, & \textit{if episode terminates at } t+1 \\ R_t + \gamma \overline{Q}_i\left(S_t^i, \underset{a' \in A}{\mathrm{argmax}} \, Q_i\left(S_{t+1}^i, a', w_t^i\right), \overline{w}^i\right), & \textit{otherwise} \end{cases} \tag{7}$$

Additionally, gradient descent is performed on $\left(\overline{G}_t - Q_i\left(S_t^i, A_t^i, w_t^i\right)\right)^2$ with respect to the parameter weights w_t^i. Note that a different Q-function, $\overline{Q}_i(\cdot)$, is used to predict the future reward; this is simply the $Q_i(\cdot)$ with the old weight parameters. Such an approach is also used to stabilize the agent's learning, as noisy environments can result in a slow learning process [50]. After N_{Updt} steps, the weights w_t^i are copied to $\overline{w}^i$, as follows: $\overline{Q}_i(\cdot) = Q_i(\cdot)$.

During training, the agent i follows the greedy policy $\pi_i(S_t^i)$ meaning that it acts greedily with respect to its current knowledge. If gradient descent is performed with samples coming solely from $\pi_i(S_t^i)$, the method inevitably would reach a local maximum very soon. Thus, to avoid being trapped in a local maximum, in $\epsilon\%$ of the time, the agent takes random actions exploring the solution space, sampling it from a uniform distribution $A_t^i \sim U(A)$. In $(100 - \epsilon)\%$ of the time, the agent follows the current policy $A_t^i \sim \pi_i(S_t^i)$. To take advantage of long-term gains, after every N_{steps_reduce} steps this value is reduced by a factor $reduce\_factor \in [0, 1]$. In summary, the algorithm is presented as follows:

Algorithm 1 Proposed learning algorithm.

Initialize the action-functions. $Q_i(\cdot)$ and $\overline{Q}_i(\cdot)$ by assigning initial weights to w_t^i and $\overline{w}^i$.
Set $n1_{counter} = 0$ and $n2_{counter} = 0$.
Initialize the DES, with the trucks at their initial locations (e.g., queueing them at the shovel).
Repeat for each episode:
    Given the current truck-shovel allocation, the DES simulates the supply material being transferred from mining facies to the processors or waste dump by the trucks.
    Once the truck i dumps the material, a new allocation must be provided.
    At this point, the agent collects the information about the state S_t^i.
    Sample $u \sim U(0, 100)$
    If $u < \epsilon\%$
        The truck-agent i acts randomly $A_t^i \sim U(A)$
    Else:
        The truck-agent i acts greedily $A_t^i \sim \pi_i(S_t^i)$
    Taking action A_t^i, observe R_t and a new state S_{t+1}^i.
    Store the tuple $e_k^i = \left(S_t^i,\ A_t^i,\ R_t,\ S_{t+1}^i\right)$ in the memory buffer D_K^i
    Sample a batch of experiences $e_k^i = \left(S_t^i,\ A_t^i,\ R_t,\ S_{t+1}^i\right)$, of size $batch\_size$, from D_K^i:
    For each transition sampled, calculate the respective G_t from Equation (7).
    Perform gradient descent on $\left(Q_1^i\left(s_{t+1}, a', w_1^i\right) - \overline{G}_t\right)^2$ according to Equation (8):

$$w_{1,next}^i \leftarrow w_{1,old}^i - \alpha \left(Q_1^i\left(s_{t+1}, a', w_1^i\right) - G_t\right) \nabla_{w_1^i} Q_1^i\left(s_{t+1}, a', w_1^i\right) \tag{8}$$

      $n1_{counter} \leftarrow n1_{counter} + 1.$
      $n2_{counter} \leftarrow n2_{counter} + 1.$
      If $n1_{counter} \geq N_{Updt}$:
        $\overline{w}^i \leftarrow w_t^i.$
        $n1_{counter} \leftarrow 0.$
      If $n2_{counter} \geq N_{step\_reduce}$:
        $\epsilon \leftarrow \epsilon * reduce\_factor.$
        $n2_{counter} \leftarrow 0.$

---

3. Case Study at a Copper—Gold Mining Complex

3.1. Description and Implementation Aspects

The proposed framework is implemented at a copper–gold mining complex, summarized in Figure 5. The mining complex comprises two open-pits, whose supply material is extracted by four shovels and transported by twelve trucks to the appropriate destinations: waste dump, mill or leach pad. Table 1 presents information regarding the mining equipment and processors. The shovels are placed at the mining facies following pre-defined extraction sequences, where the destination of each block was also pre-established beforehand. The mining complex shares the truck fleet between pits A and B. The waste dump receives waste material from both mines, whereas the leach pad material only processes supply material from pit B due to mineralogical characteristics. The truck going to the leach pad dumps the material into a crusher, then transported it to the leach pad. Regarding the milling material, each pit is associated with a crusher, and the trucks haul the high-grade material extracted from a pit and deliver it to the corresponding crusher. Next, a conveyor belt transfers this material to the mill combining the material from the two sources. Both the mill and the leach pad are responsible for producing copper products and gold ounces to be sold.

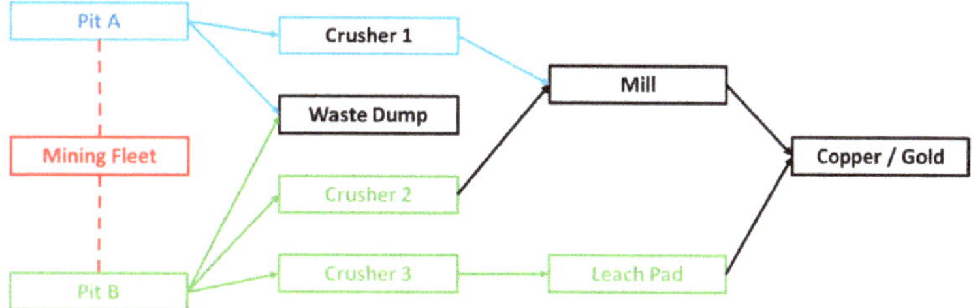

Figure 5. Diagram of the mining complex.

Table 1. Mining complex equipment and processors.

Equipment	Description
Trucks 12 in total	6 of payload capacity of 200 tons
	3 of payload capacity of 150 tons
	3 of payload capacity of 250 tons
Shovel 4 in total	2 of bucket payload of 80 tons
	1 of bucket payload of 60 tons
	1 of bucket payload of 85 tons
Mill	Capacity 80,000 ton/day, with 2 crushers.
Leach Pad	Capacity 20,000 ton/day, with one crusher.
Waste Dump	1 Waste Dump with no limitation on capacity.

The discrete event simulation, described in Section 2.1, emulates the loading, hauling and dumping operations in the mining complex. Each event is governed by uncertainties that impact the truck cycling times. Table 2 presents distributions used for the related uncertainty characterization. For simplicity, these stochastic distributions are approximated from historical data; however, a more interesting approach would have been to use the distribution directly from historical data. When the truck dumps material into a destination, a new dispatching decision must be taken by the DDQN dispatching policy. This generates samples that are used to train the DDQN dispatching policy. During the training phase, each episode lasts the equivalent of 3 days of continuous production, where the truck-agent interacts with the discrete event mine simulator environment, taking actions and collecting rewards. In total, the computational time needed for training, for the present case study, is around 4 h. For the comparison (testing) phase, the method was exposed to five consecutive days of production. This acts as a validation step, ensuring that the agents observe the mining complex's configurations which were unseen during training. The results presented show the five days of production, and the performance obtained illustrates that the method does not overfit regarding the three days of operation but maintain a consistent strategy for the additional days.

Table 2. Definition of stochastic variables considered in the mining complex.

Stochastic Variable	Probability Distribution
Loaded truck speed (km/h)	Normal (17, 4)
Empty truck speed (km/h)	Normal (35, 6)
Dumping + maneuver time (min)	Normal (1, 0.15)
Shovel bucketing load time (min)	Normal (1.1, 0.2)
Truck mean time between failures (h)	Poisson (36)
Truck mean time to repair (h)	Poisson (5)
Shovel mean time between failures (h)	Poisson (42)
Shovel mean time to repair (h)	Poisson (4)

Note that although the DDQN policy provides dispatching decisions considering a different context from the one it was trained, the new situations cannot be totally different. It is assumed that in new situations, the DDQN experiences are sampled from the same distribution observed during training. In the case where the sequence of extraction changes considerably and new mining areas as well as other destinations are prioritized, the model needs to be retrained.

Two baselines are presented to compare the performance of the proposed approach. The first one, referred to as fixed policy, is a strategy that continually dispatches the truck to the same shovel path throughout the episode. The performance comparison between the DDQN and fixed policy is denoted Case 1. The second approach, referred to as greedy policy, sends trucks to needy shovels with the shortest waiting times to decrease idle shovel time, denoted Case 2. Both cases start with the same initial placement of the trucks.

The environment is stochastic, in the sense that testing the same policy for multiple episodes generates different results. Therefore, for the results presented here, episodes of 5 days of continuous production are repeated 10 times for each dispatching policy. To assess uncertainty outcomes beyond the ones arising from operational aspects, geological uncertainty is also included in the assessment by considering 10 orebody simulations (Boucher and Dimitrakopoulos; 2009) characterizing the spatial uncertainty and variability of copper and gold grades in the mineral deposit. The graphs display results in P10, P50 and P90 percentile, corresponding to the probability of 10, 50 and 90%, respectively, of being below the value presented.

3.2. Results and Comparisons

Figure 6 presents the daily throughput obtained by running the DES over the five days of production, which is achieved by accumulating all material processed by the mill within each day. Note that here the P10, P50 and P90 are only due the equipment uncertainty. Overall, the proposed model delivers more material to the mill when compared to both cases. The DDQN method adapts the dispatching to move trucks around, relocating them to the shovels that are more in need, which constantly results in higher throughput.

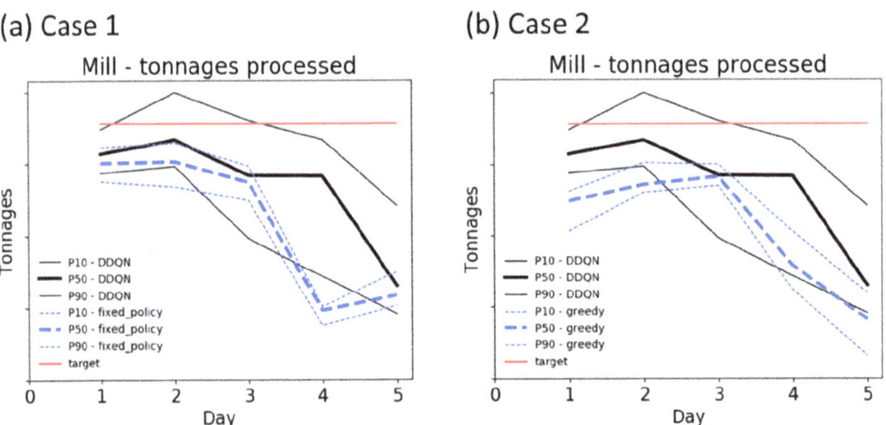

Figure 6. Daily throughput at the mill compared the DDQN policy (black line) and the respective baselines (blue line): (**a**) fixed policy and (**b**) greedy policy.

The throughput in day five drops compared to previous days, mostly due to a smaller availability of trucks as the DES considers failures in the trucks; Figure 7 presents the average number of trucks available per day. During the initial three days, the availability of trucks hovers between 10 and 12 trucks, but this rate drops in the last 2 days, which decreases the production. However, the trained policy can still provide a higher feed rate at the mill, even in this adversity. The availability of trucks on days 4 and 5 is smaller than the period for which the DDQN based method was trained, which shows an adapting capability of the dispatching approach.

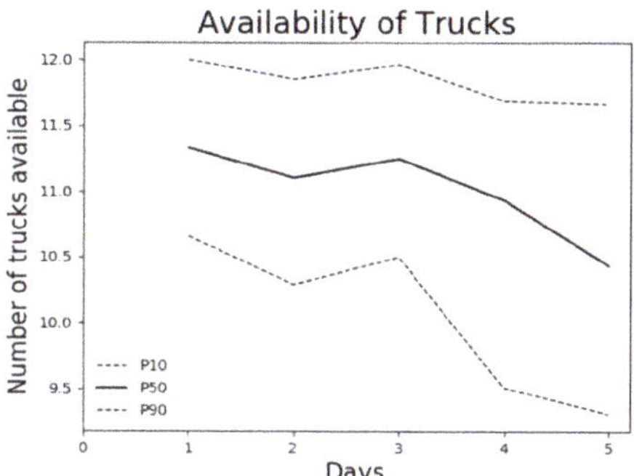

Figure 7. Availability of trucks during the five days of operation.

The framework is also efficient in avoiding queue formation. Figure 8 presents the average queue sizes at the mill and the sulphide leach. The queue at different locations is recorded hourly and averaged over each day. The plot shows that, for most of the days, the proposed approach generates smaller queues. Combined with the higher throughput obtained, this reduction in queue sizes demonstrates better fleet management. For example, during the initial three days, the DDQN approach improves the dispatching strategy by forming smaller queues at the mill. At the same time, the amount of material being delivered is continuously higher. On the 4th day, the proposed approach generates a larger queue size at the mill, which is compensated by having considerably higher throughput at this location.

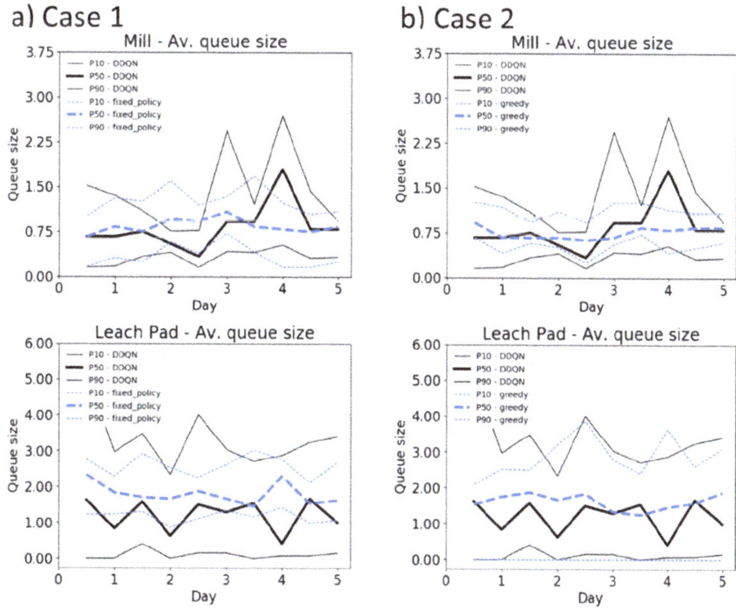

Figure 8. Queue sizes of trucks waiting at the mill (top) and Sulphide Leach (bottom) for the Deep DQN policy (black line) and the respective baseline (blue line): (**a**) fixed policy and (**b**) greedy policy.

Figure 9 displays the cumulative total copper recovered at the mining complex over the five days. Interestingly, during the first three days of DES simulation, corresponding to the training period of the DDQN approach, the total recovered copper profile between the proposed method and the baselines is similar. However, this difference is more pronounced over the last two days, which represents the situation that the trained method has not seen. This results in 16% more copper recovered than the fixed policy and 12% more than the greedy strategy. This difference in results is even larger when the total gold recovered is compared. The DDQN method generates a 20 and 23% higher gold profile in Case 1 and Case 2, respectively, Figure 10.

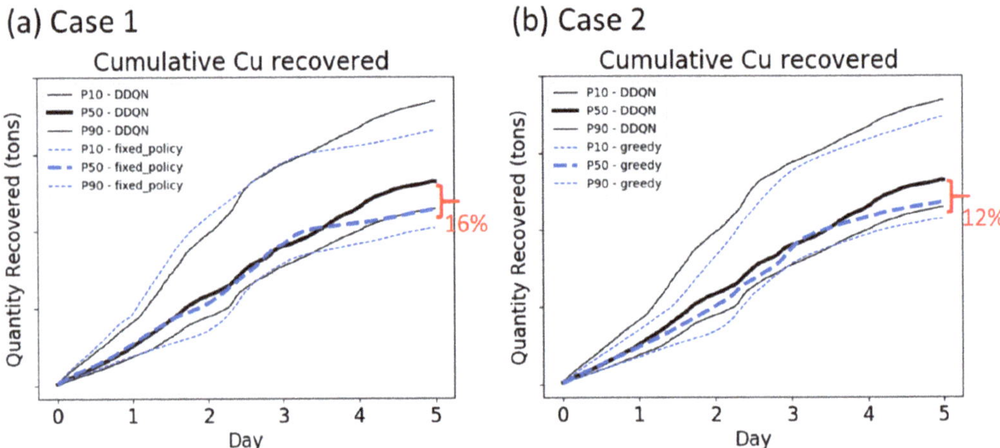

Figure 9. Cumulative copper recovered for the optimized DDQN policy (black line) and the respective baseline (blue line): (**a**) Case 1 and (**b**) Case 2.

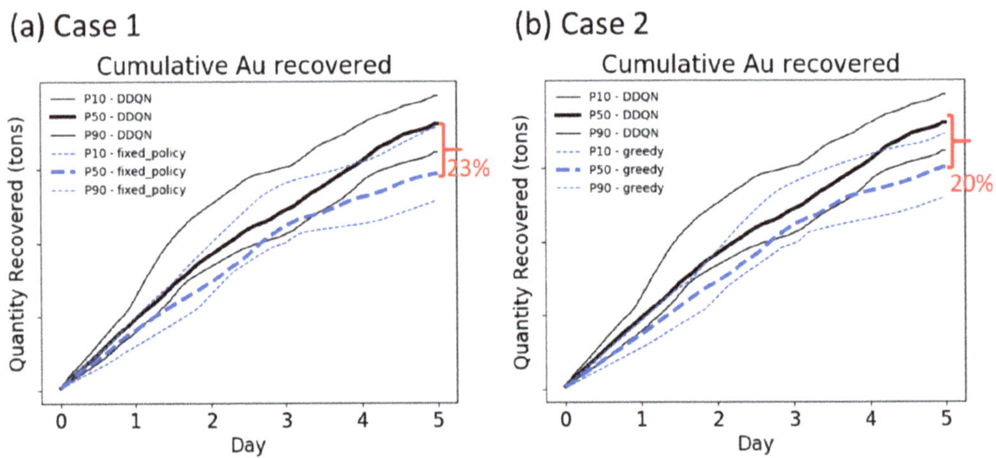

Figure 10. Cumulative gold recovered for the DDQN policy (black line) and the respective baseline (blue line): (**a**) fixed policy and (**b**) greedy policy.

4. Conclusions

This paper presents a new multi-agent truck-dispatching framework based on a reinforcement learning framework. The approach involves the interaction between a DES, simulating the operational events in a mining complex, and a truck-dispatching policy based on the DDQN method. Given a pre-defined schedule in terms of the sequence of extraction and destination policies for the mining blocks, the method improves the real-time truck-dispatching performance. The DES mimics daily operations, including loading, transportation and dumping, and equipment failures. A truck delivers the material to a processor or waste dump, and the truck-dispatcher provides it with a different shovel path. At this point, the truck receives information about the mining complex, such as other truck locations via GPS tracking, the amount of material feeding the processing plant and queue sizes at different locations. This state information is encoded into a vector, characterizing the state of the mining complex. This vector is inputted into the DDQN neural network, which outputs action values, describing the likelihood to send the truck to each shovel. Each dispatching decision yields a reward, which is received by the agent, as a performance evaluation. Initially, the truck-agent acts randomly; as the agent experiences many situations during training, the dispatching policy is improved. Thus, when new dispatching decisions are requested, an assignment is quickly obtained by the output of the DDQN agent. It differs from previous methods that solve a different optimization repeatedly during dispatching. Instead, the only requirement is to collect information regarding the state of the mining complex. With the digitalization of the mines, obtaining the required information can be done quickly.

The method is applied to a copper–gold mining complex composed of two pits, three crushers, one waste dump, one mill and one leach-pad processing stream. The fleet is composed of four shovels, and twelve trucks that can travel between the two pits. The DDQN-based method is trained for the equivalent of three days, while the results are presented for five days of production. Two dispatching baseline policies are used for comparison to assess the capabilities of the proposed method: fixed truck-shovel allocations and a greedy approach that dispatches trucks to needy shovels with the smallest queue sizes. The results show that the DDQN-based method provides the mill processing stream with higher throughput while generating shorter queues at different destinations, which shows a better fleet utilization. Over the five days of production, the proposed policy produces 12 to 16% more copper and 20 to 23% more gold than the baseline policies. Overall, the reinforcement learning approach has shown to be effective in training truck-dispatching agents, improving real-time decision-making. However, future work needs explore the development of new approaches that address the impact and adaptation of truck-dispatching decisions to changes and re-optimization of short-term extraction sequences given to the acquisition of new information in real-time and uncertainty in the properties of the materials mind.

Author Contributions: Conceptualization, J.P.d.C. and R.D.; methodology, J.P.d.C. and R.D.; software, J.P.d.C.; validation, J.P.d.C. and R.D; formal analysis, J.P.d.C. and R.D; investigation, J.P.d.C.; resources, R.D.; data curation, J.P.d.C.; writing—original draft preparation, J.P.d.C. and R.D.; writing—review and editing, R.D.; visualization, J.P.d.C.; supervision, R.D.; project administration, R.D.; funding acquisition, R.D. Both authors have read and agreed to the published version of the manuscript.

Funding: This work is funded by the National Science and Engineering Research Council of Canada (NSERC) CRD Grant CRDPJ 500414-16, NSERC Discovery Grant 239019, and the COSMO mining industry consortium (AngloGold Ashanti, AngloAmerican, BHP, De Beers, IAMGOLD, Kinross, Newmont Mining and Vale).

Institutional Review Board Statement: Not applicable.

Informed Consent Statement: Not applicable.

Data Availability Statement: Not applicable.

Conflicts of Interest: The authors declare no conflict of interest.

Appendix A.

Appendix A.1. State Definition

The definition of the state of the mining complex vector S_t^i encodes all attributes relevant to characterize the current status of the mining complex. Table A1 presents where the attributes are taken from and how it is represented in a vector format. Note that the encoding used here simply transforms the continuous attributes into values between 0 and 1, by a division of a large number. For discrete ones, a one-hot-encoding approach is used, where the number of categories defines the size of the vector, and a value of 1 is placed in the location corresponding to the actual category. This strategy attempts to avoid generating large gradients during gradient descent and facilitates the learning process. This idea can be further generalized, and other attributes judged relevant by the user can also be included.

Table A1. Attributes defining the current state of the mining complex.

Attribute in Consideration		Representation
Shovel related attributes	Destination policy of the block being currently extracted	1-hot-encoded vector (3-dimensional)
	Destination policy of next 2 blocks	1-hot-encoded (6 dimensional in total)
	Shovel capacity	1 value divided by the largest capacity
	Variable indicating if the shovel is currently in maintenance	1 value (0 or 1)
	Current distance to destination	1 value divided by a large number
	Number of trucks associated	1 value divided by a large number
	Approximated queue sizes	1 value divided by a large number
	Approximated waiting times	1 value divided by a large
Number of attributes per shovel		15
Destination related attributes	% target processed	1 value
	Amount of material received at crushers	2 values divided by a large number
	Distance to each shovel	4 values dived by a large number
	Approximated queue sizes	1 value divided by a large number
	Approximated waiting times	1 value divided by a large number
Number of attributes per destination		9
Truck related attributes	Truck capacity	1 value divided by the largest capacity
	Current number of trucks currently in operation.	1 value divided by the total number of trucks
	The last shovel visited	1-hot-encoded (4 values)
Number of attributes of each truck		
Total of attributes		102

Appendix A.2. Neural Network Parameters

Table A2. Reinforcement learning parameters.

Neural Network	Input layer = 102 nodes with ReLU activation function; Hidden layer 306 nodes with ReLU activation function; Output layer: 4 nodes without activation function.
Gradient descent	Adam optimization, with learning rate = 2×10^{-4}.
DDQN parameters	$\gamma = 0.99$ $\epsilon = 0.25$, with *reduce_factor* $= 0.98$ 10,000 episodes of training.

References

1. Chaowasakoo, P.; Seppälä, H.; Koivo, H.; Zhou, Q. Digitalization of Mine Operations: Scenarios to Benefit in Real-Time Truck Dispatching. *Int. J. Min. Sci. Technol.* **2017**, *27*, 229–236. [CrossRef]
2. Alarie, S.; Gamache, M. Overview of Solution Strategies Used in Truck Dispatching Systems for Open Pit Mines. *Int. J. Surf. Min. Reclam. Environ.* **2002**, *16*, 59–76. [CrossRef]
3. Munirathinarn, M.; Yingling, J.C. A Review of Computer-Based Truck Dispatching Strategies for Surface Mining Operations. *Int. J. Surf. Min. Reclam. Environ.* **1994**, *8*, 1–15. [CrossRef]
4. Niemann-Delius, C.; Fedurek, B. Computer-Aided Simulation of Loading and Transportation in Medium and Small Scale Surface Mines. In *Mine Planning and Equipment Selection 2004*; Hardygora, M., Paszkowska, G., Sikora, M., Eds.; CRC Press: London, UK, 2004; pp. 579–584.
5. Blom, M.; Pearce, A.R.; Stuckey, P.J. Short-Term Planning for Open Pit Mines: A Review. *Int. J. Min. Reclam. Environ.* **2018**, *0930*, 1–22. [CrossRef]
6. Afrapoli, A.M.; Askari-Nasab, H. Mining Fleet Management Systems: A Review of Models and Algorithms. *Int. J. Min. Reclam. Environ.* **2019**, *33*, 42–60. [CrossRef]
7. Temeng, V.A.; Otuonye, F.O.; Frendewey, J.O. Real-Time Truck Dispatching Using a Transportation Algorithm. *Int. J. Surf. Min. Reclam. Environ.* **1997**, *11*, 203–207. [CrossRef]
8. Li, Z. A Methodology for the Optimum Control of Shovel and Truck Operations in Open-Pit Mining. *Min. Sci. Technol.* **1990**, *10*, 337–340. [CrossRef]
9. Ta, C.H.; Kresta, J.V.; Forbes, J.F.; Marquez, H.J. A Stochastic Optimization Approach to Mine Truck Allocation. *Int. J. Surf. Min. Reclam. Environ.* **2005**, *19*, 162–175. [CrossRef]
10. Upadhyay, S.P.; Askari-Nasab, H. Truck-Shovel Allocation Optimisation: A Goal Programming Approach. *Min. Technol.* **2016**, *125*, 82–92. [CrossRef]
11. Afrapoli, A.M.; Tabesh, M.; Askari-Nasab, H. A Transportation Problem-Based Stochastic Integer Programming Model to Dispatch Surface Mining Trucks under Uncertainty. In Proceedings of the 27th International Symposium on Mine Planning and Equipment Selection—MPES 2018; Springer: Berlin/Heidelberg, Germany, 2019; pp. 255–264. Available online: https://www.springerprofessional.de/en/a-transportation-problem-based-stochastic-integer-programming-mo/16498454 (accessed on 15 May 2021).
12. White, J.W.; Olson, J.P. Computer-Based Dispatching in Mines with Concurrent Operating Objectives. *Min. Eng.* **1986**, *38*, 1045–1054.
13. Soumis, F.; Ethier, J.; Elbrond, J. Truck Dispatching in an Open Pit Mine. *Int. J. Surf. Min. Reclam. Environ.* **1989**, *3*, 115–119. [CrossRef]
14. Elbrond, J.; Soumis, F. Towards Integrated Production Planning and Truck Dispatching in Open Pit Mines. *Int. J. Surf. Min. Reclam. Environ.* **1987**, *1*, 1–6. [CrossRef]
15. Sutton, R.S.; Barto, A.G. *Reinforcement Learning: An Introduction*, 2nd ed.; MIT Press: Cambridge, UK, 2018.
16. Del Castillo, M.F.; Dimitrakopoulos, R. Dynamically Optimizing the Strategic Plan of Mining Complexes under Supply Uncertainty. *Resour. Policy* **2019**, *60*, 83–93. [CrossRef]
17. Montiel, L.; Dimitrakopoulos, R. Simultaneous Stochastic Optimization of Production Scheduling at Twin Creeks Mining Complex, Nevada. *Min. Eng.* **2018**, *70*, 12–20. [CrossRef]
18. Goodfellow, R.; Dimitrakopoulos, R. Global Optimization of Open Pit Mining Complexes with Uncertainty. *Appl. Soft Comput. J.* **2016**, *40*, 292–304. [CrossRef]
19. Pimentel, B.S.; Mateus, G.R.; Almeida, F.A. Mathematical Models for Optimizing the Global Mining Supply Chain. In *Intelligent Systems in Operations: Methods, Models and Applications in the Supply Chain*; Nag, B., Ed.; IGI Global: Hershey, PA, USA, 2010; pp. 133–163.
20. Levinson, Z.; Dimitrakopoulos, R. Adaptive Simultaneous Stochastic Optimization of a Gold Mining Complex: A Case Study. *J. S. African Inst. Min. Metall.* **2020**, *120*, 221–232. [CrossRef] [PubMed]
21. Saliba, Z.; Dimitrakopoulos, R. Simultaneous stochastic optimization of an open pit gold mining complex with supply and market uncertainty. *Min. Technol.* **2019**, *128*, 216–229. [CrossRef]
22. Both, C.; Dimitrakopoulos, R. Joint Stochastic Short-Term Production Scheduling and Fleet Management Optimization for Mining Complexes. *Optim. Eng.* **2020**. [CrossRef]
23. Whittle, J. The Global Optimiser Works—What Next? In *Advances in Applied Strategic Mine Planning*; Dimitrakopoulos, R., Ed.; Springer International Publishing: Cham, Switzerland, 2018; pp. 31–37.
24. Whittle, G. Global asset optimization. In *Orebody Modelling and Strategic Mine Planning: Uncertainty and Risk Management Models*; Dimitrakopoulos, R., Ed.; Society of Mining: Carlton, Australia, 2007; pp. 331–336.
25. Hoerger, S.; Hoffman, L.; Seymour, F. Mine Planning at Newmont's Nevada Operations. *Min. Eng.* **1999**, *51*, 26–30.
26. Chanda, E. *Network Linear Programming Optimisation of an Integrated Mining and Metallurgical Complex. Orebody Modelling and Strategic Mine Planning*; Dimitrakopoulos, R., Ed.; AusIMM: Carlton, Australia, 2007; pp. 149–155.

27. Stone, P.; Froyland, G.; Menabde, M.; Law, B.; Pasyar, R.; Monkhouse, P. Blasor–Blended Iron Ore Mine Planning Optimization at Yandi, Western Australia. In *Orebody Modelling and Strategic Mine Planning: Uncertainty and Risk Management Models*; Dimitrakopoulos, R., Ed.; Spectrum Series 14; AusIMM: Carlton, Australia, 2007; Volume 14, pp. 133–136.

28. Sitek, P.; Wikarek, J.; Rutczyńska-Wdowiak, K. Capacitated Vehicle Routing Problem with Pick-Up, Alternative Delivery and Time Windows (CVRPPADTW): A Hybrid Approach. In *Distributed Computing and Artificial Intelligence, Proceedings of the 16th International Conference, Special Sessions, Avila, Spain, 26–28 June 2019*; Springer International Publishing: Cham, Switzerland, 2019.

29. Gola, A.; Kłosowski, G. Development of Computer-Controlled Material Handling Model by Means of Fuzzy Logic and Genetic Algorithms. *Neurocomputing* **2019**, *338*, 381–392. [CrossRef]

30. Bocewicz, G.; Nielsen, P.; Banaszak, Z. Declarative Modeling of a Milk-Run Vehicle Routing Problem for Split and Merge Supply Streams Scheduling. In *Information Systems Architecture and Technology: Proceedings of 39th International Conference on Information Systems Architecture and Technology—ISAT 2018*; Świątek, J., Borzemski, L., Wilimowska, Z., Eds.; Springer: Cham, Switzerland, 2019; pp. 157–172.

31. Pillac, V.; Gendreau, M.; Guéret, C.; Medaglia, A.L. A Review of Dynamic Vehicle Routing Problems. *Eur. J. Oper. Res.* **2013**, *225*, 1–11. [CrossRef]

32. Gendreau, M.; Potvin, J.-Y. Dynamic Vehicle Routing and Dispatching. In *Fleet Management and Logistics*; Crainic, T.G., Laporte, G., Eds.; Springer US: Boston, MA, USA, 1998; pp. 115–126.

33. Secomandi, N.; Margot, F. Reoptimization Approaches for the Vehicle-Routing Problem with Stochastic Demands. *Oper. Res.* **2009**, *57*, 214–230. [CrossRef]

34. Azi, N.; Gendreau, M.; Potvin, J.-Y. A Dynamic Vehicle Routing Problem with Multiple Delivery Routes. *Ann. Oper. Res.* **2012**, *199*, 103–112. [CrossRef]

35. Torkamani, E.; Askari-Nasab, H. A Linkage of Truck-and-Shovel Operations to Short-Term Mine Plans Using Discrete-Event Simulation. *Int. J. Min. Miner. Eng.* **2015**, *6*, 97–118. [CrossRef]

36. Matamoros, M.E.V.; Dimitrakopoulos, R. Stochastic Short-Term Mine Production Schedule Accounting for Fleet Allocation, Operational Considerations and Blending Restrictions. *Eur. J. Oper. Res.* **2016**, *255*, 911–921. [CrossRef]

37. Quigley, M.; Dimitrakopoulos, R. Incorporating Geological and Equipment Performance Uncertainty While Optimising Short-Term Mine Production Schedules. *Int. J. Min. Reclam. Environ.* **2019**, *34*, 362–383. [CrossRef]

38. Bodon, P.; Fricke, C.; Sandeman, T.; Stanford, C. Combining Optimisation and Simulation to Model a Supply Chain from Pit to Port. *Adv. Appl. Strateg. Mine Plan.* **2018**, *251–267*. [CrossRef]

39. Hashemi, A.S.; Sattarvand, J. Application of ARENA Simulation Software for Evaluation of Open Pit Mining Transportation Systems—A Case Study. In Proceedings of the 12th International Symposium Continuous Surface Mining—Aachen 2014; Niemann-Delius, C., Ed.; Springer: Cham, Switzerland, 2015; pp. 213–224. Available online: https://link.springer.com/chapter/10.1007/978-3-319-12301-1_20 (accessed on 15 May 2021).

40. Jaoua, A.; Riopel, D.; Gamache, M. A Framework for Realistic Microscopic Modelling of Surface Mining Transportation Systems. *Int. J. Min. Reclam. Environ.* **2009**, *23*, 51–75. [CrossRef]

41. Sturgul, J. *Discrete Simulation and Animation for Mining Engineers*; CRC Press: Boca Raton, FL, USA, 2015.

42. Upadhyay, S.P.; Askari-Nasab, H. Simulation and Optimization Approach for Uncertainty-Based Short-Term Planning in Open Pit Mines. *Int. J. Min. Sci. Technol.* **2018**, *28*, 153–166. [CrossRef]

43. Yuriy, G.; Vayenas, N. Discrete-Event Simulation of Mine Equipment Systems Combined with a Reliability Assessment Model Based on Genetic Algorithms. *Int. J. Min. Reclam. Environ.* **2008**, *22*, 70–83. [CrossRef]

44. Law, A.M.; Kelton, W.D. *Simulation Modeling and Analysis*; McGraw-Hill.: New York, NY, USA, 1982.

45. Jaoua, A.; Gamache, M.; Riopel, D. Specification of an Intelligent Simulation-Based Real Time Control Architecture: Application to Truck Control System. *Comput. Ind.* **2012**, *63*, 882–894. [CrossRef]

46. Chaowasakoo, P.; Seppälä, H.; Koivo, H.; Zhou, Q. Improving Fleet Management in Mines: The Benefit of Heterogeneous Match Factor. *Eur. J. Oper. Res.* **2017**, *261*, 1052–1065. [CrossRef]

47. Afrapoli, A.M.; Tabesh, M.; Askari-Nasab, H. A Multiple Objective Transportation Problem Approach to Dynamic Truck Dispatching in Surface Mines. *Eur. J. Oper. Res.* **2019**, *276*, 331–342. [CrossRef]

48. Mnih, V.; Kavukcuoglu, K.; Silver, D.; Rusu, A.A.; Veness, J.; Bellemare, M.G.; Graves, A.; Riedmiller, M.; Fidjeland, A.K.; Ostrovski, G.; et al. Human-Level Control through Deep Reinforcement Learning. *Nature* **2015**, *518*, 529–533. [CrossRef] [PubMed]

49. Mnih, V.; Badia, A.P.; Mirza, M.; Graves, A.; Lillicrap, T.P.; Harley, T.; Silver, D.; Kavukcuoglu, K. Asynchronous Methods for Deep Reinforcement Learning. In Proceedings of the 33rd International Conference on Machine Learning, New York, NY, USA, 20–22 June 2016; Balcan, M.F., Weinberger, K.Q., Eds.; Volume 48, pp. 1928–1937.

50. Van Hasselt, H.; Guez, A.; Silver, D.; Van Hasselt, H.; Guez, A.; Silver, D. Deep Reinforcement Learning with Double Q-Learning. In Proceedings of the 30th AAAI Conference, Shenzhen, China, 21 February 2016.

51. Silver, D.; Schrittwieser, J.; Simonyan, K.; Antonoglou, I.; Huang, A.; Guez, A.; Hubert, T.; Baker, L.; Lai, M.; Bolton, A.; et al. Mastering the Game of Go without Human Knowledge. *Nature* **2017**, *550*, 354–359. [CrossRef]

52. Schrittwieser, J.; Antonoglou, I.; Hubert, T.; Simonyan, K.; Sifre, L.; Schmitt, S.; Guez, A.; Lockhart, E.; Hassabis, D.; Graepel, T.; et al. Mastering Atari, Go, Chess and Shogi by Planning with Learned Model. *Nature* **2020**, *588*, 604–609. [CrossRef]

53. Silver, D.; Huang, A.; Maddison, C.J.; Guez, A.; Sifre, L.; Van Den Driessche, G.; Schrittwieser, J.; Antonoglou, I.; Panneershelvam, V.; Lanctot, M.; et al. Mastering the Game of Go with Deep Neural Networks and Tree Search. *Nature* **2016**, *529*, 484–489. [CrossRef]
54. Vinyals, O.; Ewalds, T.; Bartunov, S.; Georgiev, P.; Vezhnevets, A.S.; Yeo, M.; Makhzani, A.; Küttler, H.; Agapiou, J.; Schrittwieser, J.; et al. StarCraft II: A New Challenge for Reinforcement Learning.cs.LG. *arXiv* **2017**, arXiv:1708.04782.
55. Vinyals, O.; Babuschkin, I.; Czarnecki, W.M.; Mathieu, M.; Dudzik, A.; Chung, J.; Choi, D.H.; Powell, R.; Ewalds, T.; Georgiev, P.; et al. Grandmaster level in StarCraft II using multi-agent reinforcement learning. *Nat. Cell Biol.* **2019**, *575*, 350–354. [CrossRef]
56. Paduraru, C.; Dimitrakopoulos, R. Responding to new information in a mining complex: Fast mechanisms using machine learning. *Min. Technol.* **2019**, *128*, 129–142. [CrossRef]
57. Kumar, A.; Dimitrakopoulos, R.; Maulen, M. Adaptive self-learning mechanisms for updating short-term production decisions in an industrial mining complex. *J. Intell. Manuf.* **2020**, *31*, 1795–1811. [CrossRef]
58. Kumar, A. Artificial Intelligence Algorithms for Real-Time Production Planning with Incoming New Information in Mining Complexes. Ph.D. Thesis, McGill University, Montréal, QC, Canada, 2020.
59. Goovaerts, P. *Geostatistics for Natural Resources Evaluation*; Applied Geostatistics Series; Oxford University Press: New York, NY, USA, 1997.
60. Remy, N.; Boucher, A.; Wu, J. *Applied Geostatistics with SGeMS: A User's Guide*; Cambridge University Press: Cambridge, UK, 2009; Volume 9780521514.
61. Rossi, M.E.; Deutsch, C.V. *Mineral Resource Estimation*; Springer: Dordt, The Netherlands, 2014.
62. Gómez-Hernández, J.J.; Srivastava, R.M. One Step at a Time: The Origins of Sequential Simulation and Beyond. *Math. Geol.* **2021**, *53*, 193–209. [CrossRef]
63. Minniakhmetov, I.; Dimitrakopoulos, R. High-Order Data-Driven Spatial Simulation of Categorical Variables. *Math. Geosci.* **2021**. [CrossRef]
64. Lin, L.-J. Reinforcement Learning for Robots Using Neural Networks. Ph.D. Thesis, Carnegie Mellon University, Pittsburgh, PA, USA, 1993.

Article

Partial Least Squares Regression of Oil Sands Processing Variables within Discrete Event Simulation Digital Twin

Ryan Wilson [1,*], Patrick H. J. Mercier [2,3], Bussaraporn Patarachao [2] and Alessandro Navarra [1]

1   Department of Mining and Materials Engineering, McGill University, 3610 University Street, Montreal, QC H3A 0C5, Canada; alessandro.navarra@mcgill.ca

2   National Research Council Canada, 1200 Montreal Road, Ottawa, ON K1A 0R6, Canada; patrick.mercier@corem.qc.ca (P.H.J.M.); Bussaraporn.Patarachao@nrc-cnrc.gc.ca (B.P.)

3   Corem, 1180 Rue de la Minéralogie, Quebec, QC G1N 1X7, Canada

*   Correspondence: ryan.wilson@mail.mcgill.ca

Abstract: Oil remains a major contributor to global primary energy supply and is, thus, fundamental to the continued functioning of modern society and related industries. Conventional oil and gas reserves are finite and are being depleted at a relatively rapid pace. With alternative fuels and technologies still unable to fill the gap, research and development of unconventional petroleum resources have accelerated markedly in the past 20 years. With some of the largest bitumen deposits in the world, Canada has an active oil mining and refining industry. Bitumen deposits, also called oil sands, are formed in complex geological environments and subject to a host of syn- and post-depositional processes. As a result, some ores are heterogeneous, at both individual reservoir and regional scales, which poses significant problems in terms of extractive processing. Moreover, with increased environmental awareness and enhanced governmental regulations and industry best practices, it is critical for oil sands producers to improve process efficiencies across the spectrum. Discrete event simulation (DES) is a computational paradigm to develop dynamic digital twins, including the interactions of critical variables and processes. In the case of mining systems, the digital twin includes aspects of geological uncertainty. The resulting simulations include alternate operational modes that are characterized by separate operational policies and tactics. The current DES framework has been customized to integrate predictive modelling data, generated via partial least squares (PLS) regression, in order to evaluate system-wide response to geological uncertainty. Sample computations that are based on data from Canada's oil sands are presented, showing the framework to be a powerful tool to assess and attenuate operational risk factors in the extractive processing of bitumen deposits. Specifically, this work addresses blending control strategies prior to bitumen extraction and provides a pathway to incorporate geological variation into decision-making processes throughout the value chain.

Keywords: discrete event simulation; digital twin; modes of operation; geological uncertainty; multivariate statistics; partial least squares regression; oil sands; bitumen extraction; bitumen processability

1. Introduction

With conventional oil and gas reservoirs being gradually depleted worldwide, activity in the research and exploitation of unconventional resources has grown exponentially over the past two decades. Global estimates of in-place bitumen and heavy oil resources are on the order of 5.9 trillion barrels (938 billion m^3), over 80% of which are concentrated in Canada, Venezuela and the United States [1]. Boasting the largest collection of these deposits globally with approximately 1.7 trillion barrels (270 billion m^3) of in-place resources [1], Canada is strategically positioned as an important source of unconventional petroleum products. Of this total, roughly 165 billion barrels (26.3 billion m^3) are considered technically recoverable and, thus, correspond to Canada's estimated remaining established reserves [1]. Unlike traditional light oil well drilling, which will decline over

---

Citation: Wilson, R.; Mercier, P.H.J.; Patarachao, B.; Navarra, A. Partial Least Squares Regression of Oil Sands Processing Variables within Discrete Event Simulation Digital Twin. *Minerals* **2021**, *11*, 689. https://doi.org/10.3390/min11070689

Academic Editor: Samintha Perera

Received: 30 April 2021
Accepted: 22 June 2021
Published: 26 June 2021

Publisher's Note: MDPI stays neutral with regard to jurisdictional claims in published maps and institutional affiliations.

time, forecasts show an overall 12% increase in unconventional petroleum production over the next 30 years, with peak rates reached in 2039 [2].

Bitumen and heavy oil reservoirs typically occur in unlithified sand deposits (also called "bituminous", "tar" or "oil" sands); however, heavy oil is also found within porous siliciclastic and carbonate host-rock successions due to its relative mobility [1]. These reservoirs are generally heterogeneous, containing a variety of different hydrocarbons across the American Petroleum Institute (API) gravity spectrum (from light oil to bitumen). By definition, heavy oil is marginally less dense than water (0.920 g/mL) and corresponds to API gravities in the range of 10–20°; conversely, bitumen and extra-heavy oils are denser than water with API gravities of less than 10° [1,3]. Crude bitumen and heavy oil resources are lacking in terms of lighter distillates, which significantly reduces their market value; consequently, both must undergo upgrading to increase commercial value and marketability. Upgrading to synthetic crude oil products (~20–40° API) involves the addition of hydrogen in order to attain H:C ratios similar to those of conventional crudes, as well as the removal of impurities such as nitrogen, sulphur, oxygen and heavy metals [1].

The current study is concerned with processes specific to the exploitation of oil sand deposits. Typically, economical oil sands contain on the order of 9–13% bitumen (soluble organic matter), 3–7% water and 80–85% mineral solids and insoluble organic matter [3,4]. Generally, 15–30% of the total solids are fines (mainly clays), less than 44 μm in diameter [4]. Most deposits are comprised of unconsolidated sand "bound" by a matrix of bitumen, with or without secondary cements and clays [3]. Despite broad acceptance of the origin and emplacement of bitumen reservoirs, oil sand deposits are subject to substantial levels of geological uncertainty in terms of host formation characteristics, ore composition, grade and overall processability. All of these process variables are strongly influenced by variable and complex host-formation and hydrocarbon depositional histories, in addition to post-depositional alteration processes such as biodegradation and in situ natural water washing [5,6]. Each of these contributing factors can lead to significant variability in ore feed particle size distributions, host-formation mineralogy and hydrocarbon chemistry and quality, all of which have direct impacts on downstream processing.

As with many types of mining projects, there are a variety of processes along the oil sands value chain; each of these streams may serve a particular function, but also require coordination of inputs and outputs with both up- and down-stream processes. This coordination can be difficult to maintain at times, even for systems that receive relatively stable ore feeds, but is exponentially problematic for projects dealing with heterogeneous ores. For example, blending strategies are common in mining for grade control or to minimize undesirable impurities in ore feeds; however, oil sand operations must also consider factors such as grain size distribution and mineral chemistry in order to regulate the transfer of intermediate products (e.g., slurries or froths) within hydrotransport pipelines. Furthermore, some complex ores may require additional treatment prior to, or between, conventional processing methods. For instance, heavy gas oil phases partitioned by distillation during upgrading are fed to fluid cokers and hydrocrackers to increase H:C ratios and break down long-chain molecules [7]. Similarly, problematic high chloride oil sands, often with elevated clay contents, could require ancillary control strategies (e.g., water content reduction, additives/inhibitors and blending) to reduce corrosive effects or blockages (i.e., ammonium chloride) related to hydrolysis reactions in downstream processes [8–11].

Improper process control strategies that do not suitably incorporate the geometallurgical profile of source ore feeds can result in major bottlenecks in mining and processing operations, ultimately leading to increased operating costs, decreased efficiencies and, even, potentially important reductions in project life. Several authors have emphasized the importance of establishing alternate modes of operation for mineral processing facilities [12,13]. Alternate modes are designed using mass balance and mathematical programming, and the operational decision to switch between modes is triggered by changing conditions or as critical thresholds are crossed [12,13]. The approach is particularly effective for complex mining systems dealing with heterogeneous ore feed and process variables, as demonstrated by recent studies [14–17]. Continuous online material and system process monitoring notwithstanding, a natural and powerful extension is to couple the development of operational modes with robust predictive models that benefit from earlier systematic sampling protocols. This type of integrated approach can lead to improved planning and fine-tuning earlier in the value chain, rather than being forced to make continual reactive adjustments based on process outputs (e.g., bitumen recoveries).

Given the high degree of variability inherent to oil sands mining and processing operations, it is evident that appropriate quantitative frameworks are needed in order to monitor system performance and response under changing conditions. One such computational intelligence tool is the digital twin, which is an integrated multidisciplinary probabilistic simulation of a system that uses the best available data models, updates and history to mimic the operational life of the corresponding physical system [18,19]. The development of digital twins using discrete event simulation (DES) is an effective approach because it allows for the simulation of interactions between critical parameters and processes with respect to random natural variations, e.g., geological uncertainty. Furthermore, DES models can also optimize trade-offs between available operational policies, as well as the limits that dictate the timing of their execution [15–17]. By simulating extended operating periods, potential deficiencies or bottlenecks in the coordination of system processes can be identified; strategic decisions can then be made to adjust operational policies, accordingly, thereby pre-emptively assessing and managing risk factors. The drive to improve overall system efficiencies is further amplified by increasingly stringent environmental regulations, industry best practices and pressure from community stakeholders.

This work introduces an extended framework capable of integrating predictive modelling using partial least squares (PLS) regression incorporated within a digital twin, for the evaluation of system response to geological uncertainty. A case study using data derived from Canada's oil sands is presented for a conceptual surface mining operation to assess the effect of implementing potential new ore blending schemes. The initial dataset was kindly acquired through partnerships with the National Research Council Canada (Ottawa, ON, Canada) and Syncrude Canada Ltd. (Research and Development–Edmonton, AB, Canada).

2. Background

2.1. Oil Sands Geology and Petrochemical Processing

Canada's vast oil sand resources are located almost exclusively in northeastern Alberta, within three core areas, namely the Peace River, Athabasca and Cold Lake deposit regions (Figure 1) [7,20]. Collectively, these accumulations span an area of approximately 142,000 km^2 [7,21], the largest of which is the Athabasca region, containing ~75% of the provincial reserves [20,22].

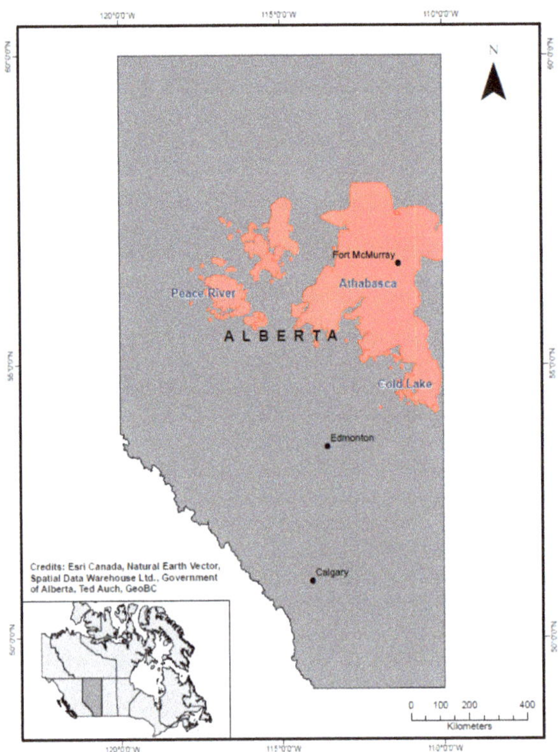

Figure 1. Location map of the Alberta Oil Sands Region (AOSR), showing the relative positions of the Peace River, Athabasca and Cold Lake oil sand deposit areas.

Generally, the Alberta oil sands have a thin overburden, and the deposits are concentrated within the early Cretaceous McMurray Formation (Mannville Group), which has variable thickness related to an original depositional surface defined by karstic features in underlying Devonian carbonates [3,4]. The present-day low API oils are the result of degraded Exshaw Shale sourced hydrocarbons [3,23,24]. Ores hosted within the McMurray Formation are subject to heterogeneities and related geological uncertainty, at both regional and individual reservoir scales. Key factors that affect the distribution and chemistry of bitumen include physical reservoir characteristics, mineralogy and mineral chemistry and fluid distribution and chemistry; these attributes are the result of dynamic host-formation and hydrocarbon depositional histories, as well as complex post-depositional alteration processes [6]. For example, though the predominant host successions are thought to have been deposited in estuarine settings, reservoirs have also been identified in fluvial and shallow marine settings, each with differing host porosities and permeabilities, mineral compositions, grain size distributions and related bitumen qualities [3,6,25,26]. Moreover, even broadly mappable geologic sequences (e.g., estuarine settings of the Middle McMurray Formation) may actually consist of multiple events overlapping in space and time, with each contributing several hierarchical heterogeneities [6,27]. Microbial degradation has been strongly linked to the quality of petroleum, having destroyed important proportions of originally emplaced conventional oil through the removal of lighter distillates [4,6]. In situ water washing, which removes the more water-soluble distillates through contact with formation waters, is considered the second most important post-depositional alteration process that affects the geochemistry, quality and bulk physical characteristics of petroleum accumulations [1,5].

Open-pit mining has traditionally been the dominant method implemented in the Alberta Oil Sands Region (AOSR), but in situ production first surpassed surface mining in 2012 and continued this trend into 2013 [1,28]. For aging open-pit mines to remain competitive with the newer in situ operations, they must be ready to implement changes to their process and, thus, adapt to their forecasted feeds. In oil sands surface mining, overburden is stripped, and conventional truck-and-shovel methods are used to excavate the ore, followed by a series of treatments to liberate the bitumen from the mineral grains for subsequent recovery and cleaning. First, the ore is crushed and mixed with water and additives to create a slurry, which is then pumped to the extraction facilities via hydrotransport pipelines [7,20]. Upon exit, the slurry is subject to water addition and gravity-settling separation processes, which produces a bitumen froth, a middlings stream and a first round of tailings. The aerated bitumen froth (comprising ~60% bitumen, 30% water and 10% fine solid particles) rises to the top of the separation vessel, meanwhile flotation cells are used to recover bitumen from the middlings [7]. The separated froth is deaerated and then sent to the froth treatment plant, where the addition of a light hydrocarbon solvent helps reduce the viscosity of the bitumen; this allows for more effective separation of any remaining impurities by centrifugation and inclined plate (gravity) settlers. The final product of this froth treatment process is called diluted bitumen (also referred to as "dilbit") and is accompanied by another tailings stream.

Depending on the choice of hydrocarbon solvent, the generated dilbit can require further upgrading (typical for naphthenic treatment) or head straight to the refinery market (possible with paraffinic treatment); in either case, the diluent is removed prior to further processing. Lighter hydrocarbon solvents yield cleaner dilbit products by reducing the viscosity of the emulsion, which allows for gravity-based removal of water and solids. Paraffinic solvents promote asphaltene (impurity) precipitation, whereas naphtha cannot do this at practical dilution rates. Upgrading converts viscous, hydrogen-deficient hydrocarbon with elevated impurity levels into high-quality synthetic crude oil products with density and viscosity attributes similar to those of conventional light sweet crude oil [7]. The process first splits the bitumen into hydrocarbon streams (i.e., light and heavy gas oil) in a vacuum distillation unit. The lighter distillates ("tops") are fed into hydrotreaters for stabilization and impurity removal (e.g., sulphur), meanwhile the heavier phases ("bottoms") are sent to fluid cokers (thermal conversion units) to remove carbon and to hydrocrackers where hydrogen is added and long-chain molecules are broken down [7].

2.2. Multivariate Statistics and Partial Least Squares (PLS) Regression

Multivariate statistics is a branch of statistics dealing with methods that examine the simultaneous effect of multiple variables [29]. Multivariate techniques extend the approach of univariate and bivariate investigations to include the analysis of covariances (or correlations) that reflect the extent of relationships between three or more variables, as well as similarities indicated by relative distances in *n*-dimensional space [29]. This area of research has expanded greatly over the past few decades due to significant technological advances in computing power and data frameworks.

Partial least squares (PLS) regression is a multivariate statistical method that combines and generalizes features from principal component analysis and multiple linear regression, with the objective of predicting a set of dependent variables from a potentially large set of independent variables [30,31]. The technique was pioneered in the 1960s by Herman Wold for use in the social sciences but has since gained traction in a variety of fields, including chemometrics, sensory evaluation and neuroimaging [30–33]. It is also becoming popular in the biological and environmental sciences with applications in soil and microbial ecology [34,35], biodiversity [36,37], paleo-climatological reconstruction [38] and ecotoxicology [39,40]. More recent studies in the geological disciplines have identified the use of near-infrared (NIR) or short-wave infrared (SWIR) reflectance measurements to build predictive models of metal concentrations in soils [41,42].

The main underlying computation for PLS is the singular value decomposition (SVD) of a matrix, which gives the best reconstruction (in a least squares sense) of the original data matrix by a matrix of lower rank (dimension reduction), while limiting the loss of significance [43]. The SVD is closely related to the well-known eigen-decomposition for non-symmetric matrices [44]. As a matter of notation, matrices are denoted by upper case bold letters, column vectors by lower case bold letters, and the superscript "T" is used to indicate transposition of either. Formally, the SVD of a given matrix, **R**, decomposes it into three matrices, comprising the left singular vectors, the singular values and the right singular vectors, as follows:

$$\mathbf{R} = \mathbf{U\Delta V^T} = \sum_{l}^{L} \mathbf{u}_l \delta_l \mathbf{v}_l^T \tag{1}$$

where **R** is the $J \times K$ correlation matrix, derived from the cross-product of the two original data tables (transpose $m \times n$ matrix of the independent variables, $\mathbf{X^T}$, and $n \times p$ matrix of the dependent variables, **Y**), as:

$$\sum_{k=1}^{n} \mathbf{x}_{ik}^T \mathbf{y}_{kj} \tag{2}$$

U is the $J \times L$ matrix of the left singular vectors (L corresponds to the rank of **R**), **Δ** is the $L \times L$ diagonal matrix of the singular values, **V** is the $K \times L$ matrix of the right singular vectors, and $\mathbf{u}_l$, δ_l and $\mathbf{v}_l^T$ are the lth left singular vector, singular value and right singular vector, respectively [43]. The non-negative singular values are sorted in decreasing order and represent the maximum covariance between each respective set of left and right singular vectors [45]. Note that both original sets of data are typically made comparable through statistical preprocessing (i.e., mean centering and rescaling each variable).

It is useful to explain the SVD from a geometric perspective as a series of orthogonal axis rotations and scaling of unit matrices about the origin. As shown in the simplified interpretation for a 2×2 matrix (Figure 2; after [46,47]), the SVD can be summarized as a linear transformation composed of three fundamental actions. These actions include: (1) rotation of the right singular vectors $\{\mathbf{v}_1, \mathbf{v}_2\}$ of matrix $\mathbf{V^T}$ within the original unit sphere; (2) scaling by the singular values $\{\delta_1, \delta_2\}$ of matrix **Δ**, which correspond to the length of the principal semiaxes of the new hyperellipse; (3) rotation of the left singular vectors $\{\mathbf{u}_1, \mathbf{u}_2\}$ of matrix **U**, oriented along the same principal semiaxes [47,48].

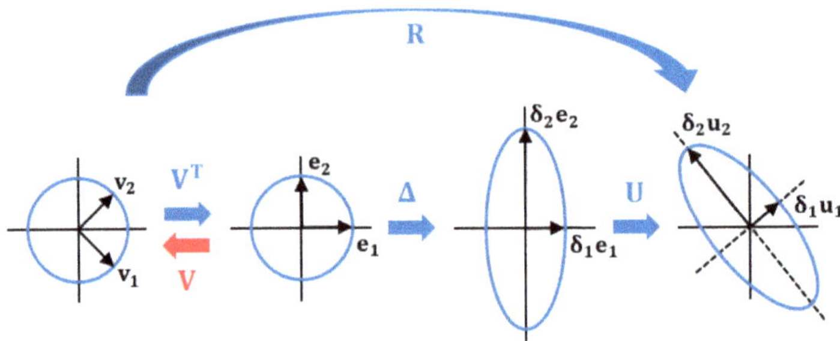

Figure 2. Geometric interpretation of the singular value decomposition (SVD) for a 2×2 matrix showing the linear transformation induced by matrix R decomposed into three actions: a rotation, a scaling and another rotation (after [46,47]).

The functional basis of PLS is to relate the information in two data tables that gather measurements on the same set of observations (i.e., samples). The method works by deriving linear combinations of the original variables through the SVD of a correlation matrix, such that covariance is maximized between each pair of the defining latent vectors (implied orthogonality) [43]. These combined variables are also referred to as latent variables, dimensions or components. In PLS regression, the SVD simultaneously decomposes matrices

X and **Y** (by virtue of the correlation matrix **R**) and iteratively computes sets of orthogonal latent variables and the corresponding regression weights [43].

Generally, this entire process is first carried out on subsets of training and validation data (i.e., measured values exist for both independent and dependent variables) to build and evaluate a regression model. The selection of training and validation splits is dependent on a number of factors, including sample population size, the nature and variability of the data and the scope of the prediction problem. Sample splits should generally be selected at random but can also be stratified when there are constraints imposed by different sample types (e.g., rock type); 80–20 training–validation splits are common in practice. Other techniques, such as k-fold cross-validation, are also widely popular to further minimize bias; one of these approaches is further detailed in Section 3.2.1. The final regression coefficients are then subsequently applied to a test dataset (i.e., for which data are only available for the independent variables) in order to predict the entire set of dependent variables.

By contrast with standard techniques, PLS regression can be used to predict a whole table of data and can also handle multicollinearity, thereby eliminating the necessity to remove certain predictor variables, which may not be linearly independent and would normally cause overfitting [43]. This is particularly important in the context of mining systems wherein a significant proportion of the data variables used for ore characterization (e.g., geological, geochemical and mineralogical) are intimately linked. For example, ~50% of the variables in the present study are strongly correlated (correlation coefficients > 0.75) with one or more other variables (Appendix A). This multicollinearity among independent variables renders the classic multiple linear regression (MLR) method inappropriate for predictive modelling in most cases due to difficulties in distinguishing the effects of individual variables [49]. This can lead to the inflation of standard errors, which may cause incorrect variable significance classifications and/or numerical instabilities related to the inversion of the covariance matrix ($\mathbf{X^T X}$) in the calculation of regression coefficients ($\mathbf{B} = (\mathbf{X^T X})^{-1} \mathbf{X^T Y}$) [49]. In PLS regression, the multicollinearity problem is bypassed by projecting the original variables to latent structures (linear combinations) and forcing orthogonality between the new variables (**t** and **u**). The different approaches of MLR and PLS are conceptualized in Figure 3. These attributes make PLS regression a powerful and highly adaptable tool because unlike other methods, it can be used on large datasets with an abundance of variables.

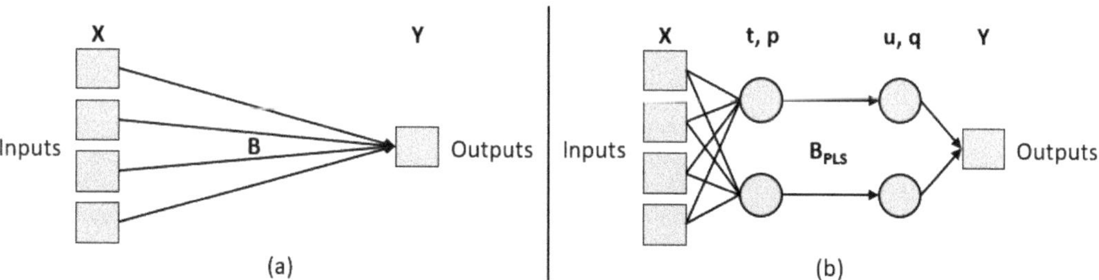

Figure 3. Simplified schematic comparison of multiple linear regression (MLR) and partial least squares (PLS) regression methods. (**a**) MLR uses all original variables to directly form a linear combination via the normal equations; (**b**) PLS first transforms the original variables by projecting to latent structures (linear combinations) that maximize covariance (orthogonality) between the new variables (**t** and **u**) (modified after [50].

To the best of the authors' knowledge, there has been very little work done using PLS regression in the areas of mining and system dynamics research to date. The most relatable study successfully developed a predictive model for the amount of kaolinite (clay mineral) in a deposit by linking earlier collected NIR spectroscopic data to confirmatory geochemical data [51]. The current work aims to extend this approach of adapting modern statistical methods for decision-making processes by integrating a predictive PLS model into a digital twin to evaluate operational risks within mining system processes. The digital twin is constructed within a discrete event simulation (DES) framework.

2.3. Digital Twins and Discrete Event Simulations

Digital twins are a form of computational intelligence tool that describe and simulate the comprehensive physical and functional aspects of integrated processes or systems and utilize the most advanced and updated information models available to quantify causal relationships between key variables and parameters [18,19]. Digital twin development using DES is particularly useful in the design and testing of alternative operational policies and associated trade-offs and can be applied to any or all lifecycle phases of an engineering project.

DES frameworks are valuable tools to expedite and support decision making within industrial systems. The key difference between mining and other industrial systems is the notion of geological uncertainty, i.e., the natural variability associated with orebodies and host geological environments. Because DES has the flexibility to incorporate random distributions, i.e., it is a form of Monte Carlo simulation [52], it can be used to evaluate the potential effects of geological uncertainty on mining system dynamics.

Paramount to the development of suitable digital twins is the establishment of alternate modes of operation that can be engaged under changing conditions or as critical thresholds are breeched [12,13]. Geometallurgical units are then defined based on the expected behaviour of classified mining blocks under each of the available modes [12,53,54]. Key performance indicators (KPIs) can also be tracked to observe system response to unexpected trends in ore feed attributes; trade-offs are optimized by adjusting available operational policies and the thresholds that control their timing [15,16,55].

Computer-based DES is a useful risk assessment tool, as it can simulate extended periods of operation to identify potential bottlenecks or other deficiencies; these risk factors can then be mitigated through implementation of operational buffers, such as stockpiling or ore blending strategies. To further support decision making (e.g., installation of new equipment), sets of validation or testing data can also be simulated in order to generate confidence intervals.

In the mining industry, DES frameworks have generally been limited to equipment selection [56], material management [57] and general transportation practices [58]. Applications related to availability and reliability data of mining equipment [59], underground mine refuge station location planning [60] and continuous mine system simulation for short-term planning and decision control [61] have also been explored to a lesser extent.

Recent mine-to-mill modeling studies include quantification of the effects of ore type spectral imaging [62] and evaluation of coupling solar radiation energy with a semi-autogenous grinding (SAG) mill [63]. As previously mentioned, some authors have developed alternate modes of operation for mineral processing via mass balance and mathematical programming [13]. Recent studies have applied this technique to mining contexts using discrete event simulation (DES), including concentrator and smelter dynamics [14,15,55], heap leach processes [16] and tailings retreatment applications [17].

There is a striking similarity between a two-mode mining system model and the RQ problem from inventory theory [15,64]. When inventory levels fall below a given "re-order point", R (i.e., critical ore level), a replenishment order of quantity, Q, is made prior to stockout. In mining systems, the critical ore level is directly proportional to the expected rate of ore consumption and the lead time required to replenish the stockpile to sufficient levels. However, this relationship does not account for potential natural variation (i.e., geological uncertainty), which could result in either surpluses or shortfalls, the latter of which is a risk for stockout (see Wilson et al., 2021 [17] for examples). This operational risk can be mitigated by raising the critical ore and/or total stockpile target levels, at the expense of higher operating (handling) and capital (equipment/storage) costs. Thus, the two-mode model is important to both risk management and multiobjective optimization (e.g., balancing throughput vs. stockpile management).

This type of framework can be formulated using commercial DES software to simulate extended operating periods and assess the system-wide response to varied stockpile management strategies. Mode A typically represents a consumption phase, whereas Mode B constitutes a phase of replenishment. While the consumption mode is generally more productive, it is not sustainable to activate indefinitely without an alternate mode of replenishment due to stockout risk [15]. The timing for switching between modes is governed by operational policy, which defines threshold crossing events (i.e., critical ore level).

When the critical ore level is raised to act as an operational buffer, this threshold no longer represents a true minimum, as the ore stock may continue to be consumed until the next replenishment mode [15]. This can result in stockout if ore levels are insufficient to last until the next planned shutdown and highlights the importance of building recourse actions into the model. The current framework incorporates blending practices as a control measure in response to geological uncertainty; ore types are blended and processed according to different set proportions (dictated by operational policy) designed to maintain consistent ore feed. In the event of stockout of one of the ore types, contingency modes are enacted until the next planned shutdown and subsequent replenishment phase (depending on mined ore supply).

The current study is an adaptation of recent DES modelling work by Navarra et al. [15,55] and Wilson et al. [17], which focused on the implementation of alternate modes of operation to balance plant ore feed types with respect to system-wide dynamics. By integrating predictive PLS regression modelling into a DES framework, this work demonstrates the potential benefits of using ore characterization data collected at an early stage to evaluate the future performance of system processes under geological uncertainty.

3. Incorporation of Quantitative Methods into Discrete Event Simulation

3.1. Case Study: Predictive and System Process Modelling of Canada's Oil Sands

An initial dataset derived from Canada's oil sands was acquired through partnerships with the National Research Council Canada (NRC) and Syncrude Canada Ltd. retrieved from the NRC Office of Energy and Research and Development (OERD) database. The set contains elemental, mineralogical and ore compositional data for a total of 60 samples collected from multiple sources in the AOSR. Of the total number of samples, 40 were sourced from various locations at the Syncrude operations, and permission has been graciously granted to include these in the present conceptual study. The remaining 20 samples come from a number of miscellaneous sources as part of smaller studies and are already available to the public domain.

Ore compositions (i.e., bitumen, water and solids contents) were analyzed by the standard Soxhlet-Dean and Stark method [65]. Separately, splits from the original sample material were prepared for subsequent analytical determinations using a micronizing procedure recently developed at NRC. In this method, ore samples are first homogenized with a spatula at room temperature; isopropanol (4 mL) and toluene (6 mL) are then added to an aliquot (~2–3 g) and micronized with agate beads in a McCrone micronizing mill. The contents are strained into a weighed petri dish (any remainder is carefully rinsed with isopropanol and toluene) and allowed to dry for 24 h in a fume hood prior to weighing. Lastly, the dried mixture is scraped and transferred for further homogenization using a mortar and pestle. Elemental compositions were determined by wavelength dispersive X-ray fluorescence (WD-XRF) using a fusion-based procedure [66] and CHNS measurements by combustion technique using an automatic elemental analyzer (Elementar Vario EL Cube) for carbon and sulphur. Mineral phase ratios were acquired by X-ray diffraction (XRD) with Rietveld refinement carried out on random orientation powder mounts. The mounts were prepared using a zero-background specimen holder (Si crystal, P-type, B-doped) with a cavity diameter of 20 mm and thickness of 0.2 mm; a glass slide was used to remove excess powder and create a flat surface. Final mineralogical compositions were based on the combined XRF, CHNS and XRD results and determined using the NRC-developed quantitative phase analysis (QPA) methodology parameterized with singular value decomposition (SVD), collectively termed SVD-QPA [67].

Basic descriptive statistics were computed to summarize the original raw dataset (Tables 1 and 2), which consists of data for 12 elements, 24 mineral phases and compounds and 4 oil sand constituents (bitumen, water, solids and proportion of fines).

The data are highly heterogeneous and reflect ores likely hosted by formations spanning mainly estuarine and shallow marine settings, with a small number possibly from fluvial settings. Because information on sample provenance was fairly limited, assumptions had to be made in order to classify sample points for the sake of this conceptual study. This actually works out reasonably well, as it highlights the importance of systematic data collection and demonstrates the powerful information that can be gained by integrating properly developed geometallurgical profiles into advanced quantitative frameworks and digital twins.

Table 1. Summary of descriptive statistics for mineral phases analyzed by X-ray diffraction (XRD) in 60 samples.

	Mineral Phase Concentrations (wt.%)					
Mineral	**Max**	**Min ***	**Median**	**Mean**	**Standard Deviation**	**Variance**
Quartz + silica	87.94	33.30	77.43	74.96	11.22	125.82
Illite	25.63	0.01	5.31	5.89	5.20	27.06
Kaolinite	24.09	−1.49	2.18	3.90	5.15	26.48
Chlorite	1.42	−1.09	0.01	0.10	0.37	0.14
Calcite	3.15	−1.06	0.01	0.09	0.50	0.25
Dolomite	8.21	−0.03	0.12	0.83	1.69	2.85
Ankerite	0.98	−0.03	0.01	0.05	0.16	0.03
Siderite	5.09	0.00	0.61	1.04	1.15	1.33
Pyrite	1.05	−0.05	0.04	0.16	0.23	0.05
Zircon	0.44	0.00	0.06	0.09	0.08	0.01
Rutile	0.62	0.03	0.13	0.15	0.09	0.01
Anatase	1.13	0.00	0.11	0.26	0.29	0.09
Ilmenite	0.05	0.00	0.01	0.02	0.01	0.00
Lepidolite	0.05	0.00	0.01	0.02	0.01	0.00
Gypsum	0.29	−0.85	0.01	0.01	0.13	0.02
Bassanite	0.30	−0.90	0.01	0.01	0.14	0.02
Anorthite	3.65	−7.14	0.12	0.22	1.22	1.49
K-feldspar	4.16	−0.35	1.06	1.19	0.86	0.74
Albite	11.38	0.00	0.34	0.96	1.75	3.08
Iron oxide + hydroxide	5.25	−1.75	0.10	0.19	1.07	1.15
Apatite	0.49	0.00	0.05	0.10	0.09	0.01
Cristobalite	0.10	0.00	0.01	0.01	0.02	0.00
Organic carbon	13.15	0.58	8.24	7.63	3.18	10.09
Organic sulphur	1.03	−0.17	0.39	0.38	0.27	0.07

* Negative values are related to the SVD-QPA methodology used for mineralogical composition reconstruction based on combined experimental results from elemental concentrations by XRF (Si, Al, K, Mg, Fe, Ti, Zr, Mg, Ca and P), carbon and sulphur contents and mineral ratios determined by Rietveld analysis of XRD powder patterns [67]. The vast majority of these are well within tolerance; the anomalous value noted for anorthite (minimum of −7.14) can be linked to a specific sample likely containing Ca-bearing smectite, which was not one of the defined phases in the QPA due to low overall Ca phases in the analyzed sample population.

Table 2. Summary of descriptive statistics for elemental and ore compositions analyzed by X-ray fluorescence (XRF) and Soxhlet-Dean and Stark extraction (respectively) in 60 samples.

Elemental Compositions (wt.%)						
Element	Max	Min	Median	Mean	Standard Deviation	Variance
Na	1.00	0.00	0.03	0.08	0.15	0.02
K	2.00	0.09	0.55	0.64	0.43	0.18
Si	42.01	26.75	38.42	37.98	2.77	7.69
Al	9.09	0.00	1.56	2.22	2.09	4.36
Fe	3.21	0.00	0.54	0.72	0.66	0.44
Mg	1.09	0.00	0.05	0.12	0.22	0.05
Ca	3.49	0.03	0.16	0.30	0.56	0.31
Ti	0.72	0.02	0.20	0.25	0.17	0.03
Zr	0.22	0.00	0.03	0.04	0.04	0.00
P	0.09	0.00	0.01	0.02	0.02	0.00
C	13.29	1.08	8.38	7.86	3.05	9.32
S	1.31	0.05	0.47	0.47	0.28	0.08

Ore Compositions (wt.%)						
Phase	Max	Min	Median	Mean	Standard Deviation	Variance
Bitumen	16.28	0.00	9.53	9.16	4.48	20.04
Water	19.57	0.41	5.63	6.65	4.37	19.06
Solids	91.25	76.79	83.97	84.42	2.86	8.16
Fines	99.46	1.36	24.80	34.22	27.01	729.80

In positing the depositional settings from which the majority of the oil sand samples came, the data were sorted based on total clay contents (illite–kaolinite–chlorite) and a cut-off level of 6 wt.% was applied; samples with less than this limit were classified as (fluvio-)estuarine and those with greater as marine. A broad inverse relationship is also evident in the data between total clay and bitumen contents, consistent with previous studies. Given that bitumen contents are generally higher in ores from fluvial and estuarine settings than those from marine settings [7], this relationship could serve as a useful check of the viability of assumed provenances. Bitumen contents in the classified (fluvio-)estuarine samples average 11.74 wt.% (standard deviation of 2.81 wt.%), which coincides with the stated range of ~9–13% for economic ores [3,4]. The average bitumen content for marine samples is 6.52 wt.% (standard deviation of 4.52 wt.%), consistent with borderline uneconomical ores [7,68]. Overall, the assumed depositional types appear fairly reasonable compared to natural deposit settings and related variations.

The dataset was also expanded to include postulated bitumen recovery data that were mostly unavailable. To this end, batch extraction unit (BEU) test data for 5 estuarine and 5 marine ore samples from previous work [69,70] were used to calculate appropriate bitumen recovery distributions for each depositional type. By applying these respective sample population means and standard deviations to a random number generator, reasonable spreads of recovery data were inferred for each sample type. This was deemed necessary in order to classify ore types based on both depositional setting (clay content) and generally related ore processability, for subsequent predictive and system process modelling (Sections 3.2.1 and 3.2.2, respectively). It is notable that the effect of fines on bitumen recovery can vary considerably depending on the type of fines and water chemistry [7]; studies have shown a depressive effect in the presence of degraded illite or smectitic clays [71], as well as ultrathin illite [69] and interstratification [72].

For the purposes of this study, all samples are being treated as though they were sourced from a single mining project, with each of 20 mining parcels (i.e., blocks) corresponding to a minimum of 3 oil sands samples. The initial concept is to develop effective predictive models using PLS regression such that bitumen recoveries could be estimated with confidence earlier in the value chain. Subsequent incorporation of these predicted data into DES frameworks would then expedite the evaluation of system response to geological uncertainty caused by heterogeneities in source ore feeds.

To demonstrate the overall concept, the classified ore types are blended according to two different schemes established through mass balancing and mathematical programming. Each of these schemes corresponds to a separate operational mode, whereby the primary blend is considered the productive phase, and the secondary blend is considered a replenishment phase. The mining parcel data, classified into proportions of each ore type, are then incorporated into a DES framework to simulate system response to ore feed availability for a designated tonnage of oil sands to be processed; bottlenecks or stockout risk for either ore type can be identified and adjustments made to the potential modes of operation (Figure 4).

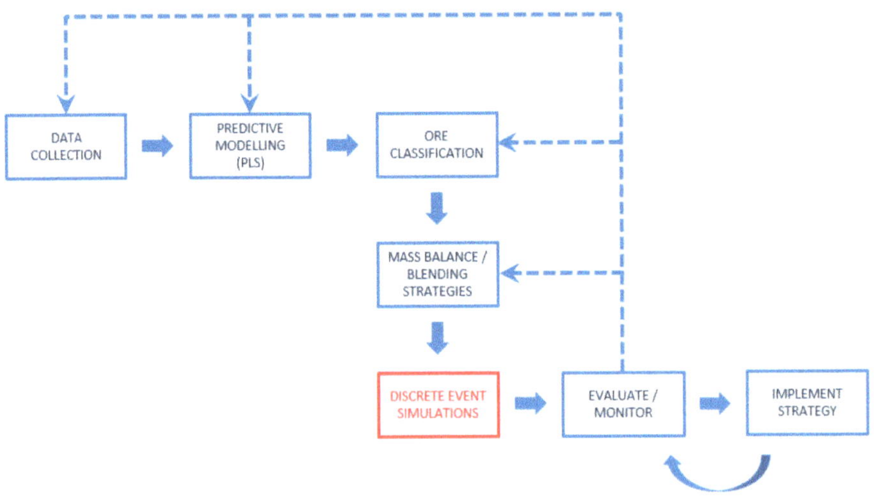

Figure 4. Generalized diagram showing the implementation of discrete event simulations (DES) in the formulation of blending control strategies and related operational modes.

Notably, since fines content is generally linked to geological setting and processability, a classification scheme based on depositional setting (and predicted recovery) actually runs in parallel to a process recently developed by Syncrude for the control of solids distribution in a bitumen froth [73]. Under this patented methodology, coarse oil sands (which normally produce high bitumen recoveries) are blended with high-fines material prior to extraction in order to improve the efficiency of pipeline transport from remote extraction sites to the froth treatment plant (Figure 5) [73]. As a result, the current framework could help quantify the effects of different ore blending strategies on downstream system processes and better guide the implementation of alternate operational modes.

This type of integrated quantitative framework allows for well-planned adjustments to process control strategies (e.g., ore blending), thereby streamlining risk-based decision making, increasing efficiencies and, likely, extending operational life through improved mine planning. The approach requires extensive sampling coupled with detailed analytical work initially, particularly in newly discovered or poorly characterized resource areas. With a sufficiently large population of sample points, robust predictive models can then be developed and implemented with confidence; at this stage, the expensive and time-consuming detailed analyses can be replaced with cheaper and faster tests earlier in the planning stages.

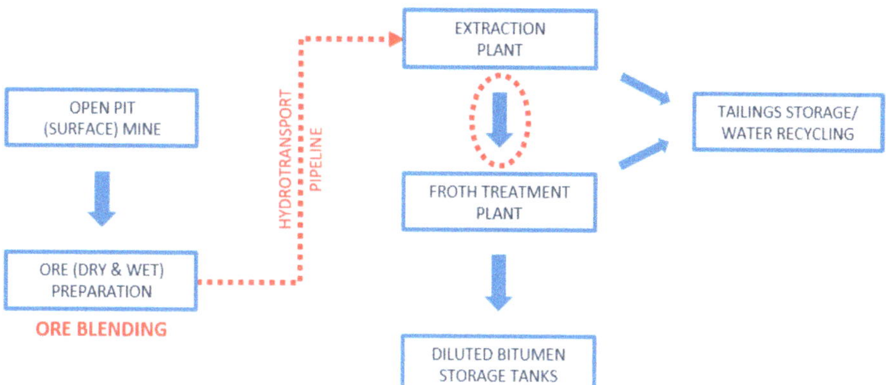

Figure 5. Generalized flowsheet for the extraction of diluted bitumen ("dilbit") from mined oil sands. Blending should occur during ore preparation (prior to processing) to improve hydrotransport along pipelines [73]. Red dotted ellipse indicates potential additional point of hydrotransport for remote extraction sites.

3.2. Sample Calculations

3.2.1. Partial Least Squares (PLS) Regression

For the predictive modelling portion of the study, elemental, mineralogical and ore composition data, coupled with depositional type, were retained for a total of 46 independent (explanatory) variables including 5 composite variables, e.g., total clays. The depositional setting variable was one-hot encoded to map its categorical data to integer values, represented as a binary vector [74]. Total bitumen recovery was reserved as the lone dependent (response) variable for the sake of this conceptual study. A PLS regression algorithm was written in Python coding and follows well-established theory after several authors [43,75,76]. The methodology begins by calculating the SVD of the correlation matrix $\mathbf{R}$, as described for Equations (1) and (2), and iteratively computing sets of orthogonal latent variables with the corresponding regression weights.

During each successive iteration, the first left and right singular vectors ($\mathbf{w}_l$ and $\mathbf{c}_l$) are used as weight vectors to calculate sets of scores ($\mathbf{t}_l = \mathbf{X}\mathbf{w}_l$ and $\mathbf{u}_l = \mathbf{Y}\mathbf{c}_l$) for $\mathbf{X}$ and $\mathbf{Y}$, respectively; loadings are then obtained by regressing $\mathbf{X}$ and $\mathbf{Y}$ against the same vector $\mathbf{t}_l$ ($\mathbf{p}_l = \mathbf{X}^T\mathbf{t}_l$ and $\mathbf{q}_l = \mathbf{Y}^T\mathbf{t}_l$) [75]. The last step of the iteration "deflates" the current data matrices (i.e., removes information related to the lth latent variable) by subtracting the outer products $\mathbf{t}\mathbf{p}^T$ and $\mathbf{t}\mathbf{q}^T$ from $\mathbf{X}$ and $\mathbf{Y}$, respectively [75]. The next component (or latent variable) can then be calculated starting from the SVD of the cross-product of the newly deflated matrices ($\mathbf{X}_{l+1}$ and $\mathbf{Y}_{l+1}$). The process continues until $\mathbf{X}$ is completely decomposed into L components and a null matrix is obtained. After each iteration, vectors $\mathbf{w}_l$, $\mathbf{t}_l$, $\mathbf{p}_l$ and $\mathbf{q}_l$ are stored as columns in their respective matrices $\mathbf{W}$, $\mathbf{T}$, $\mathbf{P}$ and $\mathbf{Q}$. The matrix of regression coefficients ($\mathbf{B_{PLS}}$) can then be calculated as:

$$\mathbf{B_{PLS}} = \mathbf{P}(\mathbf{P}^T\mathbf{P})^{-1}\mathbf{Q}^T \tag{3}$$

where $(\mathbf{P}^T\mathbf{P})^{-1}$ is in fact the Moore-Penrose pseudoinverse for the generalized case of a non-symmetric matrix [76]. Finally, the matrix of regression coefficients ($\mathbf{B_{PLS}}$) is multiplied by the original set of independent variables prior to any deflations (matrix $\mathbf{X}_0$) to obtain the predictions of the dependent variables (matrix $\hat{\mathbf{Y}}$) [43]. A number of criteria can be calculated to select the appropriate number of components to keep while limiting loss of significance, evaluate the quality of prediction and validate the model, i.e., cross-validation.

Validation is critical to the development of robust predictive models; the quality of prediction must be assessed, and model significance also determined. A common measure of prediction quality is called the residual estimated sum of squares (RESS) and is calculated as follows:

$$RESS = || \mathbf{Y} - \hat{\mathbf{Y}} ||^2 \tag{4}$$

where $|| \ ||^2$ is the squared matrix norm and decreases as prediction quality improves [43]. However, RESS alone is not the most useful metric, as it will continue to decrease until all components are added, i.e., it does not detect overfitting. An improved measure for quality of prediction is the predicted residual estimated sum of squares (PRESS), computed as follows:

$$PRESS = || \mathbf{Y} - \widetilde{\mathbf{Y}} ||^2 \tag{5}$$

where $\widetilde{\mathbf{Y}}$ represents a predicted set of values generated from cross-validation and also decreases with increasing prediction quality [43]. The selection of the optimal number of components to extract is crucial to avoid overfitting the data. Since prediction quality typically first increases then decreases upon successive component addition, a possible approach is to begin with the first component and stop as soon as the PRESS reverses direction [31]. A more intricate method is to compute the Q^2 criterion for the lth component, as follows:

$$Q_l^2 = 1 - \frac{PRESS_l}{RESS_{l-1}} \tag{6}$$

and compare against an arbitrary critical value (e.g., $1 - 0.95^2 = 0.0975$); only components with a Q_l^2 value greater than or equal to this threshold are generally kept in the model [31,33].

Because the available dataset is limited to only 60 samples, it was decided not to split the data into separate training and validation sets; instead, leave-one-out cross-validation (LOOCV), also called the "jackknife" method, was utilized. In this technique, each observation is iteratively dropped from the set, and the remaining observations then comprise a training set used to estimate the left-out observation. All estimated observations are stored in a final matrix denoted $\widetilde{\mathbf{Y}}$, which then serves as the validation set for subsequent prediction quality metrics (e.g., PRESS and Q^2 criteria) [31].

The PLS regression model was run sequentially, and a series of quality of prediction statistics were tabulated for each of 1, 2, 3, 4, 5 and 10 component scenarios (Figure 6). The Q^2 criterion indicates that only the first component should be kept in the prediction model, with a value of 0.28, as all ensuing trials resulted in values less than zero. However, not only was the coefficient of determination (R^2) quite low for the 1 component scenario (0.34), but root mean squared error (RMSE) and mean absolute residual values were also relatively high. Furthermore, the first component alone only accounts for ~88% of the total model variance, as determined by the sum of squares of the singular values. As a result, the behaviour of the PRESS statistic was tracked upon successive trials in order to identify an improved fit; ultimately, a total of 5 components was deemed appropriate for building the regression model in relation to the available dataset. This was based upon the fact that the PRESS value trended upwards over the first 4 components but dropped significantly upon addition of the fifth; this reversal also coincided with a much higher R^2 score of 0.72, improved (decreased) RMSE and residual values and an explained variance of 99.65%. Further addition of successive components (e.g., 10 components) did not greatly improve prediction accuracy or error metrics, resulted in poorer PRESS and Q^2 statistics and would likely lead to severe overfitting to the present dataset. It is also noteworthy that residuals were consistently greater for marine samples, which indicates greater variability in the predicted set for this depositional type (as expected).

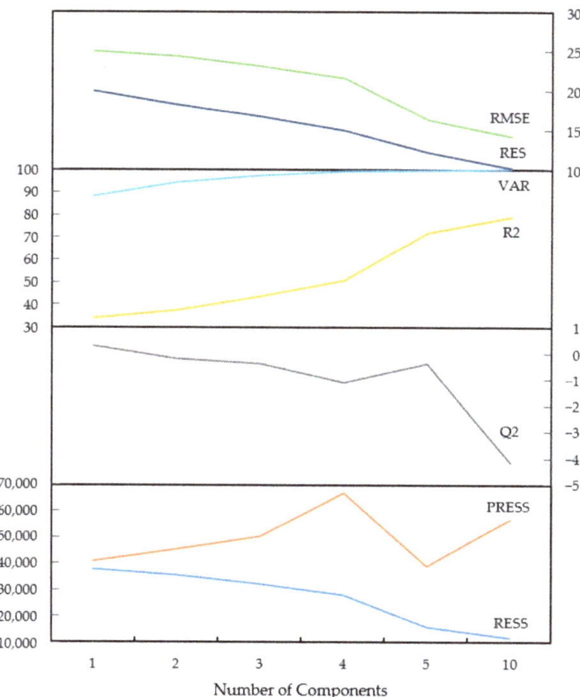

Figure 6. Stacked panel line chart for a variety of quality of prediction statistics tabulated for each of the 1, 2, 3, 4, 5 and 10 component scenarios. Abbreviation definitions: RMSE = root mean squared error; RES = mean absolute residual (all samples); VAR = percentage of model variance explained; R2 = coefficient of determination ($\times$100); Q^2 criterion (as in Equation (6)); PRESS statistic (as in Equation (5)); RESS statistic (as in Equation (4)).

Predictions from the final 5-component model are shown in Figure 7, and comparative descriptive statistics for the observed and predicted datasets are shown in Table 3. Estimated bitumen recoveries were capped at 100%, and negative values were set to zero, as crossing these thresholds is impossible in practice. The predicted values are generally quite reasonable, within ~11% for the (fluvio-)estuarine samples and ~13% for marine samples on average. This level of error (RMSE of 16.33) is not surprising on account of the assumptions made to finalize the original dataset, in addition to the significant geological variability inherent to oil sands deposits. As expected, the variability in marine sample residuals (standard deviation of 9.89%) is nearly double that of estuarine samples (standard deviation of 5.27%) and can likely be attributed to heterogeneities in clay contents and especially clay types. Overall, the PLS regression model has performed as intended and with a mere total of 60 samples from unknown and/or different mining projects altogether. This highlights the importance of rigorous sampling campaigns and characterization of appropriate geometallurgical profiles towards the development of robust predictive models, particularly for complex operations dealing with multiple and/or heterogeneous ore feeds. It is postulated that the predictive power of the present model would be greatly increased with these controls in place.

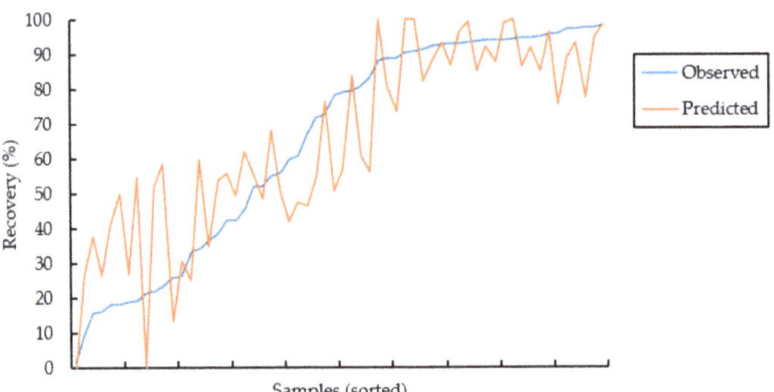

Figure 7. Plot of predicted vs. observed recoveries from the 5-component regression model. Samples are sorted according to ascending observed values to better reflect residual distances; predictions clearly improve in the middle to upper recovery ranges.

Table 3. Comparison of summary statistics for observed and predicted bitumen recoveries from PLS regression model (5 components).

Metric	Observed Values (Y)	Predicted Values ($\hat{Y}$)
Max (%)	98.29	100.00
Min (%)	1.55	0.00
Median (%)	78.80	61.27
Mean (%)	64.91	65.17
Standard Deviation (%)	31.14	26.91
Variance ($\%^2$)	969.95	724.18

Once the regression model has been finalized with the appropriate number of components, confidence intervals for the predicted values can be calculated using the "bootstrap" cross-validation method. This technique involves the random re-sampling of the original observations with replacement, i.e., each observation can be selected zero or multiple times [43]. This is repeated many times (e.g., 1000 or 10,000), and regression coefficients and corresponding predictions are computed for each bootstrapped sample set. The distribution of predicted values from all of these iterations is then used to estimate confidence limits for each variable; intervals that do not span zero (positive or negative) are considered significant [43]. Similarly, bootstrap ratios can be calculated by dividing the mean of each distribution by its standard deviation; akin to a student *t*-test, if the ratio is greater than a critical value (e.g., >2, corresponding to an alpha value of approximately 0.05), the variable is also considered significant [43].

Table 4 provides statistics computed from the distribution of 10,000 bootstrap sample sets generated from the 5-component regression model; variable significance was determined based on both bootstrap ratios and 95% confidence intervals. Of the elemental composition variables, only Na, Ca and Mg were deemed insignificant. Corresponding insignificant minerals include albite for Na; gypsum, bassanite and anorthite for Ca; chlorite for Mg; the carbonates (calcite, dolomite and ankerite) for both Ca and Mg. Interestingly, both pyrite and amorphous Fe-oxides/hydroxides were also considered insignificant (oil sands tend to contain significant heavy metals). All remaining elements and mineral phases, in addition to depositional sample type and bitumen recovery, were determined as statistically significant.

Table 4. Summary of statistics from 10,000 replications of the bootstrap cross-validation method.

| Independent (Explanatory) Variables | | | | | |
Variable	No. of Observations	Mean (wt.%)	Standard Deviation (wt.%)	Bootstrap Ratio	Lower CI (95%)	Upper CI (95%)
Sample type	60,000	0.57	0.22	2.54	0.20	1.00
Na *	60,000	0.08	0.05	1.57	0.03	0.21
K	60,000	0.64	0.13	5.12	0.45	0.95
Si	60,000	37.97	0.96	39.70	35.30	39.42
Al	60,000	2.22	0.64	3.48	1.34	3.99
Fe	60,000	0.72	0.24	2.98	0.35	1.42
Mg *	60,000	0.12	0.10	1.19	0.02	0.48
Ca *	60,000	0.30	0.31	0.99	−0.02	1.53
Ti	60,000	0.25	0.06	3.95	0.15	0.38
Zr	60,000	0.04	0.02	2.41	0.01	0.08
P	60,000	0.02	0.01	2.40	0.01	0.04
C	60,000	7.87	1.00	7.90	5.88	9.67
S	60,000	0.47	0.13	3.64	0.25	0.76
Bitumen	60,000	8.79	1.36	6.46	6.06	11.16
Water	60,000	6.64	1.42	4.69	4.25	9.97
Solids	60,000	84.42	1.11	76.29	82.27	86.83
Fines	60,000	34.27	7.91	4.33	22.22	52.30
Quartz + silica	60,000	74.94	3.46	21.64	64.62	79.82
Illite	60,000	5.89	1.66	3.55	3.43	10.21
Kaolinite	60,000	3.90	1.83	2.13	1.08	8.58
Chlorite *	60,000	0.10	0.14	0.73	−0.16	0.41
Calcite (Cal) *	60,000	0.09	0.28	0.33	−0.24	1.07
Dolomite (Dol) *	60,000	0.84	0.82	1.01	−0.03	3.67
Ankerite (Ank) *	60,000	0.06	0.07	0.85	−0.05	0.21
Siderite (Sid)	60,000	1.04	0.50	2.07	0.10	2.19
Pyrite *	60,000	0.16	0.11	1.40	−0.02	0.46
Zircon	60,000	0.09	0.04	2.41	0.02	0.17
Rutile (Rut)	60,000	0.15	0.04	3.73	0.07	0.22
Anatase (Ana)	60,000	0.26	0.10	2.64	0.12	0.49
Ilmenite	60,000	0.02	0.00	3.74	0.01	0.03
Lepidolite	60,000	0.02	0.00	3.72	0.01	0.03
Gypsum *	60,000	0.01	0.06	0.19	−0.12	0.11
Bassanite *	60,000	0.01	0.06	0.16	−0.13	0.10
Anorthite (Ano) *	60,000	0.22	0.61	0.36	−0.92	1.77
K-feldspar (Ksp)	60,000	1.20	0.38	3.17	0.39	1.96
Albite (Alb) *	60,000	0.96	0.61	1.57	0.29	2.36
Iron oxide/hydroxide (AFE) *	60,000	0.19	0.44	0.42	−0.46	1.55
Apatite	60,000	0.09	0.04	2.34	0.04	0.22
Cristobalite	60,000	0.01	0.00	2.70	0.00	0.02
Organic carbon	60,000	7.63	1.00	7.63	5.61	9.41
Organic sulphur	60,000	0.39	0.13	2.96	0.14	0.67
Total clays	60,000	9.89	3.24	3.05	5.13	19.00
Sid + AFE	60,000	1.22	0.49	2.48	0.45	2.52
Rut + Ana	60,000	0.40	0.11	3.82	0.24	0.63
Cal + Dol + Ank *	60,000	0.98	1.01	0.97	−0.05	4.39
Ano + Ksp + Alb	60,000	2.37	0.82	2.91	1.09	4.76
Dependent (response) variables					
Variable	No. of observations	Mean (wt.%)	Standard deviation (wt.%)	Bootstrap ratio	Lower CI (95%)	Upper CI (95%)
Total Recovery	60,000	64.84	28.86	2.25	1.52	109.53

* Insignificant variables determined from the distribution of 10,000 bootstrap sample sets.

The relationships between the independent variables can be observed visually by plotting the stored **X**-loadings (matrix **P**) for the first two components (Figure 8). Bitumen content is clearly most strongly linked to elemental carbon and organic carbon (as expected); it also appears in association to silicon (quartz–silica–cristobalite), sulphur (organic sulphur), titanium minerals (rutile and ilmenite) and lepidolite (Li-rich mica). The first dimension also opposes the bitumen group from the clay minerals, water content and carbonates (siderite). Notably, anatase (metastable form of TiO_2) plots opposite the other Ti-bearing phases. In the second dimension, the organic-related groups (bitumen, carbon and sulphur) clearly oppose the related silicate and oxide minerals; there is also a broad separation between silicates and carbonate + iron-bearing phases.

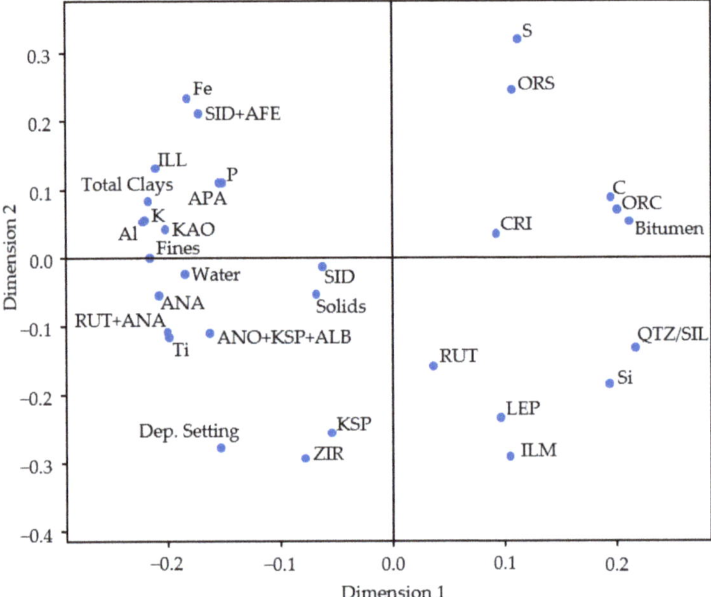

Figure 8. Plot of the **X** variable loadings (matrix **P**) for the first two components (dimensions 1 and 2).

Overall, the PLS regression model has proven to be a powerful prediction tool, capable of providing additional useful information regarding process variables that can help drive the characterization of geometallurgical profiles, sampling methodologies and other planning processes.

3.2.2. Discrete Event Simulations

For the DES portion of the study, two ore types were classified according to documented depositional setting and predicted bitumen recoveries from the 5-component PLS regression model. Ore type 1 consists entirely of marine samples (generally <75% recovery), and ore type 2 includes (fluvio-)estuarine samples (>75% recovery) as well as a few of marine type with recoveries also greater than 75%. Due to the limited nature of the dataset (only 3 samples per mining parcel), natural background noise was added to the relative proportions of ore types 1 and 2 via random number generation with a standard deviation of 5%. Two modes of operation (A and B) are considered here to balance stockpile levels against bitumen extraction rates and incoming ore feed from mining. While the conceptual mine has been operating in areas predominantly containing ore type 2 (favourable due to higher grades and recoveries) for some time, a large expansion of reserves comprising mainly ore type 1 has recently been completed. With the expansion, longer term forecasts suggest an overall deposit composition of 55–45% for ore types 1 and 2, respectively, with increased variability caused by geological heterogeneities; these values correspond to the average proportions determined from ore classification based on the predictive modelling.

In order to sustain the availability of ore type 2 and improve the economics of certain portions of the newly expanded area, the operation is evaluating possible adjustments to current blending strategies and intends to implement a secondary alternate mode. Based on the geological attributes of ore types 1 (high bitumen, low fines) and 2 (high fines), the new strategy will also serve to control the distribution of solids in ore feeds to the froth treatment plant, thereby improving amenability to transport via pipelines to the upgrader. As a result, operational Mode A will consist of an approximate 40–60 blend of ores 1 and 2. Because ore type 2 will generally be in shorter supply, a second operational Mode B, consisting of an 80–20 blend of ores 1 and 2, is needed in order to avoid stockouts, or an eventual shortage. This will ultimately stabilize feed balances, maximize equipment/infrastructure selection and utilization and allow for improved production scheduling; collectively, these factors can lead to significant reductions in operating and capital costs.

Both modes are expected to perform similarly in terms of downstream bitumen recovery processes, except that Mode B requires a pre-treatment stage to control excess chloride ions related to the marine origin and high fines content of ore type 1. Consequently, bitumen extraction rates for Mode B are set 10% lower than those for Mode A; modal parameters for each configuration are summarized in Table 5. Despite the fact that Mode A is both more productive and economical, ore stockouts would be inevitable over extended periods of usage because the weight fraction of ore type 2 (w_{2A}) is 15% higher than that of the deposit (w_{2D}). To account for the possibility of stockouts prior to a planned shutdown, contingency modes with adjusted configuration rates have been incorporated for each of Modes A and B.

Table 5. Description of operational modes in relation to deposit forecast.

		Throughput (t/h)	Ore 1 in Feed (%)	Ore 2 in Feed (%)
Algebraic Notation:		$r_{A,ACont,B,BCont}$	$w_{1A,1ACont,1B,1BCont}$	$w_{2A,2ACont,2B,2BCont}$
Mode A	Regular	30,000	40	60
	Contingency	19,500	100	0
Mode B	Regular	27,000	80	20
	Contingency	13,500	0	100
Deposit		-	55	45

Appropriate weight fractions ($w_{1A,2A,1B,2B}$) and throughput rates ($r_{A,B}$) are assessed with respect to geological estimations ($w_{1D,2D}$) using deterministic mass balancing, as follows [15,17]:

$$\left(\frac{t_A}{t_B}\right) = \left(\frac{w_{2B}w_{1D} - w_{1B}w_{2D}}{-w_{2A}w_{1D} + w_{1A}w_{2D}}\right)\left(\frac{r_B}{r_A}\right) \tag{7}$$

where t_A and t_B denote the time elapsed under Modes A and B, respectively. Average throughput between the two modes, or similarly between each mode and its corresponding contingency configuration, can then be computed as follows [15,17]:

$$r = \left(\frac{w_{1A}w_{2B} - w_{2A}w_{1B}}{\left(w_{2B}\left(\frac{r_B}{r_A}\right) - w_{2A}\right)w_{1D} - \left(w_{1B}\left(\frac{r_B}{r_A}\right) - w_{1A}\right)w_{2D}}\right)r_B \tag{8}$$

Equations (7) and (8), which ignore the risk of stockout, indicate that Mode A should be applied 1.5 times as often as Mode B, with an average throughput of 28,800 t/h. The framework aims to simultaneously maximize throughput and minimize target stockpile levels, thereby increasing production efficiency and reducing overall costs; larger stockpiles necessitate larger storage areas and equipment, as well as increased handling costs.

The current framework is designed such that mining rates exceed plant capacity, hence the plant acts as a bottleneck. To ensure stockpiles are adequately supplied to maintain consistent ore feed to the plant, ore will be mined at minimum rates of 30 kt/h under Mode A and 27 kt/h under Mode B. Target total stockpile level is a control variable that remains constant (except during extended stockout periods); however, the relative proportions of ore types 1 and 2 fluctuate contingent on the active operational mode. Mode A (productive phase) causes a relative decrease in the proportion of ore type 2, meanwhile Mode B (replenishment phase) has the opposite effect. The selection of operational mode is based on the stockpile level of the limiting ore type (in this case, ore 2) at the end of a production campaign during planned shutdowns every 4 weeks.

Under the present framework (Table 5), a naïve analysis indicates a critical threshold of 2.916 Mt for ore type 2; this level is computed as a function of campaign length (27 days) and rate of change under Mode A (108,000 t/d; plant capacity of 720,000 t/d $\times$ w_{2D} of 45% less relative critical ore 2 throughput from 40–60 blending strategy). Similarly, the analysis indicates a minimum total target stockpile level (sum of ores 1 and 2) of 4.374 Mt, determined as the maximum rate of change between ore stockpiles 1 and 2 as a function of campaign duration (under either mode). However, the digital twin is subject to the geological uncertainty of the ore, which is not taken into account by Equations (7) and (8). Unexpected fluctuations in ore feed attributes can indeed cause either overages or shortfalls for a given ore type, potentially leading to stockout towards the end of a production campaign [15]. To mitigate this risk, an operational buffer can be introduced by raising the threshold for the critical (limiting) ore type; a similar control measure would be to raise the target total stockpile level.

Because stockouts are nonetheless a real possibility, recourse actions are built into the digital twin to maintain ore feed consistency. These recourse actions depend on the timing of stockout; if an ore type is depleted during a production campaign, a contingency mode is enacted that allows the exhausted stockpile to build back up. As indicated in Table 5, Contingency Mode A only consumes ore type 1, and Contingency Mode B only consumes ore type 2. These contingency modes are much less productive than the regular configuration rates (65% for Mode A and 50% for Mode B); as a result, the duration of these segments has been limited to 1 day, which causes alternations until the next planned shutdown. If the critical ore level remains below the selected threshold at the end of a campaign, the plant will employ the alternate mode of operation to re-equilibrate stockpile levels. Time segment parameters for production campaigns, shutdowns and contingency mode duration are summarized in Table 6.

Table 6. Summary of time segment parameters.

Segment Type	Duration (Days)
Production campaign	27
Planned shutdown	1
Contingency modes	1
Regular modes	Indeterminate

The current framework was implemented, and subsequent computational results (Tables 7 and 8, Figures 9 and 10) generated, using commercial DES software (Rockwell Arena©) with Visual Basic for Applications (VBA). Extended operating periods can be simulated to assess system performance in response to geological uncertainty, with adjustments made to the critical ore and target stockpile levels as control variables. In its present configuration, the simulation model assumes that ore is mined to completion from a single parcel at a time. The framework has the flexibility to incorporate geological uncertainty by reading data from external source files. For the purposes of this study, uncertainty was introduced through Monte Carlo simulation; the proportions of ore types 1 and 2, determined from the classification of mining parcels based on depositional setting and predicted recoveries, were used to generate 100 statistical replicas through random number generation with a standard deviation of 5%. The model is configured such that 792 Mt of ore are processed within each replica, corresponding to approximately 1200 days of operation.

Table 7. Distribution of time spent in each mode type under varied target total stockpile levels.

Scenario:		1	2	3		4	5
Replications:		1	1	1	*100*	1	1
Critical Ore 2 Level (Mt):		2.916	2.916	**2.916**	*2.916*	2.916	2.916
Target Total Stockpile Level (Mt):		4.374 (1×)	6.561 (1.5×)	**8.748 (2×)**	*8.748 (2×)*	13.122 (3×)	21.870 (5×)
				Portion of time (%)			
Mode A	Regular	45.6	54.8	**59.7**	*60.1*	58.4	59.4
	Contingency	4.9	0.2	**1.6**	*1.4*	0.6	0.6
Mode B	Regular	34.8	37.3	**35**	*34.5*	37.4	36.4
	Contingency	11.1	4.2	**0.1**	*0.4*	0	0
Shutdown		3.6	3.5	**3.6**	*3.6*	3.6	3.6
Throughput (kt/h)		26.5	28.1	**28.7**	*28.7*	28.8	28.8
Replications with stockouts		-	-	-	*82*	-	-

Table 8. Distribution of time spent in each mode type under varied critical ore levels.

Scenario:		6		7		8	
Replications:		1	*100*	1	*100*	1	*100*
Critical Ore 2 Level (Mt):		5.832 (2×)	*5.832 (2×)*	**7.290 (2.5×)**	*7.290 (2.5×)*	8.748 (3×)	*8.748 (3×)*
Target Total Stockpile Level (Mt):		11.664	*11.664*	**14.580**	*14.580*	17.496	*17.496*
Mode A	Regular	58.7	*59.6*	**58.6**	*60.0*	58.6	*60*
	Contingency	0	*0.1*	**0**	*0.05*	0	*0*
Mode B	Regular	37.3	*36.3*	**37.8**	*36.3*	37.8	*36.4*
	Contingency	0.4	*0.4*	**0**	*0.05*	0	*0*
Shutdown		3.6	*3.6*	**3.6**	*3.6*	3.6	*3.6*
Throughput (kt/h)		28.8	*28.8*	**28.8**	*28.9*	28.8	*28.9*
Replications with stockouts		-	*62*	-	*5*	-	*0*

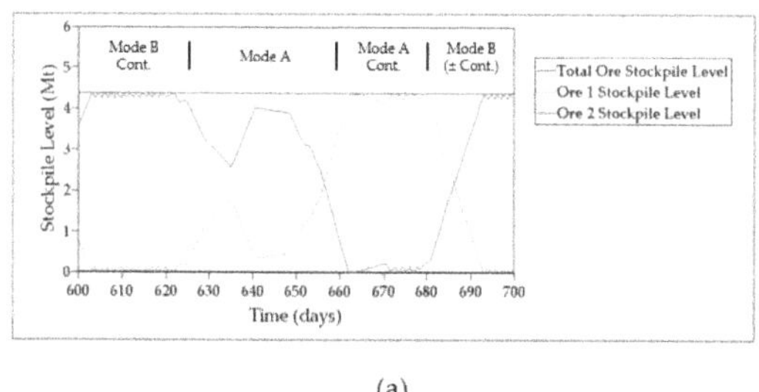
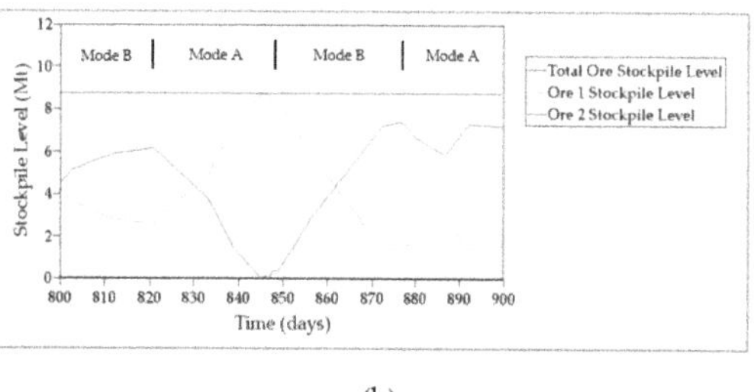

(a) (b)

Figure 9. Simulated operational dynamics of Canadian oil sands data in response to geological uncertainty, configured with a critical ore 2 level of 2.916 Mt (deterministic value) and total stockpile target levels of (**a**) 4.374 Mt under Scenario 1; (**b**) 8.748 Mt under Scenario 3. Contingency modes are depicted by the fine jagged saw-tooth pattern and result from short contingency segments of 1 day; note the abundance of these ore shortages in (**a**) compared to (**b**).

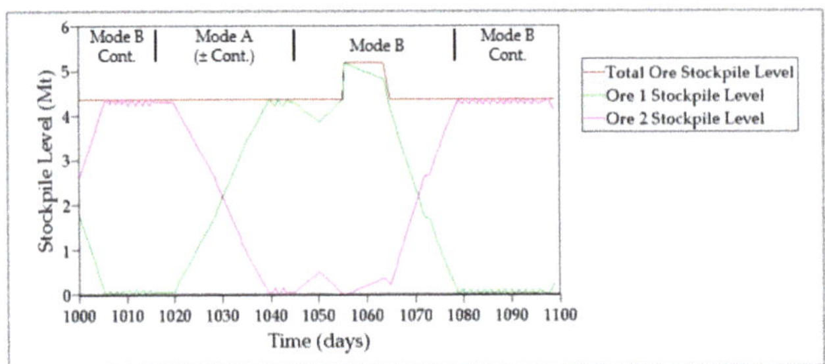

Figure 10. Simulated mining surge caused by high variability of Canadian oil sands data in response to geological uncertainty. Surges are indicated when the level of one of the ore types increases above the total stockpile target and are required to provide feed directly to the plant as recourse to a sustained stockout of the other ore type (e.g., ~1055–1065-day range).

A series of simulations were run to observe the effects of the selected control variable levels on throughput and potential stockout risk, in response to geological uncertainty. The first set of trials varied the total stockpile target levels, while holding the critical ore 2 level constant at 2.916 Mt (deterministic value). A total of 5 scenarios were considered with total stockpile levels set at 1× ("one times"), 1.5×, 2×, 3× and 5× the deterministic value (4.374 Mt); simulated results for each are summarized in Table 7.

Consistent with Navarra et al. [15] and Wilson et al. [17], the results show that naïve selection of the total stockpile target level does not perform well over extended operating periods, with Mode A being applied only 1.1× as often as Mode B for an average throughput of 26.5 kt/h. This is clearly less productive than the deterministic result of 28.8 kt/h (Mode A applied 1.5× more than Mode B), and the simulated operation also suffered from frequent sustained shortages of both ore types (Figure 9a). Increasing the total stockpile level by just 1.5× (Scenario 2) already improves overall system response; however, with Mode A applied 1.3× as often as Mode B for an average throughput of 28.1 kt/h, this is still worse than expected from Equations (7) and (8). Scenario 3, which doubled the deterministic total stockpile level to 8.748 Mt, produced the best overall results with Mode A applied 1.75× more frequently than Mode B for an average throughput of 28.7 kt/h; there was also a drastic reduction in the proportion of time spent in contingency modes (Figure 9b). Successive increases to the stockpile targets (Scenarios 4 and 5) did not show any marked changes, and system performance was actually slightly worse for both. These results suggest that in order to maximize throughput and mitigate stockout risk, the target total stockpile level is best maintained in the range of 2–3 times the selected critical ore threshold.

Using the parameter values established from Scenario 3, the framework was subsequently configured to simulate 100 replications, corresponding to approximately 120,000 days of operation. Average results from this test mirrored those of the single replication (Table 7) but highlighted repeated ore shortages as a significant operating risk under this scheme, with 82% of the replications confronted by stockouts. While not apparent from the single replication simulation, this outcome is directly related to the high variability of the dataset and is entirely possible in the context of oil sands mining, particularly when dealing with multiple and/or heterogeneous ore feed sources. Frequent and/or sustained stockout periods (especially early in a campaign) require additional consideration; as a recourse action, the possibility for mining surges has been incorporated into the framework in order to supply ore feed directly to the plant to maintain production (Figure 10).

To attenuate the significant stockout risk observed under Scenario 3, a second set of simulations were executed in which adjustments were made to the critical ore limit while keeping the total stockpile target at 2× this level. Four scenarios were tested with critical ore levels designated at 1.5×, 2×, 2.5× and 3× the deterministic value (2.916 Mt); results for each simulation trial are summarized in Table 8. While variations in the critical ore threshold had no meaningful effect on throughput rates or modal proportions, important reductions in the number and frequency of stockout periods were observed with the framework configured for 100 replications (~120,000 operating days). At twice the deterministic value (Scenario 6), the number of replications affected by ore shortages was reduced by 20% (cf. Scenario 3); at 2.5× (Scenario 7), this number decreased to just 5%. Tripling the critical value actually eliminated simulated stockout periods altogether; however, increased capital and operating costs associated with exceedingly large stockpile levels must be weighed against the risk of stockout in the decision-making process.

The time-averaged distribution of operational modes in response to geological uncertainty can be useful to evaluate the effects of varied control parameters. Figure 11a represents the naïve approach of Scenario 1, in which the deterministic values for the critical ore level (2.916 Mt) and total stockpile target level (4.374 Mt) were applied; Figure 11b depicts the enhanced framework configuration established under Scenario 7 (described above). The latter scheme is a significant improvement over the naïve setup, with an 8–9% increase in the proportion of time spent under Mode A, a much lower reliance on contingency modes (~15%) and the virtual elimination of ore stockouts. All of these factors contribute to improved production efficiencies; moreover, the enhanced configuration is also more economical based on higher consumption rates for ore type 2, which boasts higher overall grades and bitumen recoveries. Both framework applications benefit from the ability to switch between modes relatively freely in response to data variability, but the enhanced configuration is much less susceptible to operational risk caused by geological heterogeneities.

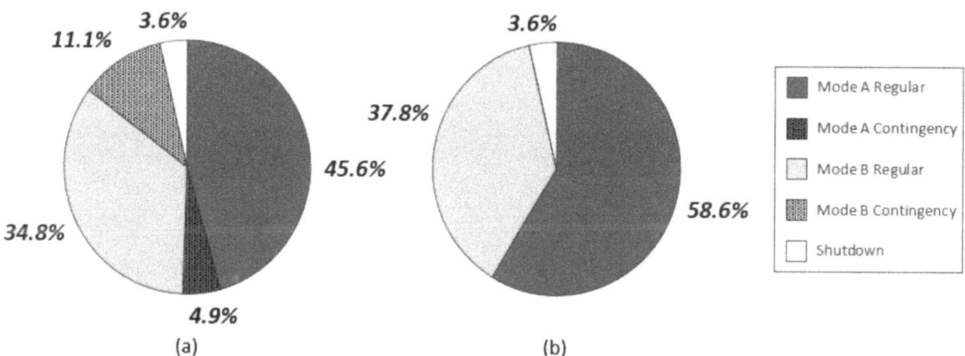

Figure 11. Time-averaged distribution of operational modes in response to geological uncertainty in the context of Canada's oil sands, for (**a**) naïve framework using the deterministic critical ore 2 threshold of 2.916 Mt and target total stockpile level of 4.374 Mt; (**b**) enhanced configuration using a critical value of 7.290 Mt (2.5×) and target total stockpile level of 14.580 Mt.

Overall, these simulation results support the flexibility of DES digital twins to integrate predictive modelling data generated through PLS regression (or other advanced methods) in order to assess the system-wide response to geological uncertainty. This quantitative framework is an extension of recent conceptual work by Navarra et al. (2019) and Wilson et al. (2021) and demonstrates its adaptation to evaluate operational risk factors associated with potential processing applications for Canada's oil sands. Simulations indicate that ore stockouts are a very real possibility due to extreme geological heterogeneities inherent to oil sands; however, the current digital twin allows for the analysis of potential adjustments to control strategies at an earlier stage, which can help drive decision making and mitigate identified risk factors. The blending control strategies described in this study would necessitate significant stockpiling infrastructure and equipment, but these implied costs could easily be offset by higher throughputs, minimized downtime and extended operational life achieved through the implementation of alternate modes of operation.

4. Discussion and Future Work

Geological variability and related uncertainty are inherent to all ore deposit types. These heterogeneities can range in intensity and generally vary both within and between deposits and/or mining districts. This can lead to unexpected changes in ore feed attributes (e.g., grade, mineralogy, grain size distribution), thereby affecting the mining and extractive processes. All of these factors have a direct impact on the interplay of critical variables and integrated coordination of processes within the overall system. Given that ore feeds are exploited from complex and heterogeneous sources, it is clear that mining systems need to be flexible, with the ability to respond to changes and communicate with all related processes (both up and downstream). The availability of alternate operational modes, each with its own set of instructions, is crucial to the sustained development of most orebodies, particularly as a project matures and evolves.

The implementation of operational buffers and other control strategies is not uncommon, but the development of suitable tools that incorporate predictive models to enhance or optimize system processes is lagging behind. Efforts are sometimes made to capture detailed information but then is not properly integrated into actual system process controls due to interdisciplinary barriers. The current study focused on ore blending strategies and overall feed management through the integration of predictive PLS regression models into a DES framework within an oil sands context. The results confirm the approach as an effective way to improve and stabilize plant throughput, despite challenges with significant geological variation of source ore feeds inherent to Canada's oil sands. Despite a small sample population and incomplete characterization (i.e., minimal depositional provenance and bitumen recovery data), the integrated quantitative framework made reasonable predictions and demonstrated how appropriate mineral and geochemical characterization could positively impact process control strategies and decision making earlier in the value chain.

As ore compositional data are routinely collected in the industry via the Soxhlet-Dean and Stark method, bitumen free solids (BFS) are readily available for geochemical and/or mineralogical analyses that can be used for predictive modelling. This work proposes that, with adequate sampling and characterization, expensive and time-consuming analytical work (e.g., organic-rich solids separation) can be replaced by faster and cheaper alternatives, such as WD-XRF and XRD executed on BFS streams [66]. The generation of robust predictive models will require extensive systematic sampling and analytical campaigns to properly characterize ore feeds and related downstream process responses; the degree of sampling is difficult to anticipate in advance and ultimately depends on data variability as well as the exact problem being addressed. Regardless, the ability to integrate reliable predictions of bitumen processability into a DES digital twin earlier in the value chain is of key importance to assess the effects of heterogeneous ore feeds on system process performance. Adjustments can then be made to operating practices (e.g., alternate modes of operation or the introduction of operational buffers) in order to mitigate the identified risk factors.

Similar to other complex mining projects, oil sands operations are host to a variety of treatments and processes. From mining to slurry and froth formation, froth treatment, upgrading and pipeline transport (each possibly comprised of multiple sub-systems), there are a large number of moving parts requiring both management and coordination. The interaction of these parts can be a major source of bottlenecks and generate severe deficiencies, which makes the oil sands context an ideal candidate for DES modelling. However, coordinated efforts between academic, government and industry partners are required in order to couple recent advances in quantitative methods with project-specific problems and data; only then can detailed flowsheets, testing and full system optimization with constraints proceed.

This conceptual study makes broad assumptions regarding ore characteristics and recoveries for demonstrative purposes but shows the suitability of the approach for multivariate ore processing systems. In practice, the ability to collect a sufficient level of data for appropriate ore characterization and build robust predictive models is challenging. This work is focused on analyzing operational risk related to oil sands mining and bitumen extraction processes with respect to geological uncertainty; in reality, overall bitumen recovery is not only based on ore characteristics, but also on changes in processing conditions, e.g., temperatures, additives and densities [77]. As such, the extensibility of the framework would allow for the eventual integration of other advanced quantitative methods, such as non-linear machine learning algorithms. The flexibility of DES digital twins to incorporate varying levels of detail makes it particularly well suited to multi-phase (re)engineering projects and can help improve confidence at each stage of development.

Author Contributions: Conceptualization, R.W., P.H.J.M., B.P. and A.N.; methodology, R.W., P.H.J.M. and A.N.; data curation, P.H.J.M.; investigation, R.W.; formal analysis, R.W.; validation, R.W.; visualization, R.W.; writing—original draft preparation, writing—reviewing and editing, P.H.J.M., B.P. and A.N.; R.W.; supervision, P.H.J.M. and A.N.; funding acquisition, P.H.J.M. and A.N. All authors have read and agreed to the published version of the manuscript.

Funding: Partial funding for this work was provided by NSERC, grant number 2020-04605, supported by the Canadian government. Financial support from the Office of Energy and Research and Development (OERD) is also acknowledged. The analytical portion of this work was performed under the High Efficiency Mining program at the National Research Council Canada (NRC).

Institutional Review Board Statement: Not applicable.

Informed Consent Statement: Not applicable.

Data Availability Statement: Restrictions apply to the availability of these data. A portion of the data was obtained from Syncrude Canada Ltd. (Research and Development–Edmonton, AB, Canada) and is available from the authors with the permission of Syncrude Canada Ltd.

Acknowledgments: The authors thank Syncrude Canada Ltd. for permission to use data on some of the ore samples analyzed here, as well as some constructive comments related to oil sands processing. These acknowledgments do not imply endorsement.

Conflicts of Interest: The authors declare no conflict of interest.

Additional Comments: The map presented in Figure 1 was created using ArcGIS® software by Esri. ArcGIS® and ArcMap™ are the intellectual property of Esri and are used herein under license. Copyright © Esri. All rights reserved. For more information about Esri® software, please visit www.esri.com.

Appendix A

Table A1. Correlation matrix for all 46 variables (including 5 composites) used in PLS regression model. Coefficients greater than 0.75 are indicated by red text.

	QTZ/SIL	ILL	KAO	CHL	CAL	DOL	ANK	SID	PYR	ZIR	RUT	ANA	ILM	LEP	GYP	BAS	ANO	KSP	ALB	AFE	APA	CRI
QTZ/SIL	1.00																					
ILL	−0.89	1.00																				
KAO	−0.91	0.81	1.00																			
CHL	−0.25	0.21	0.34	1.00																		
CAL	−0.13	0.04	−0.14	−0.04	1.00																	
DOL	−0.57	0.53	0.26	−0.08	0.57	1.00																
ANK	−0.59	0.45	0.66	0.01	−0.01	−0.08	1.00															
SID	−0.30	0.19	0.26	0.06	−0.05	0.10	0.20	1.00														
PYR	−0.06	0.20	0.01	−0.15	−0.06	0.29	−0.08	−0.23	1.00													
ZIR	−0.22	0.28	0.30	0.12	0.00	−0.04	0.13	0.04	−0.16	1.00												
RUT	0.13	−0.03	−0.12	0.08	−0.11	−0.28	−0.06	0.07	0.02	0.23	1.00											
ANA	−0.82	0.83	0.88	0.39	−0.03	0.28	0.53	0.25	−0.17	0.44	−0.20	1.00										
ILM	0.45	−0.40	−0.40	−0.18	−0.07	−0.29	−0.23	−0.10	0.13	−0.02	0.37	−0.50	1.00									
LEP	0.38	−0.33	−0.33	−0.15	−0.18	−0.33	−0.20	0.00	0.11	0.01	0.45	−0.45	0.75	1.00								
GYP	0.22	−0.36	−0.16	0.07	0.15	−0.23	−0.04	0.03	−0.20	−0.01	0.14	−0.22	0.19	0.28	1.00							
BAS	0.23	−0.37	−0.16	0.08	0.06	−0.29	−0.04	0.04	−0.19	−0.02	0.15	−0.23	0.20	0.30	0.99	1.00						
ANO	0.21	−0.35	−0.25	0.08	0.41	−0.17	−0.03	0.19	−0.50	0.04	0.06	−0.06	0.02	−0.01	0.12	0.08	1.00					
KSP	−0.19	−0.05	0.27	0.51	0.00	−0.06	0.03	0.17	−0.45	0.01	0.02	0.28	−0.13	−0.06	0.19	0.18	0.45	1.00				
ALB	−0.85	0.80	0.75	0.11	0.13	0.69	0.37	0.10	0.31	0.04	−0.20	0.60	−0.35	−0.32	−0.18	−0.20	−0.46	0.06	1.00			
AFE	−0.59	0.56	0.44	0.08	0.34	0.71	0.14	−0.35	0.24	0.06	−0.35	0.44	−0.46	−0.54	−0.20	−0.25	−0.37	−0.07	0.71	1.00		
APA	−0.60	0.61	0.40	0.07	0.31	0.71	0.07	0.11	0.35	−0.05	−0.18	0.38	−0.21	−0.19	−0.47	−0.48	−0.10	0.01	0.65	0.45	1.00	
CRI	0.35	−0.29	−0.36	−0.29	0.00	−0.17	−0.20	−0.16	0.04	−0.09	−0.02	−0.30	0.16	0.23	0.16	0.17	0.05	−0.16	−0.27	−0.20	−0.15	1.00
ORC	0.63	−0.76	−0.67	−0.32	−0.15	−0.46	−0.22	−0.17	−0.11	−0.27	0.13	−0.76	0.35	0.34	0.20	0.23	0.08	−0.18	−0.64	−0.45	−0.53	0.33
ORS	0.29	−0.40	−0.45	−0.23	0.28	0.04	−0.16	−0.09	−0.18	−0.41	0.02	−0.46	−0.04	−0.05	0.07	0.04	0.18	−0.06	−0.22	0.00	−0.26	0.14
TOTAL CLAYS	−0.94	0.95	0.95	0.32	−0.05	0.41	0.58	0.24	0.10	0.31	−0.07	0.90	−0.42	−0.35	−0.27	−0.27	−0.31	0.13	0.81	0.52	0.53	−0.35
SID+AFE	−0.77	0.64	0.61	0.13	0.24	0.69	0.30	0.61	−0.01	0.09	−0.23	0.59	−0.48	−0.45	−0.14	−0.17	−0.14	0.10	0.69	0.52	0.48	−0.32
RUT+ANA	−0.79	0.83	0.86	0.43	−0.06	0.19	0.51	0.28	−0.17	0.52	0.12	0.95	−0.38	−0.30	−0.18	−0.18	−0.04	0.29	0.55	0.33	0.32	−0.31
DOL+CAL+ANK	−0.56	0.49	0.24	−0.08	0.72	0.97	0.01	0.09	0.22	−0.03	−0.27	0.27	−0.28	−0.34	−0.16	−0.23	−0.05	−0.05	0.64	0.70	0.68	−0.16
ANO+KSP+ALB	−0.66	0.44	0.58	0.35	0.35	0.45	0.30	0.26	−0.21	0.05	−0.13	0.58	−0.33	−0.30	−0.01	−0.05	0.38	0.72	0.59	0.34	0.49	−0.26
Na	−0.85	0.80	0.75	0.11	0.13	0.69	0.37	0.10	0.31	0.04	−0.20	0.60	−0.35	−0.32	−0.18	−0.20	−0.46	0.06	1.00	0.71	0.65	−0.27
K	−0.92	0.96	0.86	0.35	0.04	0.50	0.45	0.23	0.07	0.28	−0.02	0.88	−0.42	−0.33	−0.30	−0.31	−0.22	0.23	0.79	0.53	0.60	−0.33
Si	0.96	−0.79	−0.80	−0.15	−0.23	−0.62	−0.58	−0.32	−0.05	−0.15	0.15	−0.69	0.43	0.37	0.18	0.20	0.19	−0.11	−0.80	−0.59	−0.58	0.32
Al	−0.96	0.93	0.96	0.33	0.01	0.43	0.59	0.26	0.05	0.30	−0.08	0.92	−0.43	−0.36	−0.25	−0.26	−0.21	0.22	0.81	0.51	0.55	−0.35
Fe	−0.81	0.70	0.66	0.22	0.23	0.72	0.32	0.52	0.15	0.08	−0.22	0.61	−0.47	−0.45	−0.17	−0.19	−0.22	0.08	0.75	0.60	0.54	−0.34
Mg	−0.62	0.58	0.33	0.08	0.56	0.99	−0.05	0.11	0.27	−0.02	−0.27	0.35	−0.33	−0.36	−0.22	−0.28	−0.16	0.02	0.72	0.73	0.73	−0.22
Ca	−0.40	0.27	0.08	−0.03	0.89	0.81	−0.01	0.13	0.01	−0.02	−0.20	0.18	−0.21	−0.27	0.03	−0.05	0.35	0.13	0.39	0.48	0.56	−0.10
Ti	−0.79	0.83	0.85	0.43	−0.07	0.18	0.51	0.28	−0.17	0.53	0.13	0.94	−0.37	−0.29	−0.17	−0.18	−0.04	0.29	0.54	0.32	0.32	−0.31
Zr	−0.22	0.28	0.30	0.12	0.00	−0.04	0.13	0.04	−0.16	1.00	0.23	0.44	−0.02	0.01	−0.01	−0.02	0.04	0.01	0.04	0.06	−0.05	−0.09
P	−0.60	0.61	0.40	0.07	0.31	0.71	0.07	0.11	0.35	−0.05	−0.18	0.38	−0.21	−0.19	−0.47	−0.48	−0.10	0.01	0.65	0.45	1.00	−0.15
C	0.60	−0.74	−0.66	−0.34	−0.09	−0.40	−0.22	−0.13	−0.11	−0.28	0.12	−0.76	0.33	0.33	0.20	0.22	0.09	−0.18	−0.61	−0.43	−0.49	0.32
S	0.30	−0.37	−0.47	−0.27	0.27	0.11	−0.20	−0.18	0.22	−0.46	0.06	−0.56	0.06	0.06	0.18	0.16	−0.02	−0.22	−0.11	0.06	−0.20	0.18

Table A1. *Cont.*

	QTZ/SIL	ILL	KAO	CHL	CAL	DOL	ANK	SID	PYR	ZIR	RUT	ANA	ILM	LEP	GYP	BAS	ANO	KSP	ALB	AFE	APA	CRI
Bitumen	0.73	−0.80	−0.75	−0.30	−0.18	−0.47	−0.38	−0.20	−0.11	−0.32	0.14	−0.82	0.38	0.36	0.23	0.25	0.05	−0.16	−0.67	−0.48	−0.54	0.37
Water	−0.69	0.73	0.72	0.33	0.04	0.36	0.38	0.10	0.28	0.22	−0.10	0.70	−0.31	−0.25	−0.18	−0.18	−0.18	0.14	0.68	0.43	0.56	−0.21
Solids	−0.17	0.21	0.16	0.00	0.24	0.25	0.07	0.19	−0.26	0.20	−0.09	0.30	−0.18	−0.23	−0.11	−0.14	0.18	0.06	0.10	0.15	0.03	−0.31
Fines	−0.86	0.86	0.85	0.41	0.06	0.41	0.47	0.27	−0.07	0.31	−0.15	0.91	−0.50	−0.45	−0.22	−0.23	−0.09	0.27	0.69	0.46	0.52	−0.37
Total Recovery	0.49	−0.46	−0.50	−0.17	−0.20	−0.33	−0.27	−0.11	0.08	−0.33	0.13	−0.56	0.03	0.13	0.09	0.12	−0.12	−0.24	−0.41	−0.28	−0.37	0.28

	ORC	ORS	TOTAL CLAYS	SID+AFE	RUT+ANA	DOL+CAL+ANK	ANO+KSP+ALB	Na	K	Si	Al	Fe	Mg	Ca	Ti	Zr	P	C	S	Bitumen	Water	Solids	Fines	Total Recovery
ORC	1.00																							
ORS	0.55	1.00																						
TOTAL CLAYS	−0.75	−0.45	1.00																					
SID+AFE	−0.54	−0.09	0.66	1.00																				
RUT+ANA	−0.73	−0.46	0.89	0.53	1.00																			
DOL+CAL+ANK	−0.44	0.09	0.38	0.67	0.18	1.00																		
ANO+KSP+ALB	−0.56	−0.10	0.54	0.53	0.55	0.48	1.00																	
Na	−0.64	−0.22	0.81	0.69	0.55	0.64	0.59	1.00																
K	−0.79	−0.40	0.96	0.66	0.89	0.46	0.63	0.79	1.00															
Si	0.46	0.16	−0.83	−0.79	−0.65	−0.62	−0.60	−0.80	−0.80	1.00														
Al	−0.77	−0.43	0.99	0.67	0.90	0.41	0.64	0.81	0.97	−0.85	1.00													
Fe	−0.59	−0.14	0.72	0.98	0.55	0.68	0.53	0.75	0.71	−0.82	0.72	1.00												
Mg	−0.52	0.00	0.48	0.72	0.27	0.96	0.51	0.72	0.57	−0.66	0.50	0.76	1.00											
Ca	−0.35	0.16	0.18	0.53	0.12	0.91	0.58	0.39	0.30	−0.48	0.25	0.51	0.81	1.00										
Ti	−0.73	−0.46	0.89	0.52	1.00	0.18	0.55	0.54	0.88	−0.65	0.90	0.54	0.27	0.11	1.00									
Zr	−0.27	−0.41	0.31	0.09	0.52	−0.03	0.05	0.04	0.28	−0.15	0.30	0.08	−0.02	−0.02	0.53	1.00								
P	−0.53	−0.26	0.53	0.48	0.32	0.68	0.49	0.65	0.60	−0.58	0.55	0.54	0.73	0.56	0.32	−0.05	1.00							
C	1.00	0.57	−0.74	−0.48	−0.73	−0.37	−0.53	−0.61	−0.77	0.41	−0.75	−0.53	−0.46	−0.28	−0.73	−0.28	−0.49	1.00						
S	0.52	0.90	−0.44	−0.12	−0.55	0.14	−0.20	−0.11	−0.42	0.17	−0.44	−0.11	0.06	0.16	−0.56	−0.46	−0.20	0.55	1.00					
Bitumen	0.95	0.50	−0.81	−0.58	−0.78	−0.47	−0.59	−0.67	−0.82	0.58	−0.83	−0.64	−0.53	−0.38	−0.78	−0.32	−0.54	0.94	0.49	1.00				
Water	−0.76	−0.51	0.77	0.46	0.68	0.34	0.52	0.68	0.75	−0.57	0.77	0.55	0.42	0.23	0.67	0.22	0.56	−0.76	−0.41	−0.80	1.00			
Solids	−0.41	−0.05	0.19	0.30	0.27	0.27	0.21	0.10	0.22	−0.12	0.22	0.24	0.25	0.29	0.27	0.20	0.03	−0.40	−0.18	−0.44	−0.18	1.00		
Fines	−0.82	−0.44	0.91	0.63	0.87	0.40	0.63	0.69	0.91	−0.74	0.92	0.67	0.49	0.29	0.87	0.31	0.52	−0.81	−0.50	−0.85	0.73	0.31	1.00	
Total Recovery	0.55	0.37	−0.50	−0.34	−0.52	−0.35	−0.51	−0.41	−0.51	0.41	−0.54	−0.35	−0.36	−0.35	−0.53	−0.33	−0.37	0.54	0.41	0.57	−0.53	−0.15	−0.53	1.00

References

1. Bata, T.; Schamel, S.; Fustic, M.; Ibatulin, R. *AAPG Energy Minerals Division Bitumen and Heavy Oil Committee Annual Commodity Report—May 2019*; American Association of Petroleum Geologists (AAPG): Tulsa, OK, USA, 2019.
2. Canada Energy Regulator (CER). *Canada's Energy Future 2020: Energy Supply and Demand Projections to 2050 (EF2020)*; Canada Energy Regulator: Ottawa, ON, Canada, 2020.
3. Hein, F.J. Heavy oil and oil (tar) sands in North America: An overview & summary of contributions. *Nat. Resour. Res.* **2006**, *15*, 67–84.
4. Gray, M.R.; Xu, Z.; Masliyah, J. Physics in the oil sands of Alberta. *Phys. Today.* **2009**, *62*, 31–35. [CrossRef]
5. Algeer, R.; Snowdon, L.; Huang, H.; Oldenburg, T.; Larter, S. Is water washing an important petroleum system process? In Proceedings of the AAPG Annual Convention and Exhibition, Calgary, AB, Canada, 19–22 June 2016.
6. Fustic, M.; Ahmed, K.; Brough, S.; Bennett, B.; Bloom, L.; Asgar–Deen, M.; Jokanola, O.; Spencer, R.; Larter, S. Reservoir and bitumen heterogeneity in Athabasca oil sands. *AAPG Search Discov.* **2015**, *20296*, 15.
7. Masliyah, J.H.; Czarnecki, J.; Xu, Z. *Handbook on Theory and Practice of Bitumen Recovery from Athabasca Oil Sands, Volume 1: Theoretical Basis*; Kingsley Knowledge Publishing: Cochrane, AB, Canada, 2011.
8. Gray, M.R.; Eaton, P.E.; Le, T. Inhibition and promotion of hydrolysis of chloride salts in model crude oil and heavy oil. *Pet. Sci. Technol.* **2008**, *26*, 1934–1944. [CrossRef]
9. Kaur, H.; Eaton, P.E.; Gray, M.R. The kinetics and inhibition of chloride hydrolysis in Canadian bitumen. *Pet. Sci. Technol.* **2012**, *30*, 993–1003. [CrossRef]
10. Li, X.; Wu, B.; Zhu, J. Hazards of organic chloride to petroleum processing in Chinese refineries and industrial countermeasures. *Progress Petrochem. Sci.* **2018**, *2*, 204–207. [CrossRef]
11. Londono, Y.; Mikula, R.; Eaton, P.E.; Gray, M.R. Interaction of chloride salts and kaolin clay in the hydrolysis of emulsified chloride salts at 200–350 C. *Pet. Sci. Technol.* **2009**, *27*, 1163–1174. [CrossRef]
12. Navarra, A.; Grammatikopoulos, T.; Waters, K. Incorporation of geometallurgical modelling into long–term production planning. *Miner. Eng.* **2018**, *120*, 118–126. [CrossRef]
13. Navarra, A.; Rafiei, A.; Waters, K. A systems approach to mineral processing based on mathematical programming. *Can. Metall. Q.* **2017**, *56*, 35–44. [CrossRef]
14. Navarra, A.; Wilson, R.; Parra, R.; Toro, N.; Ross, A.; Nave, J.-C.; Mackey, P.J. Quantitative methods to support data acquisition modernization within copper smelters. *Processes* **2020**, *8*, 1478. [CrossRef]
15. Navarra, A.; Alvarez, M.; Rojas, K.; Menzies, A.; Pax, R.; Waters, K. Concentrator operational modes in response to geological variation. *Miner. Eng.* **2019**, *134*, 356–364. [CrossRef]
16. Saldana, M.; Toro, N.; Castillo, J.; Hernandez, P.; Navarra, A. Optimization of the heap leaching process through changes in modes of operation and discrete event simulation. *Minerals* **2019**, *9*, 421. [CrossRef]
17. Wilson, R.; Toro, N.; Naranjo, O.; Emery, X.; Navarra, A. Integration of geostatistical modeling into discrete event simulation for development of tailings dam retreatment applications. *Miner. Eng.* **2021**, *164*, 106814. [CrossRef]
18. Boschert, S.; Rosen, R. Digital Twin—The Simulation Aspect. In *Mechatronic Futures*; Hehenberger, P., Bradley, D., Eds.; Springer International Publishing: Cham, Switzerland, 2016; pp. 59–74.
19. Glaessgen, E.H.; Stargel, D.S. The digital twin paradigm for future NASA and U.S. Air Force vehicles. In Proceedings of the 53rd Structures, Structural Dynamics, and Materials Conference: Special Session on the Digital Twin, Honolulu, HI, USA, 23–26 April 2012.
20. Kaminsky, H.A.W. Characterization of an Athabasca Oil Sands Ore and Process Streams. PhD Thesis, Department of Chemical and Materials Engineering, University of Alberta, Edmonton, AB, Canada, 2008.
21. Energy Resources Conservation Board (ERCB). *Alberta's Energy Reserves 2008 and Supply/Demand Outlook 2009–2018*; Energy Resources Conservation Board: Calgary, AB, Canada, 2009.
22. Zhao, S.; Kotlyar, L.S.; Sparks, B.D.; Woods, J.; Gao, J.; Chung, K. Solids contents, properties and molecular structures of asphaltenes from different oilsands. *Fuel* **2001**, *80*, 1907–1914. [CrossRef]
23. Barson, D.; Bachu, S.; Esslinger, P. Flow systems in the Mannville Group in the east–central Athabasca area and implications for steam–assisted gravity drainage (SAGD) operations for in situ bitumen production. *Bull. Can. Pet. Geol.* **2001**, *49*, 376–392. [CrossRef]
24. Fowler, M.; Riediger, C. Origin of the Athabasca tar sands. In *Hydrogeology of Heavy Oil and Tar Sand Deposits: Water Flow and Supply, Migration and Degradation—Field Trip Notes (GSC Open File 3946)*; Barson, D., Bartlett, R., Hein, F.J., Fowler, M., Grasby, S., Riediger, C., Underschultz, J., Eds.; Geological Survey of Canada: Calgary, AB, Canada, 2000; pp. 117–127.
25. Hein, F.J.; Langenberg, C.W.; Kidston, C.; Berhane, H.; Berezniuk, T.; Cotterill, D.K. *A Comprehensive Field Guide for Facies Characterization of the Athabasca Oil Sands, Northeast Alberta (with Maps, Air Photos, and Detailed Descriptions of Seventy–Eight Outcrop Sections): AEUB/AGS Special Report 13*; Alberta Energy Utilities Board/Alberta Geological Survey: Edmonton, AB, Canada, 2001.
26. Hein, F.J.; Cotterill, D.K.; Berhane, H. *An Atlas of Lithofacies of the McMurray Formation, Athabasca Oil Sands Deposit, Northeastern Alberta: Surface and Subsurface: AEUB/AGS Earth Sciences Report 2000–07*; Alberta Energy and Utilities Board/Alberta Geological Survey: Edmonton, AB, Canada, 2000.

27. Wightman, D.M.; Pemberton, S.G. The Lower Cretaceous (Aptian) McMurray Formation: An overview of the McMurray Area, Northeastern Alberta. In *Petroleum Geology of the Cretaceous Lower Mannville Group: Western Canada: Canadian Society of Petroleum Geologists Memoir 18*; Pemberton, G.S., James, D.P., Eds.; Canadian Society of Petroleum Geologists (CSPG): Calgary, AB, Canada, 1997; pp. 312–344.

28. Alberta Energy Regulator (AER). *Alberta's Energy Reserves 2014 and Supply/Demand Outlook 2015–2024, Statistical Series ST98–2015*; Alberta Energy Regulator: Calgary, AB, Canada, 2015.

29. Marinkovic, J. Multivariate Statistics. In *Encyclopedia of Public Health*; Kirch, W., Ed.; Springer Publishing: Dordrecht, The Netherlands, 2008; pp. 973–976.

30. Abdi, H. Partial least squares (PLS) regression. In *Encyclopedia for Research Methods for the Social Sciences*; Lewis-Beck, M., Bryman, A., Futing, T., Eds.; Sage: Thousand Oaks, CA, USA, 2003; pp. 792–795.

31. Abdi, H. Partial least square regression and projection on latent structure regression (PLS regression). *Comput. Stat.* **2010**, *2*, 97–106. [CrossRef]

32. Hoskuldsson, A. PLS regression methods. *J. Chemom.* **1988**, *2*, 211–228. [CrossRef]

33. Tenenhaus, M. *La Regression PLS: Théorie et Pratique*; Éditions Technip: Paris, France, 1998.

34. Ekblad, A. Forest soil respiration rate and delta C–13 is regulated by recent above ground weather conditions. *Oecologia* **2005**, *143*, 136–142. [CrossRef] [PubMed]

35. Allen, A.E. Influence of nitrate availability on the distribution and abundance of heterotropic bacterial nitrate assimilation genes in the Barents Sea suring summer. *Aquat. Microb. Ecol.* **2005**, *39*, 247–255. [CrossRef]

36. Maestre, F.T. On the importance of patch attributes, environmental factors and past human impacts as determinants of perennial plant species richness and diversity in Mediterranean semiarid steppes. *Div. Distr.* **2004**, *10*, 21–29. [CrossRef]

37. Palomino, D.; Carrascal, L.M. Habitat associations of a raptor community in a mosaic landscape of central Spain under urban development. *Landsc. Urban. Plan.* **2007**, *83*, 268–274. [CrossRef]

38. Seppa, H. A modern pollen climate calibration set from northern Europe: Developing and testing a tool for palaeoclimatological reconstructions. *J. Biogeogr.* **2004**, *31*, 251–267. [CrossRef]

39. Sonesten, L. Catchment area composition and water chemistry heavily affects mercury levels in perch (Perca fluviatilis L.) in circumneutral lakes. *Water Air Soil Pollut.* **2003**, *144*, 117–139. [CrossRef]

40. Spanos, T. Environmetrics to evaluate marine environment quality. *Environ. Monit. Assess.* **2008**, *143*, 215–225. [CrossRef] [PubMed]

41. Pandit, C.M.; Filippelli, G.M.; Li, L. Estimation of heavy metal contamination in soil using reflectance spectroscopy and partial least–squares regression. *Int. J. Remote Sens.* **2010**, *31*, 4111–4123. [CrossRef]

42. Kun, W.; Keyan, X.; Nan, L.; Yuan, C.; Shengmiao, L. Application of partial least squares regression for identifying multivariate geochemical anomalies in stream sediment data from Northwestern Hunan, China. *Geochem-Explor. Environ. A* **2017**, *17*, 217–230. [CrossRef]

43. Abdi, H.; Williams, L.J. Partial least squares methods: Partial least squares correlation and partial least squares regression. In *Methods in Molecular Biology: Computational Toxicology*; Reisfeld, B., Mayeno, A., Eds.; Springer: New York, NY, USA, 2013; pp. 549–579.

44. Abdi, H. Singular value decomposition (SVD) and generalized singular value decomposition (GSVD). In *Encyclopedia of Measurement and Statistics*; Salkind, N.J., Ed.; Sage: Thousand Oaks, CA, USA, 2007; pp. 907–912.

45. Melzer, T. *SVD and Its Application to Generalized Eigenvalue Problems*; University of Technology: Vienna, Austria, 2004.

46. Hern, T.; Long, C. Viewing some concepts and applications in linear algebra. In *Visualization in Teaching and Learning Mathematics*; MAA Notes, No. 19, Mathematical Association of America: Washington, DC, USA, 1991; pp. 173–190.

47. Strang, G. The fundamental theoremof linear algebra. *Am. Math Mon.* **1993**, *100*, 848–855. [CrossRef]

48. Trefethen, L.N.; Bau, D., III. *Numerical Linear Algebra*, 1st ed.; Society for Industrial and Applied Mathematics (SIAM): Philadelphia, PA, USA, 1997.

49. Siegel, A.F. Chapter 12—Multiple Regression: Predicting One Variable from Several Others. In *Practical Business Statistics*, 7th ed.; Siegel, A.F., Ed.; Academic Press: Cambridge, MA, USA, 2017; pp. 355–418.

50. Egarievwe, S.U.; Coble, J.B.; Miller, L.F. Analysis of how well regression models predict radiation dose from the Fukushima Daiichi Nuclear Accident. *Int. J. Appl. Phys. Math.* **2016**, *6*, 150–164. [CrossRef]

51. Huerta, M.; Leiva, V.; Liu, S.; Rodriguez, M.; Villegas, D. On a partial least squares regression model for asymmetric data with a chemical application in mining. *Chemometr. Intell. Lab. Syst.* **2019**, *190*, 55–68. [CrossRef]

52. Altiok, T.; Melamed, B. *Simulation Modeling and Analysis with Arena*; Academic Press: Boston, MA, USA, 2007.

53. Alruiz, O.; Morrell, S.; Suazo, C.; Naranjo, A. A novel approach to the geometallurgical modelling of the Collahuasi grinding circuit. *Miner. Eng.* **2009**, *22*, 1060–1067. [CrossRef]

54. Suazo, C.; Kracht, W.; Alruiz, O. Geometallurgical modelling of the Collahuasi flotation circuit. *Miner. Eng.* **2010**, *23*, 137–142. [CrossRef]

55. Navarra, A.; Marambio, H.; Oyarzun, F.; Parra, R.; Mucciardi, F. System dynamics and discrete event simulation of copper smelters. *Miner. Metall. Process.* **2017**, *34*, 96–106. [CrossRef]

56. Awuah–Offei, K.; Osei, B.A.; Askari–Nasab, H. Improving truck–shovel energy efficiency through discrete event modeling. In Proceedings of the Society for Mining, Metallurgy & Exploration (SME) Annual Meeting, Seattle, WA, USA, 19–22 February 2012.

57. Vagenas, N. Applications of discrete–event simulation in Canadian mining operations in the nineties. *Int. J. Surf. Min. Reclam. Environ.* **1999**, *13*, 77–78. [CrossRef]
58. Greberg, J.; Salama, A.; Gustafson, A.; Skawina, B. Alternative process flow for underground mining operations: Analysis of conceptual transport methods using discrete event simulation. *Minerals* **2016**, *6*, 65. [CrossRef]
59. Gbadam, E.; Awuah–Offei, K.; Frimpong, S. Investigation into Mine Equipment Subsystem Availability & Reliability Data Modeling Using DES. In *Application of Computers and Operations Research in the Mineral Industry*; Bandopadhyay, S., Ed.; Society for Mining, Metallurgy & Exploration, Inc. (SME): Englewood, CO, USA, 2015.
60. Tarshizi, E. Using Discrete Simulation & Animation to Identify the Optional Sizes and Locations of Mine Refuge Chambers. In *Application of Computers and Operations Research in the Mineral Industry*; Bandopadhyay, S., Ed.; Society for Mining Metallurgy & Exploration Inc. (SME): Englewood, CO, USA, 2015.
61. Shishvan, M.; Benndorf, J. Performance optimization of complex continuous mining system using stochastic simulation. In *Engineering Optimization*; Rodrigues, H.J., Mota, S.C., Miranda, G.J., Araujo, A., Folgado, J., Eds.; Taylor and Francis: Hoboken, NJ, USA, 2014; pp. 273–278.
62. Nageshwaraniyer, S.; Kim, K.; Son, Y.J. A mine–to–mill economic analysis model and spectral imaging–based tracking system for a copper mine. *J. S. Afr. I Min. Metall.* **2018**, *118*, 7–14. [CrossRef]
63. Pamparana, G.; Kracht, W.; Haas, J.; Diaz–Ferran, G.; Palma–Behnke, R.; Roman, R. Integrating photovoltaic solar energy and a battery energy storage system to operate a semi–autogenous grinding mill. *J. Clean. Prod.* **2017**, *165*, 273–280. [CrossRef]
64. Winstin, W.; Goldberg, J. The EOQ with uncertain demand: The (r, q) and (s, S) models. In *Operations Research: Applications and Algorithms (Section 16.6)*; Cengage Learning: Boston, MA, USA, 2004; pp. 895–902.
65. Bulmer, J.T.; Starr, J. *Syncrude Analytical Procedures for Oilsands and Bitumen Processing*; Alberta Oil Sands Technology and Research Authority (AOSTR): Edmonton, AB, Canada, 1974.
66. Patarachao, B.; Mercier, P.H.J.; Kung, J.; Woods, J.R.; Kotlyar, L.S.; Sparks, B.D.; McCracken, T. Optimizing XRF calibration protocols for elemental quantification of mineral solids from Athabasca oil sands. *Adv. X-ray Anal.* **2010**, *53*, 220–227. [CrossRef]
67. Couillard, M.; Mercier, P.H.J. Analytical electron microscopy of carbon–rich mineral aggregates in solvent–diluted bitumen products from mined Alberta oil sands. *Energy Fuels* **2016**, *30*, 5513–5524. [CrossRef]
68. Alberta Energy Regulator (AER). *Operating Criteria: Resource Recovery Requirements for Oil Sands Mine and Processing Plant Operations (AER Directive 082)*; Alberta Energy Regulator: Calgary, AB, Canada, 2016.
69. Mercier, P.H.J.; Patarachao, B.; Kung, J.; Kingston, D.; Woods, J.; Sparks, B.D.; Kotlyar, L.S.; Ng, S.; Moran, K.; McCracken, T. X-ray diffraction (XRD)–derived processability markers for oil sands based on clay mineralogy and crystallite thickness distributions. *Energy Fuels* **2008**, *22*, 3174–3193. [CrossRef]
70. Mercier, P.H.J.; Kingston, D.; Kung, J.; Woods, J.R.; Kotlyar, L.S.; Tu, Y.; Smith, T.; Ng, S.; Moran, K.; Sparks, B.D.; et al. *Development of an Innovative Method for Assessment of Oilsands Ore Processability by Measurement of Paramagnetic Signatures—Final Report to CONRAD Bitumen Production Research Group (ICPET Report #PET–1570–06S)*; National Research Council Canada (NRC)—Institute for Chemical Process and Environmental Technology: Ottawa, ON, Canada, 2007.
71. Wallace, D.; Tipman, R.; Komishke, B.; Wallwork, V.; Perkins, E. Fines/water interactions and consequences of the presence of degraded illite on oil sands extractability. *Can. J. Chem. Eng.* **2004**, *82*, 667–677. [CrossRef]
72. Omotoso, O.E.; Mikula, R.J. High surface areas caused by smectitic interstratification of kaolinite and illite in Athabasca oil sands. *Appl. Clay Sci.* **2003**, *25*, 37–47. [CrossRef]
73. Long, J.; Hoskins, S.; Reid, K. *United States Patent Application Publication No. US 2020/0102505 A1*; United States Patent and Trademark Office: Alexandria, VA, USA, 2020.
74. Potdar, K.; Pardawala, T.; Pai, C. A comparative study of categorical variable encoding techniques for neural network classifiers. *Int. J. Comput. Appl. Technol.* **2017**, *175*, 7–9. [CrossRef]
75. Mevik, B.-H.; Wehrens, R. The pls package: Principal component and partial least squares regression in R. *J. Stat. Softw.* **2007**, *18*, 1–24. [CrossRef]
76. Pell, R.J.; Ramos, L.S.; Manne, R. Themodel space in partial least squares regression. *J. Chemom.* **2007**, *21*, 165–172. [CrossRef]
77. Kresta, J. (Syncrude Canada Ltd., Edmonton, AB, Canada). Personal communication, 2021.

Article

Diagnosis of Problems in Truck Ore Transport Operations in Underground Mines Using Various Machine Learning Models and Data Collected by Internet of Things Systems

Sebeom Park [1], Dahee Jung [1], Hoang Nguyen [2,3] and Yosoon Choi [1,*]

1. Department of Energy Resources Engineering, Pukyong National University, Busan 48513, Korea; sebumi1v@gmail.com (S.P.); 98dahee@naver.com (D.J.)
2. Department of Surface Mining, Mining Faculty, Hanoi University of Mining and Geology, 18 Vien Str., Duc Thang ward, Bac Tu Liem Distr., Hanoi 100000, Vietnam; nguyenhoang@humg.edu.vn
3. Innovations for Sustainable and Responsible Mining (ISRM) Group, Department of Surface Mining, Mining Faculty, Hanoi University of Mining and Geology, 18 Vien Str., Duc Thang ward, Bac Tu Liem Distr., Hanoi 100000, Vietnam
* Correspondence: energy@pknu.ac.kr; Tel.: +82-51-629-6562; Fax: +82-51-629-6553

Abstract: This study proposes a method for diagnosing problems in truck ore transport operations in underground mines using four machine learning models (i.e., Gaussian naïve Bayes (GNB), k-nearest neighbor (kNN), support vector machine (SVM), and classification and regression tree (CART)) and data collected by an Internet of Things system. A limestone underground mine with an applied mine production management system (using a tablet computer and Bluetooth beacon) is selected as the research area, and log data related to the truck travel time are collected. The machine learning models are trained and verified using the collected data, and grid search through 5-fold cross-validation is performed to improve the prediction accuracy of the models. The accuracy of CART is highest when the parameters leaf and split are set to 1 and 4, respectively (94.1%). In the validation of the machine learning models performed using the validation dataset (1500), the accuracy of the CART was 94.6%, and the precision and recall were 93.5% and 95.7%, respectively. In addition, it is confirmed that the F1 score reaches values as high as 94.6%. Through field application and analysis, it is confirmed that the proposed CART model can be utilized as a tool for monitoring and diagnosing the status of truck ore transport operations.

Keywords: bluetooth beacon; classification and regression tree; gaussian naïve bayes; k-nearest neighbors; support vector machine; transport route; transport time; underground mine

Citation: Park, S.; Jung, D.; Nguyen, H.; Choi, Y. Diagnosis of Problems in Truck Ore Transport Operations in Underground Mines Using Various Machine Learning Models and Data Collected by Internet of Things Systems. *Minerals* **2021**, *11*, 1128. https://doi.org/10.3390/min11101128

Academic Editors: Rajive Ganguli, Sean Dessureault and Pratt Rogers

Received: 14 September 2021
Accepted: 12 October 2021
Published: 14 October 2021

Publisher's Note: MDPI stays neutral with regard to jurisdictional claims in published maps and institutional affiliations.

1. Introduction

Because the productivity and profits of mines can vary greatly depending on the design and planning of the production process, optimal operation methods and equipment utilization strategies are needed to maximize productivity and equipment efficiency and minimize operating costs [1–5]. The cost of transporting ore and waste accounts for over 50% of the total mine operational cost, therefore, it is crucial to design and operate the transport system efficiently [6]. Methods to improve the productivity and efficiency of the mine transport system are divided broadly into two types: methods to properly establish an operational plan so that the mine can be operated effectively, and methods to monitor and manage the site to see whether the established plan is being well implemented.

Recently, various mathematical decisions and deterministic and probabilistic simulation models have been proposed by researchers to establish an operational plan, such as optimizing the operational method and equipment allocation plan of the mine transport system and minimizing material handling costs [4,7–13]. Since the first implementation of discrete event simulation by Rist to solve problems related to ore transport in mines, many researchers have conducted research on discrete event simulation [14]. Salama and

Greberg [15] performed a simulation of a loading-haulage-dumping machine (LHD) and a truck to optimize the number of trucks used in haulage operation in an underground mine. Choi [16] developed a discrete event simulation program to simulate the shovel-truck transport system of an open-pit mine using the GPSS/H simulation language. Choi and Nieto [3] extended this to analyze the optimal transport path of a truck. Subsequently, they performed discrete event simulations of transport equipment and provided a function to visualize the simulation results. Park and Choi [17–22] developed GPSS/H-based programs and user-friendly programs to simulate truck-loader transport systems, considering various conditions such as fixed/real-time allocation, crusher capacity, and possibility of truck failure.

If the operational plan of the transport system of the mine has been properly established, it is also crucial to continuously monitor the operational status of the transport system and to verify whether the established plan is properly implemented at the site. Until now, research on monitoring and diagnosing the operating status of transport systems or equipment has been conducted by various researchers. Thompson et al. [23] provided the basis for mine maintenance management systems (MMS) by integrating data collected through onboard multi-sensors that were installed on trucks with existing mine communication and asset management systems. Park and Choi [24] developed a system that could collect truck travel time data using Bluetooth beacons and tablet computers. In addition, a method for analyzing and diagnosing the transport route status of underground mines was proposed using the collected data. Wodecki et al. [25] proposed a monitoring system that could identify major possible causes of machine failure events using the operational parameters of LHD in mines. Carvalho et al. [26] developed a system that could automatically identify the failure of a roller, one of the important components of a belt conveyor, by combining a thermal imaging camera with an unmanned aerial vehicle (UAV).

Recently, machine learning techniques have been actively utilized to monitor the transport systems and assets of mines, diagnose failures, and perform proper maintenance. Paduraru and Dimitrakopoulos [27] utilized neural networks and policy gradient reflection learning in data-driven decision-making processes to optimize material flow in large mining complexes. Ristovski et al. [28] used machine learning to predict the probability distributions of equipment activity durations used in mining operations. Xue et al. [29] and Sun et al. [30] used a machine learning model to predict truck travel time. Zhang et al. [31] used the support vector machine (SVM), a machine learning technique, to diagnose and classify the defects of the scraper conveyor in coal mine. D'Angelo et al. [32] proposed a method for real-time diagnosis of defects in rollers of belt conveyors using an object detection model based on a deep learning architecture.

Establishing operational plans, such as mine design, production forecasting, and equipment allocation, is important to ensure productivity and efficiency of mines. In addition, identifying in advance the section in which the truck travel time is expected to be abnormal is crucial because this makes it possible to prevent the occurrence of problems in the section and vehicle, as well as in the future. However, no research case has been reported thus far for monitoring and diagnosing the condition of a mine transport system using machine learning techniques. Therefore, we propose a method to evaluate the stability of transport routes and to diagnose the operational status by combining the mine production management system using a tablet computer and Bluetooth beacon with machine learning techniques. To this end, a limestone underground mine in Korea—to which a tablet computer and Bluetooth beacon was applied—was selected as a research area, and log data related to truck travel time were collected for a certain period. In addition, machine learning models were trained using the collected data. Thereafter, the stability of each section of the transport route in the study area was evaluated, diagnosed, and analyzed using the learned model.

2. Study Area and Data Collection

In this study, an underground mine (37°17′12″ N, 128°43′53″ E) owned by Seongshin Minefield in Korea was selected as the research area. Figure 1 depicts an aerial view of the study area and an underground tunnel. The mine uses the room and pillar mining method to produce 1 million tons of high-quality limestone annually. They drill with a V-Cut method using jumbo drills and crawler drills. It then produces approximately 4500 tons of limestone, with an average of 8–9 blasts per day using ammonium nitrate fuel oil (ANFO), emulite, and electric detonator (6 ms). The mined limestone is loaded into a 25–40 tons dump truck with a loader (3.0–5.6 m^3) and transported to the crusher located outside the mine. The study area operates eight loading areas and three unloading points, and three loaders and ten trucks are used to produce limestone.

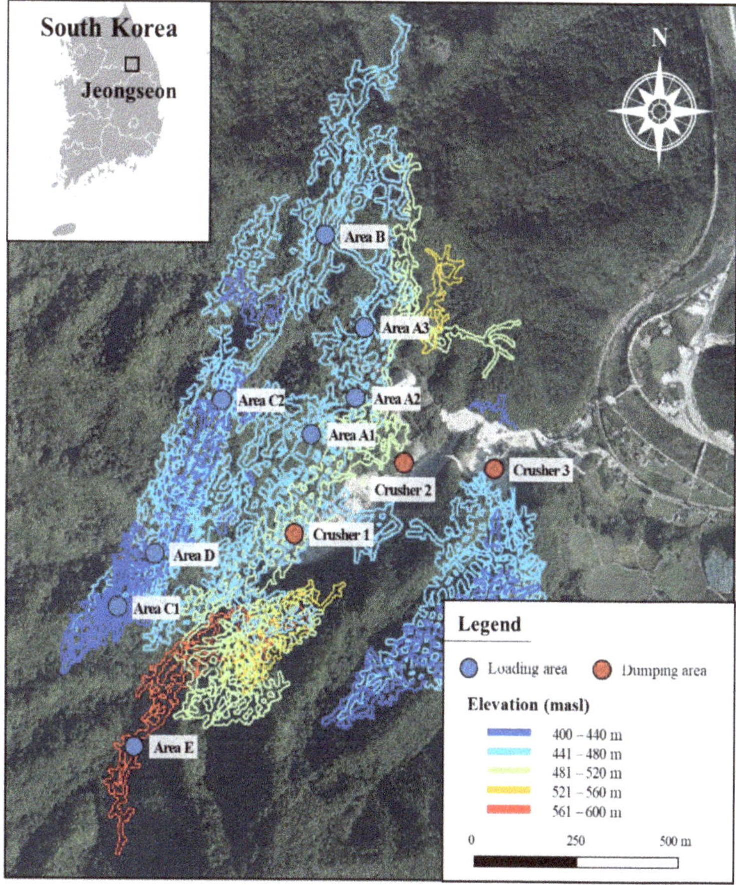

Figure 1. Map of the study area (Sungshin Minefield underground limestone mine, Jeongsun-gun, Gangwon-do, Korea) showing the loading areas and dumping areas.

The underground mine selected as the study area is equipped with a tablet computer and Bluetooth beacon-based mine production management system. This system provides functions for navigation, equipment proximity warning, production log creation, and measurement of truck travel time for each section of the underground mine [33]. The operation of the system is performed in the following order: (1) Signals are received from Bluetooth beacons installed at major points along the transport route, crusher, and loaders using a tablet computer mounted on the truck. (2) The tablet computer records the time the signal was received and the location of the truck, and (3) transmits the data stored in the internal memory to the cloud server in the area where wireless communication is possible. (4) Finally, the cloud server continuously stores and manages data transmitted from multiple trucks with tablet computers installed. For details on the operation of the system, please refer to Park and Choi [33]. Tablet computers were installed in 10 trucks used for transport operations. Bluetooth beacons were installed at loading and unloading points (8 and 3, respectively) and at major points along the transport route (11). Figure 2 shows an example of a tablet computer and Bluetooth beacon installed in the study area. Figure 3 depicts a schematic diagram showing the locations of the loading and unloading points and the Bluetooth beacon installed on the main transport route.

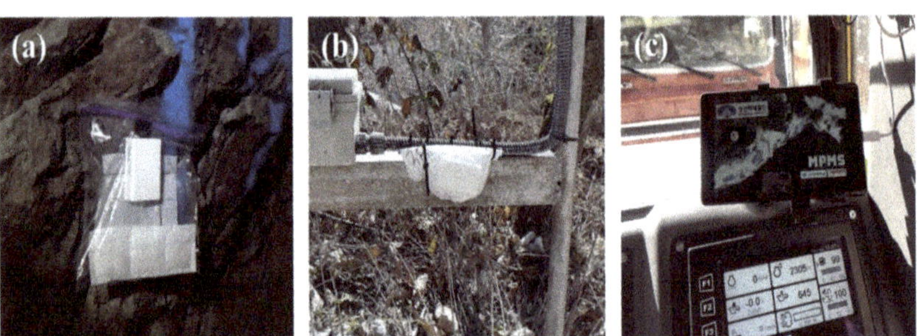

Figure 2. Example of Bluetooth beacon (Beacon i3) and Tablet PC (Galaxy A 8.0) installed for log data collection: (**a**) tunnel wall on the transport route; (**b**) near the crusher at the crusher; (**c**) windshield in the driver's seat of the truck.

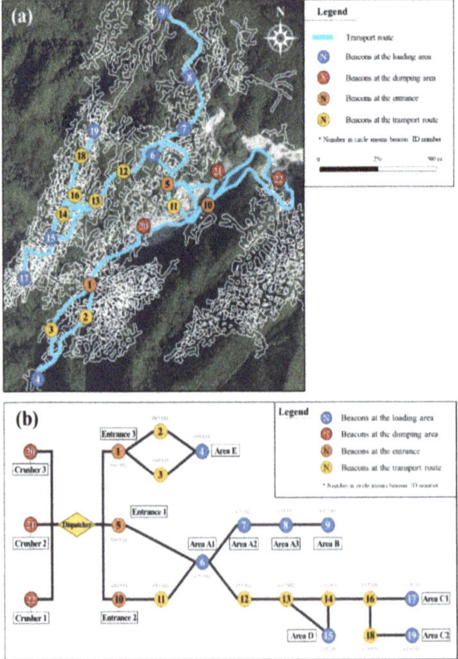

Figure 3. Transport routes between loading and unloading points and Bluetooth beacon installation points in the study area: (**a**) 2D maps; (**b**) schematic.

The purpose of this study is to calculate the truck travel time for each section based on the main points where Bluetooth beacons are installed, evaluate the stability of each transport route using a machine learning model, and diagnose the status of the transport route. The system developed by Park and Choi [33] uses a tablet computer to record the time a truck passes through the point where a Bluetooth beacon is installed; however, it cannot record the travel time of a truck traveling between the two beacons. Therefore, in this study, the truck travel time for each section was calculated using the log data analysis program developed by Park and Choi [24]. The program calls the log data files uploaded to the cloud server at once, organizes the log data, and calculates the truck travel time for each section.

In this study, log data collected from 9 November 2020 to 21 February 2021 (15 weeks) were used to evaluate and diagnose the stability of each section of the transport route using machine learning techniques. During this period, 361 log data files were uploaded to the cloud server, and 33,435 truck travel time data by section were collected.

3. Methods

The purpose of this study is to evaluate and diagnose the stability of the transport route by using the truck travel time for each section of the transport route and machine learning techniques. To achieve the purpose of the study, the research was conducted in the order of data collection for learning and verification, data processing, machine learning model selection and application of the model.

3.1. Data Preprocessing for Machine Learning Model

Factors for diagnosing the status of each section of the transport route include physical factors (location and slope of section, presence or absence of surrounding workplaces, width of transport routes, whether or not ores are loaded, etc.) and environmental factors (weather, presence or absence of groundwater, etc.) [24].

Therefore, the training data of the machine learning model for diagnosing the state of the transport path was composed of six input features and a label that judges the status of the transport path as shown in Table 1. Data types can be divided into categorical data and continuous data. The categorical data include the origin and destination (consisting of beacon IDs) of the transport route section and whether ores are loaded. Continuous data include truck travel time, average daily temperature, and daily precipitation.

Table 1. Description and data type of data set for training machine model.

Dataset	Description	Data Type
Features	Origin beacon ID	Integer (1–22)
	Destination beacon ID	Integer (1–22)
	Transport time	Seconds (sec)
	Average daily temperature	Celsius temperature (°C)
	Daily precipitation	Millimeter (mm)
	Whether ores are loaded	0: Loaded, 1: Empty
Label	Truck transport time status on transport route	0: Normal, 1: Abnormal

Coding of raw data to train the machine learning model was performed using log data related to truck travel time, which was collected from the mine production management system and weather data provided by the Korea Meteorological Administration. The status of the transport route was determined by the mine production management system installed in the research area. The truck driver judges whether the operation of the truck was normal or abnormal by considering whether any irregularity of operation occurs, such as natural causes, vehicle maintenance, tunnel closure, work interruption, accident, or excessive waiting. The truck drivers use the application of the mine production management system to input whether the operation was performed normally or abnormally when the loading, transporting, and unloading work is completed once. In this study, the case of normal operation was classified as 0, and the case of abnormal operation was classified as 1. Of the 33,435 truck travel time data collected by section, 3314 were classified as abnormal (1) by the truck driver.

The data types of input features used in this study consist of categorical data and continuous data. Because data of different dimensions are not normalized, features with small absolute values are ignored in the fault diagnosis system. Therefore, data were normalized using the min-max scaling-normalization method (Equation (1)):

$$x'_i = \frac{x_i - \min(x)}{max(x) - \min(x)} \tag{1}$$

this method can effectively prevent overfitting when training a machine learning model [34] and remove absolute differences between data items through data preprocessing, while maintaining relative differences in data within the same item. It can also improve the effectiveness of classification because it reduces the adjustment steps of parameters and improves the training speed of the model.

The dataset for training the machine learning model sets the ratio of the data classified as normal to the data classified as abnormal in the state of the transport path as 1:1, and consists of a set of 6000 data (normal: 3000, abnormal: 3000). To validate the model trained with the training dataset, the entire data was divided into a training dataset and a validation dataset. The training and validation datasets were set to 75% and 25% of the total dataset, respectively (i.e., training dataset: 4500, validation dataset: 1500).

3.2. Experimental Setup for Machine Learning Algorithms

In this study, the stability and status of each section in the underground mine was evaluated and diagnosed by using machine learning algorithms. For this, Gaussian naïve Bayes (GNB), k-nearest neighbor (kNN), support vector machine (SVM), and classification and regression tree (CART) were used.

Naïve Bayes (NB) is a set of supervised learning algorithms that apply Bayes' theorem with the "naive" assumption of independence between every pair of features [35]. Naïve Bayes can be trained efficiently in a supervised learning environment. Parameter estimation for the naïve Bayes model uses the method of maximum likelihood. In many applications, it has been confirmed that training is possible without accepting Bayesian probability or Bayesian methods. In addition, there is an advantage in the quite small amount of training data for estimating the parameters required for classification. NB can be mainly divided into Gaussian naïve Bayes (GNB) and multinomial naïve Bayes according to the type of data (i.e., continuous or categorical). GNB is an algorithm that calculates the continuous values associated with each class, often assuming that they follow a Gaussian distribution. For example, after dividing the training data including the continuous attribute x according to the class, the mean and variance of x in each class are called μ_x and σ_k, respectively. Then, assuming that a certain observation value v has been collected, the probability distribution of the values of a given class can be parameterized with μ_x and σ_k and calculated through the normal distribution equation (Equation (2)):

$$p(x = v|c) = \frac{1}{\sqrt{2\pi\sigma_c^2}} e^{-\frac{(v-\mu_c)^2}{2\sigma_c^2}} \tag{2}$$

The kNN model is one of the most intuitive and simple supervised learning models among machine learning models. The kNN does not learn in advance, but rather defers this step and then performs classification when a task request for new data is received. Therefore, it is also variously called instance-based learning, memory-based learning, or lazy learning. The idea in the kNN method is to assign new unclassified examples to the class to which the majority of its k nearest neighbors belong. It is effective to reduce the error of misclassification when the number of samples in the training dataset is large;

however, the classification accuracy depends on the value of k, the number of neighbors, and depends greatly on the distance used to calculate the closest distance to the value of k [36]. In simple kNN, the search is based on the number of class data classified closer to the new data. Figure 4 shows the classification of the data according to different k values. When the first data are found, as shown in Figure 4a, while expanding the virtual circle (in case of two-dimensional) focusing on the new data to be known, the group to which the data belong becomes the group to which the new data belong ($k = 1$). Similarly, a virtual circle is extended until three data ($k = 3$) are found, and the largest group of the three data found at this time determines the group to which the new data belong (Figure 4b).

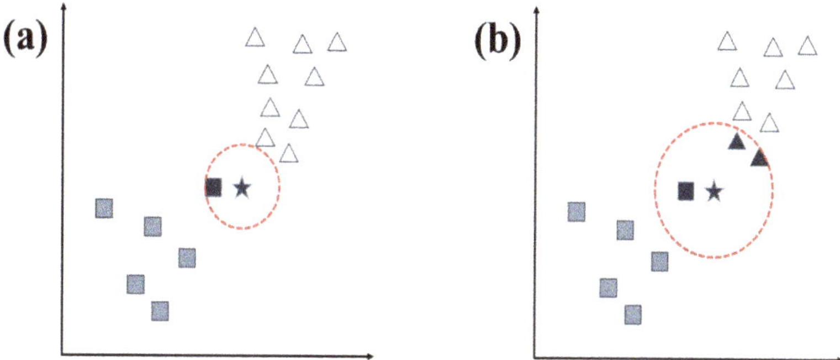

Figure 4. Example result of kNN model according to k value: (**a**) $k = 1$; (**b**) $k = 3$.

SVM was introduced by Boser et al. [37] in 1992 and has been popular in the learning community since 1996. Recently, it has been successfully applied to various problems related to pattern recognition in bioinformatics and image recognition [38]. In addition, it is sufficiently powerful to be used for both linear and non-linear regression and classification and is widely used by the public. SVM is basically a model that classifies data linearly like linear logistic regression and classifies data in three stages as shown in Figure 5. Assuming that there are two-dimensional data composed of two classes as shown in Figure 5a, there can be an infinite number of straight lines separating these classes; however, using the decision boundary selection condition of SVM, only one straight line can be selected. The selection condition is to select a hyperplane that maximizes the distance between the data points of each class that are closest to each other. First, as shown in Figure 5b, the closest points between each class are selected, and when the margin between two parallel straight lines including these points is maximized, two straight lines including these points are selected. The points used to select two straight lines are called support vectors, and when these two straight lines are determined, the central straight line located at the same distance between the two straight lines becomes the decision boundary, as shown in Figure 5c.

The optimal hyperplane can be defined as the following equation [39]:

$$y_i(\omega \cdot x_i + b) \geq \text{for } 1 \leq i \leq \text{n, } \omega \in R^d, b \in R \tag{3}$$

where x_i is an instance with its corresponding label $y_i \in (-1, 1)$, and b is an intercept term; that is, a normal vector to the hyperplane', d is the number of properties of each instance and the dimension of input vector, and n is the number of instances. A hyperplane is defined by the instances that lie nearest to it; such instances are called support vectors. By this definition, there should be no data points between the hyperplanes containing the support vectors (hard margin classification); however, this classification cannot occur in the real world. This is because real data often contain outliers that are significantly different from other instances of the same class, in addition to the possibility of errors in data entry, measurement errors, etc. Therefore, we used a definition (soft margin classification,

Equation (4)) proposed by Tuba et al. [39] for the optimal hyperplane by overcoming this problem and relaxing the conditions to use SVM for real data classification:

$$y_i(\omega \cdot x_i + b) \geq 1 - \epsilon_i, \quad \epsilon_i \geq 0, \quad 1 \leq i \leq n \tag{4}$$

here, ϵ_i is a slack variable that allows the corresponding instance to leave the margin. To find the optimal hyperplane, we must solve the quadric programming problem as follows:

$$\min \frac{1}{2}\|\omega\|^2 + C \sum_{i=1}^{n} \epsilon_i. \tag{5}$$

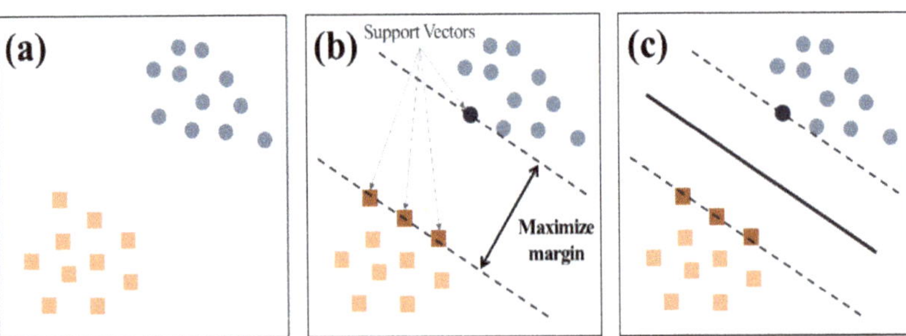

Figure 5. An example of a two-dimensional representation of a linearly separable binary classification: (**a**) two-dimensional data consisting of two classes; (**b**) selecting the closest points between each class, and selecting two straight lines for which the distance between two flat straight lines containing these points is maximum; (**c**) Select a hyperplane that is equidistant from two straight lines.

Here, C represents the parameter of the soft margin cost function, and the quality of the SVM model largely depends on the choice of this parameter; that is, the larger the value of C, the more similar the generated model obtained from the hard margin classification definition. However, because the soft margin classification method can only be applied to linearly separable data, a kernel function is used, rather than a dot product. The kernel function maps the instances into a higher-dimensional space to ensure that they can be linearly separated. There are various kernel functions, such as polynomial, Gauss (radial basis function or RBF), and sigmoid functions, but RBF is the most commonly used and can be defined as follows:

$$K(x_i, x_j) = \exp\left(-\gamma\|x_i - x_j\|^2\right) \tag{6}$$

where γ is a free parameter that significantly affects classification accuracy and this parameter defines the impact of each training instance.

CART is a decision tree (DT)-based algorithm that can be used for both classification and regression problems [40]. The data are divided into uniform labels based on the answers (yes/no) to the predictor values through an iterative procedure, and finally a binomial tree is generated. If the dependent variable is qualitative, it is called a classification tree, and if it is quantitative, it is called a regression tree. The node containing the entire dataset is called the root node. Starting from the root node, it is divided into left and right and this process is repeated until the estimation error related to the dependent variable is minimized to classify the data [41]. Because CART is inherently non-parametric, no assumptions are made regarding the underlying distribution of values of the predictor variables [42]. Therefore, CART can handle numerical data that are highly skewed or multimodal, as well as categorical predictors with either an ordinal or a nonordinal structure. In addition, it identifies the "splitting" variable based on a thorough search for all possibilities. Because efficient algorithms are used, CART has the advantage of being able to search for all possible variables with splitters, despite the existence of hundreds of possible predictors.

CART is a relatively automated machine learning method because the analyst's input is less than the complexity of the analysis.

Grid search through 5-fold cross-validation was utilized to improve the performance of the machine learning model and the reliability of the performance evaluation on the validation dataset. In general, the performance of a machine learning model depends on parameters. Various parameters exist depending on the machine learning algorithm. Therefore, to design a model with high accuracy, it is important to set the optimal parameters. 5-fold cross-validation is a method in which a dataset is divided into 5 pieces that are used one by one as a validation dataset while the rest are combined and used as a training dataset. Using this method, 100% of the data we have can be used as a validation dataset. The grid search is originally an exhaustive search based on a defined subset of the hyper parameter space [43]. That is, when creating a model, it is a search method to find the variable with the highest performance after sequentially inputting the hyperparameters set by the user. Table 2 shows the parameters and parameter tuning used in each model. The GNB predicted the accuracy of the model by setting the range of variance (var) smoothing from 10^{-9} to 1 and increasing the parameter values by approximately 1.23 times because the accuracy of the model varies depending on the var smoothing. The classification accuracy of the kNN depends on the k value, which means the number of neighbors, and the accuracy was predicted by increasing the k value by 1 from 1 to 100. Because the classification accuracy of the SVM model varies greatly depending on the parameters C and γ, the optimal pair of parameters (C: from 10 to 100, γ: from 0.1 to 1) was determined by increasing the values by 5 and 0.1, respectively. Finally, in the CART model, the accuracy of the model is determined by the minimum samples leaf (min_samples_leaf) and minimum samples split (min_samples_split). In this study, the optimal parameter was determined by setting the minimum samples leaf from 1 to 10 and increasing by 1, and for the minimum samples split, setting a range from 2 to 10 and increasing by 1.

Table 2. Values used in grid search for parameter tuning.

	GNB	**kNN**	**SVM**	**CART**
Parameter	var_smoothing	neighbors	C/γ	min_samples_leaf/ min_samples_split
Min	10^{-9}	1	10/0.1	1/2
max	1	100	100/1	10/10
Step	($\times$) 1.232847	(+) 1	(+) 5/0.1	(+) 1/1

3.3. Validation of Machine Learning Models

The parameter showing the highest learning accuracy of the machine learning model was determined using grid search through 5-fold cross-validation. Subsequently, the performance of the model was verified using the validation dataset (25% of the total data, 1500). Performance indicators that can evaluate the performance of a model generally depend on the type of supervised learning (regression or classification). In this study, the performance of the model was verified using the accuracy, precision, recall, and F1 score, which are typically used in classification problems. Accuracy refers to the number of correct predictions among all predictions, precision refers to the probability of the state actually being positive when a positive prediction is made, recall refers to the probability of correctly predicting an actual positive, and F1 score refers to the weighted average of precision and recall. The formula for each performance indicator is as follows:

$$Accuracy = \frac{TP + TN}{TP + TN + FP + FN} \tag{7}$$

$$Precision = \frac{TP}{TP + FP} \tag{8}$$

$$Recall = \frac{TP}{TP + FN} \tag{9}$$

$$F1\ score = \frac{2 \times Recall \times Precision}{Recall + Precision} \tag{10}$$

where, P (positive) and N (negative) denote whether the prediction of the model is positive (yes) or negative (no), and T (true) and F (false) imply whether the prediction is correct or wrong. When this is expressed as a matrix, it is called a confusion matrix and can be expressed as shown in Table 3.

Table 3. Confusion matrix of a classifier.

		Predicted Data	
		Negative (0)	**Positive (1)**
Actual data	Negative (0)	TN (True Negative)	FP (False Positive)
	Positive (1)	FN (False Negative)	TP (True Positive)

4. Results

4.1. Results of Data Preprocessing

For the learning and validation of machine models, 33,435 truck travel time data were collected for 15 weeks using the mine production management system installed in the study area. Except for the departure and arrival points for each section of the transport route and the transport time of the truck that can be acquired using the system, additional features such as daily average temperature, daily precipitation, ores loading of trucks, and labels to determine abnormal status of transport routes were entered. The decision to load ores can be divided into the case of empty truck or loaded truck. This was determined by judging whether the truck is headed for the loading points (empty truck) or the crusher (loaded truck) according to the sequence of beacon IDs for each section of the transport route. As a result of judging the transport route status of the data collected during the 15-week period, 3314 data out of 33,435 truck travel time data were found to be abnormally measured.

When all 33,345 data are used as training data, the normal case is much larger than the abnormal case, and a biased training result may appear. Therefore, the ratio of the data classified as normal to the data classified as abnormal was set to 1:1 to prepare the training dataset. A total of 6000 data were prepared by random sampling of 3000 data marked as abnormal and 3000 data marked as normal. Before normalizing the training dataset, the mean values, standard deviations, minimum and maximum values for truck travel time, daily average temperature, and daily precipitation were calculated (Table 4). Figure 6 shows the histogram for statistical values. The average truck travel time was 95.55 s and the standard deviation was 74.12 s. The average daily temperature was −2.71 °C and the standard deviation was 5.03 °C. The average daily precipitation was 0.24 mm and the standard deviation was 0.85 mm.

Table 4. Feature of data set for training machine learning model.

	Truck Travel Time (s)	Average Daily Temperature (°C)	Daily Precipitation (mm)
Mean	95.55	−2.71	0.24
Standard deviation	74.12	5.03	0.85
Minimum value	1.00	−14.30	0.00
Maximum value	299.00	11.40	6.90

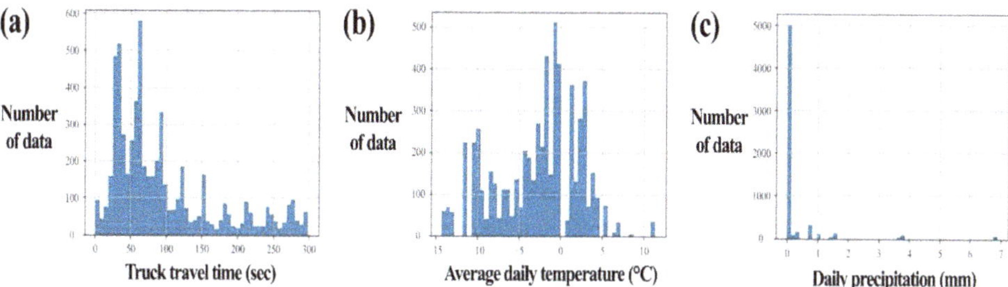

Figure 6. Feature distribution of data set for machine learning model training: (**a**) truck travel time; (**b**) average daily temperature; (**c**) daily precipitation.

4.2. Results of Model Training and Application

In this study, GNB, kNN, SVM, and CART models were used to evaluate and diagnose the stability of truck transport routes. To design the most accurate predictive model, the parameter values related to the learning accuracy of each model were optimized. For this purpose, grid search through 5-fold cross-validation was used.

The classification accuracy of the GNB model depends on the parameter var smoothing. The optimal model was determined by setting the parameter value range from 10^{-9} to 1 and increasing the parameter value by approximately 1.23 times. Figure 7 is a graph showing the accuracy of the model according to the change of the var smoothing. The accuracy of the model decreases rapidly when the parameter value exceeds 10^{-2}. The GNB showed the highest learning accuracy (0.60) when the var smoothing value was 0.000188.

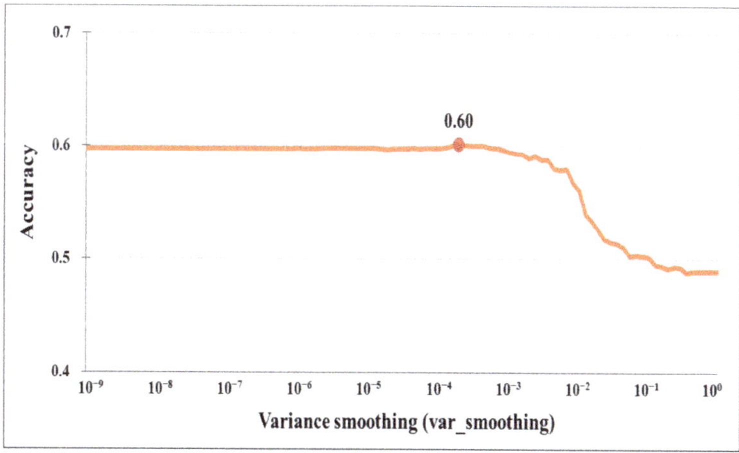

Figure 7. Variations of training accuracy in the variance smoothing (var_smoothing) change range of 10^{-9} to 1.

The accuracy of the kNN model depends on the value of k. Figure 8 shows the prediction of the accuracy of the model while increasing the k value by 1 from 1 to 100. The accuracy of the kNN model was higher as the k value was smaller (k = 1, 0.85).

The SVM model was optimized by changing the values of C and γ to determine the parameter value showing the highest training accuracy. The parameter C was set in the range from 10 to 100 and increased by 5, while γ was increased by 0.1 from 0.1 to 0.9 to calculate the accuracy of the model. Figure 9 shows the training accuracy of the model according to the change in parameter C and γ value. As the values of C and γ increased, the accuracy of the model also tended to increase. In the SVM model, when the C value was set to 100 and the γ value was set to 0.9, the model accuracy was the highest at 0.78.

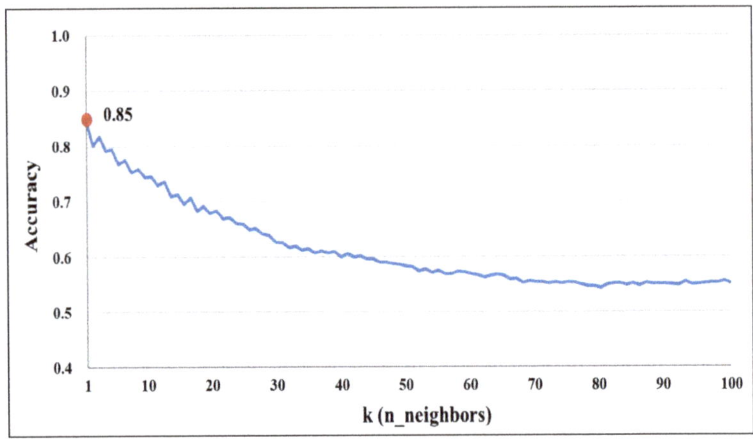

Figure 8. Variations of training accuracy in the *k* (n_neighbors) change range of 1 to 100.

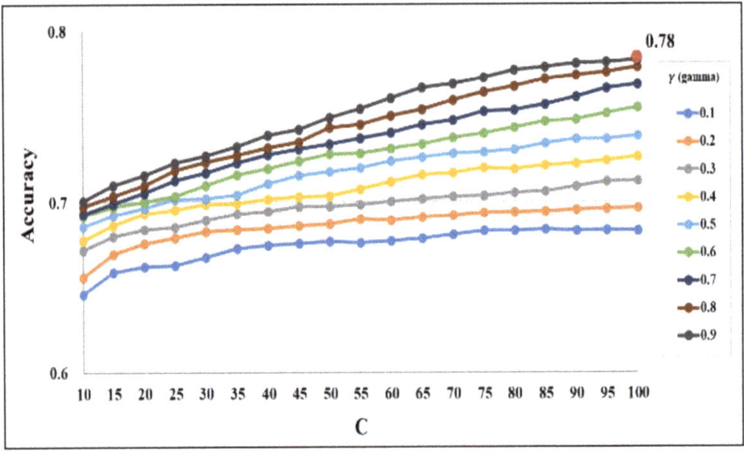

Figure 9. Variations of training accuracy in the C change range of 10 to 100 and γ change range of 0.1 to 1.

The training accuracy of CART depends on the values of minimum samples leaf and minimum samples split. In this study, the values of two parameters were optimized by increasing min_samples_leaf by 1 from 1 to 10 and increasing min_samples_split by 1 from 2 to 10. Figure 10 shows the training accuracy of the CART model depending on the changes in the two parameter values. When min_samples_leaf is 3 or less, the accuracy of the model tends to decrease as min_samples_split increases; however, when min_samples_leaf was at least 4, the accuracy did not change significantly even if min_samples_split was increased. The training accuracy of CART showed the highest accuracy (0.94) when min_samples_leaf was set to 1 and min_samples_split was set to 4.

The previously determined parameters were applied to each model, and verification was performed. The validation of the machine learning models was performed using the validation dataset (25% of the total data, 1500 pieces). Tables 5–8 shows the model verification results as a confusion matrix.

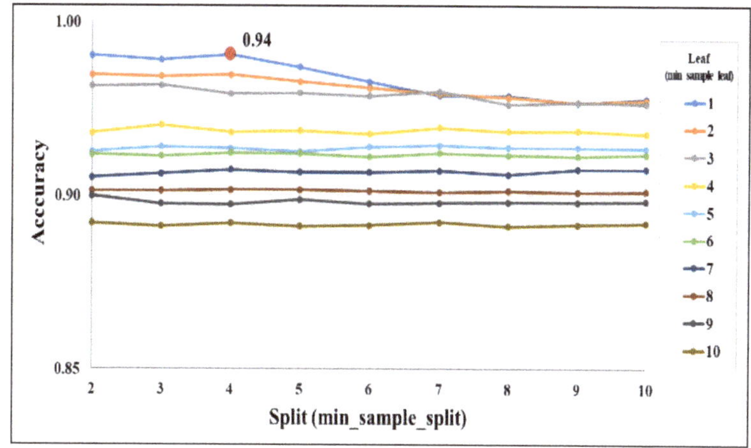

Figure 10. Variations of training accuracy in the min_samples_split change range of 2 to 10 and min_samples_leaf change range of 1 to 10.

Table 5. Confusion matrix classified using the GNB model.

Normalization $var_{smoothing} = 0.000188$		Predicted Data		
		Negative (0)	Positive (1)	Accuracy
Actual data	Negative (0)	687 (TN)	75 (FP)	0.90
	Positive (1)	496 (FN)	242 (TP)	0.33
	Accuracy	0.58	0.77	0.62
	Training accuracy			0.60

Table 6. Confusion matrix classified using the kNN model.

Normalization n Neighbors = 1		Predicted Data		
		Negative (0)	Positive (1)	Accuracy
Actual data	Negative (0)	642 (TN)	120 (FP)	0.84
	Positive (1)	122 (FN)	616 (TP)	0.83
	Accuracy	0.84	0.84	0.84
	Training accuracy			0.85

Table 7. Confusion matrix classified using the support vector machine (SVM) model.

Normalization $C = 100, \gamma = 0.9$		Predicted Data		
		Negative (0)	Positive (1)	Accuracy
Actual data	Negative (0)	655 (TN)	107 (FP)	0.86
	Positive (1)	196 (FN)	542 (TP)	0.73
	Accuracy	0.77	0.84	0.80
	Training accuracy			0.79

Table 8. Confusion matrix classified using the CART model.

Normalization Leaf = 1, Split = 4		Predicted Data		
		Negative (0)	Positive (1)	Accuracy
Actual data	Negative (0)	713 (TN)	49 (FP)	0.94
	Positive (1)	32 (FN)	706 (TP)	0.96
	Accuracy	0.96	0.94	0.95
	Training accuracy			0.94

Table 5 shows the verification results of the GNB as a confusion matrix. There were 687 cases (TN) where the section in which the truck travel time classified as normal was predicted to be normal. Conversely, there were 242 cases (TP) where the section classified as abnormal was predicted to be abnormal. In addition, it was found that there were 496 (FN) and 75 (FP) cases of predicting a section where the truck travel time was abnormal as normal and predicting a section where the truck was normal as abnormal, respectively. GNB's verification accuracy was 0.62, and when predicting the data classified as normal as normal, it showed relatively high accuracy (0.90); however, the accuracy of predicting data classified as abnormal as abnormal was very low at 0.33.

In the case of the kNN model, TN and TP, which are cases of correct prediction of the actual data among 1500 verification data, appeared 642 times and 616 times, respectively. FN and FP, which were failed predictions, appeared 122 times and 120 times, respectively (Table 6). The validation accuracy of the kNN model was found to be 0.84, and it showed a similar level of accuracy in all cases.

Table 7 shows the verification results of the SVM model as a confusion matrix. There were 655 cases (TN) where the section in which the truck travel time was classified as normal was predicted to be normal. Conversely, there were 542 cases (TP) where the section classified as abnormal was predicted to be abnormal. In addition, FN and FP, which were failed predictions, appeared 196 times and 107 times, respectively. The verification accuracy of the SVM model was 0.80, and it showed high accuracy (0.86) in the data classification problem, which was actually abnormal; however, in the problem of classifying actually normal data, the accuracy (0.73) was relatively low.

The verification results of the CART model are shown in Table 8. In fact, 713 times (TN) were predicted to be normal where the section in which the truck travel time was classified as normal, and 706 times (TP) were predicted to be abnormal where the section classified as abnormal. In addition, FN and FP that failed prediction appeared 32 times and 49 times, respectively. The verification accuracy of the CART model was very high at 0.95, and both the problem of classifying the actual normal sections (0.96) and the problem of classifying the abnormal sections (0.94).

The performance of the model was evaluated based on the confusion matrix of each model analyzed using the validation dataset. The performance assessment of the model was conducted using accuracy, precision, recall, and F1 score. Table 9 shows the performance index of each model. The prediction accuracy of the machine learning model was the highest in CART (94.6%), followed by kNN (83.9%), SVM (79.8%), and GNB (61.9%). The CART model also exhibited high precision, recall, and F1 score. Therefore, it can be said that the CART model achieves the best performance in the problem of evaluating the stability of the transport route for each section in the underground mine.

Table 9. Performance assessment indicators of machine learning (ML) models.

Performance Assessment Indicators	GNB	kNN	SVM	CART
Accuracy (%)	61.9	83.9	79.8	94.6
Precision (%)	76.3	83.7	83.5	93.5
Recall (%)	32.8	83.5	73.4	95.7
F1 score (%)	45.9	83.6	78.2	94.6

5. Discussion

5.1. Analysis of Model Accuracy for Each Transport Route Section

The prediction accuracy of each section was calculated using 1500 pieces of data used in the verification process of the CART model. Figure 11 shows the accuracy of the model for each section when operating with empty or loaded trucks. The model achieved an accuracy of at least 57.1% for all sections (45 sections) and exhibited an average accuracy of 93.3%. In the case of the route (23 sections) operating with empty trucks, the prediction accuracy of the model was found to be very high with an average of 90.9%. Except for four sections (beacon ID: 1→3, 5→21, 13→15, 16→17), all were confirmed to show an accuracy of at least 80%. In the case of operating with loaded trucks (22 sections), the prediction accuracy of the model was found to be very high, with an average of 95.9%. In addition, 20 of the 22 sections showed over 80% accuracy. The accuracy of the model for each section of transport route tended to be generally higher as the amount of data for each section included in the dataset used for learning increased (Table 10). Therefore, in the case of the section where the prediction accuracy of the model is high, it is judged that the model can be used sufficiently to evaluate whether the truck was operated normally in the section; however, in the case of a section where the accuracy is low, it is judged that the machine learning model should be improved through additional data collection for training the model is necessary.

5.2. Further Verification of the CART Model Using Unused Data

The CART model was further verified using the remaining 27,435 data not used to train the model. Table 11 shows the verification results of the CART model as a confusion matrix. There were 26,027 cases (TN) where the section in which the truck travel time was classified as normal was predicted to be normal. Conversely, there were 311 cases (TP) where the section classified as abnormal was predicted to be abnormal. There were three (FN) and 1094 (FP) cases of predicting a section where the truck travel time was abnormal as normal and predicting a section where the truck was normal as abnormal, respectively. The verification accuracy of the CART model using the remaining 27,435 data was 0.96, which was similar to the result (0.95) of the CART model trained and verified with 6000 data in Table 8. Table 12 shows the performance index of the CART model verified using the remaining 27,435 data. The prediction accuracy of the model was 96%, and the precision, recall, and F1 score were 22.1%, 99%, and 36.2%, respectively. In general, in the case of a classification problem using data with less data imbalance, it can be said that the higher the performance index, the better the model performance [44]. However, when data imbalance exists, even if precision is low, the model can be trusted when recall is high [45]. In the case of the remaining 27,435 data, there was an imbalance in the data because the normal data takes up a much larger proportion than the abnormal data. Therefore, it can be determined that the CART model is reliable when considers the value of Recall appears as 99%.

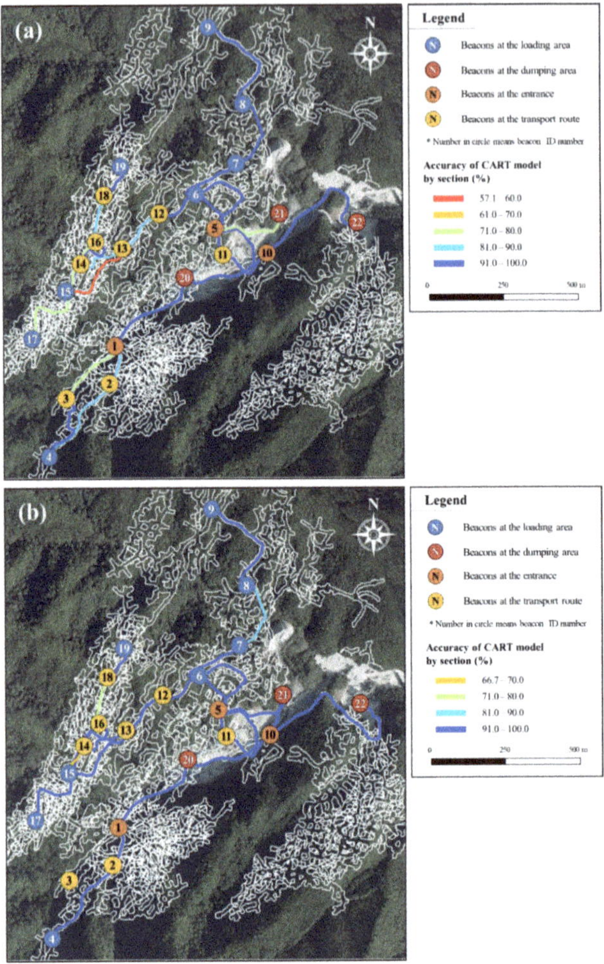

Figure 11. Prediction accuracy for each section of the CART model. (**a**) when operating with an empty truck; (**b**) when operating with loaded truck.

Table 10. Relationship between the prediction accuracy and the average number of data used in machine learning for each section.

Operation Type	Prediction Accuracy (%)	Number of Sections	Average of the Number of Data Used for Machine Learning for Each Section
Empty haul	91–100	14	105.9
	81–90	5	90.2
	71–80	3	59.3
	61–70	0	N/A
	57.1–60	1	26.0
Loaded haul	91–100	19	138.0
	81–90	1	90.0
	71–80	1	48.0
	66.7–70	1	27.0

Table 11. Confusion matrix for further verification of the CART model on the remaining 27,435 data.

Normalization Leaf = 1, Split = 4		Predicted Data		
		Negative (0)	Positive (1)	Accuracy
Actual data	Negative (0)	26,027 (TN)	1094 (FP)	0.96
	Positive (1)	3 (FN)	311 (TP)	0.99
	Accuracy	1.00	0.22	0.96

Table 12. Performance assessment indicators for further verification of CART model on remaining 27,435 data.

Performance Assessment Indicators	CART Model
Accuracy (%)	96.0
Precision (%)	22.1
Recall (%)	99.0
F1 score (%)	36.2

5.3. Practical Use at the Underground Mine Site

The proposed machine learning model can diagnose the operation status of the section by determining whether the truck travel time for each section is normal or abnormal. In this study, during the validation of the CART model, one section with high prediction accuracy and one with low prediction accuracy were selected. Then, using the log data additionally collected from the mine production management system, evaluation was performed on whether the truck was operated normally in the relevant section. For this purpose, log data collected during the 16th week (22–27 February 2021) were used for analysis. For the section of the transport route, the section from beacon ID 11 to 6 and section from beacon ID 13 to 14 were selected. In these sections, the validation accuracy of the model when validating the machine learning model was 100% and 82%, respectively.

First, in the case of sections 11 to 6 of beacon IDs, three trucks operated the section a total of 41 times in a week. By truck, truck A drove 1 time, truck B drove 25 times, and truck C drove 15 times. Log data for the section showed that truck travel time was measured within the normal range in 37 operations, and within the abnormal range in four operations. This section is a transport route for empty trucks toward the loading point, and there is no loading or dumping near the route. Therefore, most trucks have the characteristic of moving without stopping in the relevant section. After converting the log data of the relevant section (beacon ID 11→6) into the input data of the CART model, prediction was performed on whether the truck travel time was measured normally or abnormally. As a result, it was predicted that the truck travel time was measured within the normal range in the case of actual normal operation. In addition, in the case of abnormal operation, it was predicted that the operation was performed abnormally. In other words, it was found that the actual data and the prediction results by the CART model were identical. Table 13 shows the prediction results of the CART model for the 16-week data by classifying them by trucks that have operated the relevant section and is presented as a confusion matrix. Table 14 is a visualization of the confusion matrix divided by time period. This means that, during the period, trucks operated well reflecting the trend of the existing truck travel time. Furthermore, it means that there are no problems in the truck or in the transport section that will affect the truck travel time.

Table 13. Confusion matrix classified using the CART model for beacon IDs 11 to 6.

Normalization Leaf = 1, Split = 4			Predicted Data	
			Negative (0)	Positive (1)
Actual data	Truck A	Negative (0)	1 (TN)	0 (FP)
		Positive (1)	0 (FN)	0 (TP)
	Truck B	Negative (0)	22 (TN)	0 (FP)
		Positive (1)	0 (FN)	3 (TP)
	Truck C	Negative (0)	14 (TN)	0 (FP)
		Positive (1)	0 (FN)	1 (TP)

Table 14. Prediction result of the CART model by time/truck for beacon IDs 11 to 6.

Time	22 February 2021			23 February 2021			24 February 2021			25 February 2021			26 February 2021			27 February 2021		
	Truck			Truck			Truck			Truck			Truck			Truck		
	A	B	C	A	B	C	A	B	C	A	B	C	A	B	C	A	B	C
08:00					■						•	•			•		•	
09:00		•			•	• •									•		•	
10:00		•			•	•						•			•			•
11:00								•			■						•	
12:00 (Break time)																		
13:00		•			•				•		•			■	•			
14:00		•			•							•					•	
15:00		•	•							•	• •						• •	
16:00					•	■			•								•	

• TN Cases in which data that are actually normal are predicted to be normal. ■ TP: Cases in which data that are actually abnormal are predicted to be abnormal.

Next, in the section from beacon ID 13 to 14, two trucks operated a total of 58 times (Truck A: 34 times, Truck B: 24 times) during a week. In this section, 54 truck travel times were measured within the normal range, but four times were measured within the abnormal range. This section is a transport route where an empty truck goes to the loading point. However, because the loading point (Area D) is located around the route, it is a section where variations in truck travel time may occur. As a result of predicting the state of truck travel time using the CART model for the section, the prediction accuracy was found to be very low (86.2%). Normal data were predicted as normal 46 times (TN), and abnormal data were predicted as abnormal (TP) four times. In addition, it was found that there were eight (FP) cases of predicting a section where the truck was normal as abnormal (Table 15). For this section, considering that the verification accuracy has already been shown to be low, it can be confirmed that the prediction accuracy appears low even in the prediction using the 16-week data. Table 16 is a visualization of the confusion matrix divided by time period. In this section, some prediction failures of the CART model occur. To improve the accuracy of the model, additional data collection is required for training the machine learning model, and the model needs to be improved. In addition, because some data show abnormal truck travel time, this section needs to be carefully monitored. To improve the overall productivity of the mine and the efficiency of the trucking operation, and to reduce the time required to transport the ores, it is necessary to monitor and respond to these sections in advance.

Table 15. Confusion matrix classified using the CART model for beacon IDs 13 to 14.

Normalization Leaf = 1, Spli t= 4			Predicted Data	
			Negative (0)	Positive (1)
Actual data	Truck A	Negative (0)	25 (TN)	7 (FP)
		Positive (1)	0 (FN)	2 (TP)
	Truck B	Negative (0)	21 (TN)	1 (FP)
		Positive (1)	0 (FN)	2 (TP)

Table 16. Prediction result of the CART model by time/truck for beacon IDs 13 to 14.

Time	22 February 2021 Truck A	B	23 February 2021 Truck A	B	24 February 2021 Truck A	B	25 February 2021 Truck A	B	26 February 2021 Truck A	B	27 February 2021 Truck A	B
08:00	FP				TN	TN		TN	TN	TN	TN	TN
09:00			TN TN			TN TP	TN TN		TN	TN		TN
10:00		TN TN	TP		FP FP	TN	TN		TN	TN		
11:00					TN			TN	TN			
12:00 (Break time)					TN					TN	TN	
13:00	FP		TN FP					TN	TN			
14:00					TN			FP TN			TP	
15:00	FP			TN	TN	TN	TN	TP	TN	TN	TN TN	TN
16:00	FP	TN				TN						

• TN Cases in which data that are actually normal are predicted to be normal; ■ TP: Cases in which data that are actually abnormal are predicted to be abnormal; • FP: Cases in which data that are actually normal are predicted to be abnormal.

If a case in which the truck travel time is abnormally predicted is observed only in a specific truck, the possibility that the truck driver's skill is insufficient, the truck's maintenance is poor, or the maintenance period has arrived should be suspected. In addition, the manager of the mine should take appropriate action in this regard. According to what we have seen so far, the proposed CART model can predict the status of truck travel time and then monitor the problem or possibility of occurrence in the transport route or equipment in advance. In addition, it can help to analyze the cause and prepare countermeasures. Therefore, the CART model can be used as a tool for mine managers to improve the productivity and efficiency of transport operations.

5.4. Comparison between the Existing and Machine Learning-Based Methods

Various researchers are using machine learning techniques to monitor and diagnose mine operating systems, equipment, and facilities. However, hitherto, no research case has been reported on monitoring and diagnosing the condition of a mine transport system using machine learning techniques. As a similar research case related to diagnosing and predicting the status of transport routes using truck travel time data, Park and Choi [24] evaluated the stability and classified the types of transport routes using the statistics of the truck travel time for each section of the transport route. The method of collecting log data related to truck travel time is the same as that used in this study. However, Park and Choi [24] used percentiles (P10, P90) of truck travel time to evaluate the stability and condition of each section of the transport route. That is, if the newly collected truck travel time was measured in the range between percentiles P10 and P90, the status of the transport route was classified as normal; otherwise, it was classified as abnormal. The truck travel time of the mine may vary depending on the production plan, vehicle dispatch plan, tunnel maintenance and repair status, season (temperature), precipitation, and driver's driving skill. In this study, the truck travel time for each section was evaluated by considering the beacon IDs (origin and destination), temperature, precipitation, and whether the truck was loaded in addition to the transport time, and then the status of the transport route was diagnosed. Therefore, the proposed method is considered to have a higher level of reliability than the existing method used to evaluate the stability and condition of each section of the transport route by considering the statistics of the truck travel time.

6. Conclusions

In this study, we proposed a method that can utilize log data related to truck travel time and machine learning model (GNB, kNN, SVM, CART) to evaluate the stability of the underground mine transport route and to diagnose the operation status. To this end, a limestone mine that collects truck travel time data in underground mines using Bluetooth beacons and tablet computers was selected as a study area, and truck travel time data were collected for a certain period of time. In addition, learning and validation of models were performed using the collected data, and the results of monitoring and diagnosis of the transport route status in the study area were presented. As a result of performing grid search through 5-fold cross-validation using the training dataset, the accuracy (94.1%) was highest when the parameters min_samples_leaf and min_samples_split of the CART model were set to 1 and 4, respectively. In the validation of the CART model performed using the validation dataset (1500 data), data with normal truck travel time were predicted as normal 713 times, and abnormal data were predicted as abnormal 706 times. The performance of the machine learning model was judged using accuracy, precision, recall, and F1 score The accuracy of the CART model was 94.6%, and the precision and recall were 93.5% and 95.7%, respectively, and it was confirmed that the F1 score was also high at 94.6%.

The proposed CART model proposed can be used for monitoring and diagnosing the status of the transport route that constitutes the truck transport system in the underground mine. In addition, it is judged that it can be used sufficiently as a tool to improve the productivity and efficiency of mine transport operations. Because the truck travel time for each section has variability depending on the driver's driving skill, tunnel maintenance and repair status, vehicle dispatch plan, etc., the truck travel time has a significant impact on the efficiency and productivity of truck transport operations. Therefore, it is crucial to know the section in which the truck transport operation is expected to be abnormal and to prevent problems occurring in the section, the vehicle, or the future. The proposed CART model showed an average prediction accuracy of 94.1% for all sections of the study area. This means that the stability of the transport route can be evaluated and diagnosed by judging whether the newly collected truck travel time data are measured within the normal range or within the abnormal range at a relatively high level of reliability. However the prediction accuracy was relatively low in some sections. To improve the prediction accuracy of this section, additional collection of truck travel time data for training the machine learning model is required, and accordingly, the model will have to be improved.

In this study, it was confirmed that machine learning techniques can be used to diagnose and predict the condition of transport routes to maintain equipment and workplaces in underground mines. To that end, a method for diagnosing and predicting mine transport systems using machine learning techniques was proposed. We expect that the proposed method can be sufficiently applied not only in underground mines but also in open-pit mines from a methodological perspective.

Author Contributions: Conceptualization, Y.C.; methodology, Y.C. and H.N.; software, D.J.; validation, S.P.; formal analysis, S.P. and D.J.; investigation, Y.C.; resources, Y.C.; data curation, S.P. and D.J.; writing—original draft preparation, S.P. and D.J.; writing—review and editing, Y.C.; visualization, S.P.; supervision, Y.C.; project administration, Y.C.; funding acquisition, Y.C. All authors have read and agreed to the published version of the manuscript.

Funding: This work was supported by Energy & Mineral Resources Development Association of Korea (EMRD) grant funded by the Korea government (MOTIE) (Training Program for Specialists in Smart Mining).

Institutional Review Board Statement: Not applicable.

Informed Consent Statement: Not applicable.

Data Availability Statement: Data sharing not applicable.

Conflicts of Interest: The authors declare no conflict of interest.

References

1. Hartman, H.L.; Mutmansky, J.M. Unit operations of mining. In *Introductory Mining Engineering*, 2nd ed.; Wiley: New York, NY, USA, 2002; pp. 119–152.
2. Choi, Y.; Nieto, A. Optimal haulage routing of off-road dump trucks in construction and mining sites using Google Earth and a modified least-cost path algorithm. *Autom. Constr.* **2011**, *20*, 982–997. [CrossRef]
3. Choi, Y.; Nieto, A. Software for simulating open-pit truck/shovel haulage systems using Google Earth and GPSS/H. *J. Korean Soc. Miner. Energy Resour. Eng.* **2011**, *48*, 734–743.
4. Ercelebi, S.G.; Bascetin, A. Optimization of shovel-truck system for surface mining. *J. S. Afr. Inst. Min. Metall.* **2009**, *109*, 433–439.
5. Jung, D.; Baek, J.; Choi, Y. Stochastic predictions of ore production in an underground limestone mine using different probability density functions: A comparative study using big data from ICT system. *Appl. Sci.* **2021**, *11*, 4301. [CrossRef]
6. Alarie, S.; Gamache, M. Overview of solution strategies used in truck dispatching systems for open pit mines. *Int. J. Min. Reclam. Environ.* **2010**, *16*, 59–76. [CrossRef]
7. Afrapoli, A.M.; Tabesh, M.; Askari-Nasab, H. A stochastic hybrid simulation-optimization approach towards haul fleet sizing in surface mines. *Min. Technol.* **2018**, *128*, 9–20. [CrossRef]
8. Markeset, T.; Kumar, U. Application of LCC techniques in selection of mining equipment and technology. In *Mine Planning and Equipment Selection 2000*, 1st ed.; Routledge: London, UK, 2018; pp. 635–640.
9. Samanta, B.; Sarkar, B.; Mukherjee, S.K. Selection of opencast mining equipment by a multi-criteria decision-making process. *Min. Technol.* **2013**, *111*, 136–142. [CrossRef]
10. Douglas, J. *Prediction of Shovel-Truck Production: A Reconciliation of Computer and Conventional Estimates*; Stanford University: Stanford, CA, USA, 1964.
11. Smith, S.D.; Wood, G.S.; Gould, M. A new earthworks estimating methodology. *Constr. Manag. Econ.* **2000**, *18*, 219–228. [CrossRef]
12. Burt, C.N.; Caccetta, L. Match factor for heterogeneous truck and loader fleets. *Int. J. Min. Reclam. Environ.* **2008**, *21*, 262–270. [CrossRef]
13. Edwards, D.J.; Malekzadeh, H.; Yisa, S.B. A linear programming decision tool for selecting the optimum excavator. *Struct. Surv.* **2001**, *19*, 113–120. [CrossRef]
14. Sturgul, J.R. Modeling and simulation in mining—Its time has finally arrived. *Simulation* **2001**, *76*, 286–288. [CrossRef]
15. Salama, A.; Greberg, J. Optimization of truck-loader haulage system in an underground mine: A simulation approach using SimMine. In Proceedings of the MassMin 2012: 6th International Conference and Exhibition on Mass Mining, Sudbury, ON, Canada, 10–14 June 2012.
16. Choi, Y. New software for simulating truck-shovel operation in open pit mines. *J. Korean Soc. Miner. Energy Resour. Eng.* **2011**, *48*, 448–459.
17. Park, S.; Choi, Y. Simulation of shovel-truck haulage systems by considering truck dispatch methods. *J. Korean Soc. Miner. Energy Resour. Eng.* **2013**, *50*, 543–556. [CrossRef]
18. Park, S.; Choi, Y.; Park, H.S. Simulation of shovel-truck haulage systems in open-pit mines by considering breakdown of trucks and crusher capacity. *Tunn. Undergr. Space* **2014**, *24*, 1–10. [CrossRef]
19. Park, S.; Lee, S.; Choi, Y.; Park, H.S. Development of a windows-based simulation program for selecting equipments in open-pit shovel-truck haulage systems. *Tunn. Undergr. Space* **2014**, *24*, 111–119. [CrossRef]
20. Park, S.; Choi, Y.; Park, H.S. Simulation of truck-loader haulage systems in an underground mine using GPSS/H. *Tunn. Undergr. Space* **2014**, *24*, 430–439. [CrossRef]

21. Park, S.; Choi, Y.; Park, H.S. Optimization of truck-loader haulage systems in an underground mine using simulation methods. *Geosyst. Eng.* **2016**, *19*, 222–231. [CrossRef]

22. Choi, Y.; Park, S.; Lee, S.J.; Baek, J.; Jung, J.; Park, H.S. Development of a windows-based program for discrete event simulation of truck-loader haulage systems in an underground mine. *Tunn. Undergr. Space* **2016**, *26*, 87–99. [CrossRef]

23. Thompson, R.J.; Visser, A.T.; Heyns, P.S.; Hugo, D. Mine road maintenance management using haul truck response measurements. *Min. Technol.* **2013**, *115*, 123–128. [CrossRef]

24. Park, S.; Choi, Y. Analysis and diagnosis of truck transport routes in underground mines using transport time data collected through bluetooth beacons and tablet computers. *Appl. Sci.* **2021**, *11*, 4525. [CrossRef]

25. Wodecki, J.; Stefaniak, P.K.; Zimroz, P.D.S.R.; Sliwinski, M.S.P.; Andrzejewski, M.S.M. Condition monitoring of loading-haulage-dumping machines based on long-term analysis of temperature data. In Proceedings of the 16th International Multidisciplinary Scientific GeoConference SGEM 2016, Albena, Bulgaria, 28 June–6 July 2016.

26. Carvalho, R.; Nascimento, R.; D'Angelo, T.; Delabrida, S.G.C.; Bianchi, A.; Oliveira, R.A.R.; Azpúrua, H.; Uzeda Garcia, L.G. A UAV-based framework for semi-automated thermographic inspection of belt conveyors in the mining industry. *Sensors* **2020**, *20*, 2243. [CrossRef]

27. Paduraru, C.; Dimitrakopoulos, R. Responding to new information in a mining complex: Fast mechanisms using machine learning. *Min. Technol.* **2019**, *128*, 129–142. [CrossRef]

28. Ristovski, K.; Gupta, C.; Harada, K.; Tang, H.K. Dispatch with confidence: Integration of machine learning, optimization and simulation for open pit mines. In Proceedings of the 23rd ACM SIGKDD International Conference on Knowledge Discovery and Data Mining, Halifax, NS, Canada, 13–17 August 2017.

29. Xue, X.U.E.; Sun, W.; Liang, R. A new method of real time dynamic forecast of truck link travel time in open mines. *J. China Coal Soc.* **2012**, *37*, 1418–1422.

30. Sun, X.; Zhang, H.; Tian, F.; Yang, L. The use of a machine learning method to predict the real-time link travel time of open-pit trucks. *Math. Probl. Eng.* **2018**, *2018*, 4368045. [CrossRef]

31. Zhang, Y.; Ma, X.; Zhang, Y.; Yang, J. Support vector machine of the coal mine machinery equipment fault diagnosis. In Proceedings of the 2013 IEEE International Conference on Information and Automation (ICIA), Yinchuan, China, 26–28 August 2013. [CrossRef]

32. D'Angelo, T.; Mendes, M.; Keller, B.; Ferreira, R.; Delabrida, S.; Rabelo, R.; Azpurua, H.; Bianchi, A. Deep learning-based object detection for digital inspection in the mining industry. In Proceedings of the 18th IEEE International Conference on Machine Learning And Applications (ICMLA), Boca Raton, FL, USA, 16–19 December 2019. [CrossRef]

33. Park, S.; Choi, Y. Bluetooth beacon-based mine production management application to support ore haulage operations in underground mines. *Sustainability* **2021**, *13*, 2281. [CrossRef]

34. Yao, Y.; Wang, J.; Long, P.; Xie, M.; Wang, J. Small-batch-size convolutional neural network based fault diagnosis system for nuclear energy production safety with big-data environment. *Int. J. Energy Res.* **2020**, *44*, 5841–5855. [CrossRef]

35. Mitchell, T.M. *Machine Learning*; McGraw-Hill: New York, NY, USA, 1997.

36. Pandya, D.H.; Upadhyay, S.H.; Harsha, S.P. Fault diagnosis of rolling element bearing with intrinsic mode function of acoustic emission data using APF-KNN. *Expert Syst. Appl.* **2013**, *40*, 4137–4145. [CrossRef]

37. Boser, B.E.; Guyon, I.M.; Vapnik, V.N. A training algorithm for optimal margin classifiers. In Proceedings of the 5th Annual Workshop on Computational Learning Theory, Pittsburgh, PA, USA, 27–29 July 1992.

38. Yélamos, I.; Escudero, G.; Graells, M.; Puigjaner, L. Performance assessment of a novel fault diagnosis system based on support vector machines. *Comput. Chem. Eng.* **2009**, *33*, 244–255. [CrossRef]

39. Tuba, E.; Stanimirovic, Z. Elephant herding optimization algorithm for support vector machine parameters tuning. In Proceedings of the 9th International Conference on Electronics, Computers and Artificial Intelligence (ECAI), Targoviste, Romania, 29 June–1 July 2017. [CrossRef]

40. Breiman, L.; Friedman, J.H.; Olshen, R.A.; Stone, C.J. *Classification and Regression Trees*; Wadsworth International Group: Belmont, CA, USA, 1984; pp. 151–166.

41. Hasanipanah, M.; Faradonbeh, R.S.; Amnieh, H.B.; Armaghani, D.J.; Monjezi, M. Forecasting blast-induced ground vibration developing a CART model. *Eng. Comput.* **2016**, *33*, 307–316. [CrossRef]

42. Lewis, R.J. An introduction to classification and regression tree (CART) analysis. In Proceedings of the 2000 Society for Academic Emergency Medicine (SAEM) Annual Meeting, San Francisco, CA, USA, 22–25 May 2000.

43. Syarif, I.; Prugel-Bennett, A.; Wills, G. SVM parameter optimization using grid search and genetic algorithm to improve classification performance. *Telkomnika* **2016**, *14*, 1502. [CrossRef]

44. Luque, A.; Carrasco, A.; Martín, A.; de las Heras, A. The impact of class imbalance in classification performance metrics based on the binary confusion matrix. *Pattern Recognit.* **2019**, *91*, 216–231. [CrossRef]

45. Lekhtman, A. Data Science in Medicine—Precision and Recall or Specificity and Sensitivity? Available online: https://towardsdatascience.com/should-i-look-at-precision-recall-or-specificity-sensitivity-3946158aace1 (accessed on 8 October 2021).

Article

Modeling Mine Workforce Fatigue: Finding Leading Indicators of Fatigue in Operational Data Sets

Elaheh Talebi [1],*, W. Pratt Rogers [1], Tyler Morgan [2] and Frank A. Drews [3]

[1] Department of Mining Engineering, University of Utah, Salt Lake City, UT 84112, USA; pratt.rogers@utah.edu
[2] Numerite LLC, Seattle, WA 98011, USA; tyler@tylermorgan.me
[3] Department of Psychology, University of Utah, Salt Lake City, UT 84112, USA; frank.drews@psych.utah.edu
* Correspondence: elaheh.talebi@utah.edu

Abstract: Mine workers operate heavy equipment while experiencing varying psychological and physiological impacts caused by fatigue. These impacts vary in scope and severity across operators and unique mine operations. Previous studies show the impact of fatigue on individuals, raising substantial concerns about the safety of operation. Unfortunately, while data exist to illustrate the risks, the mechanisms and complex pattern of contributors to fatigue are not understood sufficiently, illustrating the need for new methods to model and manage the severity of fatigue's impact on performance and safety. Modern technology and computational intelligence can provide tools to improve practitioners' understanding of workforce fatigue. Many mines have invested in fatigue monitoring technology (PERCLOS, EEG caps, etc.) as a part of their health and safety control system. Unfortunately, these systems provide "lagging indicators" of fatigue and, in many instances, only provide fatigue alerts too late in the worker fatigue cycle. Thus, the following question arises: can other operational technology systems provide leading indicators that managers and front-line supervisors can use to help their operators to cope with fatigue levels? This paper explores common data sets available at most modern mines and how these operational data sets can be used to model fatigue. The available data sets include operational, health and safety, equipment health, fatigue monitoring and weather data. A machine learning (ML) algorithm is presented as a tool to process and model complex issues such as fatigue. Thus, ML is used in this study to identify potential leading indicators that can help management to make better decisions. Initial findings confirm existing knowledge tying fatigue to time of day and hours worked. These are the first generation of models and future models will be forthcoming.

Keywords: machine learning; mine worker fatigue; random forest model; health and safety management

Citation: Talebi, E.; Rogers, W.P.; Morgan, T.; Drews, F.A. Modeling Mine Workforce Fatigue: Finding Leading Indicators of Fatigue in Operational Data Sets. *Minerals* **2021**, *11*, 621. https://doi.org/10.3390/min11060621

Academic Editors: David Cliff and Saiied Aminossadati

Received: 15 April 2021
Accepted: 2 June 2021
Published: 10 June 2021

Publisher's Note: MDPI stays neutral with regard to jurisdictional claims in published maps and institutional affiliations.

1. Introduction

Heavy industries such as mining, which require rotational shift schedules of their personnel, are exposed to fatigue risk. This risk manifests itself in health and safety dangers presented by fatigued individuals operating heavy equipment. Fatigue is often a contributing factor to many health and safety incidents in mines, but, in addition, fatigue can also affect cognition adversely, with a negative impact on the operational performance of mine sites. These risks need improved modeling, which can enable a better understanding and better management. Improved models can eventually lead to more progressive and dynamic fatigue management with a positive impact on operational safety and efficiency.

Bauerle et al. (2018) recently discussed the limitations and lack of studies on fatigue in the mining industry [1,2]. However, several devices and technologies have been developed to identify and reduce fatigue-related risk. These tools are appealing as a risk control approach that monitors behavioral and task performance indicators that potentially indicate increases in fatigue risk [3]. Moreover, in mine operations, many real-time operational data sets exist and have great potential to provide far more analytical insight to model future undesirable events such as fatigue.

This paper presents a method that uses operational data sets to model workers' fatigue. The goal is to better understand the factors, tracked in operational technology systems, which could be used as predictors for fatigue events. The primary questions of this paper are: (1) Are there indicators within operational and other common data sets at mines that can be used to model fatigue events? (2) When these data sets are integrated and analyzed on common dimensions, is there potential value in analyzing the data with advanced computational tools such as machine learning algorithms? The approach presented in this paper is different from previous studies of mining fatigue because we use a machine learning model to identify predictor elements of workers' fatigue. The proposed model and future iterations may be useful in identifying environmental, operational and managerial events that lead to fatigue events in mine workers. This approach, when fully developed, has the potential to enhance safety and health management systems by quantifying areas of managerial focus.

The first step of the data analysis is assessing the preliminary relationships of the data. Based on the literature, there are some hypotheses around potential variables affecting fatigue in operators, which are tested in the initial data analysis section. First, does the average production or operational patterns of the mine influence the number of fatigue events? Is there any relation between time, week, month or year and the number of fatigue events? What are the differences between night and day shifts in terms of fatigue? Can the distribution of the fatigue events by shift and hour give us insights into the fatigue events? Lastly, are there any variables from weather data that cause a higher number of fatigue events?

2. Literature Review

Fitness for duty in mining is influenced by an individual's physical and psychological fitness, such as drug- and alcohol-induced impairment, fatigue, physical fitness, health and emotional wellbeing, including stress. Among these factors, fatigue is a strong driver of fitness for duty in mining, which significantly is caused by excessive work hours or insufficient rest periods associated with shiftwork [4,5]. Hence, while fatigue is identified as an issue that mine sites must address, studying factors that are contributing to or ameliorating fatigue issues is important. Fatigue in the workplace often results in a reduction in worker performance. Fatigue must be controlled and managed since it causes significant short-term and long-term risks. In the short term, fatigue can result in reduced performance, diminished productivity, human error and deficits in work quality. These effects might result in lower levels of alertness, coordination, judgment, motivation and job satisfaction, which cause increased severe health and safety issues including accidents and injuries [6–9]. Fatigue can also cause long-term negative health implications. These outcomes will result in future mental and physical morbidity, mortality, occupational accidents, work disability, excess absenteeism, unemployment, reduced quality of life and disruptive effects on social relationships and activities [10,11].

Based on a study by Drews et al. (2020), fatigue in the mining industry is different from other industries due to mining-specific environmental factors. Some of these factors are repetitive and monotonous tasks, involving long work hours, shiftwork, sleep deprivation, dim lighting, limited visual acuity, hot temperatures and loud noise [2]. However, Drews et al. (2020) also mention the high monotony of equipment operation in mining haulage as a key contributor to fatigue. Various psychological and physiological issues have effects on the fatigue of workers, which makes fatigue measurement and management difficult. Drews et al. (2020) extended a conceptual model of fatigue, which added sleep efficiency to a previously proposed model of fatigue [2]. This model shows that distal and proximal factors have effects on fatigue including clinical factors such as, life events and stressors, personality factors, previous shift conditions and sleep efficiency. Their study was based on data collected with haulage operator focus groups. Participants discussed factors that contributed to their fatigue, such as diet, shift schedule, travel time to work, sleep amount and quality, domestic factors, physical fitness and the presence of sickness. Another finding from the study is that operators have a clear awareness of fatigue's impacts on their performance and how to reduce the impact through nutrition, physical fitness, etc. [2]. Even considering this, other studies show that there is no clear approach to control, monitor and mitigate the fatigue of workers by health and safety management during mine operations [2]. Some technologies can monitor drivers of fatigue, such as tracking eye movement and head orientation (PERCLOS monitoring system) or hard hats with electroencephalogram (EEG) activity tracking. Each of these technologies has its advantages and disadvantages [2]. Each can detect fatigue when worker fatigue occurs; however, these systems do not necessarily prevent or mitigate fatigue [2]. Moreover, users of these technologies, such as the PERCLOS system, expressed privacy concerns regarding the system's constant monitoring and mentioned a high number of false alarms from the equipment, thus being a nuisance [2]. Bauerle et al. (2018) mentioned that, despite the complexity and uncertainty regarding the fatigue of miners, some real solutions could be developed for improving fatigue-related issues with fatigue assessment interventions, looking beyond sleep and physical work, and shift work effects [1]. In the same vein, Drews et al. emphasized that health and safety management should take a socio-technical systems perspective, since a sole focus on technological solutions may create an illusion of safety, while not necessarily improving safety performance. Moreover, these approaches require user acceptance and high levels of trust in order not to have an adverse impact on their functionality [2]. Successfully modeling fatigue will require a multi-faceted approach and a variety of data inputs from the mining system.

In addition to the health and safety implications on workers, fatigue can result in damage or loss of expensive mine equipment such as haul trucks. Therefore, the mining industry has long focused on measuring operational risk losses for the purpose of capital allocation and the process of managing operational risks. Operational risk results from insufficient or failed internal processes, people, control, systems or external events, including equipment health, individual health and safety and worker fatigue [12]. To manage the health of equipment, organizations have deployed early warning systems through equipment monitoring and modeling technology. These technologies depend on understanding either machine design or empirical modeling methods to determine normal equipment behavior and detect any signs of abnormal behavior [13]. These technologies learn the dynamic operational behavior of equipment using equipment sensor data and create a predictive model. The predictive model output, which is the equipment's performance, will be compared with actual measurements from sensor signals to detect any abnormalities or failures [13].

The entire mine workplace could benefit from new technologies to collect and analyze real-time safety data such as fatigue monitoring data. A critical issue is the ability to use this information to react prior to an incident. The development of new technologies can assist safety managers in providing timely measures to predict an increase in risk, resulting in the prevention of serious incidents [14]. To manage the operational safety and health in mines, it is necessary to have safety indicators. There are two different types of safety indicators: lagging and leading indicators [15]. Lagging indicators evaluate safety and health using incident and illness rates, while leading indicators measure workplace activities, conditions and safety and health-related events [16]. In the case of fatigue, lagging indicators are evident after fatigue has occurred, while leading indicators are measurements that could prevent fatigue, such as sleep patterns or caffeine intake, and steps that help to lower fatigue when it is not so high. Since lagging indicators have a reactive and delayed nature, managers need to develop appropriate leading indicators to measure workplace safety and health risk [16]. Leading indicators have a predictive value regarding unsafe workplace conditions or behavior that is followed by an incident [17–20]. There are three main uses of leading indicators: monitoring the level of safety, deciding where and how to take action and motivating managers to take action [21,22]. Passive leading indicators (PLIs) are measurements that can provide an indication of the probable safety performance [14]. On the other hand, active leading indicators (ALIs) are dynamic and more subject to active change in a short period of time [14,23]. To have predictive values, ALIs must be recorded in a timely manner in order to obtain accurate measurements and observations.

ALIs are continually being advanced as new technology is introduced into production systems. Internet of things (IoT), big data, artificial intelligence (AI) and machine learning (ML) are being used to enhance the safety, efficiency and quality of the operations [24–26]. In high-risk environments such as mines, internet of things can be used to raise safety and decrease the probability of human error and disasters [24–26]. In addition, IoT can be a relatively inexpensive and effective approach for hazard recognition and sending safety notifications [14].

Machine learning (ML) has been demonstrated to be a predictive tool to support management to make better decisions [16,27]. In spite of the abundant leading indicators, the use of ML to predict leading indicators is rare [16,27]. ML is flexible to operate, without any statistical assumptions, and has the ability to identify both linear and non-linear relationships within the phenomenon investigated [16,24,28]. Poh et al. (2018) used ML to predict safety leading indicators on construction sites [16]. They used a data set that was collected from a construction contractor to identify the input variables and develop a random forest (RF) model to forecast the safety performance of the project [16]. They mentioned that the occurrence and severity of incidents is not random, which means that there is a pattern describing the incidents, and they can be predictable [16]. This pattern can be used to explain the complexity of the leading indicators and long-term data collection helps to elucidate the interactions of safety indicators over time [16,29].

The literature suggests that finding leading indicators to predict fatigue in the mining industry is necessary [2]. Due to the complexity of fatigue, applying computational intelligence methods such as machine learning (ML) algorithms on the real-time data captured from current and future IoT technologies can benefit mine operations in modeling fatigue. Such a model could identify ALIs and predictive elements of workers' fatigue. Poh et al. collected data sets for the purpose of modeling safety. However, their study was limited to only safety data, likely neglecting other possible predictive factors. A comprehensive study incorporating a wider range of data sets will extend possible independent features in the model to identify the best predictive factors. If these factors can be developed as leading indicators of fatigue, enhanced safety and health decisions can be made earlier in the fatigue cycle.

3. Methodology

3.1. Data Set Characterization

The presented study uses 3.5 years of data at a single, large, operating surface mine. Table 1 provides an overview of the data sets, the types of information encoded in the data and the range of dates covered by each data set.

Table 1. Data sets' details.

Data Source	Key Factors	Date Range
Fatigue monitoring	Operator drowsiness, micro-sleeps, etc.	2014–2017
Time and Attendance	Hours worked, shift worked, etc.	2014–2017
Fleet management system (production and status)	Production cycles, faulty equipment, delayed equipment, etc.	2014–2017
Equipment health alarms and events	Notification of equipment abuse, use of equipment, etc.	2014–2017
Weather conditions	Temperature, wind speed, wind direction, change, precipitation, relative humidity, etc.	2014–2017

The site utilizes a PERCLOS monitoring system, which has been in place since 2014. This system uses cameras to track, monitor and model the eye movements of haul truck operators [2]. The system detects certain eye movements and can determine if the eyes are closed, blinking rapidly and other factors that indicate fatigue. If the system cannot detect that the operator's eyes are open for more than 3 s, it alerts the operator using seat buzzes and vibration. In addition to a local alarm, the system also sends a message or alarm to the dispatcher, supervisor and the company supporting the system.

Data captured from the system are categorized based on type of the events: micro-sleep with a stable head posture, other eye closure (drowsiness), eyewear interference (clear lenses), eyewear interference (sun glasses), normal driving, bad tracking, glance down, glance away, driver leaning forward, camera covered, testing, IR pods covered, no driving, video error, seat position change, partial distraction, other. Based on the study by Drews et al. (2020), micro-sleep and drowsiness are signs of operator fatigue [2]. The study assumes that the PERCLOS system is functioning and properly collaborated. Much work has been done establishing the PERCOLS technology. Testing the viability of this technology is beyond the scope of this paper. The literature shows that the fatigue events captured by these systems are important indicators of fatigue [1,2]. Therefore, for the purpose of our study, micro-sleeps and drowsiness have been used to demonstrate a fatigue event, and other types of alarms are assumed to be system errors or because of negative behaviors such as distracted driving. These are labeled in the PERCLOS systems as "other eye-closure (drowsiness)" and "micro-sleep with stable head". The operational difference between these two categories is having a stable head posture at the time of fatigue or not. In the case that the operator's head is moving downwards, the fatigue event is labeled as "other eye closure (drowsiness)". On the other hand, when the operator has a stable head posture at the time of fatigue, it is labeled "micro-sleep with stable head".

More details of fatigue events are shown in Table 2. The average number of events per day and the number of days with these fatigue events are provided for comparison. The data show more drowsiness compared to micro-sleep, representing 60% of the fatigue events that were captured by the system. The % of days with fatigue shows that on 98% and 99% of the days, there was at least one micro-sleep and drowsiness fatigue event, respectively. Therefore, fatigue is a critical daily hazard for those working in mines.

Table 2. Count of days and percentage of total days by fatigue type.

Fatigue Event Type	Average Number of Events per Day	Days with Fatigue Events	Percentage of Days with Fatigue	Percentage of Fatigue Events
Micro-Sleep with Stable Head	13	1313	98%	40%
Other Eye Closure (Drowsiness)	20	1327	99%	60%

The surface mine maintains a fleet management system (FMS), which tracks the production and status of equipment. The FMS data are made available in a business intelligence (BI) database. Status event data provide details on the state of an asset. Status event coding can be used to determine if a piece of equipment is down for maintenance, in a production activity or in standby mode. This information is valuable to compare against event rates, as well as show breaks and delays. Other information in the BI database includes the load cycle data. A production cycle shows the load of a shovel or truck. Detailed steps within a load, such as loading, dumping, running empty, running loaded, etc., are shown. The most important data for this study are the production rate by shift/hour, which can be used to normalize the data as well as understand the activity levels of haul truck drivers.

Time and attendance data are provided via the hours worked by hourly employees. The mine uses a swipe-in/swipe-out time keeping system, the data from which are processed and loaded into a time and attendance database. The data set was used to measure shifts and hours consecutively worked by haul truck operators.

Mobile machinery such as haul trucks generates large amounts of equipment health data. The data are produced by hundreds of sensors and are used to track the location, production cycles, equipment status and equipment health alarms. The sensors can be valuable predictors of production achievements and operator behavior. The surface mine utilizes an equipment health database to capture and model the health and use of their large capital assets. These databases track in detail how a given piece of equipment is being operated at any given time. The sensors can detect if an operator is operating outside of the safe boundaries of the machine and create an alarm. These alarms vary by severity and location and generate massive amounts of data.

Lastly, weather data are gathered from a local weather station in the mine. This data set includes information for the weather at the mine site. Over 10 variables are captured at 10 min intervals. Each interval contains information regarding temperature, temperature change, wind speed, precipitation and air pressure.

Data Pre-Processing

In this step, data need to be pre-processed to make them appropriate for the application of the chosen modeling approach. Initial data analyses are performed to identify possible patterns of data with the identified fatigue events. This analysis informs the next modeling step by identifying an appropriate approach to predict fatigue events with the data sets.

Fatigue data provided from the fatigue monitoring system were reviewed and divided in different categories. Among them, drowsiness and micro-sleeps were identified as the fatigue events occurring among workers, so they are considered to be the dependent variables of the model. All other data, including weather, production cycles, equipment health alarms and time and attendance data, are modeled as predictors and criterion variables.

Each data set had to be cleaned and missing data removed prior to input to the model. The process of cleaning data entails removing incorrect, duplicate, incomplete and corrupted data. Updating data types is also a common cleaning activity. A list of all variables used in the model is given in Table 3. After all data engineering, data are prepared for two distinct models: shift-based and hourly-based models. Data sets were thus grouped by shift ID and hour of data time.

3.2. Initial Data Analysis

As stated above, the primary questions posed by this study are: Are there new indicators within existing mining data sets that can be used to model fatigue events? In addition, what are potential patterns when these data sets are analyzed? In this section, the available data sets are presented to explore how they can be used to test the hypothesis of the research. Modern machine learning approaches require various levels of data engineering to facilitate statistical analysis. This section presents the process and logic used to identify key variables and the direction for further data engineering used in the development of the ML model. More specifically, the analyses presented here cover the distribution of the fatigue events, average production compared to fatigue events, number of fatigue events during night and day shifts and temperature versus fatigue events.

Fatigue is first examined by analyzing its frequency distribution by shift, which suggests non-normal distribution, as illustrated in Figure 1. This figure visualizes the distribution of the fatigue events per shift, which is seemingly close to Poisson distribution, with a mean of approximately 17 events per shift. Calculation of the probability of having 0 and >52 events per shift shows, respectively, very low probabilities $p = 0.013$ and $p = 0.0097$. However, the probability of having 7–8 events per shift, which is the mode of the distribution, is estimated to be $p = 0.052$. The next question is why some shifts have a higher number of fatigue events compared to other shifts. Therefore, to find the potential variables that drive this difference, aggregated data by shift are included in the model.

Table 3. List of the variables based on the data source.

Data Source	Variables	Data Type and Example Data
Time and Attendance	Shift ID	Integer (1 to 4140)
	Shift of Day (shift type)	Categorical Integer (0 and 1)
	Crew Name	Categorical Integer (1 to 4)
	Days On	Integer (0 to 4)
	Year	Integer (2014 to 2017)
	Month	Integer (1 to 12)
	Week	Integer (1 to 54)
	Day	Integer (1 to 31)
	Day of week	Integer (1 to 7)
	Day of year	Integer (1 to 365)
	Hour of day	Float (0 to 24)
	Shift is end of month	Categorical Integer (0 and 1)
	Shift is start of month	Categorical Integer (0 and 1)
	Shift is end of quarter	Categorical Integer (0 and 1)
	Shift is start of quarter	Categorical Integer (0 and 1)
	Shift is end of year	Categorical Integer (0 and 1)
	Shift is start of year	Categorical Integer (0 and 1)
Fleet management system (production and status)	Mine Production Factor	Integer (1335 to 589,201)
	Mine Loaded Travel Distance	Integer (37,884 to 37,797,788)
	Mine Measured Production	Integer (0 to 430,812)
	Mean Measured Production (broken down by fleet, creating 8 variables)	Float (0 to 413.83)
	Mine Load Capacity Percentage	Float (0 to 1)
	Mean Load Capacity Percentage (broken down by fleet, creating 8 variables)	Float (0 to 1)
	Mean Loaded Travel Distance	Float (3735.2 to 13,711.66)
	Mean Loaded Travel Lift	Float (272.25 to 1083.29)
	Mean Loaded Travel Lift Distance	Float (3735.2 to 13,711.66)
	St Dev Loaded Travel Distance	Float (604.55 to 16,118.27)
Weather	Mean Barometric Pressure	Float (0 to 25.1)
	Mean Precipitation	Float (0 to 439.1)
	Mean Temperature (2 m)	Float (-6.8 to 34.8)
	Min Barometric Pressure	Float (0 to 25.01)
	Min Precipitation	Float (0 to 29.71)
	Min Temperature (2 m)	Float (-8.4 to 30.44)
	Max Barometric Pressure	Float (0 to 25.1)
	Max Precipitation	Float (0 to 756.9)
	Max Temperature (2 m)	Float (-4.435 to 36.82)
	Sum Precipitation	Float (0 to 5269.17)
Equipment health alarms and events	Both Alarm Count	Integer (0 to 632)
	Electrical Alarm Count	Integer (0 to 892)
	Lockout Alarm Count	Integer (0 to 35)
	Maintenance Alarm Count	Integer (0 to 1094)
	Mechanical Alarm Count	Integer (0 to 1753)
	None Alarm Count	Integer (0 to 2608)
	Normal Alarm Count	Integer (0 to 121)
	Operational Alarm Count	Integer (0 to 819)
	Undetermined Alarm Count	Integer (0 to 1282)
	Scheduled Down Count	Integer (0 to 85)
	Unscheduled Down Count	Integer (0 to 141)
	Operational Delay Count	Integer (0 to 1126)
	Operational Down Count	Integer (0 to 80)
	Ready Non-Production Count	Integer (0 to 977)
	Ready Production Count	Integer (0 to 1322)
Fatigue monitoring system	Drowsiness and Micro-Sleep Fatigue Events Count (Normalized)	Float (0 to 1)

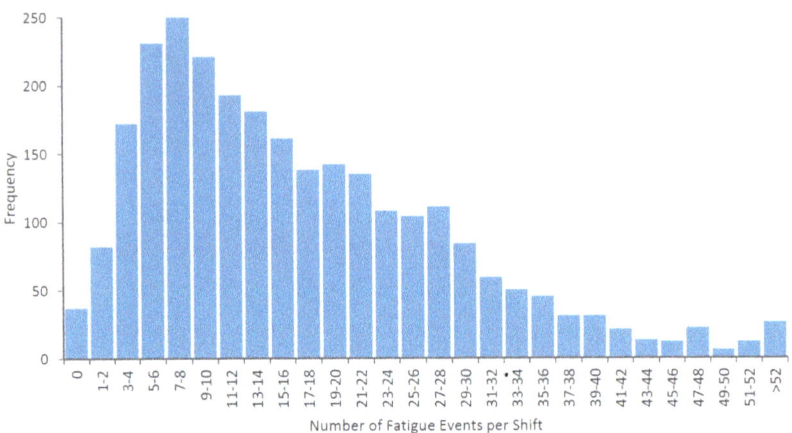

Figure 1. Fatigue events per shift frequency.

In order to analyze the effect of shift time on fatigue, Figure 2 shows the average hourly production and hourly number of fatigue events per person (including drowsiness and micro-sleep). Shift change times (7 am/pm) are indicated by substantial reductions in fatigue events due to the relatively high levels of activities associated with shift changes. In addition, the results illustrate that fatigue counts increase from the beginning of a night shift until the shift end; however, during day shifts, the fatigue levels of the operators peak at around 1 pm. Regarding the relationship between the numbers of fatigue events and hourly production, the findings suggest no clear relationship. Figure 2 suggests that the time of day and shift type could be included as additional variables in the model. This figure also suggests a negative relationship between production and fatigue. Production rates, disruptions and aggregate levels, to a certain extent, affect the operational behavior of the site. A higher number of cycles or longer cycles have the potential to influence how engaged operators are, which could provide an interesting additional measure to predict fatigue. Information about production cycles and delays will be modeled against fatigue to further explore this potential relationship.

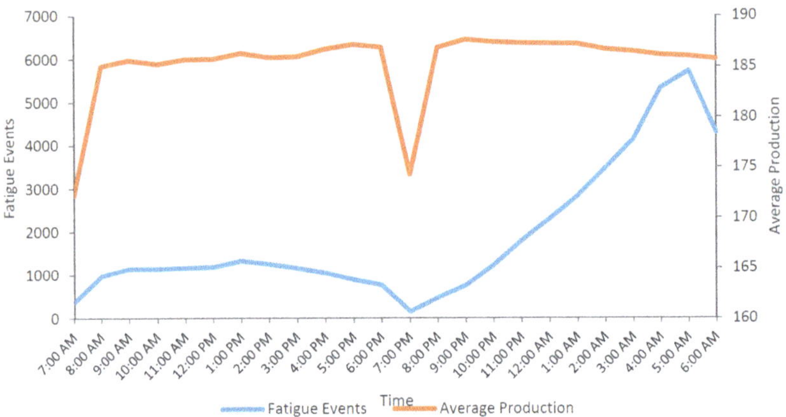

Figure 2. Hourly fatigue events and average hourly production.

To illustrate the relationship between hourly data and the frequency of fatigue events, their distribution is provided in Figure 3. This right-skewed distribution shows that more than 50% of the hours contain at least one fatigue event. This suggests that further exploration is needed to identify the variables contributing to the range of hourly fatigue events. Therefore, a second model with hourly aggregated data is developed, which will be introduced in the model section. In addition, Figure 4a shows that night shifts contain significantly more events compared to day shifts. Moreover, the average event counts by month indicate a seasonality effect, with lower rates of fatigue in spring and higher rates in summer and winter (Figure 4b). To summarize, the above explorations demonstrate that some variables, such as shift type, time of day and worked hours, have effects on fatigue. At the same time, the findings suggest that advanced approaches will be required to model fatigue events.

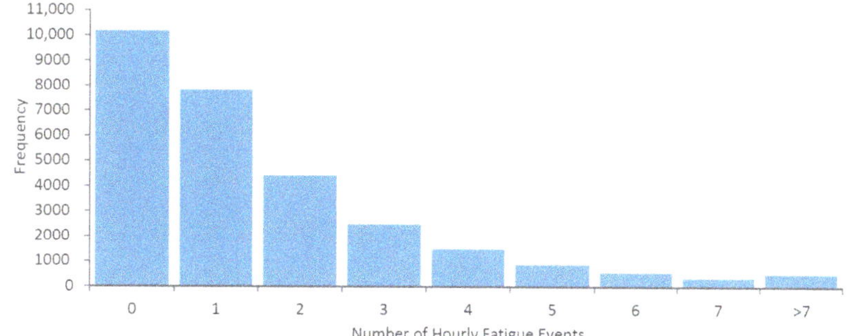

Figure 3. Hourly fatigue events frequency.

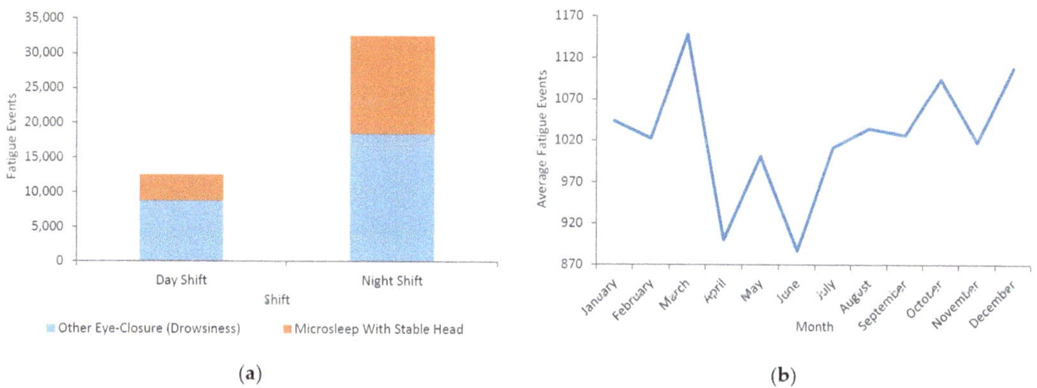

(**a**)

(**b**)

Figure 4. (**a**) Fatigue events per shift; (**b**) Average monthly fatigue events.

Next, we conduct an exploration of the influence of environmental variables on operator fatigue. Figure 5 illustrates the monthly average ambient temperature and monthly fatigue events per person, without any clear pattern. Thus, there appears to be no obvious correlation between temperature and fatigue events in this plot. Therefore, for further exploration, weather data are added as independent variables to the model.

The main purpose of the above analyses was to explore relationships between fatigue events and variables contained in the existing data sets. From our initial data analyses, fatigue appears to have some relationship with variables such as weather, shift type, time of day, etc. These analyses introduce more variables for the purpose of the modeling and data aggregating methods. The full list of variables is shown in Table 3. However, these preliminary analyses are not able to identify a pattern of fatigue based on these variables, although they are able to provide a critical insight into the data. The literature shows that fatigue is a complex issue and different psychological and physiological variables influence fatigue in workers [2]. Considering the limitations of the above analytical approaches, we use machine learning (ML) approaches as an alternative to explore the data set to elucidate relationships that are not easily identifiable. Because the above analyses show that shift type and hour of day appear to have significant effects on the fatigue of haul truck drivers, data were aggregated by shift and hour to create two different models. One approach involves fatigue prediction using the shift-based data, and the other uses hourly data to predict fatigue. The next section presents the modeling approach.

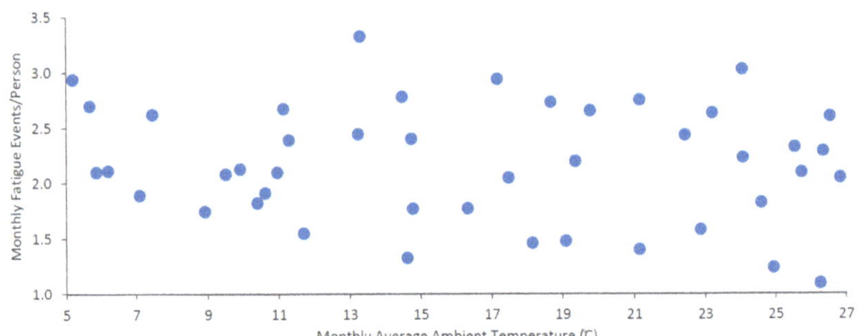

Figure 5. Monthly fatigue events per person vs. average temperature.

3.3. Machine Learning Model

Figure 6 presents the procedure and methods of the modeling steps involved in the development of the machine learning model. The process involved the following steps:

- Data collection;
- Data pre-processing;
- Data engineering;
- Training model;
- Testing model;
- Model evaluation;
- Making predictions.

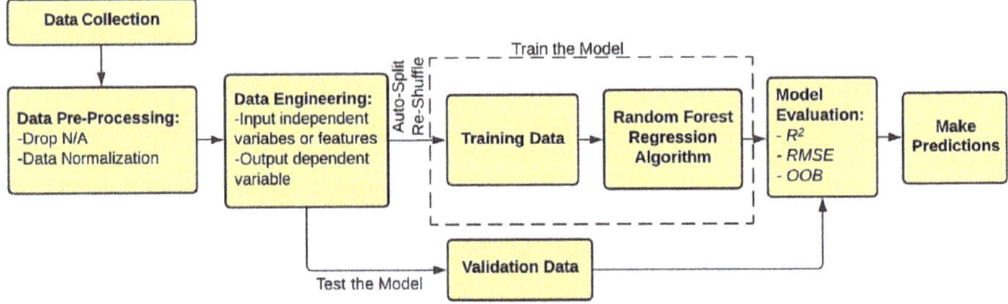

Figure 6. Procedure diagram for the prediction of fatigue using random forest regression algorithm.

3.3.1. Random Forest Regression Algorithm

The machine learning model selected for this analysis is a random forest (RF) regression algorithm. Random forest algorithms were chosen for their tendency to generalize well to a wide variety of problems, their rapid speed of training and because they are a key feature of many well-known machine learning solutions. Another key benefit of using random forests is the tooling that has been built in recent years to help researchers to gain insights into what has long been thought of as the black box of machine learning. These new analytical tools allow researchers to see the features that the model relies upon the most in order to make predications and determine how marginal changes in these features impact the predicted outcomes [30,31].

When data are not linearly scattered, a regression tree, which is a type of decision tree, can be used. In this type of decision tree, each leaf presents a threshold value (TV) for each feature of the model. For the purpose of finding the best decision tree, the model tries to find the best threshold value for each feature (independent variable) by finding the minimum sum of square residuals (SSR). SSR is the sum of the squared difference of each prediction value and actual value (Figure 7). For models with more than one feature, the decision tree root is the feature with the lowest SSR. Figure 8 represents an example of a random forest regression decision tree with five features.

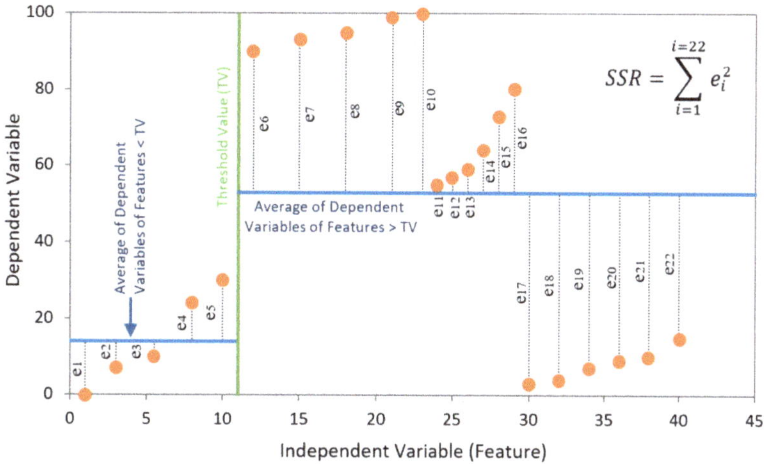

Figure 7. Finding TV and SSR for random data (*e* or error is difference of each predicted value (average value) and actual value, and *i* is the feature number).

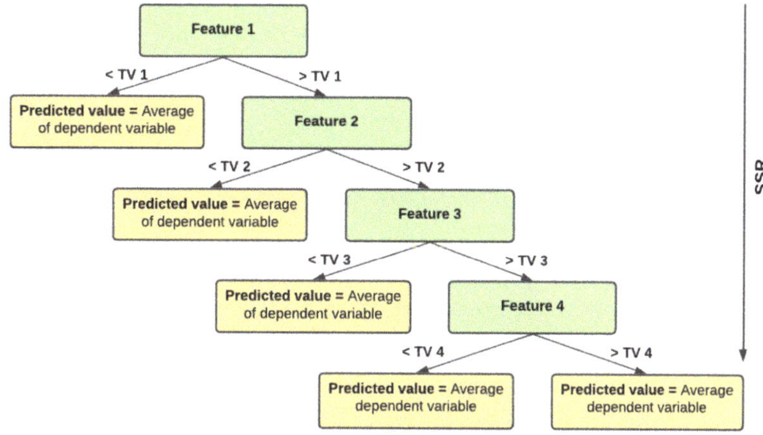

Figure 8. Schematic random forest regression decision tree.

A random forest regression algorithm is an ensemble of randomized regression trees. The random forest algorithm creates bootstrap samples from the original data. Bootstrapping is a procedure that resamples a single data set to create many simulated samples. For each of the bootstrap samples, the algorithm increases a classification or regression tree. This algorithm chooses a random sample of the predictors and selects the best split among variables. Then it predicts new data by aggregating the predictions of the trees. Models can estimate the error rate based on the training data by each bootstrap iteration [30,31].

3.3.2. Model

Two models were created using available data subsets as dependent variables. For the shift-based model, the dependent variable was the number of fatigue events in a 12 h shift, which was normalized to scheduled haulage hours in the shift (labor hours). For the hourly-based model, the dependent variable was the number of fatigue events in an hour, which was normalized to scheduled haulage hours (labor hours). All independent variables in these models, also known as features, are representations of the mine's operation as represented in the data sets. These features contain values such as the average production, average temperature and equipment alarm (see Table 3).

For this model, data were divided into two sets: 80% constituted the training data set and 20% constituted the validation data set. The goal of these models was to determine the features that can predict fatigue in such a way that minimizes *RMSE*. In these models, only data subsets with micro-sleeps and drowsiness containing fatigue were modeled. From 151,432 possible events, only 44,953 contained micro-sleep and drowsiness in the data sets to train and validate the random forest algorithm. After exploratory data analysis, this study focused on refining models to predict the fatigue of the operator.

The independent variables (features) in these models were minimally engineered. Then, possible sample counts, means, sums, mins and maxes were used without combining multiple fields from the underlying tables. The goal of these models was to predict fatigue as well as the possibility of including all available feature sets, such as the hour of the day, shift, month of the year, ambient temperature, wind speed, precipitation, etc. Data for these models were constrained to the number of days contained in the fatigue data, which was a dependent variable. Thus, the models were created using data from 1 January 2014 to 9 August 2017.

3.3.3. Evaluating Model Performance

One way to evaluate the model performance is out-of-bag error or *OOB*. The out-of-bag set includes data not chosen in the sampling process when initially building a random forest. The out-of-bag (*OOB*) error is the average error for each calculated prediction from the trees not contained in the respective sample. Here, we used the Random Forest python package, which can generate two optional information values, a value of the importance of the predictor variables (feature importance) and a value of the internal structure of the data (the proximity of different data points to one another) [30].

Next, the performance of the model was evaluated using the root mean squared errors (*RMSE*) and coefficients of determination (R^2). The coefficient of determination is the best method to compare models that are trained using different dependent variables. Both *RMSE* and coefficients of determination are important means of measuring performance between models trained to predict the same dependent variable. The reason that R^2 should be used when comparing models trained on different dependent variables is that the coefficient of determination is normalized to the mean of the dependent variable for each model.

3.3.4. Model Generalization

When creating machine learning models, it is important to ensure that the predictions are generalizable to data that the model was not trained on. A model that has a very low training error but a very high validation error is considered not to generalize well. This scenario is known as overfitting [32]. The most common method to ensure that a model has not been overfit is splitting data into training and validation sets. The model learns its parameters from the training data set. The performance of the trained model is then determined by how well it predicts the outcomes of the validation data set. The hyperparameters of the model can then be tuned by the developer, and the model is retrained to improve its performance against the validation data set. Hyperparameters are the values that define the model and cannot be learned from data; they are set by the developer of the machine learning algorithm (number of estimators, max number of features, etc.). The number of estimators and the max number of features for the best model here are 1000 and sqrt (number of features), respectively. For each model here, the data sets were split into training and validation sets. In this study, due to a lack of sufficient data for a double hold-out (test set), there is only a validation set.

3.3.5. Feature Importance

Feature importance is the process of ranking the individual elements of a machine learning model according to their relative importance to the accuracy of that model [33,34]. Feature importance is a means of determining the features that have the greatest magnitude of effect in a model. Features that have a high feature importance value have a greater impact on the model. Feature importance refers to a technique to assess the scores of independent variables to a predictive model. It indicates the relative importance of each independent variable (feature) when making a model prediction. These scores can be used to better understand the data and model and reduce the number of input features. The relative scores of feature importance can highlight which features are more useful to predict fatigue and, conversely, which features are the least helpful to predict fatigue. This may be used as the basis for gathering more or different data. Moreover, it shows that the model has been fit to the most important features. In addition, feature importance can be used to improve a predictive model. It can be used to eliminate the features with the lowest scores or retain those with the highest scores. Therefore, it can help to select features and speed up the modeling process.

4. Results

Differences between the two models were found after the analyses. The hourly-based model does not perform as well as the shift-based model according to their R^2 and $RMSE$. The best model used the shift-based data to predict fatigue. Below, we discuss feature importance and drop column tools to examine the feature set of the shift-based model.

In Table 4, the results of the best performed model are displayed. The best-performing model predicted fatigue events across the site, with an R^2 value of 0.36 and $RMSE$ value of 0.006. All other models were deemed to have values too low to warrant further exploration using this feature set. The best model used the shift-based data to predict fatigue. Below, we discuss feature importance and drop column tools to examine the feature set of the shift-based model.

Table 4. Refined model performance results.

Model	Root Mean Squared Error (*RMSE*)		Coefficient of Determination R^2		
	Training	Validation	Training	Validation	*OOB*
Shift-based model	0.002	0.006	0.93	0.36	0.47

4.1. Feature Importance of Best-Performing Model

Generally, feature importance provides a score that identifies the value of each feature in creating the random forest model. Features that have a greater effect on key decisions have higher relative importance. Table 5 shows the most important features and their values for the fatigue event prediction model (shift-based model) with the best performance.

Table 5. Permutation importance of features for shift-based model (most important features).

Data Category	Dependent Variables	Feature Importance Score
Time and Attendance	Shift type (day or night shift)	0.1650
Equipment health alarms and events	Unscheduled downtime count	0.0588
Fleet management system (production and status)	Mine load capacity percentage	0.0297
	Mine measured production	0.0293
	Mine production factor	0.0248
Time and Attendance	Year	0.0245
Weather	Mean temperature (2 m)	0.0235
Equipment health alarms and events	None alarm count	0.0230
Fleet management system (production and status)	Mine loaded travel distance	0.0226
	Mean measured production of haul truck (CAT 793D)	0.0226
Weather	Maximum temperature (2 m)	0.0223
Equipment health alarms and events	Ready production count	0.0222
	Mechanical alarm count	0.0215
Fleet management system (production and status)	Mean load capacity percentage of haul truck (CAT 793D)	0.0213
	Mean measured production of haul truck (CAT 793C)	0.0211
	Mean loaded travel distance	0.0209
	Mean measured production of haul truck (CAT 793B)	0.0209
	Mean load capacity percentage of haul truck (CAT 793C)	0.0208
	Mean load capacity percentage of haul truck (CAT 793B)	0.0207
Equipment health alarms and events	Scheduled down count	0.0206

The shift type (day/night shift) variable has the strongest effect on the model. Next, the amount of unscheduled downtime of the equipment of the whole mine for a shift affected the model. "Unscheduled downtime" is when a piece of equipment goes down for maintenance reasons in an unplanned situation. Other factors that have effects on fatigue are production variables. These outcomes corroborate with the initial data analyses regarding the effects of day shifts and night shifts on the fatigue of workers. It also demonstrates that production and equipment alarm variables such as equipment downtime can aid in predicting the occurrence of fatigue events. Moreover, weather variables such as maximum and average temperature can increase the rate of fatigue events among workers.

4.2. Drop-Column Feature Importance

With large data sets, there is always a risk of having variables that are covariates or have co-dependences. Random forest tools recognize that this risk exists and include mechanisms to address it, which can assess the individual effects of each feature on the model. Co-dependencies stem from the fact that the trees are not independent since they are sampled from the same data in the process of making the RF model. It is important to see how the model works without individual features and how each feature impacts the model, whether positively or negatively. Instead of carrying out different iterations, random forest algorithms have a built-in tool which runs models with fewer features and tracks the models' performance. This is achieved by dropping out each column (or feature) from the model, retraining the entire model and then comparing the score with the base score. Negative values show features that improve the model when removed. Positive values show features that weaken the model when removed. Values that are close to zero tend to indicate features that are correlated with other features; thus, removing them makes little difference in the model's ability to find relationships using the correlated variables. In Table 6, the ten most and ten least important features are displayed.

Table 6. Drop-column importance for the best model.

Dependent Variables	Feature Importance Score
Shift type (day or night)	0.2922
Unscheduled downtime count	0.0317
Mechanical alarm count	0.0235
Day on	0.0225
Day of week	0.0205
Mean measured production of haul truck (CAT 797F)	0.0139
Shift is end of year	0.0129
Electrical alarm count	0.0129
Mine measured production	0.0127
Undetermined alarm count	0.0125
...	...
None alarm count	−0.0022
Mean loaded travel distance	−0.0025
Mean load capacity percentage of haul truck (CAT 793D)	−0.0031
Year	−0.0034
Mean Temperature (2 m)	−0.0073
Mean load capacity percentage of haul truck (CAT 793B)	−0.0073
Mean loaded travel lift distance	−0.0074
Maintenance alarm count	−0.0083
Mean loaded travel lift	−0.0119
Mine load capacity percentage	−0.0325

As shown in Table 6, shift type has the strongest effect on the model, followed by some production and alarm variables, as indicated by their feature importance score. This score shows that dropping, for example, shift type, from the features, causes the performance of the model (R^2) to drastically decrease by 0.2922. On the other hand, eliminating the mine load capacity percentage increases the performance of the model by 0.0325.

4.3. ICE Plot

Another tool to visualize how marginal changes in features affect the predictions of the model is an individual conditional expectation (ICE) plot. An ICE plot identifies the dependence of the prediction on a feature for each instance independently. It generates one line per instance, which can be compared to one line overall in partial dependence plots. A partial dependence plot (PDP) is the average of the lines of an ICE plot. The value of a line or model score is compared when all other features are kept the same. The result is a set of points for an instance with a feature value from the grid and the respective predictions [35]. ICE plots for the four top features are displayed in Figure 9 and ICE plots of ten top features are provided in Appendix A. They show how the models' predictions change depending on marginal changes to the top features. For instance, they illustrate that the prediction difference in the model for shift of day decrease from the day shift to the night shift. The ICE plot of unscheduled downtime count shows that the prediction difference of the model for a small amount of unscheduled downtime of the equipment is not high, but it starts to increase after 40 counts of unscheduled downtime of the equipment. Moreover, the prediction difference of the model for mine measured production is small, but it increases after 300,000 tons of production. Therefore, looking at the marginal changes in the top features offers insights into how marginal changes affect the model prediction.

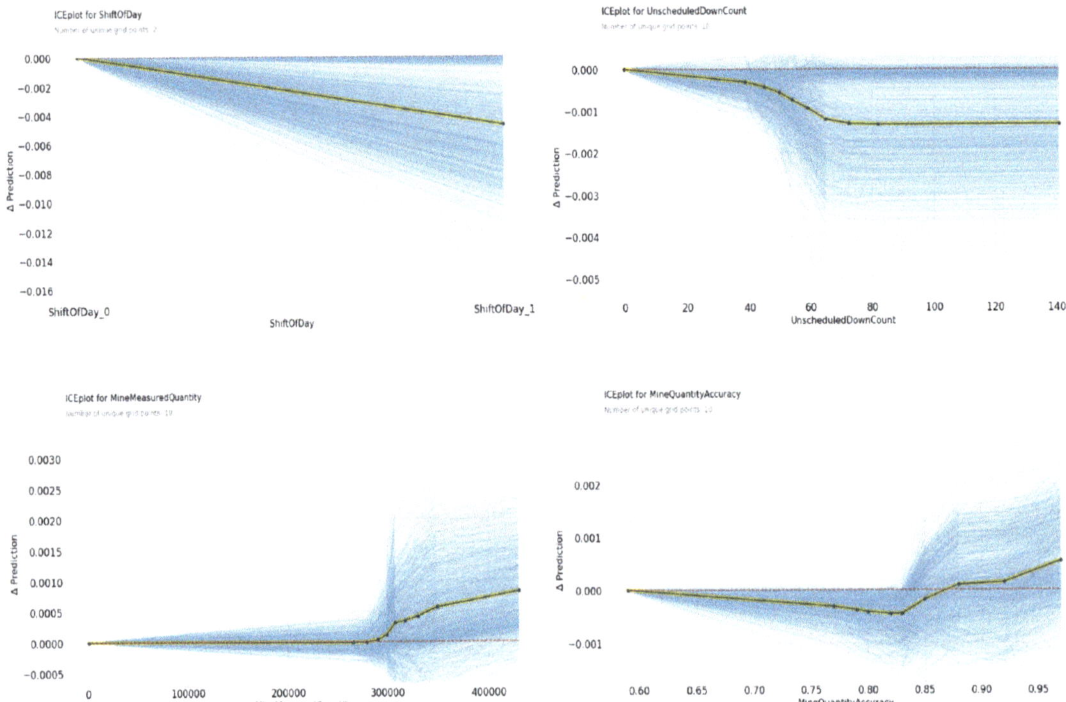

Figure 9. ICE plots of the top three features (Blue lines identify the dependence of the prediction on a feature for each instance independently, and yellow line is PDP which shows the average of the blue lines).

4.4. Comparison

Table 7 shows the top features from two different generated models. Of the models that were run, the best performance was demonstrated by the shift-based fatigue model that is used to predict fatigue events based on shift data. This model achieved an R^2 value of 0.36, which is reasonably high for the prediction of outcomes that are the result of very complex interactions. Fatigue is a complex issue and can occur for different psychological and physiological reasons; therefore, it is difficult to predict it with high accuracy.

Table 7. Comparison of the top features for hourly-based and shift-based models.

Rankings	Shift-Based Model	Hourly-Based Model
1	Shift type (day or night shift)	Mean temperature (2 m)
2	Unscheduled downtime count	Hour of day
3	Mine load capacity percentage	Mean measured production of haul truck (CAT 793B)
4	Mine measured production	Mean measured production of haul truck (CAT 793D)
5	Mine production factor	None alarm count
6	Year	St Dev loaded travel distance
7	Mean temperature (2 m)	Mean barometric pressure
8	None alarm count	Maintenance alarm count
9	Mine loaded travel distance	Undetermined alarm count
10	Mean measured production of haul truck (CAT 793D)	Mine production factor

Another model, which is based on the hourly aggregated fatigue occurrence, identifies that the time of day helps to predict the fatigue as expected, since this is one of the top features in the hourly-based model. Moreover, ambient temperature has a notable effect on fatigue, which is evidenced by the hourly-based model; however, it is obvious that temperature is linked to the time of day. More work is needed to assess the potential role of air conditioning in this. In addition to time and weather factors, some production and equipment health alarm variables have effects on the fatigue of haul truck drivers, as the hourly-based model shows (see Table 7).

5. Discussion

The model output identifies the variables that have the greatest impact on all fatigue events. Table 8 illustrates the most important features and their data sources. The results confirm our existing understanding of fatigue and offer some interesting insights into additional factors that potentially cause fatigue. While it is not surprising that shift type causes fatigue, it is interesting that maintenance processes such as unscheduled downtime and production rates, as well as other operational variables, can affect fatigue among haul truck drivers. Having identified these additional predictors for fatigue, these indicators can be used by managers to prioritize safety management efforts. The ICE plots show how marginal changes to specific variables affect the model. Therefore, they can be potentially used as thresholds for KPIs. For example, if the mine is approaching a value of 40 for unscheduled downtime, a higher risk of fatigue is indicated.

Table 8. Top features by data classification.

Data Category	Feature Rank	Feature
Time and attendance	1	Shift of day (day or night)
	6	Year
Fleet management system (production and status)	3	Mine load capacity percentage
	4	Mine measured production
	5	Mine production factor
	9	Mine loaded travel distance
	10	Mean measured production of haul truck (CAT 793D)
	14	Mean load capacity percentage of haul truck (CAT 793D)
	15	Mean measured production of haul truck (CAT 793C)
	16	Mean loaded travel distance
	17	Mean measured production of haul truck (CAT 793B)
	18	Mean load capacity percentage of haul truck (CAT 793C)
	19	Mean load capacity percentage of haul truck (CAT 793B)
Equipment health alarms and events	2	Unscheduled down count
	8	None alarm count
	12	Ready production count
	13	Mechanical alarm count
	20	Scheduled downtime count
Weather	7	Mean temperature (2 m)
	11	Maximum temperature (2 m)

In many fields of science, it is difficult to consider models that achieve R^2 values of high magnitude. Since fatigue is a complex issue, finding a comprehensive model with a high R^2 is challenging. However, the methodology and future iterations could provide beneficial insights. The finding that 36% of fatigue events can be explained by shift type, weather and operational data indicates that 64% of the variance can be attributed to factors that we currently are not modeling. Therefore, the next step in fatigue modeling would be exploring additional contributors to operator fatigue. In this study, the mine's data have been aggregated according to shift or hour, but future models could examine fatigue in a more individualistic way. Deeper integration of the data sets upon individual operators could be one way of accomplishing this. Additional factors such as an individual's habits and sleep patterns could also provide another level to the model and would give a more detailed view of the fatigue of the workers.

From the perspective of health and safety management, the most important features found in this study can be considered potential leading indicators (ALIs) to reduce fatigue. The surprising finding of unscheduled equipment downtime events is an aspect that needs to be explored further. Process disruption's impact on fatigue was one finding that was consistent with the study by Drews et al. (2020) [2]. More research from a health and safety perspective is needed to understand why some of the alarm and production variables of different fleets have a greater effect on fatigue. However, fitness for duty could be one reason behind the different fatigue events for different fleets. Mining companies can use these indicators to anticipate increases in fatigue and to potentially mitigate fatigue. These model outcomes can be utilized to implement health and safety policies, training programs and mitigation practices. If mine operations can identify the times and shift types that are more susceptible to fatigue, specific strategies could be implemented, such as mandatory break times for the operators and supervisory support during this time. Management can also train the operators to be more alert at specific times of the day and during specific shifts. They also can train them to be more aware of how fitness can decrease fatigue. The models' output shows that ambient temperature has also significant effects on the fatigue of haul truck drivers. This also must be studied further to understand the degree to which this factor influences specific individuals' fatigue states.

Moreover, the hourly-based model results provide an understanding of the effects of the variables that impact fatigue for health and safety management. It demonstrates that a leading indicator to predict fatigue is the time of day. Therefore, special attention and planning is required for those times with a higher risk of fatigue. All of these outcomes can be considered when prioritizing tasks by health and safety management.

6. Limitations and Future Work

This study shows the application of machine learning in health and safety management using operational data sets of mining operations. The findings of this study confirm that fatigue is caused by a wide variety of factors and many are likely very difficult to quantify, but there may be a small but impactful percentage of factors that can be quantified. Fatigue prediction is a matter of predicting the complex interactions between human behavior and the ever-changing work environments at mines. In the social sciences, it is very common to see situations where a low R^2 value captures relationships that quantify a relatively high amount of variance in a complex relationship [36]. Individual worker data can be added to the model to increase the accuracy of the prediction model, since only operational data and weather data are utilized in these models.

In all of the models developed, the training scores are substantially better than the validation scores. This is most often attributable to overfitting of the model, but in this case, it is likely largely due to the difficultly in generalizing a model that can predict fatigue due to the complex psychological and physiological factors associated with fatigue. This line of research will become more important as the fitness for duty of equipment operators takes on greater significance in scheduling operator work shifts.

Even when using a fairly simple model with a small data set, the best-performing model in this study is able to achieve excellent results. Many refinements were made to the models during this study, but there are many avenues of exploration that could yield even stronger predictive models. Some key areas to explore in future models could include:

- Looking at individual fatigue events instead of the aggregated fatigue events;

- Using a machine learning method that can model more complex relationships, such as a neural network;
- Increasing the size of the training data set—this could be accomplished by adding more data either from the same mine or from another mine;
- Creating common naming conventions between data sets so that they can be linked by location, operator and equipment;
- Adding more complex features such as the sleep pattern, health condition, fitness or diet of the operator;
- Adding features that represent information collected during time periods prior to when the fatigue occurred, such as downtime or production on the previous day;
- Adding some features related to the working schedule of the operator in terms of fatigue at the time and the day or week before;
- Exploring more details of each feature to reduce the number of features that have a lower impact on fatigue.

Author Contributions: Conceptualization, E.T. and W.P.R.; methodology, E.T., W.P.R. and T.M.; software, E.T. and T.M.; validation, W.P.R., E.T. and F.A.D.; formal analysis, E.T.; investigation, E.T.; resources, W.P.R.; data curation, E.T. and T.M.; writing—original draft preparation, E.T.; writing—review and editing, E.T., W.P.R. and F.A.D.; visualization, E.T.; supervision, W.P.R.; project administration, W.P.R.; funding acquisition, W.P.R. All authors have read and agreed to the published version of the manuscript.

Funding: Funding for the project was provided by the National Institute for Occupational Safety and Health (NIOSH) with grant number 75D30119C05500.

Data Availability Statement: The data presented in this study are not publicly available due to confidentiality.

Acknowledgments: We would like to express our gratitude to the National Institute for Occupational Safety and Health (NIOSH) for the funding, as well as the mining company, its management and all of the people supporting this research.

Conflicts of Interest: On behalf of all authors, the corresponding author states that there is no conflict of interest.

Appendix A

The ICE plots of the top ten features from the shift-based model are presented in Figure A1.

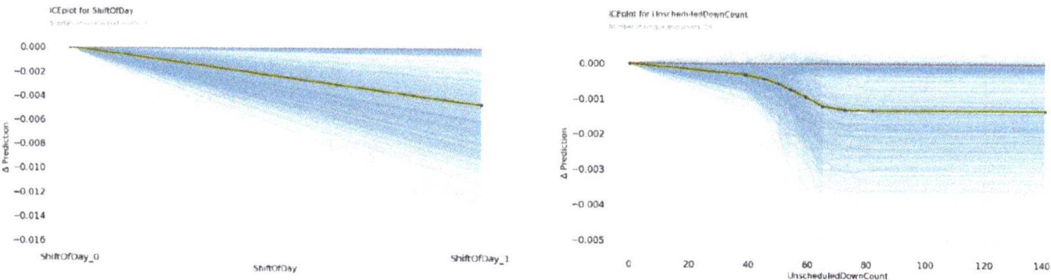

Figure A1. *Cont.*

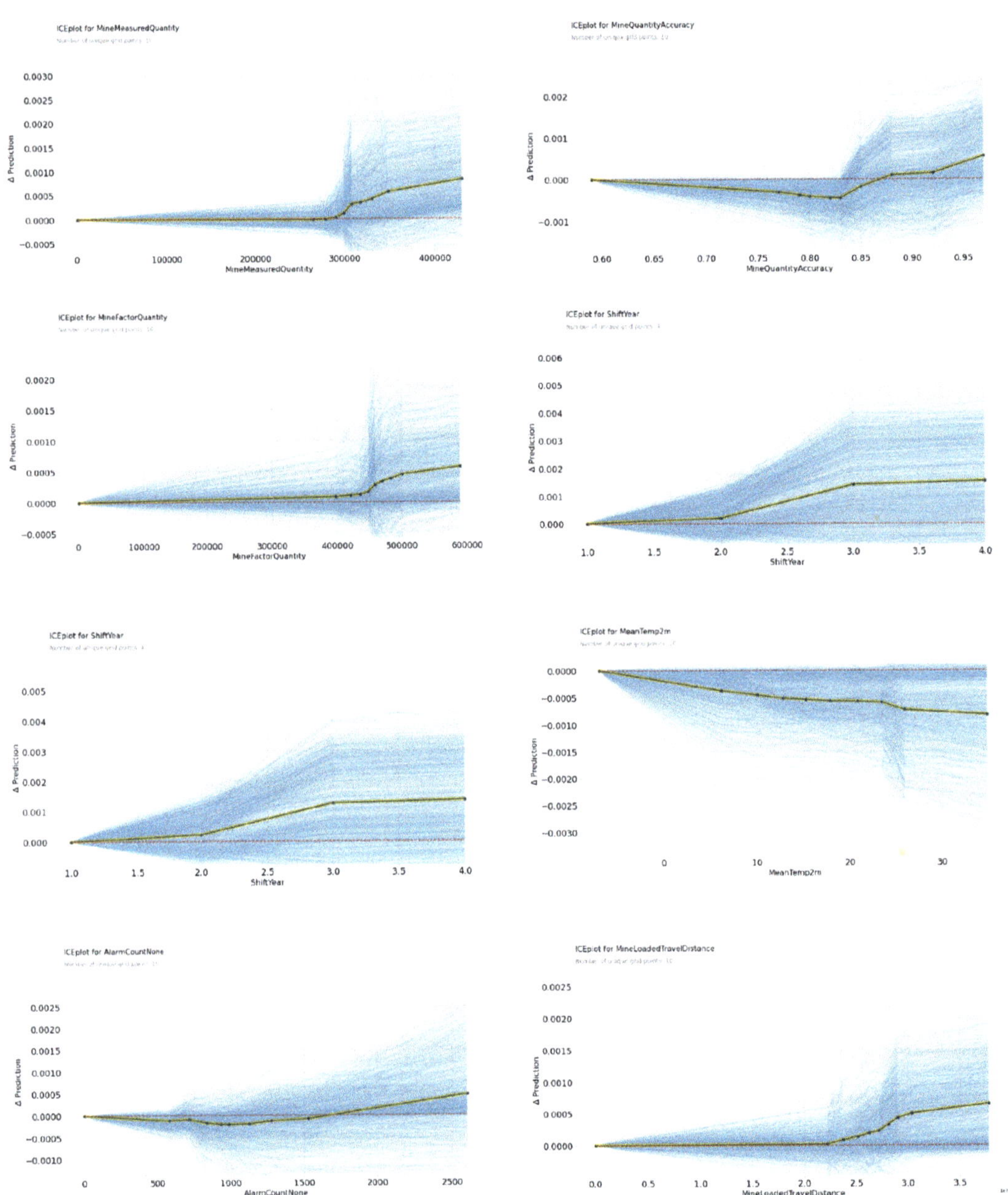

Figure A1. ICE plots of the top ten features (Blue lines identify the dependence of the prediction on a feature for each instance independently, and yellow line is PDP which shows the average of the blue lines).

References

1. Bauerle, T.; Dugdale, Z.; Poplin, G. Mineworker fatigue: A review of what we know and future directions. *Min. Eng.* **2018**, *70*, 1–19.
2. Drews, F.A.; Rogers, W.P.; Talebi, E.; Lee, S. The Experience and Management of Fatigue: A Study of Mine Haulage Operators. *Min. Metall. Explor.* **2020**, *37*, 1837–1846. [CrossRef]
3. Dawson, D.; Searle, A.K.; Paterson, J.L. Look before you (s)leep: Evaluating the use of fatigue detection technologies within a fatigue risk management system for the road transport industry. *Sleep Med. Rev.* **2014**, *18*, 141–152. [CrossRef] [PubMed]
4. Briggs, C.; Nolan, J.; Heiler, K. *Fitness for Duty in the Australian Mining Industry: Emerging Legal and Industrial Issues*; 186487421X; Australian Centre for Industrial Relations Research and Teaching: Sydney, NSW, Australia, 2001; pp. 1–52.
5. Parker, T.W.; Warringham, C. Fitness for work in mining: Not a 'one size fits all' approach. In Proceedings of the Queensland Mining Industry Health & Safety Conference, Brookfield, QLD, Australia, 4–7 August 2004.
6. Hutchinson, B. "Fatigue Management in Mining—Time to Wake up and Act", Optimize Consulting. 2014. Available online: http://www.tmsconsulting.com.au/basic-fatigue-management-in-mining/ (accessed on 25 January 2020).
7. Pelders, J.; Nelson, G. Contributors to Fatigue of Mine Workers in the South African Gold and Platinum Sector. *Saf. Health Work* **2019**, *10*, 188–195. [CrossRef]
8. Cavuoto, L.; Megahed, F. Understanding fatigue and the implications for worker safety. In Proceedings of the ASSE Professional Development Conference and Exposition, Atlanta, GA, USA, 26 June 2016.
9. Talebi, E.; Roghanchi, P.; Abbasi, B. Heat management in mining industry: Personal risk factors, mitigation practices, and industry actions. In Proceedings of the 17th North American Mine Ventilation Symposium, Montreal, QC, Canada, 28 April–1 May 2019.
10. Talebi, E.; Sunkpal, M.; Sharizadeh, T.; Roghanchi, P. The Effects of Clothing Insulation and Acclimation on the Thermal Comfort of Underground Mine Workers. *Min. Metall. Explor.* **2020**, *37*, 1827–1836. [CrossRef]
11. Yung, M. Fatigue at the Workplace: Measurement and Temporal Development. Ph.D. Thesis, University of Waterloo, Waterloo, ON, Canada, 2016.
12. Tufano, P. Who manages risk? An empirical examination of risk management practices in the gold mining company. *J. Financ.* **1996**, *51*, 1097–1137. [CrossRef]
13. Wegerich, S.W.; Wolosewicz, A.; Xu, X.; Herzog, J.P.; Pipke, R.M. Automated Model Configuration and Deployment System for Equipment Health Monitoring. U.S. Patent No. 7,640,145 B2, 29 December 2009.
14. Hinze, J.; Thurman, S.; Wehle, A. Leading indicators of construction safety performance. *Saf. Sci.* **2013**, *51*, 23–28. [CrossRef]
15. Poh, C.Q.X.; Ubeynarayana, C.U.; Goh, Y.M. Safety leading indicators for construction sites: A machine learning approach. *Autom. Constr.* **2018**, *93*, 375–386. [CrossRef]
16. Hallowell, M.R.; Hinze, J.W.; Baud, K.C.; Wehle, A. Proactive construction safety control: Measuring, monitoring, and responding to safety leading indicators. *J. Constr. Eng. Manag.* **2013**, *139*, 1–8. [CrossRef]
17. Guo, B.H.W.; Yiu, T.W. Developing Leading Indicators to Monitor the Safety Conditions of Construction Projects. *J. Manag. Eng.* **2016**, *32*, 1–14. [CrossRef]
18. Toellner, J. Improving Safety & Health Performance: Identifying & Measuring Leading Indicators. *Prof. Saf. J.* **2011**, *46*, 42–47.
19. Grabowski, M.; Ayyalasomayajula, P.; Merrick, J.; McCafferty, D. Accident precursors and safety nets: Leading indicators of tanker operations safety. *Marit. Policy Manag.* **2007**, *34*, 405–425. [CrossRef]
20. Hopkins, A. Thinking About Process Safety Indicators. In Proceedings of the Oil and Gas Industry Conference, Manchester, UK, 4–6 November 2007.
21. Hale, A. Why safety performance indicators? *Saf. Sci.* **2009**, *47*, 479–480. [CrossRef]
22. Costin, A.; Wehle, A.; Adibfar, A. Leading indicators—a conceptual IoT-based framework to produce active leading indicators for construction safety. *Safety* **2019**, *5*, 86. [CrossRef]
23. Kononenko, I.; Kukar, M. *Machine Learning and Data Mining*; Woodhead Publishing: Cambridge, UK, 2007; pp. 1–36.
24. Thibaud, M.; Chi, H.; Zhou, W.; Piramuthu, S. Internet of Things (IoT) in high-risk Environment, Health and Safety (EHS) industries: A comprehensive review. *Decis. Support Syst.* **2018**, *108*, 79–95. [CrossRef]
25. Molaei, F.; Rahimi, E.; Siavoshi, H.; Afrouz, S.G.; Tenorio, V. A Comprehensive Review on Internet of Things (IoT) and its Implications in the Mining Industry. *Am. J. Eng. Appl. Sci.* **2020**, *13*, 499–515. [CrossRef]
26. Shahmoradi, J.; Talebi, E.; Roghanchi, P.; Hassanalian, M. A comprehensive review of applications of drone technology in the mining industry. *Drones* **2020**, *4*, 34. [CrossRef]
27. Tixier, A.J.P.; Hallowell, M.R.; Rajagopalan, B.; Bowman, D. Application of machine learning to construction injury prediction. *Autom. Constr.* **2016**, *69*, 102–114. [CrossRef]
28. Lingard, H.; Hallowell, M.; Salas, R.; Pirzadeh, P. Leading or lagging? Temporal analysis of safety indicators on a large infrastructure construction project. *Saf. Sci.* **2017**, *91*, 206–220. [CrossRef]
29. Cheng, T. When Artificial Intelligence Meets the Construction Industry. Available online: https://www.gxcontractor.com/equipment/article/13031944/when-artificial-intelligence-meets-the-construction-industry (accessed on 25 May 2018).
30. Liaw, A.; Wiener, M. Classification and regression by random Forest. *R News* **2002**, *2*, 18–22.
31. Breiman, L. Random Forests. *Mach. Learn.* **2001**, *45*, 5–32. [CrossRef]
32. Segal, M.R. *Machine Learning Benchmarks and Random Forest Regression*; Technical Report for Center for Bioinformatics and Molecular Biostatistics; University of California: San Francisco, CA, USA, April 2003.

33. Heaton, J.; McElwee, S.; Fraley, J.; Cannady, J. Early stabilizing feature importance for TensorFlow deep neural networks. In Proceedings of the 2017 International Joint Conference on Neural Networks (IJCNN), Anchorage, AK, USA, 14–19 May 2017; pp. 4618–4624.
34. Menze, B.H.; Kelm, B.M.; Masuch, R.; Himmelreich, U.; Bachert, P.; Petrich, W.; Hamprecht, F.A. A comparison of random forest and its Gini importance with standard chemometric methods for the feature selection and classification of spectral data. *BMC Bioinform.* **2009**, *10*, 213. [CrossRef]
35. Molnar, C. Interpretable Machine Learning, A Guide for Making Black Box Models Explainable. Available online: https://christophm.github.io/interpretable-ml-book/ (accessed on 25 March 2021).
36. Minitab Blog Editor. Regression Analysis: How Do I Interpret R-Squared and Assess the Goodness-of-Fit? Available online: https://blog.minitab.com/en/adventures-in-statistics-2/regression-analysis-how-do-i-interpret-r-squared-and-assess-the-goodness-of-fit (accessed on 20 February 2021).

Article

Effectiveness of Natural Language Processing Based Machine Learning in Analyzing Incident Narratives at a Mine

Rajive Ganguli [1,*], Preston Miller [2] and Rambabu Pothina [1]

1 Department of Mining Engineering, University of Utah, Salt Lake City, UT 84112, USA; rambabu.pothina@utah.edu
2 Teck Red Dog Operations, Anchorage, AK 99503, USA; Preston.Miller@teck.com
* Correspondence: rajive.ganguli@utah.edu

Abstract: To achieve the goal of preventing serious injuries and fatalities, it is important for a mine site to analyze site specific mine safety data. The advances in natural language processing (NLP) create an opportunity to develop machine learning (ML) tools to automate analysis of mine health and safety management systems (HSMS) data without requiring experts at every mine site. As a demonstration, nine random forest (RF) models were developed to classify narratives from the Mine Safety and Health Administration (MSHA) database into nine accident types. MSHA accident categories are quite descriptive and are, thus, a proxy for high level understanding of the incidents. A single model developed to classify narratives into a single category was more effective than a single model that classified narratives into different categories. The developed models were then applied to narratives taken from a mine HSMS (non-MSHA), to classify them into MSHA accident categories. About two thirds of the non-MSHA narratives were automatically classified by the RF models. The automatically classified narratives were then evaluated manually. The evaluation showed an accuracy of 96% for automated classifications. The near perfect classification of non-MSHA narratives by MSHA based machine learning models demonstrates that NLP can be a powerful tool to analyze HSMS data.

Keywords: mine safety and health; accidents; narratives; machine learning; natural language processing; random forest classification

Citation: Ganguli, R.; Miller, P.; Pothina, R. Effectiveness of Natural Language Processing Based Machine Learning in Analyzing Incident Narratives at a Mine. *Minerals* **2021**, *11*, 776. https://doi.org/10.3390/min11070776

Academic Editor: Yosoon Choi

Received: 22 May 2021
Accepted: 14 July 2021
Published: 17 July 2021

Publisher's Note: MDPI stays neutral with regard to jurisdictional claims in published maps and institutional affiliations.

1. Introduction

Workers' health and safety is of utmost priority for the sustainability of any industry. Unfortunately, occupational accidents are still reported in high numbers globally. According to the recent estimates published by the International Labour Organization (ILO), 2.78 million workers die from occupational accidents and diseases worldwide [1]. In addition, 374 million workers suffer from non-fatal accidents, and lost work days represent approximately 4% of the world's gross domestic product [2,3]. It is, therefore, not surprising that researchers are constantly investigating factors that impact safety [4,5], or finding innovations and technology to improve safety [6,7].

As to the U.S. mining industry, for years 2016–2019, the National Institute for Occupational Safety and Health (NIOSH), a division of the US Centers for Disease Control and Prevention (CDC) reports 105 fatal accidents and 15,803 non-fatal lost-time injuries [8]. To bring down the rate of serious injuries and fatalities, the industry analyzes incident reports to conduct root cause analysis and identify leading indicators. Unfortunately, as noted by the International Council on Mining and Metals, a global organization of some of the largest mining companies of the world, the vast trove of incident data is not analyzed as much as it could be due to lack of analytics expertise at mine sites [9]. With the advances in natural language processing (NLP), there is now an opportunity to create NLP-based tools to process and analyze such textual data without requiring human experts at the mine site.

Natural language processing (NLP) has been explored as a tool to analyze safety reports since the 1990s [10,11]. This paper, intended for a mining industry audience, presents in this section, a brief history of NLP and its use in analyzing safety reports. NLP is the automated ability to extract useful information out of written or spoken words of a language. Exploring its application to safety is logical, as safety reports are valuable information. If causation and associated details can be automatically extracted from the safety reports, NLP can be used to quickly gain insight into safety incidents from historical reports that are filed away in the safety management databases. Additionally, with smartphone-based work site observations apps becoming popular, NLP tools can be useful in providing real time insights as incidents and observations are reported in real time. For example, in a confidential project, one of the authors of this paper advised an industrial site about a hazardous practice at the operation using an NLP analysis of data collected using a smartphone-based application. This hazard became apparent after evaluating the data because several employees had noted the practice in their worksite observations.

The efforts to apply NLP to extract causation from safety reports received a major boost when the Pacific Northwest National Laboratory (PNNL) put together a large team in the early 2000s to apply NLP and analyze aviation safety reports from the National Aeronautics and Space Administration's (NASA) aviation safety program [12]. The "meaning" of a sentence depends not just on the words, but also on the context. Therefore, PNNL used a variety of human experts to develop algorithms to extract human performance factors (HPF) from report narratives. HPF definitions were adopted from NASA [13]. The PNNL approach consisted of artificial intelligence (AI) after the text was preprocessed using linguistic rules. The linguistic rules, developed by human experts, considered specific phrases and sentence structures common in aviation reports. When automated, these rules were able to identify causes of safety incidents on par with human experts. The PNNL team, however, noted the reliance of the algorithms on human experts with domain-specific knowledge.

New developments have reduced human involvement in text analysis [14]. These developments include identifying linguistic features such as parts of speech, word dependencies, and lemmas. A million-sentence database (or "corpus" to use NLP terminology) may only contain 50,000 unique words once words such as 'buy' and 'bought' (one is a lemma of the other) are compressed into one; though that is also a choice for the human expert. After vectorization, each sentence in the database is a vector of length 50,000, with most elements being zero (a twelve-word sentence will only have ones in twelve places). When the relative order of words in a sentence is taken into account, common phrases can be identified easily. Thus, after preprocessing with NLP techniques, classical statistics and machine learning techniques can be applied to classify text. Baker et al., 2020 [15] used a variety of NLP and machine learning techniques to classify incident reports and predict safety outcomes in the construction industry. Tixier et al., 2016 developed a rule based NLP algorithm that depends on a library of accident related keywords to extract precursors and outcomes from unstructured injury reports in the construction industry [16]. In another study that was conducted on narratives from Aviation Safety Reporting System (ASRS), NLP-based text preprocessing techniques along with k-means clustering classification were used to identify various safety events of interest [17]. Baillargeon et al., 2021 [18] used NLP and machine learning techniques to extract features of importance to the insurance industry from public domain highway accident data. In an analysis conducted on infraction history of certain mine categories, ML-based classification and regression tree (CART) and random forest (RF) models were used on Mine Safety and Health Administration (MSHA) database narratives in predicting the likely occurrence of serious injuries in near future (the following 12-month period) [19].

The application of NLP-based machine learning to mining industry safety data is relatively new. Yedla et al., 2020 [20] used the public domain (MSHA) database to test the utility of narratives in predicting accident attributes. They found that vectorized forms of narratives could improve the predictability of factors such as days away from work.

Other researchers used NLP to analyze fatality reports in the MSHA database [21]. Using co-occurrence matrices for key phrases, they were able to identify some of the common causes of accidents for specific equipment.

2. Importance of this Paper

In safety-related research, it is typical to demonstrate NLP and machine learning capabilities on public domain databases. Models are first developed on a public domain database, after which its capabilities are demonstrated on an independent subset of the same database. Since modeling and subsequent demonstration of model capabilities happen on the same dataset, there is no certainty that these approaches or models would be effective on databases created by other sources. For example, every entry in an MSHA database is made by a federal employee. Would a federal employee describe an incident the same way as a mining company employee? If yes, then there exists a specific language for mine safety that is shared by safety professionals. This 'language', if it exists, can be leveraged to make NLP-based machine learning of mine safety data very effective.

This paper advances the use and application of NLP to analyze mine safety incident reports by demonstrating that machine learning models developed on public domain mine safety databases can be applied effectively on private sector safety datasets. Therefore, it demonstrates that there is a language of safety that spans organizations. Furthermore, this paper identifies key attributes of specific categories of incidents. This knowledge can be used to improve algorithms and/or understand their performance.

More generally, the paper advances the field of mine safety research. Currently, data-mining-based mine safety researchers focus only on categorical or numerical data. Therefore, gained insights are limited to statistical characterization of data (such as average age, or work experience) or models based on these data [4]. If narratives are available with incident data (as they often are), this paper will encourage researchers to evaluate them to glean more insights into the underlying causes.

3. Research Methodology

3.1. MSHA Accident Database

The MSHA accident database [22] has 57 fields used to describe safety incidents including meta-data (mine identification, date of incident, etc.), narrative description of the incident, and various attributes of the incidents. Some of the data is categorical such as body part injured and accident type. More than eighty-one thousand (81,298) records spanning the years 2011 to early 2021 were used in this research. Any operating mine in the United States that had a reportable injury is in the database. Thus, the database reflects many types of mines, jobs, and accidents.

Accidents are classified in the database as belonging to one of 45 accident types. Examples include "Absorption of radiations, caustics, toxic and noxious substances", "Caught in, under or between a moving and a stationary object", and "Over-exertion in wielding or throwing objects". Looking at these definitions, it appears that MSHA defined them to almost answer the question "What happened?" Thus, the category is simply the high level human summary of the narrative, i.e., the category is the "meaning" of the narrative. In this paper, the MSHA accident type is considered a proxy for the meaning of the narrative. Narratives are typically five sentences or less.

3.2. Random Forest Classifier

The random forest (RF) technique was used to classify the narratives based on accident types. Random forests are simply a group of decision trees. Though described here briefly, those unfamiliar with decision trees are referred to Mitchell, 1997 [23], a good textbook on the topic and the source for the description below. A decision tree is essentially a series of yes or no questions applied to a particular column ("feature") of the input data. The decision from the question (for example, miner experience > 10, where miner experience is a feature in the data set) segments the data. Each question is, thus, a "boundary" splitting the data into two subsets of different sizes. The segmented data may be further segmented by applying another boundary, though the next boundary may be on another feature. Applying several boundaries one after the other results in numerous small subsets of data, with data between boundaries ideally belonging to a single category. The maximum number of decision trees applied in the longest pathway is called the "tree depth". The method works by applying the sequence of boundaries to a sample, with the final boundary determining its class. Note that while one boundary (also called "node") makes the final decision on the class for one sample, some other boundary may make the decision for another sample. It all depends on the path taken by a particular sample as it travels through the tree. When the final boundary does not result in a unanimous class, the most popular class in the subset is used as the final decision of the class.

Boundaries are set to minimize the error on either side of the boundaries. The combination of a given data set and given boundary criteria will always result in a specific tree. In an RF, a decision tree is formed by randomly selecting (with replacement) the data. Thus, while a traditional decision tree will use the entire modeling subset for forming the tree, a decision tree in an RF will use the same amount of data, but with some samples occurring multiple times, and some not occurring at all. Thus, the same data set can yield multiple trees. In the RF technique, multiple trees formed with a random selection of data are used to classify the data. One can then use any method of choice to combine predictions from the different trees. This method of using a group of trees is superior to using a single decision tree.

In this paper, an RF classifier was applied to model the relationship between a narrative and its accident type. A non-MSHA database would contain narratives, but not any of the other fields populated by MSHA staff. Since the goal of the project is to test it on non-MSHA data, no other field in the database was used to strengthen the model. Half of the records were randomly selected to develop the model. It was tested on the remaining half of the records to evaluate its performance on the MSHA data. In the final step, the model was tested on non-MSHA data. There is no standard for what proportion of data to use for training and testing subsets, though it is expected that the subsets be similar [24]. A 50–50 split is a common practice [25,26]. RF models were developed using the function RandomForestClassifier () in the SCIKIT-LEARN [27] toolkit. As is common practice in machine learning [28], the authors did not code the RF but used a popular tool instead.

Modeling starts by making a list of non-trivial words in the narratives. As is typical in NLP, the narratives were pre-processed before the list of non-trivial words is made. Pre-processing consisted of:

- Changing case to lower case.
- Removal of specific words: This consisted of the removal of acronyms common in MSHA databases, and a custom list of "stop words". Stop words are words such as stray characters, punctuation marks, and common words that may not add value. These are available from several toolkits. The stop words list available from NLTK [29] was modified and used in this paper.
- Lemmatizing: This was done using the lemmatizer in the spacy [30] toolkit. Lemmatizing is the grouping of similar words, or rather, identifying the foundational word. This is done so that related words are not considered separately. For example, consider the two sentences, "He was pushing a cart when he got hurt" and "He got hurt as he pushed a cart". The lemmatizer would provide "push" as a lemma for both pushing and pushed, and push would replace pushed and pushing in the narrative.

The combined length of all narratives was 1.72 million words, consisting of 31,995 unique words or "features". The list of unique features is called the vocabulary. The input data set is then prepared by selecting the top 300 most frequently occurring words ("max features"). Essentially, the vocabulary is cut from its full length to just the words occurring most frequently. These words are used to vectorize each narrative such that each narrative is represented as a vector of size 300. The value at a given location in the vector would represent the number of occurrences of that word in that narrative. The top 5 words were: fall, right, left, back, and cause.

The output for the narrative consisted of a 1 or a 0, indicating whether it belonged ("1") to a particular category of accident or not ("0"). "Max features" is a parameter in RF modeling, and was set to 300 after trial and error exercises. Similarly, the number of trees ("n_estimators") was set to 100. Another parameter is "max_depth" (maximum depth of tree). This parameter was not set. Whenever a parameter is not specified, the tool uses default values. In the default setting for tree depth, data is continually segmented till the final group is all from the same class. According to the user guide of the tool, the main parameters are the number of trees, and max features. The rest of the parameters were not set, i.e., default values were used. The interested reader can visit the provided links for technical details about the toolkits in the footnotes, including the default values. The tool combines the outputs of the various trees by averaging them to obtain the final classification.

Among the 45 accident types are some whose names start with the same phrase. For example, there are four over-exertion (OE) types, all of which start with the phrase over-exertion. They are (verbatim): Over-exertion in lifting objects, over-exertion in pulling or pushing objects, over-exertion in wielding or throwing objects, and over-exertion NEC. Accident categories whose names begin with the same phrase are considered to belong to the same "type group", with the phrase defining the grouping.

NEC stands for "not elsewhere classified," and is used within some type groups. When it exists, it is often the largest sub-group as it is for everything that is not easily defined. There are 11 types that start with "Fall", including two that start with "Fall to". Five types start with "Caught in". Six start with "Struck by". These accident type groups contain 26 of the 45 accident types, but 86% of all incidents (35,170 out of 81,298). Table 1 shows the four type groups that were modeled in this paper. Separate models were developed for some of the sub-groups to get an understanding of these narrowly defined accidents. These were:

- Over-exertion in lifting objects (OEL).
- Over-exertion in pulling or pushing objects (OEP).
- Fall to the walkway or working surface (FWW).
- Caught in, under or between a moving and a stationary object (CIMS), and
- Struck by flying object (SFO).

Table 1. The four type groups of accidents modeled in the paper.

Type Group: Caught in	Type Group: Fall	Type Group: Over-Exertion	Type Group: Struck
Caught in, under, or between a moving and a stationary object	Fall down raise, shaft or manway	Over-exertion in lifting objects	Struck by concussion
Caught in, under, or between collapsing material or buildings	Fall down stairs	Over-exertion in pulling or pushing objects	Struck by falling object
Caught in, under, or between NEC	Fall from headframe, derrick, or tower	Over-exertion in wielding or throwing objects	Struck by flying object
Caught in, under, or between running or meshing objects	Fall from ladders	Over-exertion NEC	Struck by powered moving object
Caught in, under, or between two or more moving objects	Fall from machine		Struck by rolling or sliding object
	Fall from piled material		Struck by... NEC
	Fall from scaffolds, walkways, platforms		
	Fall on same level, NEC		
	Fall onto or against objects		
	Fall to lower level, NEC		
	Fall to the walkway or working surface		

Thus, a total of nine RF models were developed; four for the four type groups, and five for the specific types. Table 2 shows the characterization of the training and testing subsets that went into developing the models. It is apparent that each category was represented about the same in the two subsets.

Table 2. Various accident categories in the training and testing subsets. Each subset has 40,649 samples.

Subset	Type Group: OE	Type Group: Caught in	Type Group: Struck by	Type Group: Fall	OEP	OEL	FWW	CIMS	SFO
Training	8909	4563	10,216	4802	1290	2838	2130	3337	1586
Testing	8979	4524	10,226	4926	1275	2961	2130	3310	1590

In classification exercises, it is common to develop a single model to classify a data set into multiple categories, rather than develop models for each category individually. The reason for developing nine models instead of one is discussed in the next section.

4. Results

4.1. Performance within MSHA Data

Table 3 shows a summary of the modeling within the MSHA test set. To understand the table, consider the OE type group. Of the 40,649 records in the test set, 8979 records were from this type. The success of an RF model can be determined by identifying the OE type as OE type and/or by classifying a non-OE type (31,670 records) as not belonging to OE. This is shown below through a simple computation.

Table 3. Results of RF models in the MSHA test set.

Metrics	Type Group: OE	Type Group: Caught in	Type Group: Struck by	Type Group: Fall	OEP	OEL	FWW	CIMS	SFO
Records from Category	8979	4524	10,226	4926	1275	2961	2130	3310	1590
Overall Success % from Category	92%	96%	90%	95%	98%	96%	96%	95%	97%
Accurately Predicted	81%	71%	75%	71%	37%	59%	34%	55%	25%
False Positive	4%	1%	5%	2%	<1%	<1%	<1%	2%	<1%

- Total samples (n_samples): 40,649
- Total samples in target category (n_target): 8979
- Total samples in other categories (n_other): n_samples − n_target = 31,670
- Samples from target category predicted accurately (n_target_accurate): 7248
- Samples from other category predicted wrongly as target (false_predicts): 1331
- Samples from other category predicted correctly as other (other_accurate): 31,670 − 1331 = 30,339
- Percentage of targets accurately predicted: 100 × n_target_accurate/n_target = 100 × 7248/8979 = 81%
- False positive rate: false_predicts/n_other = 1331/31,670 = 4%
- Total correct predictions (total_correct): n_target_accurate + other_accurate = 7248 + 30,339 = 37,587
- Overall success rate (%) = 100 × total_correct/_samples = 100 × 37,587/40,649 = 92%

The overall success was 92%, i.e., a very high proportion of narratives were classified correctly as belonging to OE type group, or as not belonging to OE type group. Though it is an indicator of overall success, this type of evaluation is not particularly useful, as classifying a narrative as "not belonging to OE" is not helpful to the user. It is more useful to look at how successful RFs were in correctly identifying narratives from the accident type in question (OE type group in this example). As shown in the table and in the example computation, 81% of these 8918 (7248) were accurately identified. The false positive rate was 4%, i.e., 1331 of the 31,670 non-OE records were identified as OE. The low positive rate implies that if a narrative was classified as belonging to the OE type group, it was highly likely to belong to that type. The success in the other type groups was lower, and ranged from 71% to 75%, with false positives ranging from 1% to 5%. Thus, one could expect RF to accurately identify about 75% of the narratives in the MSHA database from the four type groups, with a good false positive rate.

The success rate takes a dramatic downturn with the individual models. Only 25% to 59% of narratives belonging to the individual types are correctly classified though with a negligible false positive rate. The negligible false positive implies that when the model classifies the narrative as belonging to a specific category, it is almost guaranteed to be in that category. The low number of records in the individual categories is one part of the explanation of the poor performance, as models would be less powerful if they are trained on fewer records. For example, only about 3% of the records were from the OEP category. This means that 97% of the data seen by the OEP model was not relevant to identifying OEP. An additional explanation is obtained from trigram analysis of the narratives that belong to these accident types. Trigrams explore the sets of three words that occur consecutively the most. Trigram analysis was conducted using the NLTK collocations toolkit.

Table 4 shows the tri-word sequences that occur the most frequently in the OE accident types. They are listed in order of frequency. The overlap between the tri-words is immediately apparent. Back, shoulders, knee, abdomen, and groin are injured most in these types of accidents. The overlap between OEP and OEL would cause accidents to be misclassified as belonging to the other category. This issue is also evident in the Fall accident types (Table 5), where losing balance, slipping, and falling seem to be the major attributes. Even the two types "Caught in" and "Struck by" have some overlap (Table 6). Caught in makes it apparent that it is the fingers that are predominantly injured in this type of accident. SFO highlights that eyes and safety glasses are impacted when someone is struck by a flying object.

Table 4. Results of trigram analysis on OE accident types.

Type Group: OE	OE Lifting	OE Pulling
feel pain back	feel pain back	feel pain back
pain low back	pain low back	feel pain shoulder
feel pain low	feel pain low	feel pain right
feel pain right	feel low back	feel pain low
feel pain shoulder	feel pain shoulder	feel pain left
feel pain left	feel pain right	feel pain groin
feel pain knee	feel pain left	feel pain abdomen

Table 5. Results of trigram analysis on Fall accident types.

Fall	FWW
lose balance fall	lose balance fall
slip fall ground	slip fall right
cause lose balance	slip fall left
foot slip fall	slip fall ground
slip fall backward	cause lose balance
step lose balance	place restrict duty
lose balance cause	slip fall ice

Table 6. Results of trigram analysis on Caught in and Struck by accident types.

Caught in …	CIMS	Struck by …	SFO
right index finger	right index finger	piece rock fell	wear safety glass
left index finger	left index finger	rock fall strike	safety glass eye
right middle finger	left ring finger	cause laceration require	eye safety glass
left ring finger	right middle finger	left index finder	behind safety glass
right ring finger	right ring finger	strike left hand	go safety glass
left middle finger	pinch index finger	right index finger	safety glass face
pinch index finger	left middle finger	wear safety glasses	safety glass left

The success rate for classification was dramatically lower when a single RF model was developed to classify the narratives into separate categories. OEP, OEL, FWW, CIMS, SFO had success rates of only 23%, 33%, 19%, 29%, and 17% respectively compared to 37%, 59%, 34%, 55%, 25% respectively. Multiple models for multiple categories would require that multiple models be applied to the same data, resulting in multiple predictions of category. It would be possible then for a particular narrative to be categorized differently by the different models. In such situations, one could determine the similarity between the narrative and the narratives from the multiple categories in the training set to resolve the conflicting classifications. The features (words) of the category within the training set are the foundation behind the model for the category. For example, the words in the "Struck by" category in the training set play a key role in what RF trees are formed in the "Struck by" model. Thus, when a test narrative is classified as "Struck by" by one model, and "Caught in" by another, one could find the similarity between words in the test narrative, and the words in the two categories of the training data, "Struck by" and "Caught in", to resolve the conflict. This is demonstrated in the next section.

4.2. Performance on Non-MSHA Data

The nine RF models were applied to data from a surface metallic mine in the United States that partnered in this project. The data consisted of narratives that described various safety incidents. Injury severity ranged from very minor incidents to lost time accidents. Narratives were typically longer than MSHA narratives (about twice the length), and formats were sometimes different (such as using a bulleted list). They usually had more details about the incident. The narratives were written by a staff member from the safety department. Narratives from the 119 unique incidents logged in 2019 and 2020 were analyzed. Some narratives were duplicated in the database. Duplicates of narratives were ignored. Each model was applied to the 119 narratives separately.

The RF models classified 76 out of the 119 narratives (Table 7) with a high degree of success. 17 narratives were classified by multiple models, but not misclassified (explained later). Forty-three (43) narratives were ignored by all nine models, i.e., they were not classified as belonging to a particular category. The classifications were manually evaluated by the authors to see if they would match the MSHA Accident Types. In many cases, the MSHA database contained an accident that was not only similar to the narrative being manually evaluated but was also classified into the same accident type as the narrative in question. Therefore, the manual validation was easy. A narrative was deemed as accurately classified if it was also classified as such by the authors. The 43 narratives that were not classified by any of the nine models could possibly belong to one of the 19 MSHA accident types not modeled in this paper. The overall success rate was 96%.

Table 7. Performance of RF models on non-MSHA data.

Metrics	OE	OEP	OEL	Fall	FWW	Caught in	CIMS	Struck by	SFO	Overall
Number	26	1	4	14	3	9	7	27	2	93
Validation	85%	100%	100%	100%	100%	100%	100%	100%	100%	96%

The OE category is quite broad and, therefore, one would expect some narratives to be wrongly classified as OE. Therefore, it is not surprising that 4 out of the 26 classified as OE did not belong in that category. One narrative involved an employee who had a pre-existing soreness in the wrist. The 'incident' was simply the employee reporting to the clinic. Two incidents involved employees backing into or walking into a wall or object while working. The fourth incident involved chafing of the calves from new boots. Some of these incidents would perhaps have been also classified differently had models been developed for the other accident types.

Table 8 shows examples of some of the narratives and the automated classifications. Examples are shown for the narrowest categories as they would normally be the most challenging to identify. Table 9 shows how the overlapping occurred in the 17 narratives. Three narratives were classified as both Fall and FWW, while seven were categorized as both "Caught in" and CIMS. Since nine models were used in parallel, it was possible for each narrative to be categorized into nine different categories. Yet, no narrative was categorized as belonging to three or more different categories. Except for one, these overlaps should be expected. For example, OEL is a subset of OE. Therefore, a narrative classified as OEL by the OEL model is expected to be also classified as OE by the OE model. The overlap between a type group and one of its sub-type is a confirmation that models are working properly. It is good that there was no overlap between OEL and OEP. The overlap between "Caught in" and "Struck by" was surprising as they are different categories. The narrative that was classified as both "Caught in" and "Struck by" is (verbatim): "while installing a new motor/pump assy. using portable a cherry picker, the cherry picker tipped over and the assembly caught the employee leg and ankle between the piping and the motor assembly." Tools and equipment that tip over and cause injury have been reported in the "Struck by" category in the MSHA database. A limb caught in between two objects is reported in the "Caught in" category in the MSHA database. Thus, the RF models were correct in their classification of the narrative. However, the overlap in classification presents a good opportunity to demonstrate how one could use "similarity scores" to resolve the overlap. The steps of the process, to resolve conflicting classifications of "Caught in" and "Struck by" are:

1. Consider the non-trivial words in the problem narrative: "instal new motor/pump assy.use portable cherry picker cherry picker tip assembly catch leg ankle piping motor assembly". This list of non-trivial words was obtained after pre-processing. Note that "instal" is not a typo but a product of the lemmatizer.

2. Consider the word frequencies of the training set when the accident category was "Caught in". There were 4894 unique words in the 4563 narratives from that category. The top 5 words were finger (0.036), hand (0.021), right (0.015), pinch (0.0148), and catch (0.0143) with the number in parenthesis indicating the proportion of times the word occurred within that category of narratives.

3. Similarly, consider the list of words in the "Struck by" category. There were 7758 unique words in the 10,216 narratives. The top 5 words were strike (0.019), left (0.014), right (0.014), cut (0.013), and fall (0.012).

4. Now obtain the similarity score between the narrative and a category by weighing each word of the narrative by the proportion of occurrence within the category. This makes sense as the frequency of occurrence of a word in a category is an indicator of its importance to the category. For example, if "leg" gets "Caught in" less frequently than "Struck by", it will occur in lower proportion in "Caught in" than in "Struck by". The words in the "Struck by" list occurred 16 times in the narrative for a total similarity score of 0.0168. There are 13 unique words in the 16 occurrences. The top 3 contributors were "leg", "/" and "install" with scores of 0.004, 0.0027, and 0.0023 for each occurrence in the narrative.

5. Similarly, obtain the total similarity score for all the other categories. For "Caught in", the score is 0.0338. The top 3 contributors in the narrative were "catch" (0.014), "tip" (0.0045), and "install". It is insightful to note how much more "catch" contributed as

a top word than "leg" did as a top word. Clearly, "catch" is a bigger determiner of "Caught in" than leg is of "Struck by".

6. The decision as to which category the narrative belongs is the one with the highest similarity score. In this case, the narrative is deemed to be of the category "Caught in".

Table 8. Examples from the partner mine HSMS, and the automated classifications. Narratives are shown verbatim, but some text has been deleted (identified by . . .) to not disclose sensitive information.

Accident Type	Narrative
OEP	Employee pulled a heavy bag with helper and felt sharp pain in mid back area
OEL	. . . Employee strained lumbar back while carrying a portable generator...
FWW	The operator . . . . began the pre-shift walk around, but did not notice the slick ground conditions. The operator was not wearing any type of traction device, and slipped and landed on their side/back.
SFO	.. While doing so a small piece of shrapnel from shank guard struck mechanic in the left inner thigh and was lodged into skin . . .
CIMS	While moving a turbo charger rotor, employee pinched finger between the rotor shaft and the crate . . .

Table 9. Counts of overlapping accident types.

Overlapping Types	Count
Fall, FWW	3
Caught in, Struck by	1
OEL, OE	3
OEP, OE	1
Struck by, SFO	2
Caught in, CIMS	7

5. Discussion

Two thirds of the narratives in the partner database could be successfully classified (96% accuracy) without any human intervention. The narratives that are not automatically classified could belong to categories not modeled in this paper. At this time, they were not manually analyzed to determine their nature. The nearly absent overlap in predictions for distinct accident types is encouraging as that allows the multiple-model-for-multiple-category approach to work. That is further strengthened by the low false positive rates for the distinct categories, i.e., when a particular model for a distinct category (say OEP) claims that a narrative belongs to that category, the classification is most likely valid. The similarity score approach is presented to resolve cases where a narrative is classified into multiple categories due to the use of multiple models.

The classifications done in the paper were not an empty computational exercise thanks to how MSHA classified the accidents. An increase in narratives being classified as SFO would tell management that foreign matter was entering the eyes of their employees. This is the same as humans reading the narratives, understanding them, and reaching that conclusion. Thus, in some sense, the RF models picked up what the narratives "meant". The high classification success rate also meant that there were specific ways safety professionals describe incidents and that NLP tools can extract that language.

These tools have excellent applicability to help the mining industry reach the industry goal of preventing serious injury and fatalities. On noting an increase in SFO classifications, management can deploy eye protection related interventions. An increase in OEL incidents could result in more training about safe lifting. The safety "department" in most mines means a single person with no mandate or expertise to analyze data. These types of tools can assist mines to analyze data without human intervention. As mines deploy smartphone-based apps to collect employee reports on worksites, the volume of information will explode. However, these tools will help mines process that data and identify hazards before they become incidents.

The detection rate for the narrowest of categories needs to be improved. Improving this would be the most logical next step for this research. A reason why NLP tools were not always effective may be how incidents are described in the narratives. A limitation of the approach is that it is dependent on the terminology and the writing style. For example, "roof bolter" related incidents may not be detected by NLP in narratives when the writer uses the term "pinner" to refer to a bolter (though the diligent NLP developer would notice the frequent occurrence of "pinner" in narratives involving "roof"). "Pinner" is a common term for roof bolters in certain parts of the US. Terminology aside, writing style can vary dramatically depending on the region and the English language abilities of the writer. Considering all of these, the MSHA database may not be a great resource for English based NLP tools in other parts of the world. Regardless, organizations (or nations) developing their own NLP tools could provide training to standardize the writing of safety narratives, so that data is generated to assist automation.

The extremely low false positive rate for the narrowest accident types is a wonderful argument for considering these tools. The overall false positive rate across all accident types is quite low, which is good.

6. Conclusions

Natural language processing based random forest models were developed to classify narratives in the MSHA database depending on accident types. Nine models were developed. Four of the models, i.e., Over-exertion, Fall, "Caught in" and "Struck by", looked at type groups, i.e., groups of particular accident types. Five models looked at specific accident types within these broad groups. They were: Over-exertion in lifting objects, Over-exertion in pulling or pushing objects, Fall to the walkway or working surface, "Caught in", under or between a moving and a stationary object, and Struck by flying object. All models had high overall success rates (typically 95% or higher) in classification on MSHA data when considering both false positive and false negative rates. The success in detecting an accident type within a narrative was higher for type groups (71–81%) than for individual categories (25–59%). Detection was done with low false positive rates for type groups (1–5%), and extremely low false positive rate (<1%) for individual categories.

When a single model was developed to classify narratives into multiple categories, it did not perform as well as when a separate model was developed for each category. A similarity score based method was developed to resolve situations where a particular narrative may be classified differently according to different models.

When applied to non-MSHA data, the developed models were successful in classifying about two-thirds of the narratives in a non-MSHA database with 96% accuracy. The narratives that are not classified by the models could belong to accident types not modeled in this paper. In classifying the non-MSHA narratives with near perfect accuracy, the paper demonstrates the utility of NLP-based machine learning in mine safety research. It also demonstrates that there exists a language for mine safety, as models developed on narratives written by MSHA personnel apply to narratives written by non-MSHA professionals. They also demonstrate that natural language processing tools can help understand this language automatically.

Author Contributions: Conceptualization, R.G.; data curation, P.M. and R.P.; formal analysis, R.G., P.M., and R.P.; funding acquisition, R.G.; investigation, R.G., P.M., and R.P.; methodology, R.G., P.M., and R.P.; validation, R.G., P.M., and R.P.; visualization, R.G. and P.M.; writing—original draft, R.G.; writing—review & editing, P.M. and R.P. All authors have read and agreed to the published version of the manuscript.

Funding: This project was partially supported by the National Institute of Occupational Safety and Health.

Data Availability Statement: Not Applicable.

Acknowledgments: This project was partially supported by the National Institute of Occupational Safety and Health. Their support is gratefully acknowledged.

Conflicts of Interest: The authors declare no conflict of interest. The funders had no role in the design of the study; in the collection, analyses, or interpretation of data; in the writing of the manuscript, or in the decision to publish the results.

References

1. ILO. Safety and Health at the Heart of the Future of Work: Building on 100 Years of Experience. International Labour Organization, April 2019 Issue. Available online: https://www.ilo.org/wcmsp5/groups/public/---dgreports/---dcomm/documents/publication/wcms_686645.pdf (accessed on 10 April 2021).
2. Hämäläinen, P.; Takala, J.; Boon, K.T. Global estimates of occupational accidents and work-related illnesses. In Proceedings of the XXI World Congress on Safety and Health at Work, Marina Bay Sands, Singapore, Workplace Safety and Health Institute, Marina Bay Sands, Singapore, 3–6 September 2017.
3. Takala, J.; Hämäläinen, P.; Saarela, K.; Yun, L.; Manickam, K.; Jin, T.; Heng, P.; Tjong, C.; Kheng, L.; Lim, S.; et al. Global estimates of the burden of injury and illness at work in 2012. *J. Occup. Environ. Hyg.* **2014**, *11*, 326–337. [CrossRef] [PubMed]
4. Jiskani, I.M.; Cai, Q.; Zhou, W. Distinctive model of mine safety for sustainable mining in Pakistan. *Min. Metall. Explor.* **2020**, *37*, 1023–1037. [CrossRef]
5. Talebi, E.; Rogers, W.P.; Morgan, T.; Drews, F.A. Modeling Mine Workforce Fatigue: Finding Leading Indicators of Fatigue in Operational Data Sets. *Minerals* **2021**, *11*, 621. [CrossRef]
6. Basu, A.J.; Kumar, U. Innovation and technology driven sustainability performance management framework (ITSPM) for the mining and minerals sector. *Int. J. Surf. Min. Reclam. Environ.* **2004**, *18*, 135–149. [CrossRef]
7. Aznar-Sáncheza, J.A.; Velasco-Muñoza, J.F.; Belmonte-Ureña, L.J.; Manzano-Agugliaro, F. Innovation and technology for sustainable mining activity: A worldwide research assessment. *J. Clean. Prod.* **2019**, *221*, 38–54. [CrossRef]
8. NIOSH. NIOSH Mine and Mine Worker Charts. Available online: https://wwwn.cdc.gov/NIOSH-Mining/MMWC (accessed on 15 June 2021).
9. ICMM. 2012. Available online: http://www.icmm.com/en-gb/guidance/health-safety/indicators-ohs (accessed on 5 April 2021).
10. Garcia, D. COATIS, an NLP system to locate expressions of actions connected by causality links. In *Knowledge Acquisition, Modeling and Management. EKAW 1997*; Plaza, E., Benjamins, R., Eds.; Lecture Notes in Computer Science (Lecture Notes in Artificial Intelligence); Springer: Berlin/Heidelberg, Germany, 1997; Volume 1319, pp. 347–352. [CrossRef]
11. Kaplan, R.; Berry-Rogghe, G. Knowledge-based acquisition of causal relationships in text. *Knowl. Acquis.* **1991**, *3*, 317–337. [CrossRef]
12. Posse, C.; Matzke, B.; Anderson, C.; Brothers, A.; Matzke, M.; Ferryman, T. Extracting information from narratives: An application to aviation safety reports. In Proceedings of the IEEE Aerospace Conference Proceedings, Big Sky, MT, USA, 5–12 March 2005. [CrossRef]
13. Maille, N.P.; Ferryman, T.A.; Rosenthal, L.J.; Shafto, M.G.; Statler, I.C. What Happened, and Why: Towards an Understanding of Human Error Based on Automated Analyses of Incident Reports—Volume I. NASA ONERA, 2015. Available online: https://ntrs.nasa.gov/api/citations/20060023334/downloads/20060023334.pdf?attachment=true (accessed on 15 April 2021).
14. Jurafsky, D.; Martin, J.H. Speech and Language Processing: An Introduction to Natural Language Processing, Computational Linguistics, and Speech Recognition. 2020. Available online: https://web.stanford.edu/~{}jurafsky/slp3/ed3book_dec302020.pdf (accessed on 18 January 2021).
15. Baker, H.; Hallowell, M.R.; Tixier, A.J.-P. AI-based prediction of independent construction safety outcomes from universal attributes. *Autom. Constr.* **2020**, *118*, 103146. [CrossRef]
16. Tixier, A.J.-P.; Hallowell, M.R.; Rajagopalan, B.; Bowman, D. Automated content analysis for construction safety: A natural language processing system to extract precursors and outcomes from unstructured injury reports. *Autom. Constr.* **2016**, *62*, 45–56. [CrossRef]
17. Rose, R.; Puranik, T.G.; Mavris, D.N. Natural language processing based method for clustering and analysis of aviation safety narratives. *Aerosp.* **2020**, *7*, 143. [CrossRef]
18. Baillargeon, J.T.; Lamontagne, L.; Marceau, E. Mining actuarial risk predictors in accident descriptions using recurrent neural networks. *Risks* **2021**, *9*, 7. [CrossRef]
19. Gernard, J.M. Machine learning classification models for more effective mine safety inspections. In Proceedings of the 2014 International Mechanical Engineering Congress and Exposition IMECE2014, Montreal, QC, Canada, 14–20 November 2014.
20. Yedla, A.; Kakhki, F.D.; Jannesar, A. Predictive modeling for occupational safety outcomes and days away from work analysis in mining operations. *Int. J. Environ. Res. Public Health* **2020**, *17*, 1–17.
21. Raj, V.K.; Tarshizi, E.K. *Advanced Application of Text Analytics in MSHA Metal and Nonmetal Fatality Reports*; SME Annual Meeting & Expo: Phoenix, AZ, USA, 2020.
22. MSHA (Mine Safety and Health Administration). Mine Data Retrieval System. Available online: https://www.msha.gov/mine-data-retrieval-system (accessed on 31 January 2021).
23. Mitchell, T.M. Machine Learning. In *Machine Learning*; McGraw-Hill: New York City, NY, USA, 1997; Volume 45.

24. Ganguli, R.; Dagdelen, K.; Grygiel, E. *Systems Engineering. Mining Engineering Handbook*; Darling, P., Ed.; Society for Mining, Metallurgy and Exploration, Inc.: Englewood, CO, USA, 2011.
25. Röger, C.; Ismayilova, I. Predicting ambient traffic of a vehicle from road abrasion measurements using random forest. In Proceedings of the Conference 13th International Workshop on Computational Transportation Science (IWCTS'20), Seattle, WA, USA, 3 November 2020; pp. 1–7.
26. Weedon, M.; Tsaptsinos, D.; Denholm-Price, J. Random forest explorations for URL classification. In Proceedings of the 2017 International Conference On Cyber Situational Awareness, Data Analytics And Assessment (Cyber SA), London, UK, 19–20 June 2017; Institute of Electrical and Electronics Engineers, Inc.: New York City, NY, USA, 20 June 2017. ISBN 9781509050604. [CrossRef]
27. Scikit-Learn. sklearn.ensemble.RandomForestClassifier. Available online: https://scikit-learn.org/stable/modules/generated/sklearn.ensemble.RandomForestClassifier.html (accessed on 15 January 2021).
28. Humphries, G.R.W.; Magness, D.R.; Huettmann, F. *Machine Learning for Ecology and Sustainable Natural Resources Management*; Springer: Berlin/Heidelberg, Germany, 2018. [CrossRef]
29. NLTK. Natural Language Tool Kit. Available online: https://www.nltk.org/ (accessed on 15 January 2021).
30. Explosion Spacy. Industrial-Strength Natural Language Processing. Available online: https://spacy.io/ (accessed on 15 January 2021).

Article

Selected Artificial Intelligence Methods in the Risk Analysis of Damage to Masonry Buildings Subject to Long-Term Underground Mining Exploitation

Leszek Chomacki [1], **Janusz Rusek** [2,*] **and Leszek Słowik** [1]

[1] Building Research Institute, 00611 Warsaw, Poland; l.chomacki@itb.pl (L.C.); l.slowik@itb.pl (L.S.)
[2] Faculty of Mining Surveying and Environmental Engineering, AGH University of Science and Technology, 30059 Cracow, Poland
* Correspondence: rusek@agh.edu.pl; Tel.: +48-12-617-22-74

Abstract: This paper presents an advanced computational approach to assess the risk of damage to masonry buildings subjected to negative kinematic impacts of underground mining exploitation. The research goals were achieved using selected tools from the area of artificial intelligence (AI) methods. Ultimately, two models of damage risk assessment were built using the Naive Bayes classifier (NBC) and Bayesian Networks (BN). The first model was used to compare results obtained using the more computationally advanced Bayesian network methodology. In the case of the Bayesian network, the unknown Directed Acyclic Graph (DAG) structure was extracted using Chow-Liu's Tree Augmented Naive Bayes (TAN-CL) algorithm. Thus, one of the methods involving Bayesian Network Structure Learning from data (BNSL) was implemented. The application of this approach represents a novel scientific contribution in the interdisciplinary field of mining and civil engineering. The models created were verified with respect to quality of fit to observed data and generalization properties. The connections in the Bayesian network structure obtained were also verified with respect to the observed relations occurring in engineering practice concerning the assessment of the damage intensity to masonry buildings in mining areas. This allowed evaluation of the model and justified the utility of the conducted research in the field of protection of mining areas. The possibility of universal application of the Bayesian network, both in the case of damage prediction and diagnosis of its potential causes, was also pointed out.

Keywords: mining exploitation; masonry buildings; damage risk analysis; artificial intelligence; Bayesian network; Naive Bayes; Bayesian Network Structure Learning (BNSL)

check for
updates

Citation: Chomacki, L.; Rusek, J.; Słowik, L. Selected Artificial Intelligence Methods in the Risk Analysis of Damage to Masonry Buildings Subject to Long-Term Underground Mining Exploitation. *Minerals* **2021**, *11*, 958. https://doi.org/10.3390/min11090958

Academic Editor: Rajive Ganguli

Received: 9 August 2021
Accepted: 28 August 2021
Published: 1 September 2021

Publisher's Note: MDPI stays neutral with regard to jurisdictional claims in published maps and institutional affiliations.

1. Introduction

The mining process of underground resources significantly disturbs the structure of the rock mass. This leads to negative effects manifested on the surface of mining areas. Usually, these take the form of large-scale continuous deformations [1,2], mining tremors [3,4] and local discontinuous deformations [5,6]. All these phenomena are a potential threat both in the context of safety [7,8], and often they are the cause of a significant reduction in the utility of buildings [9,10]. Regarding both of these issues, there have been efforts for many years to optimize the possibility of extraction of resources with the lowest possible degree of degradation of the existing buildings on the surface of the mining area. Today, apart from the issues related to mining technology, the assessment of potential mining damage is one of the most important problems conditioning the possibility of conducting underground operations. It is a very complex socio-economic problem, which concerns both owners or managers of buildings and mine officials. From a practical point of view, it is very important to be able to reliably predict the expected damage to buildings before mining activities start, as well as to diagnose the causes of the damage during the occurrence of effects. An additional difficulty is the uncertainty occurring in the process of collecting information on the technical condition of buildings over a long period of exploitation and imprecise impact forecasts for the planned exploitation. Awareness of the complexity of this problem led the authors to undertake research using methods classified as artificial intelligence (AI) [11,12]. Particular attention was paid to methods operating on the basis of probabilistic notation of uncertainty, which include Naive Bayes Classifier and Bayesian networks [13,14].

The research was carried out on a database created of masonry buildings, for which information was established on the intensity of long-term impacts at the location of each structure and on the structural and material features of buildings, the quality of maintenance and the diagnosed range and intensity of damage. On this basis, two prognostic models, understood as decision-making systems for damage risk assessment, were created and compared. The first was the learned and tested structure of the Naive Bayes Classifier (NBC) [15]. The second, methodologically more complex model, was the Bayesian network structure [16]. During the construction of this model, an advanced technique involving the optimal extraction of the Bayes network structure from data (TAN-CL Chow-Liu's Tree Augmented Naive Bayes) [17] was used. Thus, a method belonging to the field of Bayesian networks structure learning from data (BNSL) [18] was used.

The research methodology adopted in this paper to establish a meaningful decision system for damage risk assessment is currently under development, especially in the areas of medical science [19], biology [20], genetics [21,22]. This type of methodology is also used in civil engineering issues, especially in the context of safety [23], risk [24] and reliability [14] assessment. In addition, a great number of scientific studies indicate the effectiveness of this type of approach in issues related to hazards arising from the impact of random phenomena of natural origin (floods [25], earthquakes [26], tsunami [27]). However, recently one can also encounter implementation of this approach in the interdisciplinary area combining mining and civil engineering [13,28]. The separation of a model for damage risk assessment in buildings, in addition to the benefits in terms of optimizing the planning of the mining process, can be used in a wide spectrum of socio-economic issues and the digitalization of the construction industry—cf. Figure 1. It seems particularly important to implement such a tool in the intensively increasing trend of BIM, as an important component for the area of AEC (Architecture-Engineering-Construction) [29] and FM (Facilities Management) [30,31]. On the other hand, integration of building issues with the mining process allows us to include the problem of damage risk assessment into the activities related to the development of the Industry 4.0 [32].

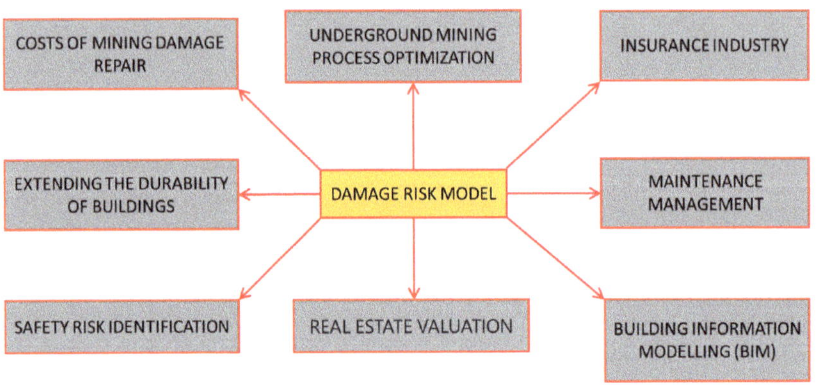

Figure 1. Diagram of the interaction of damage risk issues in various mining and civil engineering fields (own source).

Evaluating the issue addressed in this paper only through the narrow prism of the development of digitalization and the need for optimal planning of the mining exploitation process, it is clear that there is a need to create a tool to improve the work of both mining and civil engineers involved in the process related to the underground exploitation of resources. An additional advantage of such a tool, based on AI methodology, is the ability to update the model with access to new data resources. This is especially the case with Bayesian networks, which, in addition to allowing inference in the mode of prediction of the range and intensity of damage and diagnosis of its causes [13,28]. This significantly extends the possibilities of implementing such a tool in other industrial sectors that coexist with the mining industry.

2. Indicating the Innovation of the Methodological Approach Used for the Research

AI methods are currently used in many areas of science and technology, especially where, due to the complexity of the issues, it is necessary to apply heuristic approaches. Although individual tools belonging to this group are still being developed and improved, the effectiveness of many of them has already been confirmed. The validity of AI methods has also been justified by research conducted within the interdisciplinary field of mining and civil engineering [9,13,28,33,34]. Among the available AI methods, the methods that allow representation of formal uncertainty during inference are particularly useful for describing the risk of building damage. To date, the most popular methods that allow a mathematical representation of formal uncertainty are systems based on fuzzy logic [34] and Bayesian inference principle using probabilistic formalism [35]. Currently, there are also hybrids resulting from the combination of these two approaches [36,37].

From the point of view of the problem undertaken, there were two main criteria for the choice of methodology for the construction of the damage risk model. The first criterion was the use of a notation that would allow mathematical treatment of uncertainty during the construction and subsequent operation of such a system. The second criterion was dictated by the practice of making assessments, in which sometimes a prediction of damage intensity is required for predicted mining impacts, and sometimes a diagnosis of the causes of damage that has occurred. Therefore, it was decided to reduce the group of potential AI methods to those based on probabilistic notation. This form of uncertainty description is used every day by engineers and is found in international standards guidelines [38].

At this point, it should be indicated that other methods whose high efficiency in relation to the analysis of building damage has been confirmed by numerous studies. The main tool for failure analysis in building structures, from a mechanical point of view, is the Finite Elements Method (FEM). With regard to the issues of mining impacts, it is applicable and confirms its effectiveness, both in static issues related to the impact of land subsidence [39–41], and in dynamic issues related to the impact of mining tremors [4,42]. However, this type of approach cannot be effectively applied when it is necessary to forecast the intensity of damage for a large number of buildings. An additional aspect that hinders this type of approach is often the differences between the structural arrangement of individual buildings and the lack of transparency of their spatial static arrangements. For this reason, with regard to the undertaken problem, with full awareness of the advantages of the FE methodology, it was decided to undertake research based on in-situ data and apply advanced statistical methods, which include machine learning tools.

In this paper, two methods using probabilistic uncertainty notation are used: Naive Bayes Classifier (NBC) and Bayesian Network (BN) methodology. The NBC methodology was used to obtain a reference model for the methodologically more complex approach using Bayesian networks. The results obtained for NBC provided a reference basis for assessing the quality of the established BN structure, which was considered the target model for describing the issue addressed in this paper.

Although Bayesian networks have already been successfully applied in issues of risk assessment of the occurrence of various types of negative natural or anthropogenic phenomena [43,44], the main problem is to determine the appropriate structure of such a model. Applications of Bayesian networks, whose structure is mostly arbitrarily determined by an expert, are encountered in mining and civil engineering problems. This approach is efficient but is limited by finite human perception. This makes it impossible to build expert systems for processes or phenomena with a large number of variables. The risk problem related to the phenomenon of damage to existing buildings affected by mining exploitation depends on dozens of factors, the influence of which cannot be neglected in the analysis. Therefore, the application of a methodology based on the detection of the BN structure from data is indispensable in this case. In this regard, the issue is still open and requires testing various methods involving BN structure learning from data (BNSL—Bayesian Network Structure Learning). Implementations of BNSL methods are not often seen in engineering problems, especially civil engineering and mining. For this reason, the issue addressed in this paper is considered original and innovative in the above engineering fields.

3. Characteristics of the Information Collected in the Building Database and Description of Mining Impacts

Fulfilling the set research purposes, which consisted in building and verifying the NBC classifier and the BN structure, at the beginning required collecting data on the behavior of buildings subjected to the influence of mining exploitation. The most relevant factor here was the observed damage in the buildings before the mining operation and the actualization of the damage after it.

During the passage of underground exploration on the ground surface, deformation occurs. In general, vertical (w [mm]) and horizontal (u [mm]) displacements occur. However, in order to relate the deformation of the terrain to the problem of the threat of buildings, detailed measures are introduced, which are derived from vertical and horizontal displacements. These measures are horizontal deformations (ε [mm/m]), inclinations (T [mm/m]) and curvatures (R [km] of terrain. The values of these parameters may be established on the basis of model tests or as a result of geodetic measurements [45].

In order to explain the meaning of variables used in the research, the process of formation of a mining basin was presented and interpreted schematically in Figure 2. However, the characteristic damage to buildings on the convex and concave margins of a mining basin is illustrated in Figure 3.

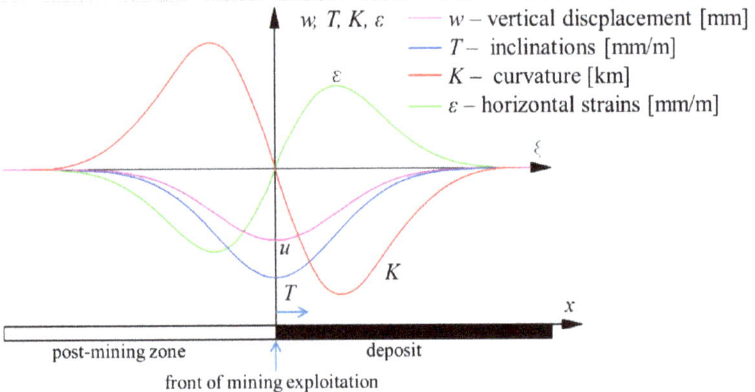

Figure 2. Distribution of deformation indices over the excavation: (*u*) vertical displacement; (*T*) inclination; (*K*) ground curvature; (ε) horizontal ground strain (own source).

Finally, a group of 207 buildings was qualified for further analysis. Next, "in-situ" field research was carried out, during which information on the buildings was collected and finally archived in the database. The "in-situ" field research for the selected group of buildings concerned the determination of, among others, the geometrical parameters, applied structural solutions, existing protection against mining influence and the range and intensity of the existing damage. An example of two representative buildings from among all those qualified for the survey is presented in Figure 4.

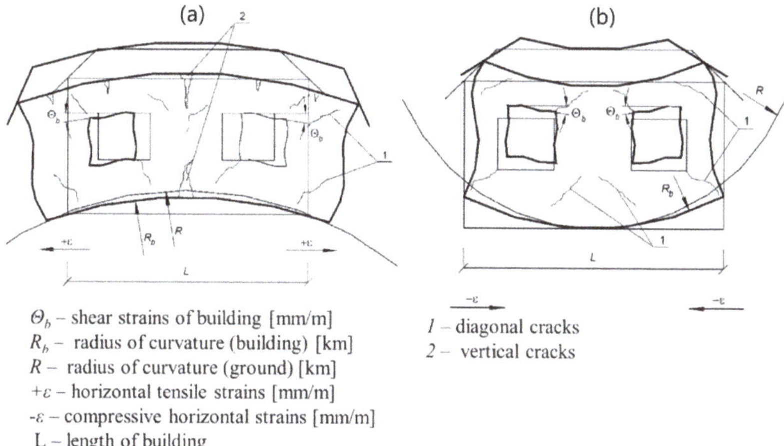

Figure 3. Diagram of characteristic damage to buildings in a mining basin: (**a**) convex edge of the basin; (**b**) concave edge of the basin; (1) diagonal cracks, (2) vertical cracks [46].

Figure 4. Two examples of the 207 masonry buildings qualified for the study: (**a**) example of a partially renovated building; (**b**) example building after full renovation.

Investigations of the state of damage to buildings taking into account the range and intensity, as well as the threat to the safety of the structure and users, made it possible to classify each case into one of four categories [47].

The description of the accepted building damage categories can be presented as follows:

- damage category 1: no structural damage, possible occurrence of damage in the form of insignificant cracks on the plaster of walls and ceilings.
- damage category 2: more intensive damage to non-structural elements and finishing elements, such as cracking or local separation of ceiling soffits, trimming of ceiling and wall plaster, cracking of elevation and interior wall plaster.
- damage category 3: damage in structural elements, the range, intensity and location of which, in the case of further ground deformation influences, may lead to the local loss of load-bearing capacity or stability of structural elements
- damage category 4: damage threatening the local load-bearing capacity of its elements (which could have already been subjected to temporary protective works), or buildings in which there is large natural wear of structural elements, manifested by extensive and advanced erosion of masonry or concrete and reinforcement.

The distribution of damage intensity to buildings in the period 2011–2017 according to the adopted categorization is presented in Figure 5.

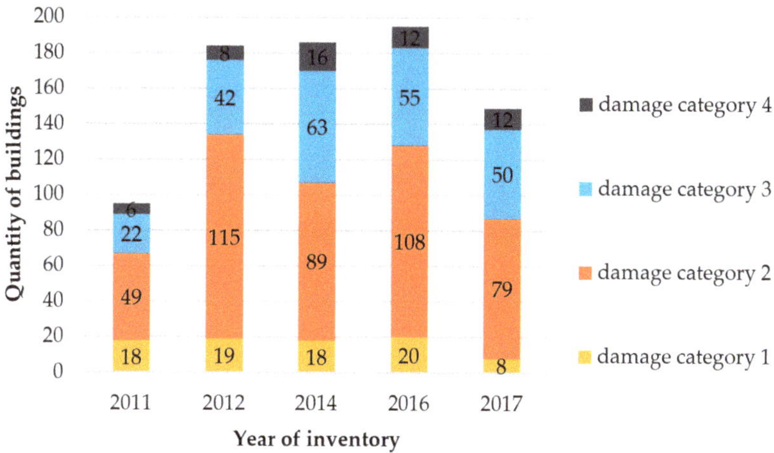

Figure 5. Distribution of building damage categories determined during the inventory in the years 2011–2017.

In the period of research, i.e., in the years 2011–2017, the development in the study was subjected to the influence of coal mining, carried out in the system with roof collapse, whose characteristic parameters are summarized in Table 1.

Table 1. Summary of basic parameters of mining exploitation in the study area.

Deck	Wall	Height [m]	Depth [m]	Period of Exploitation
503	4	2.6–3.3	625–720	2011–2013
510 wg	30a and 31a	2.0–2.4	725–805	2013–2015
503	5 i 6	2.0–2.3	670–680	2015–2017

In the process of creating the database, information was collected on the occurring values of the horizontal ground deformations (ε—cf. Figure 2) in the locations of particular buildings. The basis for determining the values and directions of strains were the results of surveying measurements conducted by the mine. In turn, approximation of values and directions of strains to the location of each building was performed using dedicated modeling methods based on Budryk–Knothe theory [48]. The quantity of buildings, which were affected by horizontal tensile strain ε^+, together with their values with the accuracy of 0.5 mm/m, is presented in Figure 6. Whereas the quantity of buildings, which were affected by horizontal compressive strain ε^- is presented in Figure 7.

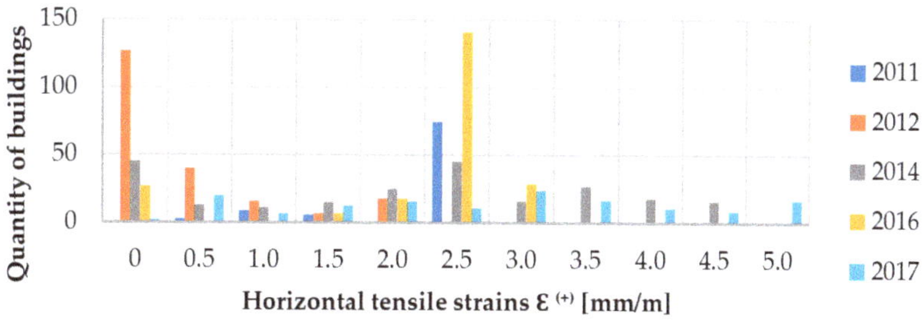

Figure 6. Distribution of data on the occurrence of tensile horizontal strain ε^+ [mm/m] at the location of buildings from the period 2011–2017.

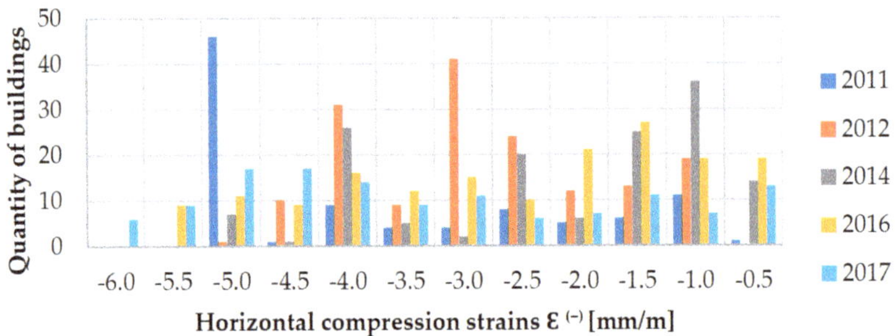

Figure 7. Distribution of data on the occurrence of compressive horizontal strain ε^- [mm/m] at the location of buildings from the period 2011–2017.

Based on the horizontal deformation impacts collected above, the resulting mining terrain categories were determined [46]. In this form, the intensity of influences from continuous deformation of the mining area was classified and included in the database.

Finally, taking into account multiple inspections of the technical state for the same buildings, a study material of 594 design cases was collected. A synthetic summary of building data and meaningful mining impacts is presented in Table 2.

Table 2. List of variables with indication of their discretization (categories).

Variable Type	Variable	Code	No. of Categories
	Length	Geo1	8
	Width	Geo2	6
	Building area	Geo3	10
	Number of overground storeys	Geo4	5
	Volume	Geo5	11
Geometry	Length of a series of compact buildings	Geo6	12
	Dilation method	Geo7	3
	The shape of the building's body	Geo8	4
	Basement	Geo9	3
	Variable level of foundation	Geo10	2
	Variable building height	Geo11	2
	Foundation type	Con1	3
	Basement wall material	Con2	3
	Material of the walls of the ground floor and above	Con3	2
Construction	Ceiling above the basement	Con4	5
	Ceiling above the ground floor and above	Con5	3
	Lintels	Con6	3
	Protections for mining influence	Con7	4
	Protections—supplementary data	Con8	3
	Year of construction	Otd1	8
Other technical data	Natural wear (technical condition)	Otd2	5
	Repair factor	Otd3	2
	Static (deformation) resistance category	Otd4	3
Mining impacts	Mining threat category of the terrain	MC	3
Damage	Damage category before impacts	Dmg1	4
	Damage category after impacts	Dmg2	4

4. Characteristics of the AI Methods Used in the Research

As part of the research, it was decided to choose supervised learning as the optimal method to achieve the purpose [49]. Within this area, the NBC (Naive Bayes Classification) and BNs (Bayesian Networks) methods were qualified for further research. This choice was dictated by the fact that these methods allow notation of risk in a probabilistic form, which is in accordance with the functioning nomenclature in this area at an international level [50]. An additional advantage of these methods is the ability to capture uncertainty and, in the case of BNs, additionally incompleteness of information concerning the input variables. The last very important advantage from the utilitarian point of view, and concerning only BNs, is the possibility of inference in any direction. In the problem of damage risk, this proves that this model can be used both in the case of predicting the intensity of damage as well as to diagnose its causes.

Finally, it was concluded that the target damage risk model would be created using the BN method. The NBC method, on the other hand, would serve as a reference basis for verifying the quality of the model described by the extracted optimal BN structure.

4.1. NBC—Naive Bayes Classification

The NBC method determines the probability of occurrence of particular classes/ labels/categories of the so-called decision variable depending on a given set of input variables. On the basis of the probability value, the classification result is determined by means of ranking. This result is called the classifier indication. Unlike BN, the assumption of mutual independence of particular input variables is used here. Taking n input variables described as x_1, x_2, x_3, ... , x_n and the output variable y described by the number of classes: $c_1, c_2, c_3, ... , c_k$ the mathematical form of the inference process expressed as [51] is obtained (1):

$$P(C_k|x_1, x_2, x_3, \ldots, x_n) = \frac{P(C_k) \prod_{j=1}^{n} P(x_j|C_k)}{P(x_1, x_2, x_3, \ldots, x_n)} \tag{1}$$

The assumption of independence is often overly optimistic (naive), but it allows for significant simplification of the computational procedure.

The schematic diagram of the NBC network structure is presented in Figure 8.

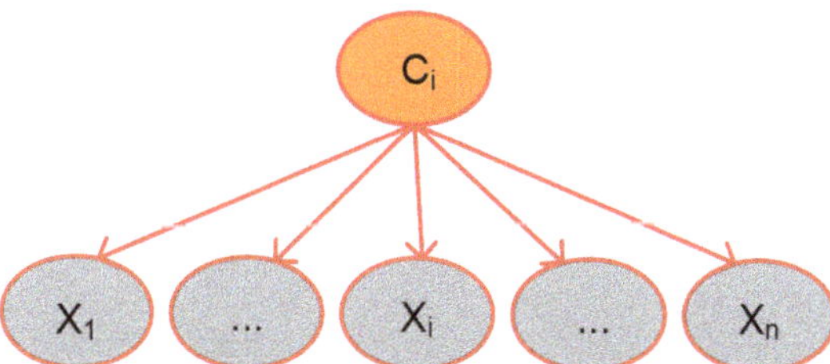

Figure 8. Schematic diagram of the NBC network structure.

Two procedures are used to build an NBC classifier from the learning dataset: Maximum Likelihood Estimation (MLE), which maximizes the conditional likelihood $P(c|x)$ understood here as the verifiable claim of the existence of each class for the learning data [51] (2):

$$h_{MLE} = arg\ max P(C_k|x_j) \tag{2}$$

or Maximum a Posteriori Estimation (MAP), which maximizes the posteriori probability of occurrence of each class for the learning set [51] (3):

$$h_{MAP} = arg\ max P(x_j|C_k) \tag{3}$$

The main advantages of the NBC classifier include high learning speed with relatively high classification accuracy. The quality of classification is not strongly determined by the number of learning data, which is also considered to be an advantage of this approach. However, due to the assumption of mutual independence of all input variables, it can be assumed that the model structure may not reflect the real relationships between the analyzed variables in the issue of describing the risk of building damage. Therefore, in the framework of the present research, the NBC model created will play a comparative role, giving an idea about the effectiveness of a more complex model in the form of a separate BN structure.

4.2. BN—Bayesian Network

The Bayesian network (BN) can be interpreted as a Directed Acyclic Graph (DAG). The graph structure (G) encodes information about the interrelationships between the variables $X = \{X_1, \ldots, X_N\}$, k which is represented by graph edges (E) and nodes (V). In a meaningful sense, the fixed BN represents the joint probability distribution over the set of all random variables, which can be represented as [52] (4):

$$P(X|G, \Theta) = \prod_{i=1}^{N} P(X_i | \Pi_{X_i}, \Theta_{X_i})$$

(4)

where:

$G = G(X, E, V)$—mathematical notation for describing the acyclic directed graph structure
$X = \{X_1, \ldots, X_N\}$—the set of all variables that belong to the nodes of the graph
$X_i = \{x_i^{(1)}, \ldots, x_i^{(k_i)}\}$—states of the i-th variable of X
E—set of all edges
V—set of all nodes
$\Pi_{X_i} = \{x_i^{(q_1)}, \ldots, x_i^{(q_i)}\}$—the set of parents, i.e., all nodes of the graph that determine the state of the node X_i
$\theta = \{\theta_{X_1}, \ldots, \theta_{X_N}\}$—the set of all parameters of conditional relations between particular nodes X_i, and a set of their parents Π_{X_i}

In the case of discrete variables, the latent parameters of the model $\theta_{X_j} = \{\theta_{ijk}\}$ are represented in terms of a multinomial Conditional Probability Table (CPT) whose elements are expressed as [51] (5):

$$\theta_{ijk} = P\left(X_j = x_j^{(i)} \middle| \Pi_{X_j} = \pi_j^{(k)}\right)$$

(5)

According to relation (4), the joint distribution $P(X|G, \Theta)$ is decomposed based on the conditional local distributions $P(X_i | \Pi_{X_i}, \Theta_{X_i})$, described over each random variable X_i relative to its corresponding set of conditional variables so-called parents Π_{X_i}. This formulation is possible due to the concept of conditional independence introduced by Pearl [53]. This allows for a significant reduction in the number of links that do not show cause-and-effect relationships. The introduction of the proposed linkage reduction allows significant simplification of the calculations related to the modelling of the joint probability distribution and simplifies the subsequent interpretation of the structure by the human user.

A diagram of an exemplary BN structure is presented in Figure 9. The provided diagram illustrates in a simplified manner the coding within CPT and the meaning interpretation of the nodes of the so-called parents.

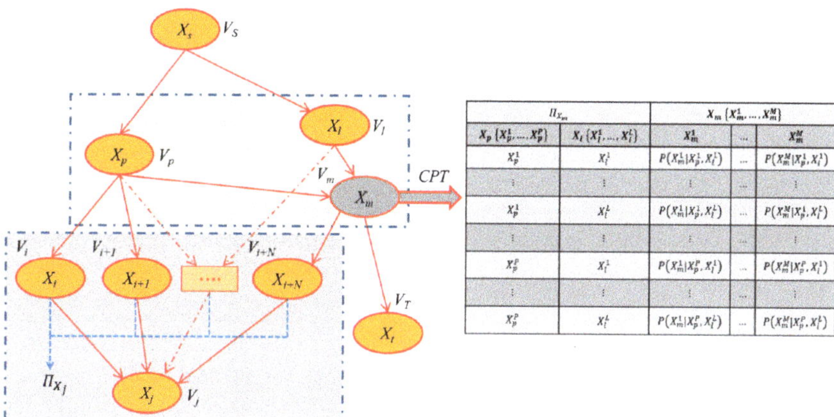

Π_{X_m}		$X_m \{X_m^1, \ldots, X_m^M\}$		
$X_p \{X_p^1, \ldots, X_p^P\}$	$X_l \{X_l^1, \ldots, X_l^L\}$	X_m^1	...	X_m^M
X_p^1	X_l^1	$P(X_m^1 \mid X_p^1, X_l^1)$	...	$P(X_m^M \mid X_p^1, X_l^1)$
⋮	⋮	⋮	...	⋮
X_p^1	X_l^L	$P(X_m^1 \mid X_p^1, X_l^L)$	...	$P(X_m^M \mid X_p^1, X_l^L)$
⋮	⋮	⋮	...	⋮
X_p^P	X_l^1	$P(X_m^1 \mid X_p^P, X_l^1)$	...	$P(X_m^M \mid X_p^P, X_l^1)$
⋮	⋮	⋮	...	⋮
X_p^P	X_l^L	$P(X_m^1 \mid X_p^P, X_l^L)$	...	$P(X_m^M \mid X_p^P, X_l^L)$

Figure 9. Schematic diagram of Bayesian Network structure (BN) as a *Directed Acyclic Graph* (DAG) with tables of conditional probabilities for a selected node (CPT).

The procedure for learning a BN from data consists of two interconnected steps: *Structure Learning* and *Parameter Learning* [54], which can be written as (6):

$$\underbrace{P(Model|D) = P(G, \Theta|D) =}_{learning} \quad \underbrace{P(G|D)}_{structure\ learning} \cdot \underbrace{P(\Theta|G, D)}_{parameter\ learning} \tag{6}$$

The self-contained extraction of BN structure from the data is much more difficult than the implementation of an arbitrarily determined model, e.g., based on expert knowledge. This approach is mostly used where it is required to extract relationships among a large number of variables used to describe a given process. In such situations, determining the network structure from the data based on expert knowledge is impossible. This is dictated by the limited human perception when it comes to analyzing multivariate problems.

With respect to learning the BN structure from data, the unknowns are both the network structure (G) and the parameters (θ) of the multinomial probability distribution tables (CPTs). In general, there are three different approaches in learning BN structure from data: constraint-based structure learning, score-based structure learning, and hybrid algorithms [35].

The risk of damage to buildings is described by numerous factors with subtle contributions, as demonstrated by years of research described, among others, in [10]. With these considerations in view, it is important that as many of the variables as possible are included in the model when extracting the BN structure. In turn, the basic criterion is that the probability distribution represented by the BN has the highest possible agreement with the information contained in the learning dataset. With this in mind, the research conducted analyses through a number of available score-based and constraint-based algorithms. As a result of these analyses, the optimal form of the DAG network was obtained for the learning method using Chow-Liu's tree Augmented Naive Bayes (TAN-CL) algorithm [17].

5. Results

In order to select the optimal method of building a damage risk assessment model, the assembled database was adapted for analysis. Then the calculation stage was carried out to obtain classifiers to assess the risk of damage to masonry buildings. In line with the previous justification regarding the choice of research methodology, the NBC and BN approaches were used for further analysis. At the same time, as part of the BNs methodology, an approach was used based on teaching the structure of BN from data (BNSL).

The classifiers were built in the R [55] development environment with the use of the following packages: *bnlearn* [54], *bnclassify* [17], *caret* [56] oraz *naivebayes* [57].

5.1. Preparation of Data for Analysis

At the initial stage of data preparation for further analysis, extreme cases were rejected, the relative frequency of which for each of the variables did not exceed 5%. The data set filtered in this way was used for the stage in which the training and testing of individual models commenced.

Moving on to the learning stage, the data set was divided into training and test sets in the proportion of 80:20. Additionally, in order to maintain the completeness of the patterns for the learning and testing processes, the stratified sampling approach was applied [58]. In general, it forces the presence of patterns of the same category in both the training set and the test set. Thus, the information is complete for both the learning process and the subsequent testing.

Ultimately, the number of separated sets was 478 cases for the training set and 116 cases for the test set.

The training set was used for learning, as required for each method included in the research. The test set, which did not participate in the learning process, was used as unbiased to evaluate the created models in the context of generalization properties.

5.2. Interpretation of the Results and the Adopted Method of Their Verification

In order to effectively compare the results of individual methods, a universal measure of the classification correctness assessment was used, namely the confusion matrix. An example of such a matrix in a binary (dichotomous) classification is presented in Table 3.

Table 3. Confusion matrix for a binary classifier.

	Actual Positive	**Actual Negative**
Predicted positive	True positives TP	False positives FP
Predicted negative	False negatives FN	True negatives TN

The basic comparative parameter here is the overall accuracy, which is the quotient of the sum of correctly classified cases and their total number [59] (7):

$$ACC = \frac{TP + TN}{TP + FP + FN + TN} \tag{7}$$

It is advisable that the chosen method should also be characterized by the highest possible precision and sensitivity:

- Positive Predictive Value (PPV) [59] (8):

$$PPV = \frac{TP}{TP + FP} \tag{8}$$

- True Positive Rate (TPR) [59] (9):

$$TPR = \frac{TP}{TP + FN} \tag{9}$$

As part of building models from the AI group, a very important feature is the generalization of the knowledge obtained during the learning process, which can be verified on the testing set. In this sense, knowledge generalization is defined as the ability of a model to predict the right response for non-learning cases.

In order to compare the generalization abilities of individual models, the relative difference in the accuracy of the classification for the training set and the test set ΔACC was calculated in relation to the results obtained on the training set.

Finally, after creating the NBC model and after extracting the structure for the BN method, they were compared in terms of the quality of classification and generalization properties. These results, together with a detailed discussion, are presented in Sections 5.3 and 5.4 of this work. As part of the presentation of the obtained results, error matrices were used, taking into account the division into training and test sets. In accordance with the formulas (7)–(9), these matrices also summarize the results on the accuracy of classification as well as the average precision and sensitivity.

5.3. The Results Obtained for the NBC Method

The construction of the NBC classifier was carried out using four packages in the R environment. The best classification accuracy was characterized by the classifier built using the *naivebayes* package [57] and this model was taken into account in the further part of the research.

In the selected package, the implemented algorithm detects and assigns classes to individual variables, which allows the use of different distributions for each of them [60]. A multinomial distribution was assigned to 23 variables. For the remaining four variables, with dichotomous values, Bernoulli distribution was proposed. In turn, the parameters for the conditional probability distribution tables (CPT) were determined by the maximum likelihood method (MLE—p. 4.1).

At the stage of building the NBC classifier, it is also necessary to use the *Laplace* smoothing parameter. It is characterized by the fact that for its lower values, the accuracy of classification increases, but its effectiveness deteriorates significantly in atypical cases [9]. Based on the multiple analyses carried out, it was found that when this parameter equal to the value of $p_L = 10$ was used, good classification accuracy was obtained while maintaining appropriate generalization properties.

The created model was assessed in the context of the correctness of the classification on the training and test sets as well as the generalization properties, in accordance with the criteria specified in Section 5.2. The results in the form of a confusion matrix are presented in Table 4.

Table 4. Confusion matrix of the NBC classifier—number of cases, precision, sensitivity and accuracy of classification.

Training Set—478 Cases							
Damage State Category after Impacts (Dmg2)		Observed				Σ	PPV
		1	2	3	4		
Predicted	1	28	33	12	0	73	38.36%
	2	12	216	1	8	237	91.14%
	3	1	3	140	6	150	93.33%
	4	0	1	0	17	18	94.44%
Σ		41	253	153	31	478	avg. PPV 79.32%
TPR		68.29%	85.38%	91.50%	54.84%	avg. TPR 75.00%	ACC 83.89%
Test Set—116 Cases							
Damage State Category after Impacts (Dmg2)		Observed				Σ	PPV
		1	2	3	4		
Predicted	1	7	6	5	0	18	38.89%
	2	4	58	1	3	66	87.88%
	3	0	2	20	7	29	68.97%
	4	0	0	0	3	3	100.00%
Σ		11	66	26	13	116	avg. PPV 73.93%
TPR		63.64%	87.88%	76.92%	23.08%	avg. TPR 62.88%	ACC 75.86%

Table 4 shows that the constructed model is characterized by a good classification accuracy of 83.89% for the training set. The results for the test set are satisfactory and the classification accuracy is 75.86%. Using the previously defined relative measure of ΔACC, the generalization abilities of the model were assessed as satisfactory ($\Delta ACC = 9.57\%$).

5.4. The Results Obtained for the BN Method

The assumption of the possible mutual influence of individual variables allowed the analyses to be carried out in accordance with the BNs methodology.

The BN approach results in a network structure that depends on the selected classifier training method. The results for selected eight methods of learning the network structure were analysed. Some of the methods studied qualified for the constraint-based approach, and some for the score-based structure learning approach. These methods are available in the bnlearn [54] and bnclassify [17] packages.

The best results were obtained using the Chow-Liu's tree Augmented Naive Bayes (TAN-CL) learning method. The chosen method of training the TAN-CL network is the result of a combination of two methods. The Tree Augmented Naive Bayes (TAN) method [61], which approximates the interactions between variables using a tree-shaped structure, with the *Chow-Liu* junction detection algorithm [62].

The controlling parameter in the construction of the model is the measure of the fit of the model acting as an objective function for score-based optimization. The impact of three selected functions was analysed: Log-Likelihood (loglik), Akaike Information Criterion (AIC) and Bayesian Information Criterion (BIC). Ultimately, the best results were obtained for the AIC criterion.

The created model was assessed in the context of the correctness of classification and generalization properties. For this purpose, the results obtained from the simulation of the model response for the training and test set were used. As in the case of the NBC classifier, the representation of the results in the form of a matrix of errors was used, which is summarized in Table 5.

Table 5. Confusion matrix of the BN classifier—number of cases, precision, sensitivity and accuracy of classification.

Training Set—478 Cases							
Damage State Category after Impacts (Dmg2)		Observed				Σ	PPV
		1	2	3	4		
Predicted	1	40	15	0	0	55	72.73%
	2	1	220	29	5	255	86.27%
	3	0	15	118	3	136	86.76%
	4	0	3	6	23	32	71.88%
Σ		41	253	153	31	478	avg. PPV 79.41%
TPR		97.56%	86.96%	77.12%	74.19%	avg. TPR 83.96%	ACC 83.89%
Test set—116 cases							
Damage State Category after Impacts (Dmg2)		Observed				Σ	PPV
		1	2	3	4		
Predicted	1	11	2	0	0	13	84.62%
	2	0	58	4	0	62	93.55%
	3	0	5	22	3	30	73.33%
	4	0	1	0	10	11	90.91%
Σ		11	66	26	13	116	avg. PPV 85.60%
TPR		100.00%	87.88%	84.62%	76.92%	avg. TPR 87.35%	ACC 87.07%

Table 5 shows that the created BN network is characterized by a high classification accuracy of 83.89% for the training set. The results for the test set are also high and the classification accuracy is 87.07%. On the other hand, using the previously defined relative measure ΔACC, it can be concluded that the BN model has high generalization properties of the acquired knowledge from the learning stage (ΔACC = 3.79%). In this case, it proves an advantage in terms of the separated structure of BN (DAG), as well as correctly determined parameters in the learning process (θ).

5.5. Comparison of Established NBC and BN Models

Considering the fact that both models achieved a high degree of correct classification for the training set, the main criterion for selecting a model to assess the risk of damage was the verification of generalization properties. This was done by analysing the classification accuracy of both models for the test set. In order to detail the verification process, the obtained results were additionally analysed in terms of precision (PPV) and sensitivity (TPR). The list of reliable criterion values is summarized in Table 6 and subjected to a graphic interpretation, which is illustrated in Figure 10.

Table 6. Comparison of classification accuracy (ACC), mean precision (avg. PPV) and sensitivity (avg. TPR) for selected artificial intelligence methods.

Model Building Method	ACC for Training Set	ACC for Testing Set	Average PPV for Testing Set	Average TPR for Testing Set
NBC	83.89%	76.72%	74.55%	64.80%
BN	83.89%	87.07%	85.60%	87.35%

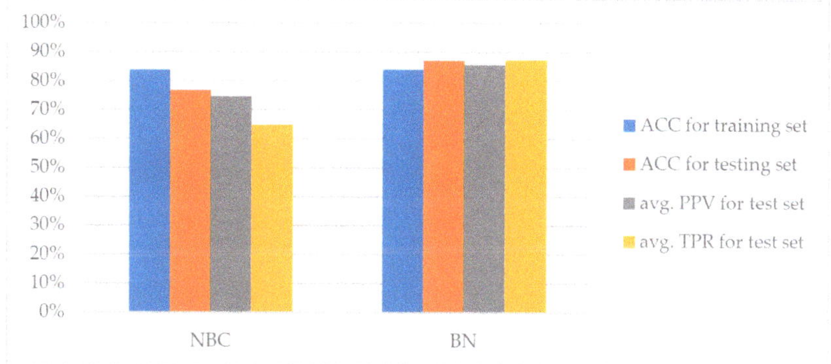

Figure 10. Comparison of the criteria values used to verify the effectiveness of the created NBC and BN models.

As a result of the comparative analysis of the NBC and BN models, it was found that the separated structure of BN allows for the description of the modelled process, obtaining better results than in the case of the NBC model. As shown in Tables 3 and 6, the BN network obtained better results in relation to each of the verified criteria (ACC, PPV and TPR). Additionally, by simulating both models on the data from the test set, it was shown that the BN network has better generalization properties than NBC. This is evidenced by the value of the defined measure ΔACC, which for the BN model reached the value of ΔACC = 3.79%, and for the NBC model ΔACC = 9.57%. In this case, the smaller the value means the smaller the difference in the correctness of the classification for the training set and the test set. Thus, a lower value of the ΔACC measure indicates better generalization properties.

Taking into account the results of the comparative analysis, it was found that a more effective tool for modelling the risk of damage to masonry buildings is the separated structure of the BN. In addition, implicit evidence was obtained that in order to model the risk of damage, it is necessary to take into account the relationship between individual variables, which is not taken into account in the NBC method.

5.6. Analysis of Connections Occurring in the Separated Structure of the Bayesian Network

One of the advantages of the BN approach is the possibility of representing the model in the form of a graph structure (*DAG*), which increases the interpretability of the model itself, and also supports the user in making decisions (*Decision Support Tool*) [63].

During the process of learning the BN structure from the data, any external interference in the relationships between the variables was abandoned. And so, for the TAN-CL learning method, the definition of constraints (in the form of lists defining blocked or forced connections) [54] was abandoned, thus giving full autonomy to the adopted method of teaching the BN network structure. Figure 11 shows the structure of the Bayesian network, indicating the direction of inference, and presents the variables taken into account in the decision-making process.

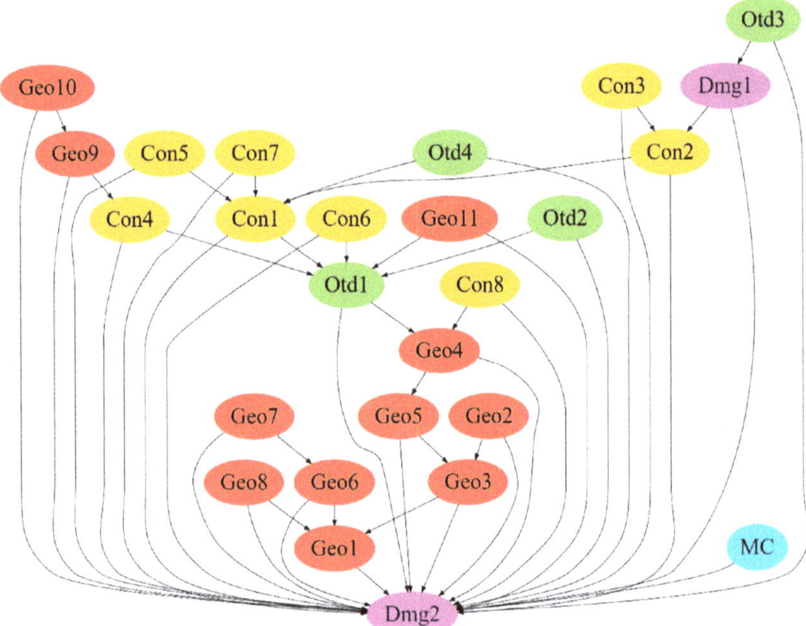

Figure 11. Network structure for the BN classifier obtained using the TAN-CL learning methods.

Based on the qualitative assessment of the relationships occurring in the separated BN structure, created by applying the TAN-CL learning algorithm (Figure 11), it was found that there are numerous cause-effect relationships:

- geometry variables (Geo) are linked together, as are the structure variables (**Con**),
- the variable on mining impacts MC is not related to other variables and has an impact on the output variable Dmg2,
- out of 48 connections, 45 were positively assessed, and three connections were neutral.

To sum up, it is estimated that the network structure is coherent and logical, and the obtained connections between the variables mostly coincide with those observed in engineering practice.

6. Conclusions

This work presents an example of the use of selected tools from the group of artificial intelligence (AI) methods to assess the risk of damage to masonry buildings located in the mining area of active mining facilities. After taking into account a number of criteria resulting from the practice of making these types of assessments for the protection of the development of mining areas, two AI tools belonging to the group of supervised learning methods were selected for analysis: NBC—Naive Bayes Classification and BN—Bayesian Networks.

Ultimately, using the "in-situ" data collected over many years of inspections of the technical condition of buildings in the mining area, 574 cases were collected and recorded in the form of a database. These data were used to train and test the NBC and BN models. At the same time, as part of building the BN model, it was necessary to isolate the unknown structure of connections between the variables describing the process under study. This task was finally carried out using the TAN-CL algorithm, which belongs to the group of methods for teaching the structure of BN from data (BNSL).

The obtained results were subjected to detailed individual and collective assessment. On this basis, it was found that the BN methodology was more effective than the simpler NBC approach. Thus, it has been shown implicitly that in order to describe a complex process which is the risk of damage to buildings, it is necessary to involve dependencies between individual variables, and thus to use BNSL methods.

The paper shows that better results in the context of mapping the information contained in the original data set were obtained for the extracted DAG structure of the Bayesian network compared to the simpler NBC model. This indicates the need to take into account the interrelationships between individual variables that are not taken into account in other AI methods, including the NBC model. Moreover, during the process of extracting the DAG structure of the Bayesian network, the connection between the variable describing the damage and the variable indicating the intensity of mining impacts was spontaneously separated.

The authors of the paper have currently undertaken research in the context of determining the significance of the relationships between the individual variables. Establishing the significance between individual nodes of the Bayesian network is necessary to complete the description of the damage process and to enable a more effective application of such a model in practice.

It should be emphasized that as in the case of all AI methods based on supervised learning, the reliability of the results obtained is strictly dependent on the information contained in the model data. This also means that a lot of emphasis should be placed already at the stage of collecting and archiving the data saved in the database. The authors dealt with the problem closely related to the issues of decision making under uncertainty. For this reason, based on specific criteria (Chapter 2), two presented methods based on the Bayesian inference formalism were distinguished for analysis. However, the choice of method depends absolutely on the type of problem. And so, for example, in the analysis of structural reliability assessment, other heuristic models, e.g., artificial neural networks, can be successfully used as a supporting tool for the FORM or SORM methods (First and Second Order Reliability Methods) [64,65].

As mentioned in chapter 1, the methodology of Bayesian networks is characterized by a very wide range of applications, an example of which can be found in issues related to threats of natural origin (e.g., floods, earthquakes, tsunamis, etc.).

The implementation of this type of tool may be implemented in the near future within the developing BIM concept. In conjunction with the IoT technology [66], it will allow for permanent monitoring of building structures along with the simultaneous assessment of the risk from the impact of the industrial environment, which also includes the impact of mining activities. In turn, automatic data archiving and updating the damage risk model will contribute to a more detailed understanding of this phenomenon, which may bring great socio-economic benefits.

Author Contributions: L.C.—conceptualization, methodology, software analysis, validation, formal analysis, investigation, resources, data collection, writing—original draft preparation, writing—review and editing, visualization. J.R.—conceptualization, methodology, formal analysis, investigation, writing—original draft preparation, writing—review and editing, visualization, supervision.

L.S.—conceptualization, formal analysis, supervision, project administration, funding acquisition. All authors have read and agreed to the published version of the manuscript.

Funding: This research received no external funding.

Data Availability Statement: The data presented in this study are available on request from the corresponding author. The data are not publicly available because they were taken from studies carried out for private enterprises.

Conflicts of Interest: The authors declare no conflict of interest.

References

1. Dudek, M.; Tajduś, K. FEM for prediction of surface deformations induced by flooding of steeply inclined mining seams. *Geomech. Energy Environ.* **2021**, *28*, 100254. [CrossRef]
2. Tajduś, K. The nature of mining-induced horizontal displacement of surface on the example of several coal mines. *Arch. Min. Sci.* **2014**, *59*, 971–986. [CrossRef]
3. Tajduś, K.; Tajduś, A.; Cała, M. Seismicity and rock burst hazard assessment in fault zones: A case study. *Arch. Min. Sci.* **2018**, 747–765. [CrossRef]
4. Pachla, F.; Tatara, T. Dynamic Resistance of Residential Masonry Building with Structural Irregularities. In *Seismic Behaviour and Design of Irregular and Complex Civil Structures II*; Springer: Berlin, Germany, 2020; pp. 335–347.
5. Lian, X.; Zhang, Y.; Yuan, H.; Wang, C.; Guo, J.; Liu, J. Law of Movement of Discontinuous Deformation of Strata and Ground with a Thick Loess Layer and Thin Bedrock in Long Wall Mining. *Appl. Sci.* **2020**, *10*, 2874. [CrossRef]
6. Chen, S.; Chen, B.; Yin, D.; Guo, W. Characteristics of Discontinuous Surface Deformation Due to Mining in Hard, Thick Bedrock: A Case Study. *Geotech. Geol. Eng.* **2019**, *37*, 2639–2645. [CrossRef]
7. Straub, D.; Schneider, R.; Bismut, E.; Kim, H.J. Reliability analysis of deteriorating structural systems. *Struct. Saf.* **2020**, *82*, 101877. [CrossRef]
8. Drobiec, Ł.; Niemiec, T.; Kawulok, M.; Słowik, L.; Chomacki, L. The method of strengthening the church building in terms of the planned mining exploitation. *MATEC Web Conf.* **2019**, *284*, 05003. [CrossRef]
9. Rusek, J. The Point Nuisance Method as a Decision-Support System Based on Bayesian Inference Approach. *Arch. Min. Sci.* **2020**, *65*, 117–127.
10. Firek, K.; Rusek, J. Partial least squares method in the analysis of the intensity of damage in prefabricated large-block building structures. *Arch. Min. Sci.* **2017**, *62*. [CrossRef]
11. Carleo, G.; Cirac, I.; Cranmer, K.; Daudet, L.; Schuld, M.; Tishby, N.; Vogt-Maranto, L.; Zdeborová, L. Machine learning and the physical sciences. *Rev. Mod. Phys.* **2019**, *91*, 45002. [CrossRef]
12. Pan, Y.; Zhang, L. Roles of artificial intelligence in construction engineering and management: A critical review and future trends. *Autom. Constr.* **2021**, *122*, 103517. [CrossRef]
13. Rusek, J.; Tajduś, K.; Firek, K.; Jędrzejczyk, A. Score-based Bayesian belief network structure learning in damage risk modelling of mining areas building development. *J. Clean. Prod.* **2021**, *296*, 126528. [CrossRef]
14. Cai, B.; Liu, Y.; Liu, Z.; Chang, Y.; Jiang, L. *Bayesian Networks for Reliability Engineering*; Springer: Singapore, 2020.
15. Li, B.; Li, H. Prediction of tunnel face stability using a naive Bayes classifier. *Appl. Sci.* **2019**, *9*, 4139. [CrossRef]
16. Scanagatta, M.; Salmerón, A.; Stella, F. A survey on Bayesian network structure learning from data. *Prog. Artif. Intell.* **2019**, *8*, 425–439. [CrossRef]
17. Mihaljević, B.; Bielza, C.; Larrañaga, P. bnclassify: Learning Bayesian Network Classifiers. *R J.* **2019**, *10*, 455. [CrossRef]
18. Zhang, Z.; Cui, P.; Zhu, W. Deep Learning on Graphs: A Survey. *IEEE Trans. Knowl. Data Eng.* **2020**. [CrossRef]
19. Cherny, S.S.; Nevo, D.; Baraz, A.; Baruch, S.; Lewin-Epstein, O.; Stein, G.Y.; Obolski, U. Revealing antibiotic cross-resistance patterns in hospitalized patients through Bayesian network modelling. *J. Antimicrob. Chemother.* **2020**, *76*, 239–248. [CrossRef]
20. Liber, Y.; Cornet, D.; Tournebize, R.; Feidt, C.; Mahieu, M.; Laurent, F.; Bedell, J.-P. A Bayesian network approach for the identification of relationships between drivers of chlordecone bioaccumulation in plants. *Environ. Sci. Pollut. Res.* **2020**, *27*, 41046–41051. [CrossRef] [PubMed]
21. bint E Ajmal, H.; Madden, M.G. Dynamic Bayesian Network Learning to Infer Sparse Models from Time Series Gene Expression Data. *IEEE/ACM Trans. Comput. Biol. Bioinform.* **2021**, 1. [CrossRef] [PubMed]
22. Qu, L.; Wang, Z.; Huo, Y.; Zhou, Y.; Xin, J.; Qian, W. Distributed Local Bayesian Network for Gene Regulatory Network Reconstruction. In Proceedings of the 2020 6th International Conference on Big Data Computing and Communications (BIGCOM), Deqing, China, 24 July 2020; pp. 131–139.
23. Kabir, S.; Papadopoulos, Y. Applications of Bayesian networks and Petri nets in safety, reliability, and risk assessments: A review. *Saf. Sci.* **2019**, *115*, 154–175. [CrossRef]
24. Zhou, Y.; Li, C.; Zhou, C.; Luo, H. Using Bayesian network for safety risk analysis of diaphragm wall deflection based on field data. *Reliab. Eng. Syst. Saf.* **2018**, *180*, 152–167. [CrossRef]
25. Wu, Z.; Shen, Y.; Wang, H.; Wu, M. Urban flood disaster risk evaluation based on ontology and Bayesian Network. *J. Hydrol.* **2020**, *583*, 124596. [CrossRef]

26. Zhang, S.; Li, C.; Zhang, L.; Peng, M.; Zhan, L.; Xu, Q. Quantification of human vulnerability to earthquake-induced landslides using Bayesian network. *Eng. Geol.* **2020**, *265*, 105436. [CrossRef]
27. Song, M.J.; Cho, Y.S. Modeling maximum tsunami heights using bayesian neural networks. *Atmosphere* **2020**, *11*, 1266. [CrossRef]
28. Rusek, J.; Firek, K.; Słowik, L. Extracting structure of bayesian network from data in predicting the damage of prefabricated reinforced concrete buildings in mining areas. *Eksploat. Niezawodn.* **2020**, *22*. [CrossRef]
29. Noghabaei, M.; Heydarian, A.; Balali, V.; Han, K. Trend Analysis on Adoption of Virtual and Augmented Reality in the Architecture, Engineering, and Construction Industry. *Data* **2020**, *5*, 26. [CrossRef]
30. Hilal, M.A.; Maqsood, T.; Abdekhodaee, A. *A Hybrid Conceptual Model for BIM Adoption in Facilities Management: A Descriptive Analysis for the Collected Data BT—Collaboration and Integration in Construction, Engineering, Management and Technology*; Ahmed, S.M., Hampton, P., Azhar, S., D. Saul, A., Eds.; Springer International Publishing: Berlin, Germany, 2021; pp. 327–332.
31. Amos, D.; Au-Yong, C.P.; Musa, Z.N. *Overview of Facilities Management and the Public Healthcare System in Ghana BT—Measurement of Facilities Management Performance in Ghana's Public Hospitals*; Amos, D., Au-Yong, C.P., Musa, Z.N., Eds.; Springer: Berlin, Germany, 2021; pp. 9–19. ISBN 978-981-33-4332-0.
32. Ghobakhloo, M. Industry 4.0, digitization, and opportunities for sustainability. *J. Clean. Prod.* **2020**, *252*, 119869. [CrossRef]
33. Rusek, J. Support Vector Machines and Probabilistic Neural Networks in the Assessment of the Risk of Damage to Water Supply Systems in Mining Areas. Available online: https://depot.ceon.pl/handle/123456789/12466 (accessed on 9 August 2021).
34. Malinowska, A.; Hejmanowski, R.; Rusek, J. Estimation of the Parameters Affecting the Water Pipelines on the Mining Terrains with A Use of An Adaptive Fuzzy System. *Arch. Min. Sci.* **2016**, *61*. [CrossRef]
35. Scutari, M.; Vitolo, C.; Tucker, A. Learning Bayesian networks from big data with greedy search: Computational complexity and efficient implementation. *Stat. Comput.* **2019**, *29*, 1–14. [CrossRef]
36. Zhang, L.; Wu, X.; Qin, Y.; Skibniewski, M.J.; Liu, W. Towards a Fuzzy Bayesian Network Based Approach for Safety Risk Analysis of Tunnel-Induced Pipeline Damage. *Risk Anal.* **2016**, *36*, 278–301. [CrossRef] [PubMed]
37. Rostamabadi, A.; Jahangiri, M.; Zarei, E.; Kamalinia, M.; Alimohammadlou, M. A novel Fuzzy Bayesian Network approach for safety analysis of process systems; An application of HFACS and SHIPP methodology. *J. Clean. Prod.* **2020**, *244*, 118761. [CrossRef]
38. International Organization for Standardization Switzerland for PN ISO 2394: 2000. *General Principles on Reliability for Structures*; International Organization for Standardization: Geneva, Switzerland, 2000.
39. Qin, J.; Zhu, J.; Hua, D.; Yuan, Y.; Yang, S. FEM analysis on masonry houses subjected to ground deformation. *Ind. Constr.* **2002**, *32*, 41–44.
40. Abdallah, M.; Verdel, T. Behavior of a masonry wall subjected to mining subsidence, as analyzed by experimental designs and response surfaces. *Int. J. Rock Mech. Min. Sci.* **2017**, *100*, 199–206. [CrossRef]
41. Capanna, I.; Aloisio, A.; Di Fabio, F.; Fragiacomo, M. Sensitivity assessment of the seismic response of a masonry palace via non-linear static analysis: A case study in l'aquila (italy). *Infrastructures* **2021**, *6*, 8. [CrossRef]
42. Tatara, T. Influence of rigid lintels for dynamic response of low masonry buildings due to mining-related surface vibrations. *Czas. Tech. Bud.* **2006**, *103*, 143–157.
43. Han, L.; Zhang, J.; Zhang, Y.; Ma, Q.; Alu, S.; Lang, Q. Hazard Assessment of Earthquake Disaster Chains Based on a Bayesian Network Model and ArcGIS. *ISPRS Int. J. Geo-Inform.* **2019**, *8*, 210. [CrossRef]
44. Mitra, D.; Bhandery, C.; Mukhopadhyay, A.; Chanda, A.; Hazra, S. *Landslide Risk Assessment in Darjeeling Hills Using Multi-criteria Decision Support System: A Bayesian Network Approach BT—Disaster Risk Governance in India and Cross Cutting Issues*; Pal, I., Shaw, R., Eds.; Springer: Berlin, Germany, 2018; pp. 361–386. ISBN 978-981-10-3310-0.
45. Tajduś, K. Analysis of horizontal displacement distribution caused by single advancing longwall panel excavation. *J. Rock Mech. Geotech. Eng.* **2015**, *7*, 395–403. [CrossRef]
46. Kawulok, M. *Diagnozowanie budynków zlokalizowanych na terenach górniczych*; Instytut Techniki Budowlanej: Warsaw, Poland, 2021.
47. Kawulok, M. Osąd eksperta w ochronie istniejących obiektów budowlanych na terenach górniczych. *Przegląd Górniczy* **2015**, *71*, 38–43.
48. Hejmanowski, R. Modeling of time dependent subsidence for coal and ore deposits. *Int. J. Coal Sci. Technol.* **2015**, *2*, 287–292. [CrossRef]
49. Qin, T. *Machine Learning Basics BT—Dual Learning*; Qin, T., Ed.; Springer: Berlin, Germany, 2020; pp. 11–23. ISBN 978-981-15-8884-6.
50. ISO ISO 2394:2015. *General Principles on Reliability for Structures*; International Organization for Standardization: Geneva, Switzerland, 2015.
51. Murphy, K. *Machine Learning A Probabilistic Perspective*; Massachusetts Institute of Technology: Cambridge, MA, USA, 2012.
52. Scutari, M.; Graafland, C.E.; Gutiérrez, J.M. Who learns better Bayesian network structures: Accuracy and speed of structure learning algorithms. *Int. J. Approx. Reason.* **2019**, *115*, 235–253. [CrossRef]
53. Pearl, J. *Probabilistic Reasoning in Intelligent Systems: Networks of Plausible Inference*; Elsevier: Amsterdam, The Netherlands, 2014.
54. Scutari, M.; Ness, R. Package 'bnlearn'. 2019. Available online: https://cran.r-project.org/web/packages/bnlearn/bnlearn.pdf (accessed on 9 August 2021).
55. R Core Team. R: A Language and Environment for Statistical Computing. 2019. Available online: http://r.meteo.uni.wroc.pl/web/packages/dplR/vignettes/intro-dplR.pdf (accessed on 9 August 2021).

56. Kuhn, M.; Wing, J.; Weston, S.; Williams, A.; Keefer, C.; Engelhardt, A.; Cooper, T.; Mayer, Z.; Ziem, A.; Scrucca, L.; et al. Package 'caret' R Topics Documented. 2020. Available online: https://cran.r-project.org/web/packages/caret/caret.pdf (accessed on 9 August 2021).
57. Majka, M. Package Naivebayes: High Performance Implementation of the Naive Bayes Algorithm in R. 2019. Available online: https://cran.r-project.org/web/packages/naivebayes/naivebayes.pdf (accessed on 9 August 2021).
58. Ramezan, C.A.; Warner, T.A.; Maxwell, A.E. Evaluation of sampling and cross-validation tuning strategies for regional-scale machine learning classification. *Remote Sens.* **2019**, *11*, 185. [CrossRef]
59. Chicco, D.; Jurman, G. The advantages of the Matthews correlation coefficient (MCC) over F1 score and accuracy in binary classification evaluation. *BMC Genom.* **2020**, *21*, 1–3. [CrossRef] [PubMed]
60. Majka, M. Introduction to naivebayes package Main functions. *Cran. R-Proj.* **2020**, 1–15. Available online: https://cran.r-project.org/web/packages/naivebayes/vignettes/intro_naivebayes.pdf (accessed on 9 August 2021).
61. Long, Y.; Wang, L.; Sun, M. Structure extension of tree-augmented naive bayes. *Entropy* **2019**, *21*, 721. [CrossRef] [PubMed]
62. Russell, S.; Norvig, P. *Artificial Intelligence: A Modern Approach*; Prentice Hall: Upper Saddle River, NJ, USA, 2009; ISBN 0131038052.
63. Olivier, J.; Aldrich, C. Use of decision trees for the development of decision support systems for the control of grinding circuits. *Minerals* **2021**, *11*, 595. [CrossRef]
64. Papadopoulos, V.; Giovanis, D.G. Reliability Analysis. In *Stochastic Finite Element Methods: An Introduction*; Springer International Publishing: Berlin, Germany, 2018; pp. 71–98. ISBN 978-3-319-64528-5.
65. Xiao, N.C.; Zuo, M.J.; Zhou, C. A new adaptive sequential sampling method to construct surrogate models for efficient reliability analysis. *Reliab. Eng. Syst. Saf.* **2018**, *169*, 330–338. [CrossRef]
66. Scuro, C.; Sciammarella, P.F.; Lamonaca, F.; Olivito, R.S.; Carni, D.L. IoT for structural health monitoring. *IEEE Instrum. Meas. Mag.* **2018**, *21*, 4–14. [CrossRef]

Article

Coupling NCA Dimensionality Reduction with Machine Learning in Multispectral Rock Classification Problems

Brian Bino Sinaice [1,*], Narihiro Owada [2], Mahdi Saadat [1], Hisatoshi Toriya [1], Fumiaki Inagaki [1], Zibisani Bagai [3] and Youhei Kawamura [4]

1   Department of Geosciences, Geotechnology and Materials Engineering for Resources, Graduate School of International Resource Sciences, Akita University, Akita 010-8502, Japan; msaadat014@gmail.com (M.S.); toriya@gipc.akita-u.ac.jp (H.T.); fumiaki.inagaki@gipc.akita-u.ac.jp (F.I.)
2   Technical Division, Faculty of International Resource Sciences, Akita University, Akita 010-8502, Japan; jixidahetian@gmail.com
3   Department of Geology, University of Botswana, Private Bag UB 0022, Gaborone, Botswana; bagaizb16@gmail.com
4   Division of Sustainable Resources Engineering, Faculty of Engineering, Hokkaido University, Sapporo 060-8628, Japan; kawamura@eng.hokudai.ac.jp
*   Correspondence: bsinaice@rocketmail.com

Citation: Sinaice, B.B.; Owada, N.; Saadat, M.; Toriya, H.; Inagaki, F.; Bagai, Z.; Kawamura, Y. Coupling NCA Dimensionality Reduction with Machine Learning in Multispectral Rock Classification Problems. *Minerals* **2021**, *11*, 846. https://doi.org/10.3390/min11080846

Academic Editors: Rajive Ganguli, Sean Dessureault, Pratt Rogers and Amin Beiranvand Pour

Received: 20 May 2021
Accepted: 3 August 2021
Published: 5 August 2021

Abstract: Though multitudes of industries depend on the mining industry for resources, this industry has taken hits in terms of declining mineral ore grades and its current use of traditional, time-consuming and computationally costly rock and mineral identification methods. Therefore, this paper proposes integrating Hyperspectral Imaging, Neighbourhood Component Analysis (NCA) and Machine Learning (ML) as a combined system that can identify rocks and minerals. Modestly put, hyperspectral imaging gathers electromagnetic signatures of the rocks in hundreds of spectral bands. However, this data suffers from what is termed the 'dimensionality curse', which led to our employment of NCA as a dimensionality reduction technique. NCA, in turn, highlights the most discriminant feature bands, number of which being dependent on the intended application(s) of this system. Our envisioned application is rock and mineral classification via unmanned aerial vehicle (UAV) drone technology. In this study, we performed a 204-hyperspectral to 5-band multispectral reduction, because current production drones are limited to five multispectral bands sensors. Based on these bands, we applied ML to identify and classify rocks, thereby proving our hypothesis, reducing computational costs, attaining an ML classification accuracy of 71%, and demonstrating the potential mining industry optimisations attainable through this integrated system.

Keywords: hyperspectral imaging; multispectral imaging; dimensionality reduction; neighbourhood component analysis; artificial intelligence; machine learning

1. Introduction

The adoption of advanced automated technology into various industries has proven to be highly effective in improving sustainability and efficiencies. This is greatly due to the optimisation of system designs, data collection methods and the overall implementation of automation. With this said, the mining industry has been no stranger to this growing trend. This industry strives for the improvement of safety regulations via increasing the distance between miners and the environment [1]. This is where automated technology plays its part, by improving site data collection methods followed by high accuracy analysis methods [2]. One of such improvements has been demonstrated by researchers [2,3], where they employed hyperspectral signatures of rocks and a neural network to classify rocks based on their spectral signatures. With such studies have proved the advantages of using these technologies in terms of safety and improved data analysis, they have highlighted their main disadvantages. Though the hundreds of spectral bands in hyperspectral imaging provide a multitude of highly detailed data [4], this data suffers from what is referred to as the 'dimensionality curse'. This is defined as the inability to visualise such depth possessing data structures [5].

Moreover, Tong et al. [6] highlight that though deep neural networks acquire high accuracy results, executing them is computationally costly and time-consuming, deeming them highly difficult to employ in rapid on-site investigations [7]. To counter these shortcomings, this paper is a proposal whose attempt is to improve the application of rock spectral imaging. By converting from hyperspectral imaging to multispectral imaging [8], this will be performed through dimensionality reduction (DR) via Neighbourhood Component Analysis (NCA). Lastly, employing different Machine Learning (ML) models whose purpose is to access the attainable rock discrimination capabilities based on the NCA selected bands as summarised in Figure 1.

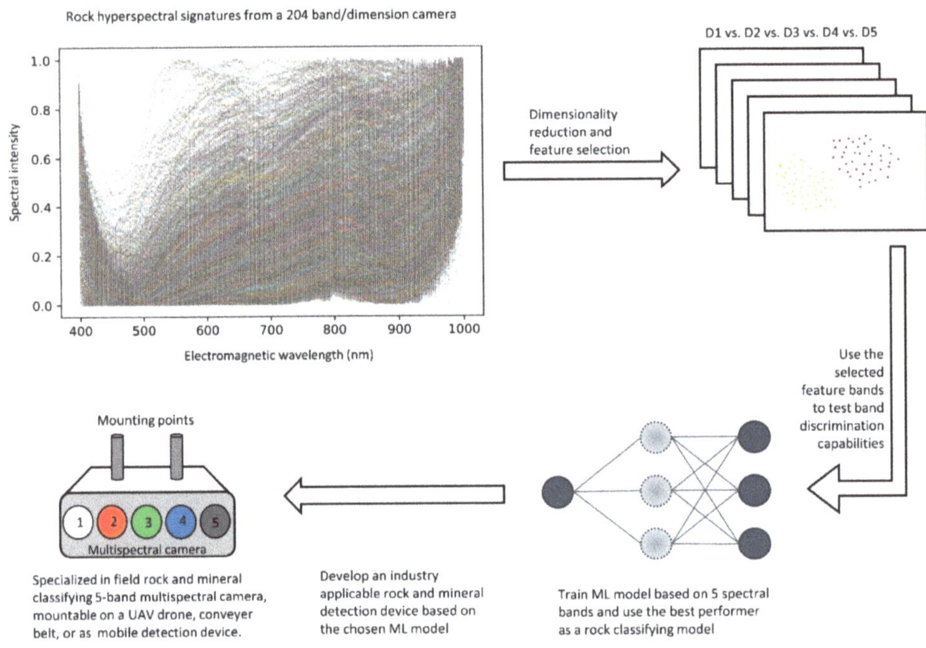

Figure 1. Our proposed system design, consisting of hyperspectral imaging, dimensionality reduction via Neighbourhood Component Analysis, Machine Learning classification of rocks and minerals to test system viability based on the selected features, and finally employing those features together with the ML model in developing a unmanned aerial vehicle drone-mountable multispectral camera.

The benefits of employing this proposed system within the mining industry are endless. For instance, multispectral signatures of rocks are more than viable in the discrimination of rocks [9] for the purpose of determining the correct blasting procedures based on the type and state of rocks. Moreover, mining engineers are always faced with tasks, such as determining adequate slope angles within open pit mines, determining dilution ratios in the processing of ore, and determining waste rock quantities, amongst other standards [10]. These all depend on rock information, this rock information being attainable via rock multispectral signatures [9] without the need to employ expensive hyperspectral imaging, hence achieving cost efficiency, fast data collection times, and quicker analysis rates.

Other than this, with NCA, engineers should hypothetically be able to convert from using highly detailed hyperspectral data together with its often-redundant datasets [11], to using more specialised multispectral datasets. This specialisation can be set such that the multispectral bands focus on detecting specific phenomena, these being specific rocks, minerals, ores, tailings, dam metal contaminants, and more. Moreover, NCA can be used in determining the most important criteria [12] in mining, ore processing, quarrying, geological, and geotechnical assessments. This is through assigning feature weights [12] to datasets, such as rock hardness, presence of clay minerals, weathering intensity and water content, amongst others. These weights consequently allow for the elimination of redundant data based on the intended applications of such data [13].

Having determined the specific bands viable to discriminate a certain rock and/or mineral database, or equivalent via NCA, the selected features or multispectral bands can be accessed via various machine learning (ML) models to determine the distinguishing capabilities of such models. Their performances are usually judged based time required to train, global accuracies and sub-class precisions [14,15]. Thereafter, engineers should potentially be able to commission the construction of a multispectral sensing device built using the best performing ML model. Consequently, making this integrated system a novel method in which specific rocks, minerals and other phenomena can be classified via a specialised multispectral sensor. Advantages of such would include improved remote sensing, which in turn improves workplace safety, traceability of data and results, and the overall optimisation of the classification system. As one can imagine, our proposed method is not only limited to mining industry applications, and it has the potential to be employed in multitudes of other industries, such as in agriculture, forensics, biology, and banking, amongst others.

To understand the technicalities of this proposed integrated system (Figure 1), we will explain each of these technologies and how they have been previously applied by other researchers in Section 2. Having defined the ideal number of spectral bands whose position within the electromagnetic spectrum will be determined by NCA, these said bands can potentially be employed in multiple areas within the mining industry. One of such potential applications is the development of 5-band-multispectral sensing cameras mountable on unmanned aerial vehicle (UAV) drones, because current industry standards for in-situ classifications related to the state of the environment are usually limited to 5-bands. Therefore, our paper will aim to satisfy the current 5-band multispectral production camera and drone trends, yet still demonstrate the different ways in which this integrated system can be taken advantage of in mining, rock and mineral engineering industries and/or studies.

2. Methodology for Coupling NCA Dimensionality Reduction with Machine Learning

2.1. Hyperspectral Imaging

Hyperspectral imaging, as defined by researchers [2,14], refers to the collection of hundreds of pixel-scale imagery information pertaining to a subject from within the electromagnetic spectrum. The collection of such data, which in our case was within the Visible-Near-Infrared-Range (VNIR), numerically translates to the 400–1000 nm electromagnetic range [7]. Hyperspectral imaging is a graduation from multispectral imaging, meaning that within the same spectral range, hyperspectral imaging has a higher resolution, thereby facilitating the extraction of detailed spectral signatures [2,7,14]. Since its discovery, it has seen various applications in fields, such as soil sciences, hydrology, geology and the mining industry [7,15]. When an image is captured using hyperspectral camera, such as our 204 band Specim IQ capturing camera, information pertaining to the subject's interaction with light is recorded [16]. This makes each of the 204 VNIR spectral bands receive a specific signal within each of the approximately 3 nm wide spectral bands. It should be mentioned that camera specifications may differ in terms of the number of spectral bands per spectral range provided by a certain manufacturer. This, in essence, affects the width of each spectral band, it however does not affect the underlying signatures exhibited by specific rocks and minerals.

Having said this, it is evident that analysing hyperspectral data requires sophisticated analysis software. This is because this type of data is computationally costly to analyse, due to the depth of information bands it possesses—often referred to as dimensionalities [5], hence the term dimensionality-curse [5,17,18]. To counter this phenomenon, a method referred to as DR needs to be applied to reduce or eliminate redundant information. Doing so requires a selection of the most representative spectral bands, able to distinguish rocks within our database without affecting or altering their inherent spectral signature differences.

2.2. Dimensionality Reduction

DR techniques have in the last couple of decades been a topic of interest for researchers working in computational statistics [19]. It is a key technique in data analysis, aimed at revealing expressive structures and unexpected relationships in multivariate data [20]. It should, however, be noted that, in general, it is not possible to preserve all pairwise relationships between data points in the DR process [12]. DR is used for many purposes; it is beneficial as a visualisation tool to present multivariate data in a humanly accessible form (lower dimensions). Moreover, DR can be applied as a method of feature extraction, and as a preliminary transformation applied to data, such as our rock hyperspectral database prior to the usage of other analysis tools like clustering and classification [21].

There are many criteria that can be used to sort the various methods of DR. With our objective being a classification task, the aim of our DR is, therefore, to project high-dimensional data points in a low-dimensional subspace whilst keeping most of the 'intrinsic information' contained in the original data preserved. This, in principle, keeps the within-class-sample compactness and between class-sample distinguishability [22]. The success of which means that low-dimensional presentation of original data may provide enough information for classification.

2.2.1. Supervised vs. Unsupervised Methods

Several DR techniques that reduce the size of the data table, while minimising loss of information have been studied, all of which can describe the essence of the primary data generated. Among these numerous methods, principal component analysis (PCA), linear discriminant analysis (LDA) and maximum margin criterion (MMC) are the most famous ones because of their simplicity and effectiveness [23]. Due to the nature of our data, we found that, geometrically, feature extractors based on maximum margin criterion (MMC) maximise the (average) margin between classes after dimensionality reduction. This would not improve our research as our goal is to use machine learning for this task [23]. On the other hand, the linear discriminant analysis (LDA) method operates by finding a linear combination of input features. However, the performance of LDA is degraded when encountering limited available low dimensional spaces and singularity problems, which is one of the disadvantages of LDA [23]. Lastly, PCA is a linear dimensionality reduction technique that transforms a set of correlated variables into a smaller number of uncorrelated variables called principal components, while retaining as much of the variation in the original dataset as possible [23–25].

In addition, sometimes the performance of these methods is limited, as these methods are often unsupervised. Therefore, these methods only use the global structure of the sample, while ignoring the local structure, which are extremely important in helping to improve the discrimination of the sample in the projection space. To improve on this, we have employed a supervised NCA method. The assumption being, with this supervised method, the outcome of interest informs the DR solution—this occurs because this method naturally considers the local structures and their labels [24].

2.2.2. Why Use NCA

While PCA is one of the most commonly used approaches for DR, the method does not reduce the number of variables [25]. The analyst chooses the number of components to include in analyses based on a prior defined criterion. For example, looking at the screen plot, selecting components with eigenvalues above one, or selecting the number of components that explain a prespecified proportion of the variance in the data [23]. Because PCA forces orthogonality between components, it imposes a rigid structure [23,24].

NCA, on the other hand, performs better both in terms of classification performance in the projected representation and in terms of visualisation of class separation as compared to the standard unsupervised methods. Moreover, regarding NCA, one can substantially reduce the storage, search, running and time spent on waiting costs at the test phase by forcing the learned distance metric to be low rank. This, therefore, favours its potential application in real-time field analyses [26].

NCA is a distance-based feature weighting, non-parametric supervised method, it works by automatically selecting the most significant features [11,25]. To calculate the correlation between features and target, a Mahalanobis distance-based fitness function is used. The weighting of features is carried out as follows; initial weights are assigned randomly, thereafter, weights are updated using the stochastic gradient descent or ADAM optimisation method and Mahalanobis distance-based function, hence positive weights are generated for each feature [11,25,26].

Though Goldberger et al. [11] applied their NCA algorithm for face recognition, they too mention that the NCA algorithm learns a training set distance metric, and can improve k-NN classifications, hence achieving very good performance. Koren & Carmel [20] further support the employment of an NCA model by saying it provides a linear transformation model that optimises the performance of k-NN in the learnt low-dimensional space. These said advantages influenced our desire to employ NCA in distinguishing rocks from within our rock hyperspectral database.

However, researchers [11,20,26,27] note that unlike the common PCA method, which is both convex and has an analytical solution, another key difference distinguishing the two is that NCA is a non-convex optimisation problem. This means every time one runs NCA, they may get a different solution, and like K-Means and other non-convex algorithms, it is advisable to run it more than once and take the best solution. Hence, our paper presents the best NCA bands from having run NCA multitudes of times and selected features that express themselves most frequently. Researchers [20,26] explain this by noting that this occurs as NCA components are not ordered nor dependent on the chosen target dimensions.

This, however, is not a drawback as run-times are extremely short. Moreover, once the number and specific band positions have been specified, subsequent classification tasks require significantly less storage, fewer test times, and the redundant bands are eliminated along with their datasets. These chosen components (spectral bands) are, therefore, assumed to be the most sufficient in determining the rocks within, or related to, a said database. These said sample signatures may include mine, laboratory or environmental rock spectral signatures, such as in our case.

2.3. Why Machine Learning?

Though NCA provides the opportunity to ignore redundant data-heavy bands, it does not provide any information related to the retained classification accuracy, hence the need to employ an ML algorithm(s). We use ML for multitudes of data-related tasks or problems. It has grown as a subdomain of Artificial Intelligence (AI) that comprises models capable of deriving useful information from data and utilising that information in self-learning that aids in making good classifications or predictions [7,8]. ML has gradually gained popularity, due to its accuracy and reliability [28]. Improved hardware and software components of machine vision systems have aided in building ML algorithms that process data faster and give reliable decisions in very little time [13]. Since we are dealing with a classification problem with labelled data, we employed and compared several supervised ML algorithms. Supervised learning requires learning a model from labelled training data that helps in making classification or predictions about the future data [13,29]. Supervised, in essence, indicates samples sets in which the desired output is known. In other words, the labelling of data is done to guide the machine to look for the exact desired pattern.

3. Practical Experiments

3.1. Capturing Rock Hyperspectral Signatures

To craft this proposed system, the approach involved a series of steps to be followed to get to the ultimate goal of rock classification from integrating hyperspectral imaging, NCA and ML. To develop, test and propose this system for rock engineering and classification problems, we employed 32 different igneous rocks belonging to eight rock lithologies (four samples per lithology), namely, granite, diorite, gabbro, granodiorite, rhyolite, andesite, basalt and dacite. These samples were specimens of Akita Mining Museum, each with several representative samples, such as shown in Figure 2.

Figure 2. Images of some of the 32 rock samples from eight rock lithologies used in building our rock hyperspectral database.

To capture their spectral signatures, we used a Specim IQ hyperspectral camera (VNIR 400–1000 nm, 204 bands) to record the pixel-by-pixel signatures from rocks; the main components of the spectral data extraction setup are illustrated in Figure 3. As van der Meer [3] has pointed out, gathering this data entails standardising the spectral signature recording process. This was done by initialising the camera with a white reference board (provided by the manufacturer, Specim), the purpose of which is to filter out the noise and verify subsequent data is recorded under the same standardised conditions. The experimental setup utilises tungsten-halogen lamps to illuminate the stage as they have high output capabilities throughout the VNIR, which coincides with the camera capturing range.

Having captured the depth possessing, high dimensionality hyperspectral imagery data of all rocks, this data is then converted to numerical data using hyperspectral analysing software. Converting to numerical data entails the extraction of spectra from specific pixel blocks. This is performed after having assigned a data extraction area to allow automatic selection of pixel blocks with spectra to be considered for analysis (Figure 4). Since the Specim IQ camera acquires images of 512 × 512 pixels from 204 bands, the software randomly and automatically extracts 20 × 20-pixel information from these 204 band images (Figures 4 and A1). Meaning, each spectrum is an average of the spectral reflectance information from a 20 × 20-pixels (400 pixels) area; hence, each block becomes an average spectral strength with a depth of 204 bands (wavelength). The selection area boundaries are set by the user to ensure the software extracts only relevant data.

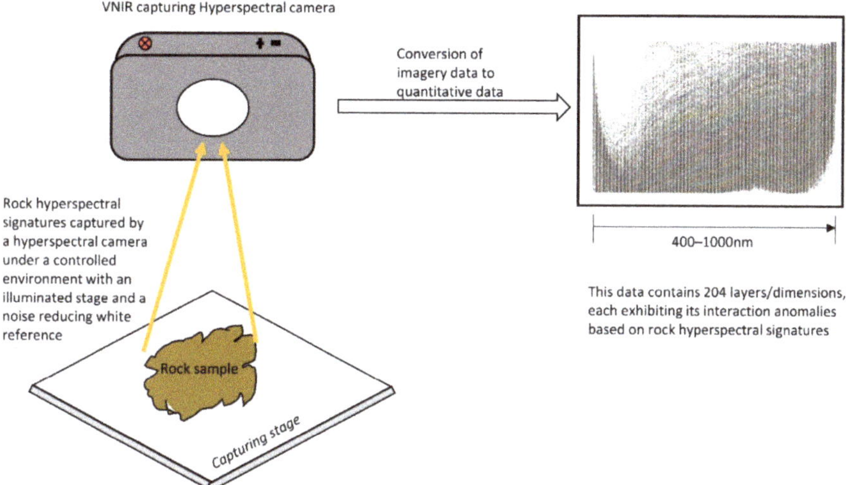

Figure 3. A hyperspectral signature capturing setup is used in collecting and building a rock and/or mineral spectral database. The curves are spectra in different spatial pixels of eight rock lithologies.

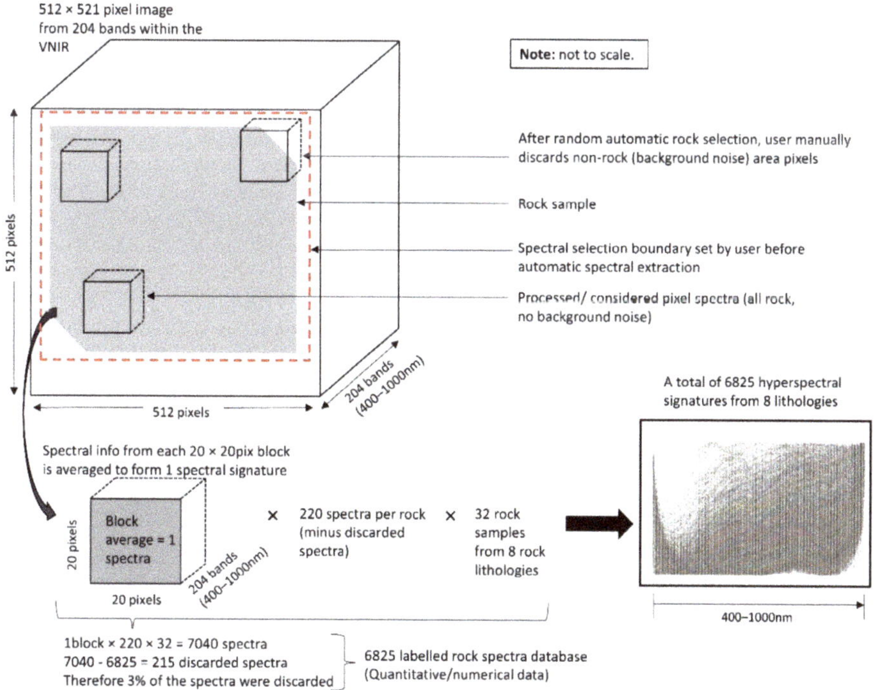

Figure 4. Automatic random extraction of rock pixel spectra captured via a Specim IQ camera that acquires 512 × 512 pixel images from 204 bands within the Visible-Near-Infrared-Range. Extracted spectra are used to build a rock hyperspectral database.

To extract 100% of the captured image spatial area of the 512 × 512 pixels = 262,144 total pixels, a total of approximately 655 spectra (the exact number is 655.36 pixels), each with a spatial area of 20 × 20pixels = 400 pixels would have to be extracted. This is derived from dividing 262,144 pixels by 400 pixels to get 655. However, since the spectral extractor used in this study extracts 220 pixels per image, this results in approximately 30% [262,144/(220 × 400)] of the image area being used for analysis. This 30% (made up of 220 spectra minus manual elimination of non-rock spectra) of the extracted whole image area, however, can be placed anywhere on the image area using the spectral extraction boundary controlled by the user. Therefore, should a perfect 88,000 pixels (220 × 20 × 20) area be defined by the user, 100% of the selected rock spectral information without background noise (see Figures 2 and 4) would be extracted. However, this was not the case in this study, as the extraction boundaries were randomly set judging from the area in which actual rock resides within each image.

Performing manual elimination (by the user) of unwanted spectra from the automatically extracted (by the software) 220 results in a lesser number of extracted spectra than the initial 220. As a result, from the 32 rock samples, we now have a total of 6825 [(220 spectra × 32 samples) minus unwanted background noise)] viable representative spectra from eight rock lithologies, each with a 204-band depth having been extracted for analysis. This, in essence, means the quantitative dataset has a matrix of size 204 × 6825 = 1,392,300 spectral information, which is used as input data in subsequent procedures. This data goes through a preprocessing stage where each dataset is assigned a relevant label; hence, a hyperspectral rock database was built based on the eight igneous rock lithologies. It should be noted that this process can be performed on any rocks, minerals or the combination or which. The choice or type of data used to develop a database depends entirely on the purpose in which rock or mineral classification is intended to be based upon. This, as a result, enables the AI coupled system to be highly specialised in classifying that which is within or related to the database.

3.2. Selecting the Appropriate Feature Bands

As previously mentioned in Section 2.2, the common problem that may arise during a DR process is to define how many features to select for analysis. This is often dictated by the purpose of employing such a DR method. In this paper, based on the currently available industry produced spectral imaging devices, such as the 'DJI P4 multispectral drone' used in agricultural applications and environmental monitoring, our objective was to identify the appropriate rock classifying multispectral bands. From these bands, it would then be easier to develop a UAV drone-mountable multispectral sensing camera specialised in classifying rocks and minerals.

To achieve this, we convert heavy hyperspectral imagery classification data to less heavy multispectral classification data to meet weight restrictions, industry standards and production costs of developing such a device. This conversion is performed in consideration of the rocks and/or minerals from which subsequent multispectral data collection, such as from a UAV drone-mounted multispectral camera, is to be recorded and classified for a plethora of rock engineering purposes.

Transitioning from high dimensional hyperspectral to low dimensional multispectral data is not without challenges. The selection of a suitable method according to the type of data is a big issue that often needs to be addressed. It is essential to find a suitable mechanism to attain the highest level of accuracy when comparing the outputs of different DR techniques. Since it is well documented that supervised methods generally outperform unsupervised methods, we employed the NCA DR technique as it is a supervised and highly acclaimed method. As a way of determining the significance of employing the 204 spectral bands with all 'redundant features', we used our NCA algorithm to eliminate and record the attainable output accuracies in classifying the rocks. This was based on the full 204 feature bands, down to 100, 50, 25, 10, and finally, the current UAV drone-mountable multispectral 5-band feature classification bands. NCA DR eliminates redundant information by assigning each dimensionality from within the hyperspectral signatures a feature weight.

As [11,25–27] researchers have mentioned before, finding the relevant and important features is a problematic task. It entails domain knowledge, and human expertise to extract the most relevant features for future processing and selection of ML models for classification [8,13,28,29]. Employing NCA, however, makes this process easier as the algorithm assigns feature weights to each of the dimensions, thereby highlighting the most relevant features/bands for such a database. Having employed NCA to select the most relevant features that contribute most to the prediction (dependent) variable, the final step entails exporting the selected features into an ML model. The model is then trained, thereafter, we can determine the classification capabilities based solely on these feature bands.

3.3. Post-NCA Classification via ML

To commence with the post-NCA classification task, we begin by preprocessing our data based on the number of spectral bands intended to be used in the rock and mineral classification task. For the initial training and classification, 100% of all the 204 spectral band signature data is employed for classification—this acts as a control task. Thereafter, depending on NCA feature weights, only the high-feature-weight possessing spectral bands are employed in succeeding classifications, which in essence means discarding the rest of the data that is deemed redundant. By doing this, we decrease data storage costs, as well as take a step towards developing a field applicable multispectral band camera. The classification was performed for 204-bands, 100, 50, 25, 10 and 5-bands using various ML models, thereby allowing for classification accuracy checks for the various band reductions.

4. Experimental and Analytical Results
4.1. Findings Based on Hyperspectral Imaging

As a way of visualising the characteristic rock and light interactions at a pixel level from within the VNIR, Figure 5 hyperspectral signatures are typical illustrations used to visualise these inherent reflectance signatures. Each anomaly represents a given 20×20 pixels block as an average spectral reflectance strength from the image scene. Based on Figure 5, one can appreciate the differences in spectral reflectance strength signatures attainable from different pixels within the same hyperspectral image. Moreover, the way different rock sample variants of the same rock exhibit different signatures combined to form hyperspectral signatures, hence, each of the eight rock lithologies shows dispersed hyperspectral signatures. Taylor [30] employed VNIR spectroscopy on their 'Mineral and Lithology Mapping of Drill Core Pulps' problem and concluded that spectrometry, like XRD, provides an evaluation of quantitative mineralogy that is very reliable. Hence, we are confident in hyperspectral imaging is very useful in our rock identification problem, as has been hypothesised. We see these inherent differences in the spectral signatures exhibited by the rocks in our database (Figure 5).

Patterns can be drawn from hyperspectral signatures, enabling one to distinguish individual rocks and/or minerals. However, it is difficult to extract a certain anomaly from each of the eight hyperspectral rock signatures and deem it the most representative spectral signature of a particular rock and/or mineral. This can be said when for example, examining the general spectral patterns of granite with those of diorite. Their anomaly shapes seem rather similar in terms of resembling 'check marks', with some of them displaying comparable reflectance intensities even; the same can be said when comparing gabbro signatures with those of andesite (Figure 5). Having seen the advantages and disadvantages of hyperspectral signatures employed as a means for rock and/or mineral classification, one can acknowledge that there is a need to employ a method by which significant data is given priority over redundant data. This allows for better comparisons and distinguishability of rocks and/or minerals via their spectral signatures, which is where NCA improves on this method of rock discrimination.

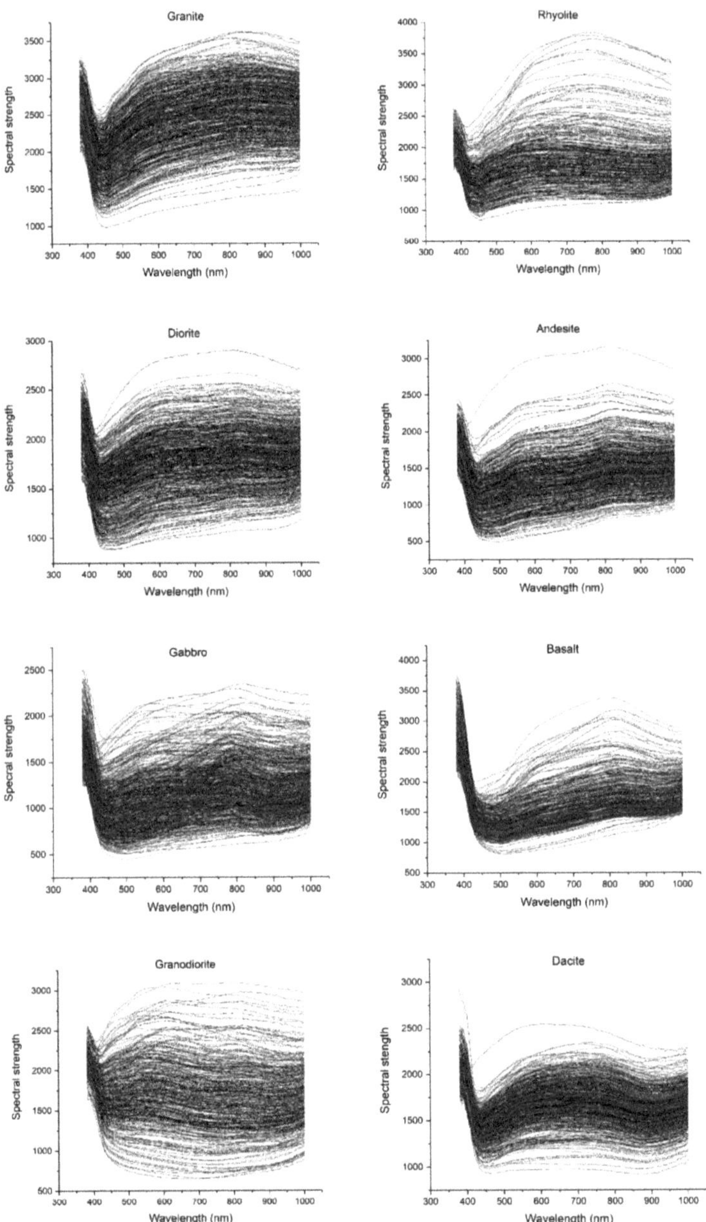

Figure 5. Reflectance hyperspectral signatures of eight rock lithologies employed in the construction of a hyperspectral database. Each anomaly represents the interaction between each 20 × 20 pixels 2D area with a depth of 204 bands within a rock's hyperspectral image with light, captured via a Visible-Near-Infrared-Range hyperspectral camera.

4.2. Findings Based on NCA

NCA is a method that seeks to identify and down-scale global unwanted variability within the data. The method changes the feature space used for data representation by a global linear transformation which assigns large weights to relevant dimensions, which are the most discriminatory spectral bands. Consequently, low weights are assigned to irrelevant dimensions, which we can, thus, refer to as less discriminatory spectral bands [26]. These relevant dimensions are estimated using a subset of points that are known to belong to the same although unknown class, also referred to as chunklets [31]. These chunklets are obtained from equivalence relations by applying a transitive closure within the algorithm. This transformation is, therefore, intended to reduce clutter, so that in the new feature space, the inherent structure of the data can be more easily unravelled [31,32].

Based on Figure 6, our NCA algorithm flawlessly reduced the dimensionality space of the hyperspectral signatures. We are, therefore, able to compare the different projection graphs of each of the 5-bands against one another in 2D spaces, hence mapping or visualising the manner in which the rocks plot at these chosen high classification dimensionalities. Results from the NCA algorithm in Figure 6 show that there are multitudes of spectral bands which one would refer to as relevant as they possess a substantial feature weight relative to the rest. Depending on the computational resources an entity or individual possesses, the number of spectral bands one desires to employ for future classifications having done away with redundant bands, is upon the user. Having said this, we used Figure 6 to select the most rated bands as we can indeed see the redundancy in some of the feature bands.

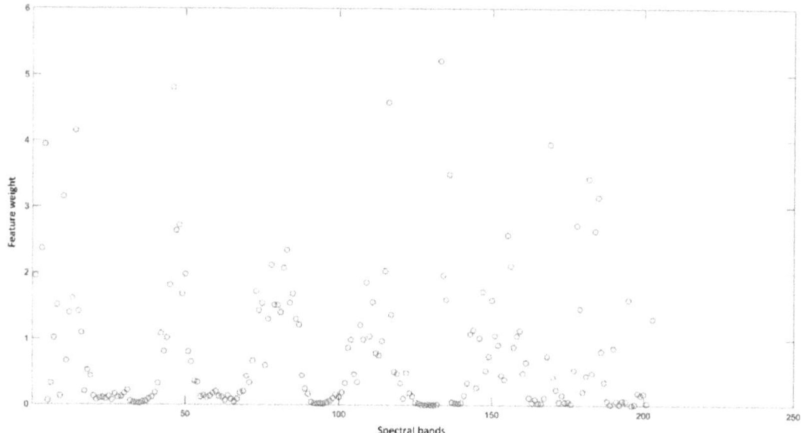

Figure 6. Neighbourhood Component Analysis feature selection which assigns higher weights to the most discriminatory hyperspectral bands by eliminating redundancy in data, with the top five feature-bands being 14, 46, 116, 133 and 169.

As stated in Sections 1 and 3, our intended use of the most representative feature bands requires 5-bands which, according to Figure 6, are located at positions 14, 46, 116, 133 and 169 from the 204 feature bands of the VNIR. From these selected feature band positions, we can then convert these positions into electromagnetic wavelength bands. Doing so, we get 441 nm, 535 nm, 741 nm, 791 nm and 897 nm as the most discriminatory spectral bands for our rock database. It should be noted that, considering each of these spectral bands are approximately 3 nm wide, a system designed to classify rocks based on these five spectral bands would have an error of $+/-3$ nm, as stated in Section 2.1. Having said this, we can safely say our NCA algorithm flawlessly assigns feature weights to high dimensionality hyperspectral data. This allows the user to select the number of spectral bands they wish to employ based on NCA assigned feature weights.

DR is the transformation of data from a high-dimensional space into a low-dimensional space so that the low-dimensional representation retains some meaningful properties of the original data, ideally close to its intrinsic dimension. From this statement and having selected the now five multispectral bands to employ in future classification problems related to our data, our NCA breaks down the hyperspectral signatures. This allows for visualisation in 2D spaces whose X and Y planes are the NCA-defined spectral bands with the highest feature weights, as shown in Figure 7.

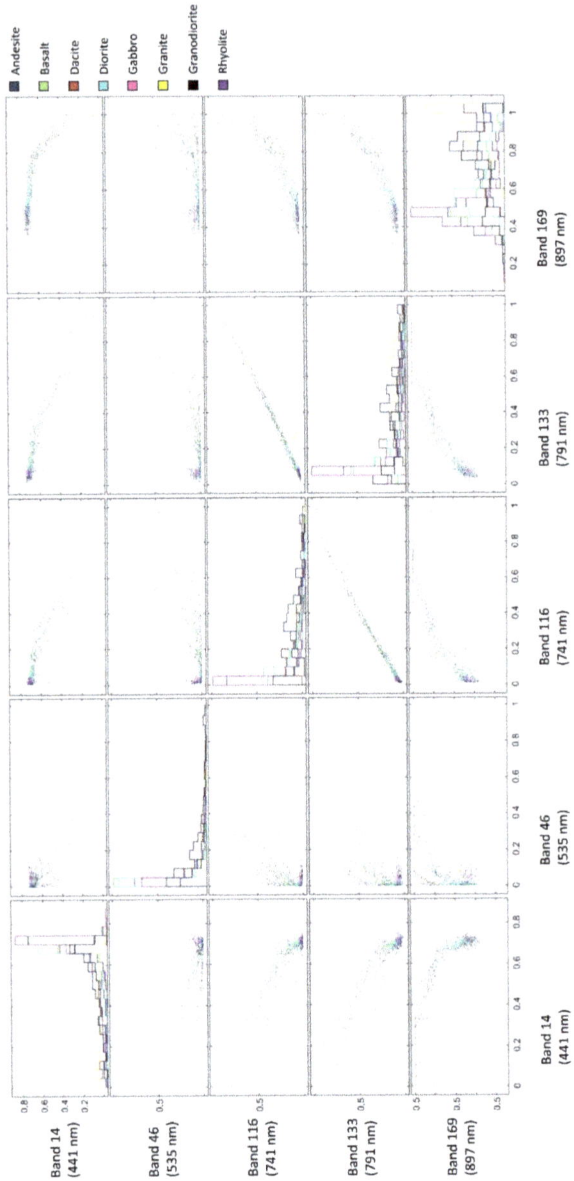

Figure 7. Projection of complex rock hyperspectral signature data points in dimensionality reduced (via Neighbourhood Component Analysis feature selection) 2D planes showing the relative reflectance spectral strength relationships between five different spectral bands for eight rock lithologies.

It should be noted that there are numerous ways in which these (Figure 7) dimensionally reduced band-by-band projections can be interpreted. Starting with the relative reflectance spectral strength scatter plots (normalised to 1), where each point represents the relative reflectance spectral strength of each rock sample's previously extracted 20×20 pixels averages. The area within which every point plot within the scatter plots is governed by the relative spectral strength between its (previously extracted 20×20 block averages) spectral reflectance strength within its respective spectral band (for example, band 14), in relation to the other band (hence 2D), as well as in relation to the other seven rocks (eight in total). Having said this, the point with the highest relative spectral reflectance strength in these three categories (respective band, other band and other rocks) would plot at the right most region of the scatter plot. Having plotted these points, the relative frequency histogram sums and summarises the frequency densities of the points from 4 scatter plots into 1. Since these histograms are also relative, the most densely populated rock points are normalised to 1. From interpreting these relative scatter and histogram plots, we can assess the spectral reflectance strengths of rocks in lower, dimensionality reduced, 2D planes.

Here, is an example of interpretations deducible from Figure 7 where band 14 occupies the x-axis, and bands 46, 116, 133 and 169 occupy the y-axis. From band 14's histogram (left most), we can make the following assumptions based on the number of rock spectral reflectance strength points summarising the scatter plots directly below it. Gabbro has the highest relative density of points (hence, the highest peak), meaning this rock has the highest concentration of points located within the further right small patch area of band 14. This is supported by the four scatter plot projections at bands 46, 116, 133 and 169, where we can see a similar dense cluster of gabbro data points (pink colour) located at this said location of the relative scatter plot projections (Figures 7 and A2). Within the same patch of area, we see that the next densely populated points belong to basalt and andesite, where basalt shows a slight edge over andesite. Below these point frequency densities, we find diorite, followed by rhyolite, granodiorite, granite and dacite as the least dense.

However, we see a difference in the density of points for the histogram bar on the immediate left to the previously described. The frequency of points starts from diorite as the most densely populated within this small area. This is followed by basalt, dacite, gabbro, granodiorite, a tie between andesite and rhyolite, and granite as the least dense. Within the same band 14 projection against other bands, we can see a relatively equal frequency density of data points within the first half (left to middle) of the projection for seven rocks. This is with the exception of granite, which has a higher density of points within this wide area. Therefore, we can make similar assessments of data points for the 46, 116, 133 and 169 band, as the x-axis and draw different patterns based on the frequency and location of rock relative reflectance strength points. Having said this, we can safely say, based on Figures 7 and A2 relative scatter and frequency histograms—thus, we can make predictions on future rock identification problems related to those of our study.

Hence, should new data, related to our rock database with similar multispectral bands (441 nm, 535 nm, 741 nm, 791 nm and 897 nm), be introduced, we expect such data to exhibit similar patterns as the rocks we have assessed. This may be in terms of relative frequency density relationships, or areas within which such rock relative reflectance spectral strength points are expected to exist. From these density histograms and scatter plots, we can make one more assumption. The more the frequency of data points exist within a small area, the easier it is to identify with a naked eye such rock relative points based on the scatter plots. The opposite is true for scattered data points. As much as NCA can reduce dimensionality, visual rock identification based on Figures 7 and A2 patterns alone is time-consuming and prone to some human error. Hence, there is a need to employ objective ML models which can draw patterns faster and accurately. Moreover, ML models give feedback with regards to the best rock delineation strategies.

4.3. Classification with ML, Post-NCA

Supervised Learning uses an algorithm that requires external help. The provided input database is automatically separated into training and testing datasets. The output variable is predicted or classified from the training database. Algorithms try to learn some shapes during training of the database and implement these learnt patterns to the testing database, which provides results in relation to the learnt patterns [13,28,29]. From these output results, we can evaluate the performance of each algorithm.

As shown in Figure 8, a 5-folds-cross-validation (6825 divided 5) was used to process the data at all times, resulting in 1365 (Figure A1) samples (including all eight rock lithologies) being used in each set. This ensures that every observation set (with each of the eight rocks contributing) from the original dataset has the chance of appearing in training and test sets as the ultimate goal is to classify entire rocks (Figure 8). This method generally results in a less biased model compared to other methods, it is said to be one of the best approaches whenever there is a limited amount of input data [29]. Since the number of samples is 6825 (rows), it does not reduce, only the number of bands/features reduces (columns, from 204 down to 5). Each row represents the spectral reflectance strength of each rock signature, whilst each column represents the position of the wavelength band from which the spectral strengths have been extracted, hence forming a 2D matrix. As a result, a breakdown of the input dataset matrices is as follows; for 204 bands (full database), input dataset is 204 (columns) × 6825 (rows) matrix = 1,392,300; for 100 bands, input dataset is 100 × 6825 matrix = 682,500; for 50 bands, input dataset is 50 × 6825 matrix = 341,250; for 25 bands, input dataset is 25 × 6825 matrix = 170,625; for 10 bands, input dataset is 10 × 6825 matrix = 68,250; for five bands, input dataset is 5 × 6825 matrix = 34,125.

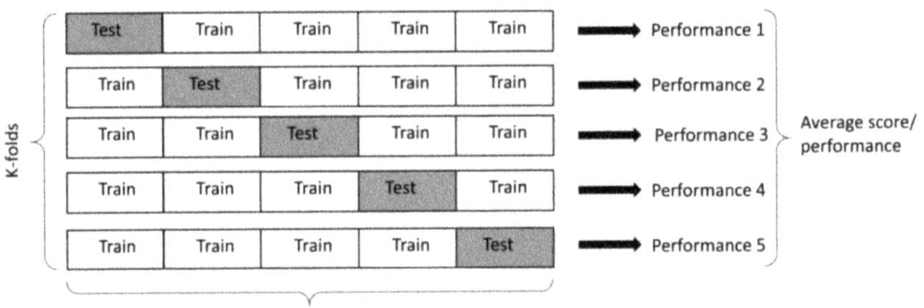

Figure 8. Database handling of training and testing sets. A 5-fold-cross-validation was used in all instances.

Using MATLAB R2020b classification learner Machine Learning toolbox, we assessed multiple ML algorithms and combined the best five classification performers in terms of training, average per class precision, and time taken to train the algorithm. These attributes are said to be the most important classification evaluation criteria. Moreover, these attributes govern industrial applicability, and the overall viability of the algorithm. Table 1 is a compilation of the top-performing ML algorithms per given number of selected spectral bands from the pre-DR 204 spectral bands, down to 100, 50, 25, 10 and our intended goal of five spectral bands. It demonstrates the differences in classification based on bands with the most feature weights. Results from the ML models in Table 1 show that the highest performing model in all pre- and postclassifications was Cubic Support Vector Machine (SVM).

Table 1. Top five machine learning classification comparisons based on predimensionality reduction from 204-bands, to postdimensionality reduction (using Neighbourhood Component Analysis) for 100, 50, 25, 10 and 5 rock spectral bands.

Number of Classification Bands Post-NCA	Machine Learning Algorithm	Global Accuracy (%)	Average per-Class Precision (%)	Training Time (s)
204-bands [1]	SVM (Cubic SVM)	90.7	90.0	28.7
	SVM (Quadratic SVM)	87.0	86.0	27.3
	SVM (Linear SVM)	79.1	76.5	13.8
	Linear discriminant	80.4	78.4	4.6
	Ensemble (Subspace discriminant)	81.2	79.3	41.5
100-bands	SVM (Cubic SVM)	89.4	88.7	37.7
	SVM (Quadratic SVM)	84.7	83.8	21.8
	Quadratic Discriminant	77.9	77.5	1.0
	SVM (Linear SVM)	76.9	76.0	5.9
	Linear Discriminant	76.7	75.6	1.1
50-bands	SVM (Cubic SVM)	86.9	85.7	39.1
	SVM (Quadratic SVM)	84.3	82.1	23.2
	Quadratic Discriminant	79.1	78.9	1.1
	SVM (Linear SVM)	76.1	75.4	5.3
	Ensemble (Subspace KNN)	75.2	75.0	34.9
25-bands	SVM (Cubic SVM)	86.3	86.2	45.1
	SVM (Quadratic SVM)	83.9	82.6	28.2
	SVM (Fine Gaussian SVM)	75.9	70.6	6.7
	Quadratic Discriminant	75.7	75.3	1.2
	Ensemble (Subspace KNN)	75.4	70.0	29.2
10-bands	SVM (Cubic SVM)	81.0	80.3	78.2
	SVM (Quadratic SVM)	78.2	76.0	40.7
	SVM (Fine Gaussian SVM)	72.7	70.1	8.2
	Ensemble (Bagged tress)	71.0	69.8	19.1
	Ensemble (Subspace KNN)	70.6	70.3	17.9
5-bands	SVM (Cubic SVM)	70.9	72.0	182.1
	Ensemble (Bagged trees)	68.6	67.0	12.3
	SVM (Quadratic SVM)	68.4	65.8	76.7
	SVM (Fine Gaussian SVM)	68.4	66.6	7.5
	KNN (Fine KNN)	67.3	66.0	5.7

[1] Predimensionality reduction.

A similar approach was applied by Galdames et al. [4] where they performed a feature selection from 2424 spectral channels to 73 spectral channels. Their study employed colour images, a VNIR sensor, as well as a SWIR (900–2500 nm) sensor. They achieved a classification performance of 99.73% using Conditional Mutual Information Maximisation to select their most important features. Considering the tools and number of bands selected at the most intrinsic bands, we would argue our methods achieves more for less. On the other hand, Mei et al. [33] employed Unsupervised Spatial-Spectral Feature Learning by 3D Convolutional Autoencoder for Hyperspectral Classification. Though with high classification capabilities, we believe this method is computationally taxing as convolutional neural networks are known to require a lot of training data and times, making CNNs invalid in our five feature bands quest.

From the results compiled in Table 1, we can appreciate the differences in accuracies acquired and elapsed times when training our ML models before and after DR. This confirms our hypothesis, which stated that with NCA, ML will maintain rapid run times and good accuracies, while maintaining without compromise, the fundamental differences in the hyperspectral signatures of rocks within our database. Global accuracy refers to the validation accuracy acquired during training. Average per-class precision refers to the individual rock classification sum averages in testing the models. Lastly, training time refs to elapsed time in training the models to classify the rocks based on the number of spectral band datasets.

As our goal was to reduce the number of hyperspectral bands to five multispectral bands capable of distinguishing rocks at a substantial, industry applicable accuracy, we assessed the highest performing Cubic SVM ML model for the 5-band classification. The results are presented in Figure 9. To assess the viability of this Cubic SVM model, Figure 9 presents two performance metrics. The first is True Positive Rates (TPR), defined as the probability that an actual positive will test positive (Equation (1)). The second is False Negative Rates (FNR), defined as the probability that a true positive will be missed by the test (Equation (2)). Both variables are highly viable in assessing the capability of the ML model in classifying each rock. Another assessment that can be drawn from Figure 9 confusion matrix is the average per-class precision of 72%. This is substantial considering the magnitude of the DR from 100% of that hyperspectral data (204-bands) to approximately 2.5% (5-bands), which we now refer to as multispectral data. We have, therefore, determined an applicable classification model for this particular problem. In addition to this, we gained a reduction in computational costs and storage requirements, ease of data management, ease of data application and visualisation, and most importantly, viability in rapid field applications.

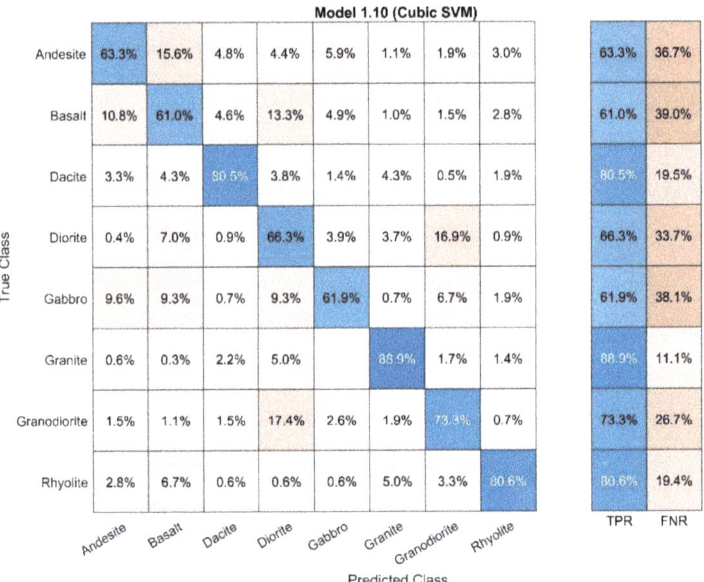

Figure 9. Confusion matrix from a Cubic SVM machine learning model used in evaluating the classification viability of post dimensionality reduction spectral bands.

In addition to the above-given assessment, Figure 9 illustrates the in depth classification capabilities of the ML algorithm post-DR for each class of rocks employed in this study. From the Figure 9 confusion matrix, 63.3% of the initial input andesite datasets (for 5-bands) were correctly (TPR) classified as andesite. On the other hand, the remaining 36.7% (FNR) was incorrectly classified as basalt (15.6%), dacite (4.8%), diorite (4.4%), gabbro (5.9%), granite (1.1%), granodiorite (1.9%), and rhyolite (3.0). Similar assessments can be made for all rocks, resulting in different ratios of both TPR and FNR. Comparing Figures 7 and A2 and nine results, we can make the following assumptions; the flatter the relative frequency histograms (Figures 7 and A2), the higher the prediction precision (Figure 9), hence granite has the highest ML prediction precision. On the other hand, the steeper the relative frequency histograms (Figures 7 and A2), the lower the ML prediction precision (Figure 9), hence basalt and gabbro have lower ML prediction precision outcomes. By developing algorithms on a particular type of rock, it is possible to improve any of the Figure 9 results to favour that specific rock, mineral or environmental phenomenon of interest. This thereby makes this system highly applicable in a magnitude of highly specialised classification problems. Doing so simply requires importing the most discriminative hyperspectral bands of any particular rock, mineral, or phenomenon, and giving them priority over other spectral bands, hence improving their succeeding ML classification outputs. However, since the goal of this paper was to classify eight igneous rock lithologies as a collective based on five multispectral bands, our system was not preprogrammed to be biased towards any of the eight igneous rock lithologies, but rather used the data as is, hence the true/unmodified results.

True Positive Rates (TPR):

$$TPR = (TP/TP + FN) \times 100 \tag{1}$$

False Negative Rates (FNR):

$$FNR = (FN/FN + TP) \times 100 \tag{2}$$

where FN is false negatives, and TP is true positives.

5. Significance of Proposed System

Therefore, given our findings, we can confirm that our proposed system, which consists of DR of rock hyperspectral data and subsequently employing specific discriminant features for our igneous rock database, performs extremely well. This, in essence, means for rock engineering, problems requiring discrimination of rocks, minerals, soils and other environmental phenomena based on their spectral signatures can indeed employ this system. By setting desired attributes founded on preknowledge of a site, such as types of rocks present within a mine site, rocks transported via a conveyer belt, or the general mapping of the environment, it is possible to maximise data collection. Based on specific multispectral bands, we can eliminate unnecessary storage, processing or classification costs associated with massive data.

With our integrated system, here are several optimisations we were able to achieve:

- Through DR, we can reduce the storage capacity required to store and handle a database, thereby reducing storage costs as we have proven there is no need to collect, store and process redundant data;
- With DR, we were able to break down hyperspectral signature data into different dimensionalities, hence the ability to plot such data in 2D planes, which as a result allows for easy visual assessment;
- As proven with post-NCA specialised multispectral imaging, we can attain respectable classification accuracies. This proves that multispectral imaging is a good enough option as it can be programmed to be highly specialised, costs less, has lower operation costs, has the flexibility of being applied in specialised multispectral imaging, such as on a UAV drone. Having said this, it is important to note that samples used in this study were clean and manually prepared before analysis, which is not the state in which rocks are found in the field, due to dirt and other matter. Therefore, classification accuracy variations in our envisioned identification of these rocks in the field, compared to the study's attained results, are likely to exist;
- Through ML, we can analyse and classify multispectral signatures produced by rocks and minerals with high accuracies. By finding the right model for a particular dataset, subsequent related data is relatively easier to classify as the training data always assists the model in future predictions as proven;
- With our proposed combined system, we have proved that any industry looking to cut spectral data (or equivalent) analysis costs whilst still retaining high classification accuracies, DR via a feature selection supervised NCA algorithm to specify

the most discriminative bands, and verifying the viability of selected bands via ML, thereafter employing these 5-bands (or more, depending on application) in future specialised classifications, could potentially be the key to achieving several system design optimisations;

- Via a post-NCA 5-band rock and mineral classification specialised multispectral camera mounted on a UVA drone, such as the 'DJI P4 Multispectral drone used in agricultural applications', there is a plethora of applications in which this specialised technology could find potential use. This, as a result, minimises purchase, operation and data interpretation costs as compared to a hyperspectral imaging system. This could aid in remote sensing from long distances without the need for physical presence, as well as rapid in situ assessments of the state of the environment via the UAV drone, possibilities are endless.
- Lastly, there is potential to employ such a post-NCA specialised multispectral camera in the frequent monitoring of mine dams. This would allow quicker assessment of contaminants based on spectral signatures produced by unexpected and/or anticipated metal contaminants. Hence, we deem this proposed system viable in all mining-related stages, from exploration, operation and closure.

6. Conclusions

This paper proposes the combination and DR of hyperspectral data via NCA to multispectral imaging, coupled with ML as a method by which subsequent spectral characteristics of rocks, minerals and the environment can be performed without unnecessary processing of redundant data. With our NCA algorithm, we proved the viability of our hyperspectral data DR from 204-bands, to 100, 50, 25, 10, and finally, the industry standard 5-band multispectral dimensionality. Thus, from NCA, we can conclude that the most viable discriminative five multispectral bands viable in the classification of igneous rocks, such as granite, diorite, gabbro, granodiorite, rhyolite, andesite, basalt and dacite, are bands with the following wavelengths—441 nm, 535 nm, 741 nm, 791 nm and 897 nm. With this DR, we were able to produce 2D data plots, which provide better interpretation, visualisation and somewhat data prediction capabilities in the form band-against-band scatter plots, as well as frequency density histograms. Therefore, it can be said that by eliminating redundant features, DR can be a useful technique employable for various datasets possessing the dimensionality curse.

The proposed method flawlessly merges with several ML models. Hence, we are provided with quantitative outputs pertaining to the classification abilities of each ML model, an example being our Cubic SVM model, which outperformed all other ML models in the classification of igneous rocks in our database. This, in essence, deems the Cubic SVM ML model the most viable as it attained a global classification accuracy of 71%, and an average per-class accuracy of 72%, which is considerable given the magnitude of the DR from 204-bands to 5-bands.

Author Contributions: Conceptualisation, B.B.S. and Y.K.; Data curation, H.T.; Formal analysis, M.S.; Funding acquisition, Y.K.; Investigation, B.B.S.; Methodology, B.B.S.; Resources, F.I.; Software, N.O.; Supervision, Y.K.; Validation, Z.B.; Visualisation, M.S. and H.T.; Writing—original draft, B.B.S.; Writing—review & editing, Z.B. and Y.K. All authors have read and agreed to the published version of the manuscript.

Funding: This Work Was Supported by JSPS 'Establishment of Research and Education Hub on Smart Mining for Sustainable Resource Development in Southern African Countries'. Grant number: JPJSCCB2018005.

Acknowledgments: This research was supported by SATREPS in collaboration between JST and JICA. We would like to express our sincere gratitude to Akita Mining Museum for permitting us to use their specimens as our research data, from which we were able to build a considerable database. Anonymous reviewers are thanked for critically reading the manuscript and suggesting substantial improvements.

Appendix A

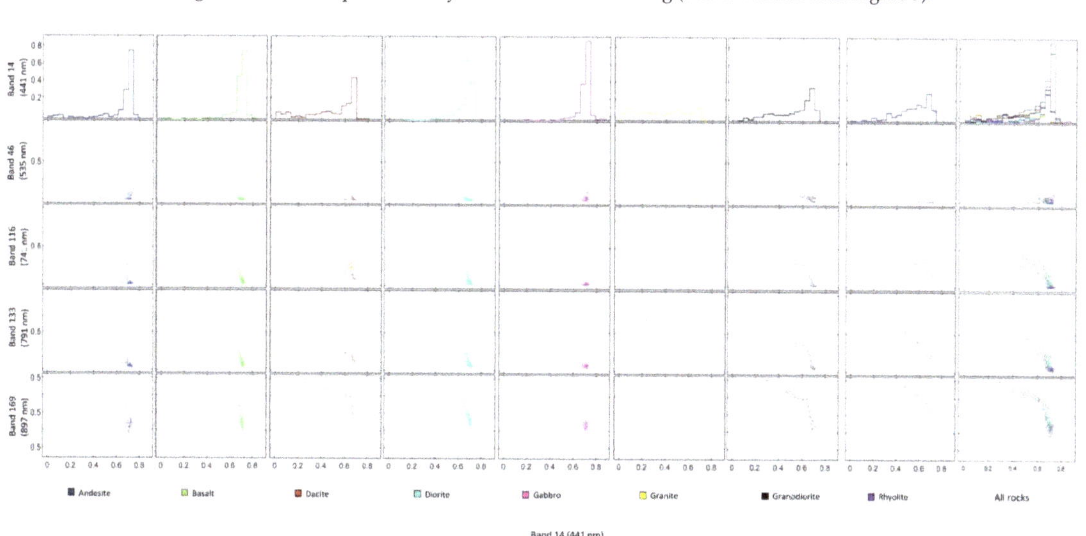

Figure A1. Summary of rock spectral reflectance strength data accumulation and preprocessing prior to dimensionality reduction via Neighbourhood Component Analysis and Machine Learning (a continuation from Figure 5).

Figure A2. Relative scatter and histogram plots of rock spectral reflectance strengths in 2D plane projections, where bands 441 nm is the x-axis, and bands 535 nm, 741 nm, 791 nm and 897 nm are the y-axis. The position of each point (spectra) is based on the relative intensities between bands, as well as between rocks.

References

1. Jia, F.; Lei, Y.; Lin, J.; Zhou, X.; Lu, N. Deep Neural Networks: A Promising Tool for Fault Characteristic Mining and Intelli-gent Diagnosis of Rotating Machinery with Massive Data. *Mech. Syst. Signal Process.* **2016**, *72*, 303–315. [CrossRef]
2. Zhang, X.; Li, P. Lithological Mapping from Hyperspectral Data by Improved Use of Spectral Angle Mapper. *Int. J. Appl. Earth Obs. Geoinf.* **2014**, *31*, 95–109. [CrossRef]
3. Van der Meer, F. The Effectiveness of Spectral Similarity Measures for the Analysis of Hyperspectral Imagery. *Int. J. Appl. Earth Obs. Geoinf.* **2006**, *8*, 3–17. [CrossRef]
4. Galdames, F.J.; Perez, C.A.; Estévez, P.A.; Adams, M. Rock Lithological Classification by Hyperspectral, Range 3D and Color Images. *Chemom. Intell. Lab. Syst.* **2019**, *189*, 138–148. [CrossRef]
5. Ruiz Hidalgo, D.; Bacca Cortés, B.; Caicedo Bravo, E. Dimensionality Reduction of Hyperspectral Images of Vegetation and Crops Based on Self-Organized Maps. *Inf. Process. Agric.* **2020**. [CrossRef]
6. Tong, Z.; Gao, J.; Zhang, H. Recognition, Location, Measurement, and 3D Reconstruction of Concealed Cracks Using Con-volutional Neural Networks. *Constr. Build. Mater.* **2017**, *146*, 775–787. [CrossRef]
7. Sinaice, B.B.; Kawamura, Y.; Kim, J.; Okada, N.; Kitahara, I.; Jang, H. Application of Deep Learning Approaches in Igneous Rock Hyperspectral Imaging. In *Proceedings of the 28th International Symposium on Mine Planning and Equipment Selec-tion—MPES 2019*; Springer Series in Geomechanics and Geoengineering; Topal, E., Ed.; Springer International Publishing: Cham, Switzerland, 2020; pp. 228–235. [CrossRef]
8. Sharma, N.; Sharma, R.; Jindal, N. Machine Learning and Deep Learning Applications—A Vision. *Glob. Transit. Proc.* **2021**, *2*, 24–28. [CrossRef]
9. Asscher, Y.; Angelini, I.; Secco, M.; Parisatto, M.; Chaban, A.; Deiana, R.; Artioli, G. Mineralogical Interpretation of Multispectral Images: The Case Study of the Pigments in the Frigidarium of the Sarno Baths, Pompeii. *J. Archaeol. Sci. Rep.* **2021**, *35*, 102774. [CrossRef]
10. Rahimi, B.; Sharifzadeh, M.; Feng, X.-T. A Comprehensive Underground Excavation Design (CUED) Methodology for Geotechni-cal Engineering Design of Deep Underground Mining and Tunneling. *Int. J. Rock Mech. Min. Sci.* **2021**, *143*, 104684. [CrossRef]
11. Goldberger, J.; Roweis, S.; Hinton, G.; Salakhutdinov, R. Neighbourhood Components Analysis. 8. Available online: https://www.cs.toronto.edu/~{}hinton/absps/nca.pdf (accessed on 5 February 2021).
12. Venna, J. Dimensionality Reduction for Visual Exploration of Similarity Structures. 82. Available online: https://www.researchgate.net/publication/27516587 (accessed on 23 March 2021).
13. Saha, D.; Annamalai, M. Machine Learning Techniques for Analysis of Hyperspectral Images to Determine Quality of Food Products: A Review. *Curr. Res. Food Sci.* **2021**, *4*, 28–44. [CrossRef] [PubMed]
14. Salles, R.D.R.; de Souza Filho, C.R.; Cudahy, T.; Vicente, L.E.; Monteiro, L.V.S. Hyperspectral Remote Sensing Applied to Uranium Exploration: A Case Study at the Mary Kathleen Metamorphic-Hydrothermal U-REE Deposit, NW, Queensland, Australia. *J. Geochem. Explor.* **2017**, *179*, 36–50. [CrossRef]
15. Kruse, F.A. Mapping Surface Mineralogy Using Imaging Spectrometry. *Geomorphology* **2012**, *137*, 41–56. [CrossRef]
16. Ganesh, U.K.; Kannan, S.T. Creation of Hyper Spectral Library and Lithological Discrimination of Granite Rocks Using SVCHR -1024: Lab Based Approach. *J. Hyperspectral Remote. Sens.* **2017**, *7*, 168–177.
17. Li, Y.; Chai, Y.; Zhou, H.; Yin, H. A Novel Dimension Reduction and Dictionary Learning Framework for High-Dimensional Data Classification. *Pattern Recognit.* **2021**, *112*, 107793. [CrossRef]
18. Li, J.; Wang, T. Dimension Reduction in Binary Response Regression: A Joint Modeling Approach. *Comput. Stat. Data Anal.* **2021**, *156*, 107131. [CrossRef]
19. Zuniga, M.M.; Murangira, A.; Perdrizet, T. Structural Reliability Assessment through Surrogate Based Importance Sampling with Dimension Reduction. *Reliab. Eng. Syst. Saf.* **2021**, *207*, 107289. [CrossRef]
20. Koren, Y.; Carmel, L. Robust Linear Dimensionality Reduction. *IEEE Trans. Vis. Comput. Graph.* **2004**, *10*, 459–470. [CrossRef]
21. Bar-Hillel, A.; Hertz, T.; Shental, N.; Weinshall, D. Learning a Mahalanobis Metric from Equivalence Constraints. *J. Mach. Learn. Res.* **2005**, *6*, 937–965.
22. Zhang, C.; Lei, Y.-K.; Zhang, S.; Yang, J.; Hu, Y. Orthogonal Discriminant Neighborhood Analysis for Tumor Classification. *Soft Comput.* **2016**, *20*, 263–271. [CrossRef]
23. Hu, H.; Feng, D.-Z.; Chen, Q.-Y. A Novel Dimensionality Reduction Method: Similarity Order Preserving Discriminant Analysis. *Signal Process.* **2021**, *182*, 107933. [CrossRef]
24. Kalia, V.; Walker, D.I.; Krasnodemski, K.M.; Jones, D.P.; Miller, G.W.; Kioumourtzoglou, M.-A. Unsupervised Dimen-sionality Reduction for Exposome Research. *Curr. Opin. Environ. Sci. Health* **2020**, *15*, 32–38. [CrossRef]
25. Jiang, Z.; Li, J. High Dimensional Structural Reliability with Dimension Reduction. *Struct. Saf.* **2017**, *69*, 35–46. [CrossRef]
26. Liu, J.; Chen, S. Discriminant Common Vecotors Versus Neighbourhood Components Analysis. 32. *Image Vis. Comput.* **2006**, *24*, 249–262. [CrossRef]
27. Kartun-Giles, A.P.; Bianconi, G. Beyond the Clustering Coefficient: A Topological Analysis of Node Neighbourhoods in Complex Networks. *Chaos Solitons Fractals X* **2019**, *1*, 100004. [CrossRef]
28. Gunduz, H. An Efficient Dimensionality Reduction Method Using Filter-Based Feature Selection and Variational Auto encoders on Parkinson's Disease Classification. *Biomed. Signal Process. Control* **2021**, *66*, 102452. [CrossRef]

29. Jain, U.; Nathani, K.; Ruban, N.; Joseph Raj, A.N.; Zhuang, Z.; Mahesh, V.G.V. Cubic SVM Classifier Based Feature Extraction and Emotion Detection from Speech Signals. In Proceedings of the 2018 International Conference on Sensor Networks and Signal (SNSP), Xi'an, China, 28–31 October 2018. [CrossRef]
30. Taylor, G.R. Mineral and Lithology Mapping of Drill Core Pulps Using Visible and Infrared Spectrometry. *Nat. Resour. Res.* **2000**, *9*, 257–268. [CrossRef]
31. Bar-Hillel, A.; Hertz, T.; Shental, N.; Weinshall, D. Learning Distance Functions Using Equivalence Relations. In Proceedings of the Twentieth International Conference on Machine Learning (ICML-2003), Washington, DC, USA, 21–24 August 2003; Available online: https://www.researchgate.net/publication/221344999_Learning_Distance_Functions_using_Equivalence_Relations (accessed on 23 March 2021).
32. Ayesha, S.; Hanif, M.K.; Talib, R. Overview and Comparative Study of Dimensionality Reduction Techniques for High Dimensional Data. *Inf. Technol. Control* **2020**, *59*, 44–58. [CrossRef]
33. Mei, S.; Ji, J.; Geng, Y.; Zhang, Z.; Li, X.; Du, Q. Unsupervised Spatial–Spectral Feature Learning by 3D Convolutional Autoencoder for Hyperspectral Classification. *IEEE Trans. Geosci. Remote. Sens.* **2019**, *57*, 6808–6820. [CrossRef]

 *minerals*

Article

Application of Deep Learning in Petrographic Coal Images Segmentation

Sebastian Iwaszenko [1,*] and Leokadia Róg [2]

[1] Department of Acoustics, Electronics and IT Solutions, GIG Research Institute, 40-166 Katowice, Poland
[2] Department of Solid Fuels Quality Assessment, GIG Research Institute, 40-166 Katowice, Poland; lrog@gig.eu
* Correspondence: siwaszenko@gig.eu

Abstract: The study of the petrographic structure of medium- and high-rank coals is important from both a cognitive and a utilitarian point of view. The petrographic constituents and their individual characteristics and features are responsible for the properties of coal and the way it behaves in various technological processes. This paper considers the application of convolutional neural networks for coal petrographic images segmentation. The U-Net-based model for segmentation was proposed. The network was trained to segment inertinite, liptinite, and vitrinite. The segmentations prepared manually by a domain expert were used as the ground truth. The results show that inertinite and vitrinite can be successfully segmented with minimal difference from the ground truth. The liptinite turned out to be much more difficult to segment. After usage of transfer learning, moderate results were obtained. Nevertheless, the application of the U-Net-based network for petrographic image segmentation was successful. The results are good enough to consider the method as a supporting tool for domain experts in everyday work.

Keywords: coal; petrographic analysis; macerals; image analysis; semantic segmentation; convolutional neural networks

Citation: Iwaszenko, S.; Róg, L. Application of Deep Learning in Petrographic Coal Images Segmentation. *Minerals* **2021**, *11*, 1265. https://doi.org/10.3390/min11111265

Academic Editors: Rajive Ganguli, Sean Dessureault and Pratt Rogers

Received: 8 September 2021
Accepted: 10 November 2021
Published: 13 November 2021

Publisher's Note: MDPI stays neutral with regard to jurisdictional claims in published maps and institutional affiliations.

1. Introduction

Coal petrography is a science that, despite the passage of many years, is developing and updating its knowledge with a view to new directions for use in the energy industry. Particular emphasis is placed on clean coal technologies as well as the recovery of critical elements from coal [1–6].

Coal is a heterogeneous substance in terms of its chemical composition. Its heterogeneity is due to the variation in the peat-forming plant material from which it was formed and the variation in the conditions, time, pressure, and temperature to which the organic material was subjected during both its biochemical and geochemical phases [7]. The basic units of the structure of coal, homogeneous in physical and chemical terms, are macerals. The study of the petrographic structure of coal is important from both a cognitive and a utilitarian point of view [7–11]. It is the petrographic constituents and their individual characteristics and features that are responsible for the property of coal and the way it behaves in various technological processes [7,8,12–15].

Knowledge of the percentage of individual petrographic constituents in coal is very important, as the petrographic constituents differ in terms of their physical and chemical properties, such as volatile matter content, elemental composition, vitrinite reflectance, and specific density, all of which affect the chemical, physical and technological properties of coal [7,13,16–18].

The knowledge of the petrographic and mineral composition of the coal deposit, and the properties resulting from this composition, should be the basis for optimizing the conditions in coal preparation plants [6,19,20]. Such an approach makes it possible to control the properties of the final product in order to obtain concentrates with precisely defined parameters and properties, usable in clean coal technologies (CCT), especially in combustion, coking, and gasification processes [21–27].

Petrographic studies are also used to determine the composition of coke mixtures and to forecast coke quality. In addition, petrographic analyses of coal are used in research on the production of liquid fuels by means of the direct hydrogenation method [13]. Another important parameter in the selection of coal for various technologies is the degree of coalification of the organic matter in coal, a measure of which is the reflectance value determined on one of the most homogeneous macerals, namely vitrinite [16,28].

The petrographic composition of hard coals and their degree of coalification, determined by analyzing huminite reflectance for low-rank coals, and vitrinite reflectance for medium- and high-rank coals, have become an important element in the classification of coals, both according to the Polish classification PN-G-97002:2018:11 *Hard coal. Types* and the international classification ISO 11760:2005 *Classification of coals and international codification system for medium and high-rank coals*, ECE/COAL/115, United Nations Publication, New York, 1998. The international classification and codification are based on both the maceral content in coal and the degree of vitrinite carbonization, while the Polish classification by type is based on the vitrinite reflectance value. Although huminite reflectance could be a useful parameter for low-rank coal, the heterogeneous geochemical characteristics of huminite macerals might also be problematic [7,29–33]. Based on the above information, it can be concluded that petrographic studies of coal are very frequently used, and thus they are becoming more and more important.

The microscopic analysis of composition is based on the identification of individual petrographic constituents observed by the operator in the microscopic image. The method is standardized and based on ISO 7404-3:1994 *Methods for the petrographic analysis of bituminous coal and antaracite–Part 3: Method of determining maceral group composition*, as well as the maceral classification of the International Committee for Coal and Organic Petrology (ICCP) [9–11]. The determination is based on the identification of macerals at a minimum of 500 points. Even though several attempts have been made to design automated systems, this analysis is still commonly based on a manual method, which is very time consuming. It requires a lot of knowledge and skill on the part of the operator.

The application of image analysis and computer vision in minerals science has a long and successful history. The images were used for the identification of the minerals based on their color and textural features [34–36]. There were also attempts at the determination of grains sizes using the computer vision approach when the boundaries were used using color [37,38] as well as texture [39,40] features. Computer vision was also used to analyze the microscopic images of minerals. Martens et al. analyzed the microscopic pictures of mortars to assess the distribution of the sand grains sizes [41]. Kazak et al. successfully applied machine learning methods in the analysis of focused ion beam scanning electron microscopy (FIB SEM) images for void space characterization of tight reservoir rocks [42]. Zhou et al. investigated the possibility of segmenting mineral grains in petrographic images using various edge detection algorithms [43]. A similar problem, but with the application of the level set method [44], was also considered in [45]. The attempts were made to use image analysis in the petrography, macerals, cleat, and lithotypes of coal samples as early as in the 1980s [46–48]. The researchers investigated different possibilities to differentiate the organic and mineral matter. Hou et al. used the time-of-flight secondary ion mass spectrometry to obtained images of the macerals and selected mineral matter [49]. Alphana et al., based on scanning electron microscope images (SEM) [50], calculated a set of color and textural features for each image. The classification process is aimed at predicting the coal quality, namely as best, good and poor. The application of radial base functions neural networks (RBFNN) outperformed the other classification methods. However, there was no attempt to identify the macerals composing the sample, and the application of SEM complicated the image acquisition. The most interesting and desirable solution should address the automation of coal petrography using the simplest image acquisition procedure possible. In this direction, O'Braien et al. proposed the usage of full maceral reflectograms for maceral identification [51]. The coal samples were submerged in resin and polished. After the resin regions were masked out in the image, the cumulative curve of reflectance was computed using the gray values of the remaining pixels. The shape of the curve reflects the composition of the coal. The method turned out to be successful, but it required appropriate preparation of sample and imaging protocol (e.g., usage of red dye for resin and green light for images acquisition). The gray level values had to be calibrated, so that the resulting curve could be interpreted in terms of maceral's containment. The idea was further enhanced by using simultaneous analysis of optic and SEM gathered images [52]. The idea of the usage of optical microscopy obtained images for automated macerals identification was also considered by other researchers [53–58]. Młynarczuk and Skiba proposed the usage of machine learning (ML) and artificial intelligence methods in maceral identification [59]. The maceral group identification is based on the color features vector computed for the square neighborhood of the selected pixel. The k nearest neighbors (kNN) and multilevel perceptron (MLP) were used as the classifier. The results were very promising, and the method was developed to identify the macerals within the inertinite group [60]. The features vector was extended to include both color and texture properties of the pixels. The results were satisfactory, but the effectiveness depended on the maceral. However, none of the methods tried to semantically segment the image. One of the attempts in this direction was made by Wang et al. [56]. In this attempt, the shapes of macerals groups were identified using a clustering procedure, namely a modified k-means algorithm. Then, the discovered objects were classified using morphological, color and texture features. It should be emphasized that in addition to the analysis of images from various types of microscopes, other methods were also used for carbon analysis, examples of which can be found in [61,62].

Semantic segmentation is a topic of much research interest nowadays [63,64]. The application of deep learning (DL) and convolutional neural networks (CNN) allowed for achieving stunning results. The DL was used as a tool for microfossils, core images, petrographic and rock images classification [65]. Attempts are being made to apply these approaches to the analysis of coal characteristics and particularly its petrography using visual information. The most fundamental characteristic of coal's run-of-mine (ROM) distinguishes between coal rocks and the accompanying gangue. Pu et al. used the VGG16 CNN for the classification of images presenting coal or gangue in different configurations (as stockpiles, during transportation, photographed in laboratory conditions, etc.) with satisfactory results [66]. Li et al. developed a solution for the identification of coal and gangue rocks on images [67]. The proposed framework processed the Gaussian pyramid of the input image. The rock grains were detected and classified as coal or gangue. The authors reported impressive accuracy, exceeding 98%, in rock type recognition. The application of semantic segmentation to maceral group identification was developed by Lei et al. [68]. The proposed network utilizes the U-Net [69] network enhanced with the attention gates. The authors used the multi-class form of the output layer of the network. The segmentation results were very good and proved the robustness of DL methods.

The identification of the macerals directly on the image by means of the direct assignment of the maceral label to every pixel would be beneficial in many ways. First of all, it will allow the determination of maceral composition. Secondly, it would provide the scientist with information allowing them to judge whether the individual parts of the image have been correctly identified. Third, the calibration should not be critical in maceral identification, because not only are the color statistics considered, but also the spatial arrangement of pixels constituting the maceral groups (for medium- and high-rank coals). Therefore, attempts were made to develop a method suitable for such coal petrographic images analysis. The presented paper provides a proposition of such a method using the deep learning approach. The highlights of the presented results include the development of the coal petrographic images database, the method of image preparation and augmentation, and the development of a U-Net [69]-based convolutional neural network for the semantic segmentation of coal petrographic images. The proposed approach is based on using single-class classification—a separate model of the same architecture was trained for each of the macerals.

2. Materials and Methods

The identification of macerals is based on the microscopic evaluation of grain morphology and color. On this basis, three groups of macerals were distinguished: liptinite, vitrinite, and inertinite (Figure 1) [9–11]. The color of liptinite changes from brown through dark grey to light grey in the microscopic image. Under incident light, depending on coal rank, the color of vitrinite changes from dark grey through light grey to almost white. On the other hand, in the same light conditions, the color of inertinite in coal is always the brightest and changes from light grey to white and bright white. The reflectance of all macerals increases with the increasing carbonization of the organic matter of the coal (Figure 2). At the vitrinite reflectance (%Rr) level of about 1.5%, the simultaneous differences in reflectance and in color between liptinite and vitrinite disappear, and with a %Rr about 2.4%, the differences between vitrinite and inertinite also disappear.

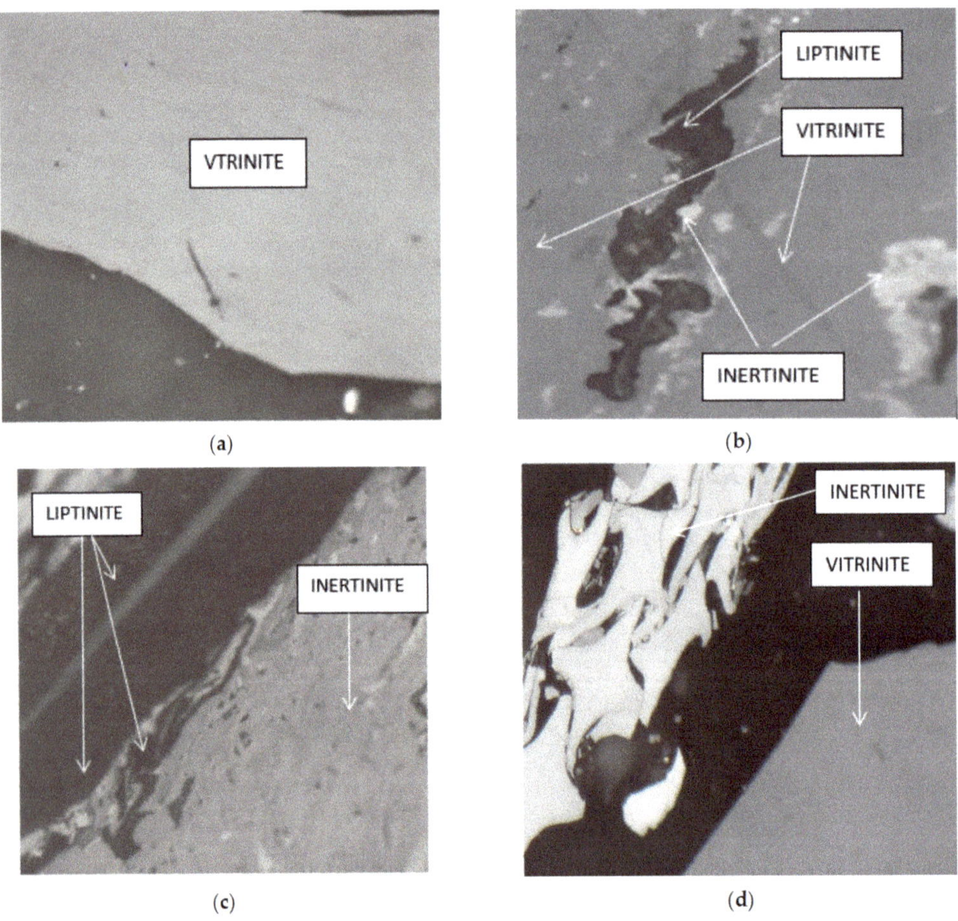

Figure 1. Maceral groups: (**a**) vitrinite; (**b**) liptinite and inertinite; (**c**) liptinite; (**d**) inertinite and vitrinite. Oil immersion, magnification 500×.

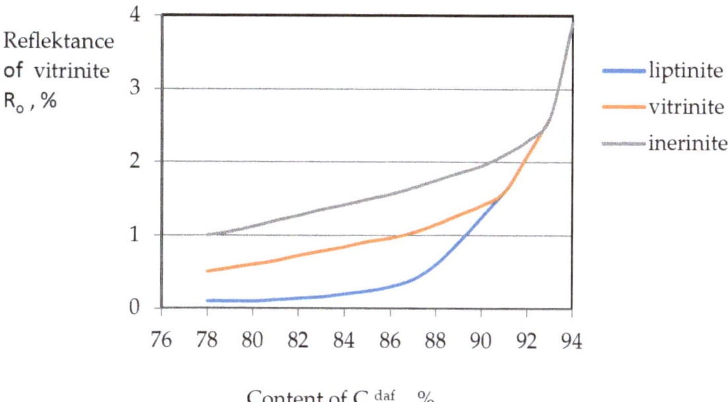

Figure 2. Variability of macerals' groups reflectivity in coals of different carbonization degree [70].

For the purposes of this study, medium-sized samples were prepared for petrographic analyses, according to the PN-ISO 7404-2:2005 *Methods for the petrographic analysis of bituminous coal and anthracite–Part 2: Method of preparing coal samples,* from selected coal samples in which vitrinite reflectance did not exceed 0.8%. Coal samples were taken from coals originating from Polish coal basins: the Upper Silesian Coal Basin and the Lublin Coal Basin. Data on the tested coal samples (rank, the origin of the samples, and maceral compositions are presented in Table 1. The microscopic specimens were prepared by the immersion of coal dust in a mixture of epoxy resin and hardener, obtained by mixing the components at a ratio of 8:1. The immersed microscopic specimens were left for at least 24 h until solidification. The solidified specimens were ground and polished using a Struers LaboForce-3 grinding/polishing machine (Struers Inc., Cleveland, OH, USA). A Zeiss Axio Imager Z 2m microscope (Carl Zeiss AG, Oberkochen, Germany) (Figure 3) was used for the study. A magnification of 500 times and white light reflected in oil immersion were used. Surfaces were selected for which photographs were taken using an Axiocam 506 color camera. The set of microscope photographs obtained showed different macerals for which a mask set was developed. In the petrographic analysis, the participation of maceral groups was most important. The results of the determination of the mineral substance are rarely used. Therefore, they were omitted in the first stage of the research. We plan to take care of this problem in the future.

The images were captured with the resolution 3072×2304 pixels with 8-bit RGB color space. For further processing, the images were cut into 512×512 parts. Then, the manual segmentation of the vitrinite, inertinite, and liptinite was performed by a domain expert. The segmentation was used as the ground truth for further processing. There were separate masks created for each of the macerals. The completed database consisted of 162 images for which the masks were created (three masks were created for each input image). The example image and masks are presented in Figure 4.

Table 1. Data on the tested coal samples.

Origin Samples	Numer of Samples	Maceral Groups (% vol.)			Vitrinite Reflectance (%)
		Vitrinite	Liptinite	Inertinite	
Upper Silesian Coal Basin	12	45–79	7–12	17–31	0.51–0.80
Lublin Coal Basin	10	58–72	6–15	18–35	0.58–0.75

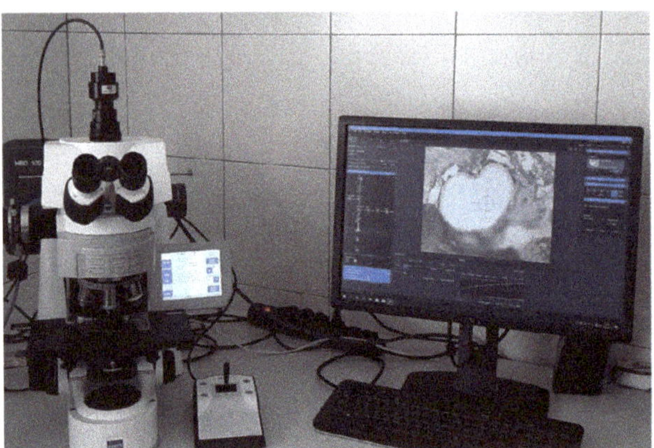

Figure 3. Microscope Zeiss Axio Imager Z 2m.

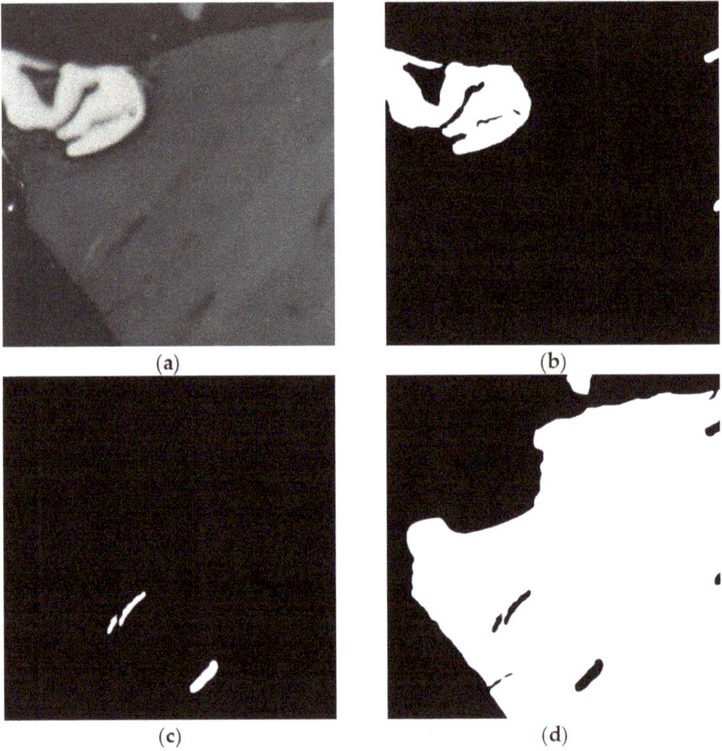

Figure 4. Example of the input image and accompanying masks: (**a**) the input image; (**b**) mask for the inertinite; (**c**) mask for the liptinite; (**d**) mask for the vitrinite.

It was decided to use a separately trained single-class model for the maceral identification. This approach has its advantages, dictated by both practical and computational considerations. In petrographic practice, the identification of macerals is used in various variants. For example, as far as the analysis of the vitrinite reflectance index is concerned, the recognition of only one maceral—collotelinite—is required. Similarly, the research carried out in order to determine the coke-forming properties requires the recognition of two types of petrographic components, namely reactive ones, which include macerals from the vitrinite and liptinite groups, and inert ones, which include macerals from the inertinite group. Using a single-class approach gives the opportunity to make it possible to obtain a network that is particularly sensitive to a specific group of macerals. Such a network is expected to be easier to train than a multiclass network and will allow for the usage of a simpler architecture without sacrificing the performance. The above is also true with respect to the preparation of a set of training images. It also provides the possibility to optimize the network architecture for each group of macerals. The usage of single-class models does not limit the common analysis of all of the maceral groups simultaneously. The outputs of the models can be combined into the result, showing all classes on a single image using for example argmax function (argmax function returns the argument for which the maximum value of output was achieved).

For the segmentation experiments, the U-Net convolutional semantic segmentation network was used [71]. The architecture of the network is presented in Figure 5.

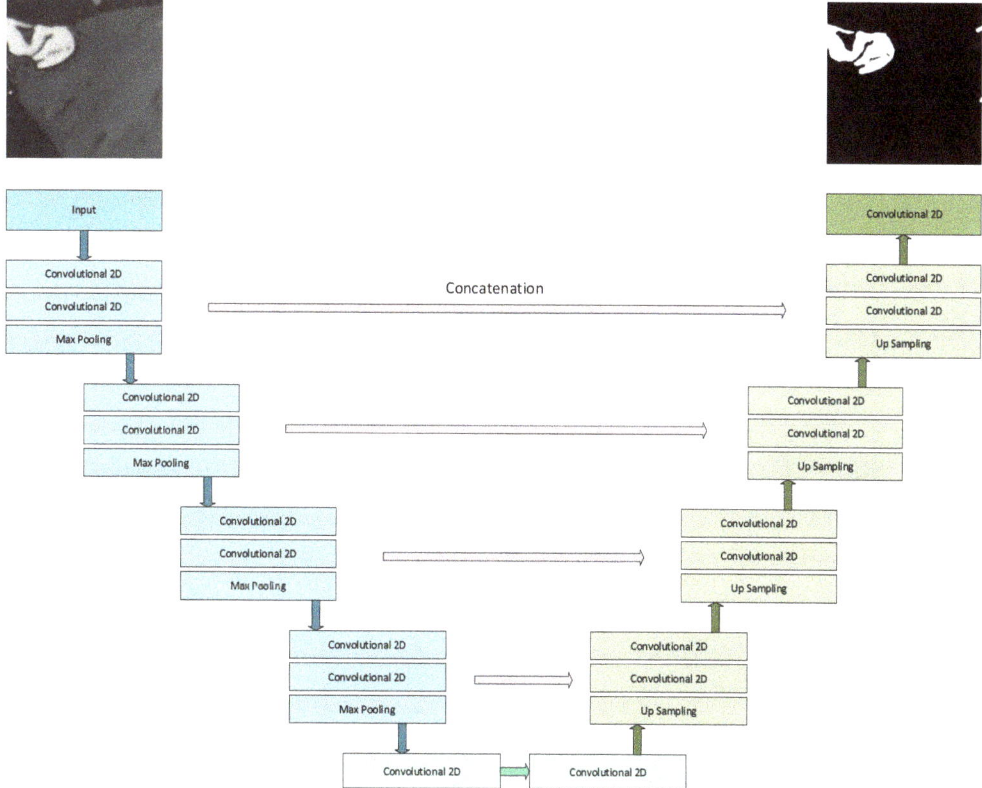

Figure 5. The architecture of U-Net network [71].

The U-Net is an example of an autoencoder network. It can be divided into three parts: the contraction part (4 blocks composed of two convolutional steps and pulling step), the bottleneck (the convolutional layer with 1024 channels), and the expansion part (4 blocks composed of two convolutional layers followed by upscaling layer). All convolutional layers use the rectified linear unit (ReLU) as the activation function. The input layer in the constructed U-Net-based network has a 512×512 resolution, which is in accordance with the input image size. The output layer was constructed with a 1×1 convolutional layer with a sigmoid activation function. The list of layers along with their shapes is presented in Table 2. The network architecture was implemented using the Tensorflow deep learning framework library.

Table 2. The shapes of layers used for the U-Net-based maceral segmentation network.

Layer	Shape	Activation
Input	$512 \times 512 \times 3$	-
Convolutional 2D	$512 \times 512 \times 16$	ReLU
Convolutional 2D	$512 \times 512 \times 16$	ReLU
Max Pooling	$256 \times 256 \times 16$	
Convolutional 2D	$256 \times 512 \times 32$	ReLU
Convolutional 2D	$256 \times 512 \times 32$	ReLU
Max Pooling	$128 \times 128 \times 32$	
Convolutional 2D	$128 \times 128 \times 64$	ReLU
Convolutional 2D	$128 \times 128 \times 64$	ReLU
Max Pooling	$64 \times 64 \times 64$	
Convolutional 2D	$64 \times 64 \times 128$	ReLU
Convolutional 2D	$64 \times 64 \times 128$	ReLU
Max Pooling	$32 \times 32 \times 128$	
Convolutional 2D	$32 \times 32 \times 256$	ReLU
Convolutional 2D	$32 \times 32 \times 256$	ReLU
Up Sampling	$64 \times 64 \times 256$	
Convolutional 2D	$64 \times 64 \times 128$	ReLU
Convolutional 2D	$64 \times 64 \times 128$	ReLU
Up Sampling	$128 \times 128 \times 128$	
Convolutional 2D	$128 \times 128 \times 64$	ReLU
Convolutional 2D	$128 \times 128 \times 64$	ReLU
Up Sampling	$256 \times 256 \times 64$	
Convolutional 2D	$256 \times 256 \times 32$	ReLU
Convolutional 2D	$256 \times 256 \times 32$	ReLU
Up Sampling	$512 \times 512 \times 32$	
Convolutional 2D	$512 \times 512 \times 16$	ReLU
Convolutional 2D	$512 \times 512 \times 16$	ReLU
Convolutional 2D	$512 \times 512 \times 1$	Sigmoid

The input images were split randomly into training and validation sets. The validation set was formed with 10% of all images. The binary cross-entropy function was used as a loss function during the network training process. During the training, the pixel-wise accuracy (PA), intersection-over-union (IoU), and mean intersection-over-union (MIoU) were also monitored as effectiveness measures. The ADAM optimizer was chosen for model learning [72]. The training of the network was performed in two stages. In the first stage, the batch size and the learning rate range were estimated. The model training was stopped after just a few epochs and the training results were analyzed. The upper limit for learning rate was established by choosing the value at which the model improved the performance in at least 4 consecutive epochs. The batch size was limited by the size of the input dataset. The bigger the batch size, the fewer steps per epoch the training procedure can make. It was assumed that the biggest batch size allowed for at least several dozen steps for the epoch. The second stage was devoted to model training. During the training, the decreasing learning rate was used. The model was trained for 50 epochs with a constant rate. If no improvement to the loss function was observed, the learning rate was decreased by 10 and the training process was repeated. The accuracy for the validation set was observed as an indicator for possible overtraining. The training was stopped once the validation set accuracy start to decrease. The presented learning procedure was used for each of the macerals. During the training process, two kinds of data modifications were performed:

1. The images which do not show the given maceral were excluded from the training set. For example, if the model was trained for vitrinite segmentation, all images where vitrinite was not present were excluded from the training set;
2. Basic images augmentation was performed. The augmentation was limited to rotation by $\pi/2$, π, $3\pi/2$ and mirroring horizontally and vertically.

The order of the images during each epoch was randomized. All input images were in RGB color space. All masks were binary. No image preprocessing except for the described augmentation was performed.

All the calculations were performed on an MS Windows workstation equipped with an Intel i7 processor running at 3.6 GHz (maximum), 32 GB RAM, and an NVIDIA GeForce GTX 1080 graphic card. All software necessary for computation was prepared with the Python programming language using the TensorFlow framework [73].

3. Results and Discussion

The learning rate during the first stage of experiments was changed from 10^{-1} to 10^{-6}. It was observed that learning rates greater than 10^{-4} caused huge changes in the loss function for consecutive epochs. The loss hardly shows any improvement. Therefore, the rate of 10^{-4} was chosen as the largest learning rate used in the calculation. After every 50 epochs during the training, the results were examined and the learning rate decreased once the loss function values started to oscillate from one epoch to another. The calculations were stopped once the validation test accuracy started to increase. At this moment, the learning rate was as low as 10^{-7}.

The training for the inertinite started from the randomly initialized network, using Xavier initializer [74]. During the 250 epochs of the training process, the learning rate was changed from 10^{-4} to 10^{-6}. The final accuracy computed for the validation set was equal to 0.9385. The values of IoU and MIoU were equal 0.79 and 0.85, respectively. The segmentation results compared with the input image and the ground truth for selected images are presented in Figure 6.

The presented results of segmentation are indeed the output values from the last layer in the used U-Net based network. The values, being the values of the sigmoid function, vary from 0 to one, which is reflected by the grayscale level in the picture. The segmentation quality can be assessed as very good, though not perfect. It may be noticed that some minor artifacts are visible on each of the presented images. The network has difficulties in recognizing the tiny structures of inertinite visible among other macerals of similar greyscale and textures. In addition to that, vast structures were correctly noticed by U-Net and marked.

The training procedure and the results obtained for vitrinite were similar to those for inertinite. The learning process also lasted 250 epochs, though the learning rate was changed in a wider range. It started at 10^{-4}, but ended with 10^{-7}. The selected results of the segmentation are presented in Figure 8. The obtained accuracy computed for the validation set was equal to 0.9176. The IoU was equal to 0.78 and the MIoU was equal to 0.75.

The analysis of the results shows that the quality of segmentation is similar to that for inertinite; however, a slightly higher level of artifacts was observed, which is also visible in the presented images (see Figure 7). The results can be considered very good and suitable for practical applications.

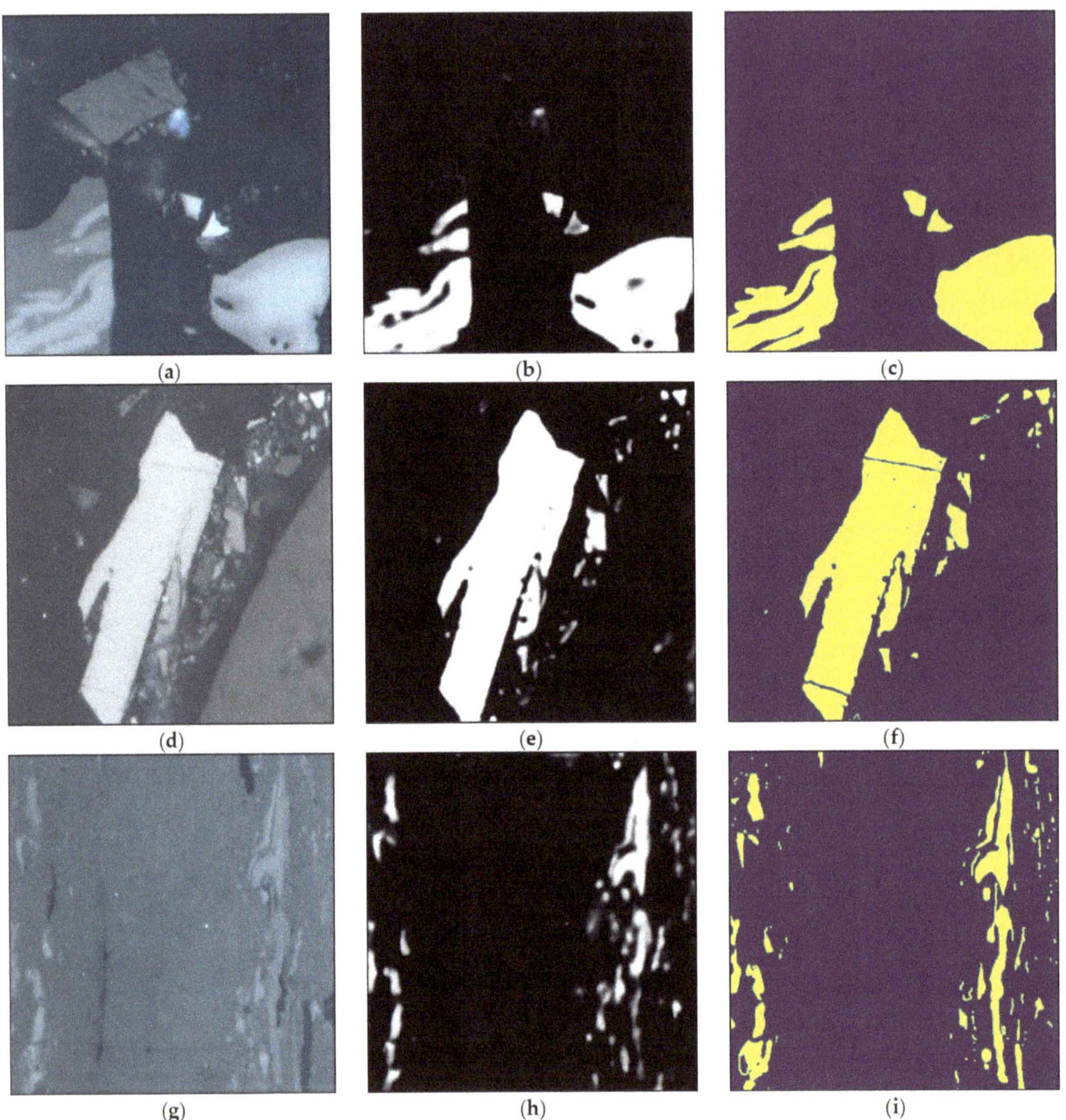

Figure 6. The results of inertinite segmentation. (**a**,**d**,**g**) Input images; (**b**,**e**,**h**) segmentation results; (**c**,**f**,**i**) ground truth.

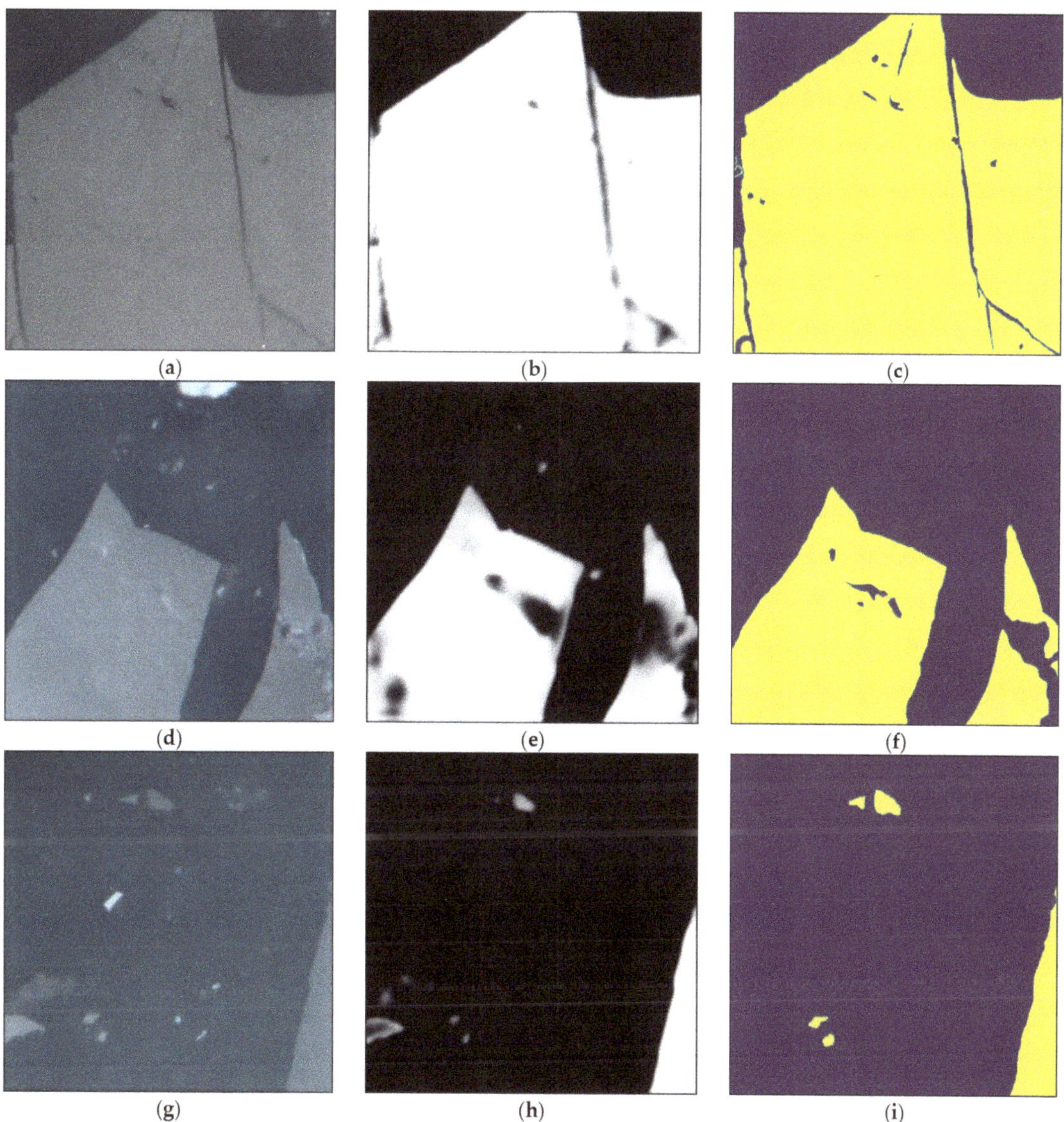

Figure 7. The results of vitrinite segmentation. (**a,d,g**) Input images; (**b,e,h**) segmentation results; (**c,f,i**) ground truth.

As expected for medium- and high-rank coals, the segmentation of liptinite turned out to be the most difficult. There was no success with the training network beginning with the randomly initialized weights. Moderately satisfactory results were obtained when the training process for liptinite used the weights from the trained model for inertinite segmentation. The application of such performed transfer learning made it possible to obtain acceptable liptinite segmentation, but the errors and artifacts are clearly visible in the resulting images. The accuracy value for the validation set was 0.9791. Such a large value, with a relatively low quality of segmentation, results from the small area covered by the liptinite on the analyzed images. The calculated values for the IoU (0.18) and the

MIoU (0.58) show that the segmentation is indeed poor, and can be treated as a rough identification of liptinite's presence. The results of segmentation are presented in Figure 8.

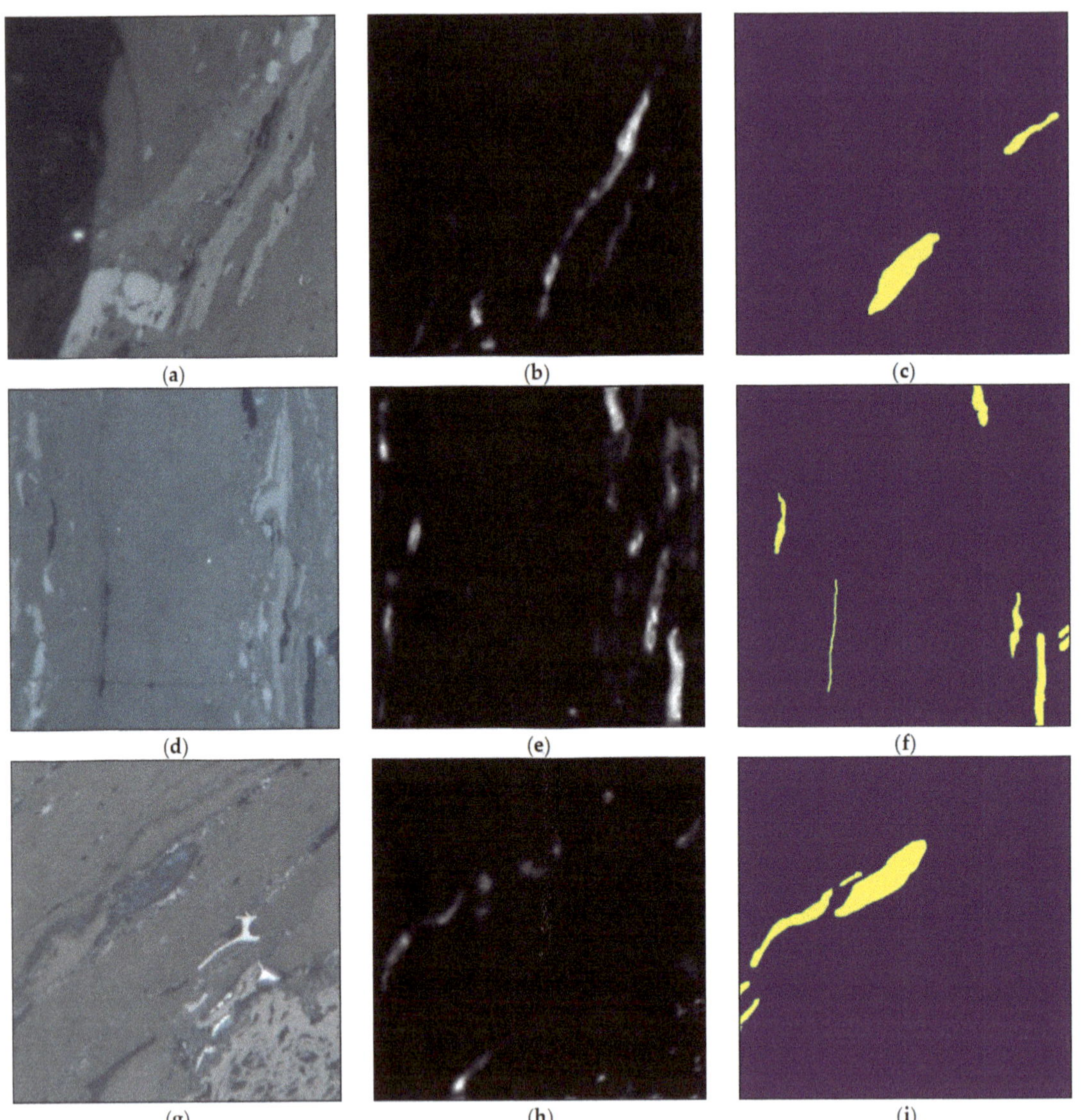

Figure 8. The results of liptinite segmentation. (**a,d,g**) Input images; (**b,e,h**) segmentation results; (**c,f,i**) ground truth.

A summary of the obtained values of accuracy, IoU, and MIoU is presented in Table 3. The quality of segmentation obtained for inertinite and vitrinite was good enough to be used as the basis for the development of an autonomous maceral identification method. The imperfections were small, not differing much from the ground truth. Moreover, during the analysis of the results, it turned out that the network was able to identify the small inertinite structures overlooked during manual segmentation. It seems reasonable to use the U-Net-based convolutional network for the segmentation of the mentioned macerals with only a little attention from a domain expert. Unfortunately, this is definitely not true for liptinite. The network was able to identify the liptinite only roughly. The result should instead be treated as approximate, possible locations of liptinite structures which have to be verified and corrected by a domain expert. The training for liptinite was also more difficult than for other macerals. It is probably caused by its more varied appearance. In addition, liptinite covered small areas in the images and was present only on relatively small numbers of them. Nevertheless, such support in assessing the maceral can be useful in practice. The obtained results can be related to others reported in the literature [56,68]; however, the comparison is not obvious as the mentioned papers do not provide the measures for the macerals' groups separately. Therefore, it is reasonable to use the mean values for the IoU and MIoU presented in Table 3 and the values of the same measures presented in [68]. The presented U-Net-based network gives better results than the non-DL methods. The results obtained by the improved U-Net (enhanced with the use of attention gates) are better than presented in the paper, though the difference is small (IoU ~ 0.8554 and MIoU ~ 0.631 for best enhanced network presented in [68]). However, it is impossible to assess how it is divided into individual macerals groups. When the liptinite, with the worst results, is omitted, the mean IoU and MIoU for inertinite and vitrinite are much greater. As the proposed models address the segmentation of each of the macerals individually, they should be treated as complementary to the model presented in [68]. Wang et al. presents the results obtained using different deep learning networks architectures, such as U-Net (standard multi-class architecture), SegNet, and DeepLab V3+ [57]. The results are provided for each of the macerals separately. The comparison is presented in Table 4.

Table 3. The values of accuracy, IoU and MIoU obtained for validation images set.

Macerals' Group	PA	IoU	MIoU
Inertinite	0.9385	0.79	0.85
Liptinite	0.9791	0.18	0.58
Vitrinite	0.9176	0.78	0.75
Mean	0.9451	0.58	0.79
Mean without liptinite	0.9280	0.73	0.80

Table 4. The IoU values obtained for various deep networks architectures.

Macerals' Group	U-Net [57]	SegNet [57]	DeepLab V3+ [57]	Simplified U-Net
Inertinite	0.57	0.66	0.71	0.79
Liptinite	0.52	0.37	0.83	0.18
Vitrinite	0.81	0.83	0.87	0.78

The proposed simplified U-Net-based network did very well in segmenting the inertinite, achieving a better result than much more sophisticated DeepLab V3+ network (the best from architectures compared in [57]). The results obtained for vitrinite are slightly worse than for the other two networks. There are very large differences in the case of liptinite. The network architecture was probably too simple to successfully cope with the most difficult to recognize maceral groups. The results obtained for two other maceral groups are optimistic. In particular, the IoU measure for the inertinite is good enough to contribute to the assumptions made and present the network's robustness. The proposed network can be efficiently used for inertinite and vitrinite identification in the petrographic images.

The discussed approaches present different means to provide the solution for maceral groups identification. The usage of different models trained for each maceral group separately gives the opportunity for finetuning. This also allows the architecture of the net to be kept relatively simple (e.g., simpler than in the original U-Net) while still providing good performance, at least for inertinite and vitrinite. The results are also encouraging in research targeted at discovering the simplest and most robust neural network structure for the efficient analysis of petrographic images.

4. Conclusions

The application of a U-Net-based CNN network for macerals segmentation on the coal petrographic optical microscope images has been presented. The set of images was manually segmented by experts and used further as the ground truth. The network was trained to segment inertinite, liptinite, and vitrinite. During the training, basic image augmentation was used (horizontal and vertical flipping, rotation by multiplicity of $\pi/2$ angle). The result show that very good results can be achieved for inertinite segmentation. The vitrinite was segmented slightly worse, but also at a very good level. The liptinite was most difficult to process. Moderately good results were obtained after the transfer learning usage. Even so, the segmentation was noticeably worse than for inertinite and vitrinite.

The obtained results show that the proposed convolutional autoencoder could effectively be used for maceral segmentation. Although the results for the liptinite were worse than those for other macerals, due to the advanced rank of analyzed coal samples, the network in most cases was able to locate the estimated maceral location. The inertinite and vitrinite segmentation are good enough to be considered as a base for autonomous petrographic processing. Although the results do not justify such a sentence in the case of liptinite, it still can be a valuable tool supporting the expert during petrographic image analysis. Data augmentation and transfer learning in particular proved their effectiveness in at least partially solving the problems in difficult cases. The comparison of the results with similar research showed that the obtained values of IoU and MIoU are better than those reported in the literature for the ML models, and are similar to those achieved by using the DL models (for inertinite and vitrinite). The segmentation of liptinite with the simplified, U-Net-based network is still a challenge and requires further research. The ML methods as well as image analysis methods are very promising, and have been utilized for coal analysis by many scientists with satisfactory results. The approach proposed here, though encouraging, fulfills only a tiny portion of the scientific challenges related to coal petrography. Further research work in this field is required.

Author Contributions: Conceptualization, S.I. and L.R.; methodology, S.I. and L.R.; software, S.I.; validation, L.R.; formal analysis, S.I.; investigation, S.I.; resources, L.R.; data curation, S.I. and L.R.; writing—original draft preparation, S.I. and L.R.; writing—review and editing, S.I. and L.R.; visualization, S.I.; supervision, L.R.; project administration, L.R.; funding acquisition, L.R. All authors have read and agreed to the published version of the manuscript.

Funding: This research was funded by Ministry of Science and Higher Education (Poland), grant number 11185011.

Data Availability Statement: Not applicable.

Conflicts of Interest: The authors declare no conflict of interest. The funders had no role in the design of the study; in the collection, analyses, or interpretation of data; in the writing of the manuscript, or in the decision to publish the results.

References

1. Dai, S.; Hower, J.C.; Finkelman, R.B.; Graham, I.T.; French, D.; Ward, C.R.; Eskenazy, G.; Wei, Q.; Zhao, L. Organic associations of non-mineral elements in coal: A review. *Int. J. Coal Geol.* **2019**, *218*, 103347. [CrossRef]
2. Dai, S.; Finkelman, R.B. Coal as a promising source of critical elements: Progress and future prospects. *Int. J. Coal Geol.* **2018**, *186*, 155–164. [CrossRef]
3. Finkelman, R.B.; Dai, S.; French, D. The importance of minerals in coal as the hosts of chemical elements: A review. *Int. J. Coal Geol.* **2019**, *212*. [CrossRef]
4. Hower, J.C. Clean Coal Technologies and Clean Coal Technologies Roadmaps-by Colin Henderson, International Energy Agency, CCC/74 and CCC/75, 2003; and Trends in Emission Standards by Lesley L. Sloss, International Energy Agency, CCC/77, 2003. *Int. J. Coal Geol.* **2004**, *4*, 270–271. [CrossRef]
5. Hower, J.C.; Ban, H.; Schaefer, J.L.; Stencel, J.M. Maceral/microlithotype partitioning through triboelectrostatic dry coal cleaning. *Int. J. Coal Geol.* **1997**, *34*, 277–286. [CrossRef]
6. Hower, J.C.; Robl, T.L.; Thomas, G.A. Changes in the quality of coal delivered to Kentucky power plants, 1978 to 1997: Responses to Clean Air Act directives. *Int. J. Coal Geol.* **1999**, *41*, 125–155. [CrossRef]

7. O'Keefe, J.M.; Bechtel, A.; Christanis, K.; Dai, S.; DiMichele, W.A.; Eble, C.F.; Esterle, J.; Mastalerz, M.; Raymond, A.L.; Valentim, B.; et al. On the fundamental difference between coal rank and coal type. *Int. J. Coal Geol.* **2013**, *118*, 58–87. [CrossRef]

8. Hower, J.C.; Eble, C.F.; O'Keefe, J.M. Phyteral perspectives: Every maceral tells a story. *Int. J. Coal Geol.* **2021**, *247*, 103849. [CrossRef]

9. ICCP, I. Handbook Coal Petr. Suppl. to 2nd Ed. ICCP. 1998. The New Vitrinite Classification (ICCP System 1994). *Fuel* **2001**, *77*, 349–358.

10. Sỳkorová, I.; Pickel, W.; Christanis, K.; Wolf, M.; Taylor, G.; Flores, D. Classification of Huminite—ICCP System 1994. *Int. J. Coal Geol.* **2005**, *62*, 85–106. [CrossRef]

11. Pickel, W.; Kus, J.; Flores, D.; Kalaitzidis, S.; Christanis, K.; Cardott, B.; Misz-Kennan, M.; Rodrigues, S.; Hentschel, A.; Hamor-Vido, M.; et al. Classification of Liptinite–ICCP System 1994. *Int. J. Coal Geol.* **2017**, *169*, 40–61. [CrossRef]

12. Parzentny, H.R.; Róg, L. Dependences between Certain Petrographic, Geochemical and Technological Indicators of Coal Quality in the Limnic Series of the Upper Silesian Coal Basin (USCB), Poland. *Arch. Min. Sci.* **2020**, *65*, 665–684. [CrossRef]

13. Kruszewska, K.; Dybowa-Jachowicz, S. *Draft of Coal Pertology*; Publishing House of the University of Silesia: Katowice, Poland, 1997.

14. Parzentny, H.; Róg, L. Evaluation the Value of Some Petrographic, Physico-Chemical and Geochemical Indicators of Quality of Coal in Paralic Series of the Upper Silesian Coal Basin and Attempt to Find a Correlation Between Them. *Gospod. Surowcami Miner.* **2017**, *33*, 51–76. [CrossRef]

15. Vasconcelos, L.D.S.E. The petrographic composition of world coals. Statistical results obtained from a literature survey with reference to coal type (maceral composition). *Int. J. Coal Geol.* **1999**, *40*, 27–58. [CrossRef]

16. Permana, A.K.; Ward, C.R.; Gurba, L.W. Maceral Characteristics and Vitrinite Reflectance Variation of The High Rank Coals, South Walker Creek, Bowen Basin, Australia. *Indones. J. Geosci.* **2013**, *8*. [CrossRef]

17. Denge, E.; Baiyegunhi, C. Maceral Types and Quality of Coal in the Tuli Coalfield: A Case Study of Coal in the Madzaringwe Formation in the Vele Colliery, Limpopo Province, South Africa. *Appl. Sci.* **2021**, *11*, 2179. [CrossRef]

18. Suárez-Ruiz, I.; Ward, C.R. Basic Factors Controlling Coal Quality and Technological Behavior of Coal. In *Applied Coal Petrology*; Elsevier: Amsterdam, The Netherlands, 2008; pp. 19–59.

19. Cutruneo, C.M.; Oliveira, M.; Ward, C.; Hower, J.C.; de Brum, I.A.; Sampaio, C.H.; Kautzmann, R.M.; Taffarel, S.R.; Teixeira, E.C.; Silva, L. A mineralogical and geochemical study of three Brazilian coal cleaning rejects: Demonstration of electron beam applications. *Int. J. Coal Geol.* **2014**, *130*, 33–52. [CrossRef]

20. Mastalerz, M.; Padgett, P.L. From in situ coal to the final coal product: A case study of the Danville Coal Member (Indiana). *Int. J. Coal Geol.* **1999**, *41*, 107–123. [CrossRef]

21. Méndez, L.; Borrego, A.; Martinez-Tarazona, M.; Menéndez, R. Influence of petrographic and mineral matter composition of coal particles on their combustion reactivity. *Fuel* **2003**, *82*, 1875–1882. [CrossRef]

22. Jelonek, I.; Jelonek, Z. Influence of petrographic properties of bituminous coal on the quality of metallurgical coke. *Sci. J. Inst. Miner. Energy Econ. Pol. Acad. Sci.* **2017**, *100*, 49–66.

23. Jelonek, I.; Mirkowski, Z. Petrographic and geochemical investigation of coal slurries and of the products resulting from their combustion. *Int. J. Coal Geol.* **2015**, *139*, 228–236. [CrossRef]

24. Bielowicz, B. Petrographic Characteristics of Coal Gasification and Combustion by-Products from High Volatile Bituminous Coal. *Energies* **2020**, *13*, 4374. [CrossRef]

25. Bielowicz, B. Petrographic Composition of Coal from the Janina Mine and Char Obtained as a Result of Gasification in the CFB Gasifier. *Miner. Resour. Manag.* **2019**, *35*, 99–116. [CrossRef]

26. Mirkowski, Z.; Jelonek, I. Petrographic composition of coals and products of coal combustion from the selected combined heat and power plants (CHP) and heating plants in Upper Silesia, Poland. *Int. J. Coal Geol.* **2018**, *201*, 102–108. [CrossRef]

27. Bielowicz, B.; Misiak, J. The Impact of Coal's Petrographic Composition on Its Suitability for the Gasification Process: The Example of Polish Deposits. *Resources* **2020**, *9*, 111. [CrossRef]

28. Róg, L. Vitrinite reflectance as a measure of the range of influence of the temperature of a georeactor on rock mass during underground coal gasification. *Fuel* **2018**, *224*, 94–100. [CrossRef]

29. Büçkün, Z.; Inaner, H.; Oskay, R.G.; Christanis, K. Palaeoenvironmental Reconstruction of Hüsamlar Coal Seam, SW Turkey. *J. Earth Syst. Sci.* **2015**, *124*, 729–746. [CrossRef]

30. Oskay, R.; Christanis, K.; Inaner, H.; Salman, M.; Taka, M. Palaeoenvironmental reconstruction of the eastern part of the Karapınar-Ayrancı coal deposit (Central Turkey). *Int. J. Coal Geol.* **2016**, *163*, 100–111. [CrossRef]

31. Mastalerz, M.; Hower, J.; Taulbee, D. Variations in Chemistry of Macerals as Refl Ected by Micro-Scale Analysis of a Spanish Coal. *Geol. Acta* **2013**, *11*, 483–493.

32. Querol, X.; Cabrera, L.; Pickel, W.; López-Soler, A.; Hagemann, H.; Fernandez-Turiel, J.-L. Geological controls on the coal quality of the Mequinenza subbituminous coal deposit, northeast Spain. *Int. J. Coal Geol.* **1996**, *29*, 67–91. [CrossRef]

33. Suarez-Ruiz, I.; Jimenez, A.; Iglesias, M.J.; Laggoun-Defarge, F.; Prado, J.G. Influence of Resinite on Huminite Properties. *Energy Fuels* **1994**, *8*, 1417–1424. [CrossRef]

34. Oestreich, J.; Tolley, W.; Rice, D. The development of a color sensor system to measure mineral compositions. *Miner. Eng.* **1995**, *8*, 31–39. [CrossRef]

35. Singh, V.; Rao, S.M. Application of image processing and radial basis neural network techniques for ore sorting and ore classification. *Miner. Eng.* **2005**, *18*, 1412–1420. [CrossRef]
36. Tessier, J.; Duchesne, C.; Bartolacci, G. A machine vision approach to on-line estimation of run-of-mine ore composition on conveyor belts. *Miner. Eng.* **2007**, *20*, 1129–1144. [CrossRef]
37. Iwaszenko, S.; Nurzynska, K. Rock Grains Segmentation Using Curvilinear Structures Based Features. In Proceedings of the Real-Time Image Processing and Deep Learning 2019. *SPIE Int. Soc. Opt. Eng.* **2019**, *10996*, 109960V.
38. Zhang, Z.; Yang, J.; Su, X.; Ding, L.; Wang, Y. Multi-Scale Image Segmentation of Coal Piles on a Belt Based on the Hessian Matrix. *Particuology* **2013**, *11*, 549–555. [CrossRef]
39. Perez, C.A.; Estévez, P.A.; Vera, P.A.; Castillo, L.E.; Aravena, C.M.; Schulz, D.A.; Medina, L.E. Ore grade estimation by feature selection and voting using boundary detection in digital image analysis. *Int. J. Miner. Process.* **2011**, *101*, 28–36. [CrossRef]
40. Iwaszenko, S.; Smoliński, A. Texture features for bulk rock material grain boundary segmentation. *J. King Saud Univ.—Eng. Sci.* **2020**, *33*, 95–103. [CrossRef]
41. Mertens, G.; Elsen, J. Use of computer assisted image analysis for the determination of the grain-size distribution of sands used in mortars. *Cem. Concr. Res.* **2006**, *36*, 1453–1459. [CrossRef]
42. Kazak, A.; Simonov, K.; Kulikov, V. Machine-Learning-Assisted Segmentation of Focused Ion Beam-Scanning Electron Microscopy Images with Artifacts for Improved Void-Space Characterization of Tight Reservoir Rocks. *SPE J.* **2021**, *26*, 1739–1758. [CrossRef]
43. Zhou, Y.; Starkey, J.; Mansinha, L. Segmentation of petrographic images by integrating edge detection and region growing. *Comput. Geosci.* **2004**, *30*, 817–831. [CrossRef]
44. Sethian, J. Fast Marching Methods and Level Set Methods for Propagating Interfaces. In Proceedings of the 29th Computational Fluid Dynamics, Rhode-Saint-Genese, Belgium, 23–27 February 1998.
45. Lu, B.; Cui, M.; Liu, Q.; Wang, Y. Automated Grain Boundary Detection Using the Level Set Method. *Comput. Geosci.* **2009**, *35*, 267–275. [CrossRef]
46. Chao, E.; Minkin, J.; Thompson, C. Application of automated image analysis to coal petrography. *Int. J. Coal Geol.* **1982**, *2*, 113–150. [CrossRef]
47. Crelling, J.C. Automated petrographic characterization of coal lithotypes. *Int. J. Coal Geol.* **1982**, *1*, 347–359. [CrossRef]
48. Kuili, J.; Jian, X.; Duohu, H. The use of automated coal petrography in determining maceral group composition and the reflectance of vitrinite. *Int. J. Coal Geol.* **1988**, *9*, 385–395. [CrossRef]
49. Hou, X.; Ren, D.; Mao, H.; Lei, J.; Jin, K.; Chu, P.K.; Reich, F.; Wayne, D.H. Application of imaging TOF-SIMS to the study of some coal macerals. *Int. J. Coal Geol.* **1995**, *27*, 23–32. [CrossRef]
50. Alpana; Mohapatra, S. *Automated Coal Characterization Using Computational Intelligence and Image Analysis Techniques*; IEEE: Piscataway, NJ, USA, 2015; pp. 176–180. [CrossRef]
51. O'Brien, G.; Jenkins, B.; Esterle, J.; Beath, H. Coal characterisation by automated coal petrography. *Fuel* **2003**, *82*, 1067–1073. [CrossRef]
52. O'Brien, G.; Gu, Y.; Adair, B.; Firth, B. The use of optical reflected light and SEM imaging systems to provide quantitative coal characterisation. *Miner. Eng.* **2011**, *24*, 1299–1304. [CrossRef]
53. Crelling, J.; Glasspool, I.; Gibbins, J.; Seitz, M. Bireflectance imaging of coal and carbon specimens. *Int. J. Coal Geol.* **2005**, *64*, 204–216. [CrossRef]
54. Lester, E.; Watts, D.; Cloke, M. A novel automated image analysis method for maceral analysis. *Fuel* **2002**, *81*, 2209–2217. [CrossRef]
55. Tiwary, A.K.; Ghosh, S.; Singh, R.; Mukherjee, D.P.; Shankar, B.U.; Dash, P.S. Automated coal petrography using random forest. *Int. J. Coal Geol.* **2020**, *232*, 103629. [CrossRef]
56. Wang, H.; Lei, M.; Chen, Y.; Li, M.; Zou, L. Intelligent Identification of Maceral Components of Coal Based on Image Segmentation and Classification. *Appl. Sci.* **2019**, *9*, 3245. [CrossRef]
57. Wang, Y.; Bai, X.; Wu, L.; Zhang, Y.; Qu, S. Identification of maceral groups in Chinese bituminous coals based on semantic segmentation models. *Fuel* **2021**, *308*, 121844. [CrossRef]
58. Vranjes-Wessely, S.; Misch, D.; Kiener, D.; Cordill, M.; Frese, N.; Beyer, A.; Horsfield, B.; Wang, C.; Sachsenhofer, R. High-speed nanoindentation mapping of organic matter-rich rocks: A critical evaluation by correlative imaging and machine learning data analysis. *Int. J. Coal Geol.* **2021**, *247*, 103847. [CrossRef]
59. Mlynarczuk, M.; Skiba, M. The application of artificial intelligence for the identification of the maceral groups and mineral components of coal. *Comput. Geosci.* **2017**, *103*, 133–141. [CrossRef]
60. Skiba, M.; Młynarczuk, M. Identification of macerals of the inertinite group using neural classifiers, based on selected textural features. *Arch. Min. Sci.* **2018**, *63*, 827–837. [CrossRef]
61. Busse, J.; de Dreuzy, J.; Torres, S.G.; Bringemeier, D.; Scheuermann, A. Image processing based characterisation of coal cleat networks. *Int. J. Coal Geol.* **2016**, *169*, 1–21. [CrossRef]
62. Maxwell, K.; Rajabi, M.; Esterle, J. Automated classification of metamorphosed coal from geophysical log data using supervised machine learning techniques. *Int. J. Coal Geol.* **2019**, *214*. [CrossRef]
63. Garcia-Garcia, A.; Orts-Escolano, S.; Oprea, S.; Villena-Martinez, V.; Garcia-Rodriguez, J. A Review on Deep Learning Techniques Applied to Semantic Segmentation. *arXiv* **2017**, arXiv:1704.06857.

64. Taghanaki, S.A.; Abhishek, K.; Cohen, J.P.; Cohen-Adad, J.; Hamarneh, G. Deep semantic segmentation of natural and medical images: A review. *Artif. Intell. Rev.* **2020**, *54*, 137–178. [CrossRef]
65. De Lima, R.P.; Bonar, A.; Coronado, D.D.; Marfurt, K.; Nicholson, C. Deep convolutional neural networks as a geological image classification tool. *Sediment. Rec.* **2019**, *17*, 4–9. [CrossRef]
66. Pu, Y.; Apel, D.B.; Szmigiel, A.; Chen, J. Image Recognition of Coal and Coal Gangue Using a Convolutional Neural Network and Transfer Learning. *Energies* **2019**, *12*, 1735. [CrossRef]
67. Li, D.; Zhang, Z.; Xu, Z.; Xu, L.; Meng, G.; Li, Z.; Chen, S. An Image-Based Hierarchical Deep Learning Framework for Coal and Gangue Detection. *IEEE Access* **2019**, *7*, 184686–184699. [CrossRef]
68. Lei, M.; Rao, Z.; Wang, H.; Chen, Y.; Zou, L.; Yu, H. Maceral groups analysis of coal based on semantic segmentation of photomicrographs via the improved U-net. *Fuel* **2021**, *294*, 120475. [CrossRef]
69. Long, J.; Shelhamer, E.; Darrell, T. Fully convolutional networks for semantic segmentation. In Proceedings of the IEEE Conference on Computer Vision and Pattern Recognition (CVPR), Boston, MA, USA, 7–12 June 2015; pp. 3431–3440. [CrossRef]
70. Van Krevelen, D.W.; Schuyer, J. *Coal. Coal Chemistry and Its Structure*; Państwowe Wydawnictwo Naukowe: Warsaw, Poland, 1959.
71. Ronneberger, O.; Fischer, P.; Brox, T. U-Net: Convolutional Networks for Biomedical Image Segmentation. In *Proceedings of the International Conference on Medical Image Computing and Computer-Assisted Intervention, Munich, Germany, 5–9 October 2015*; Springer: Berlin/Heidelberg, Germany, 2015; pp. 234–241.
72. Kingma, D.P.; Ba, J. Adam: A Method for Stochastic Optimization. *arXiv* **2017**, arXiv:1412.6980.
73. Abadi, M.; Agarwal, A.; Barham, P.; Brevdo, E.; Chen, Z.; Citro, C.; Corrado, G.S.; Davis, A.; Dean, J.; Devin, M.; et al. TensorFlow: Large-Scale Machine Learning on Heterogeneous Systems. *arXiv* **2015**, arXiv:1603.04467.
74. Glorot, X.; Bengio, Y. Understanding the Difficulty of Training Deep Feedforward Neural Networks. In Proceedings of the Thirteenth International Conference on Artificial Intelligence and Statistics, Sardinia, Italy, 13–15 May 2010; pp. 249–256.

*minerals*

Article

Model Scaling in Smartphone GNSS-Aided Photogrammetry for Fragmentation Size Distribution Estimation

Zedrick Paul L. Tungol [1,*], Hisatoshi Toriya [1], Narihiro Owada [1], Itaru Kitahara [2], Fumiaki Inagaki [1], Mahdi Saadat [1], Hyong Doo Jang [3] and Youhei Kawamura [4]

[1] Department of Earth Resource Engineering and Environmental Sciences, Akita University, Akita 010-8502, Japan; toriya@gipc.akita-u.ac.jp (H.T.); nowaddd67@gmail.com (N.O.); fumiaki.inagaki@gipc.akita-u.ac.jp (F.I.); msaadat014@gmail.com (M.S.)

[2] Center for Computational Sciences, University of Tsukuba, Tsukuba 305-8577, Japan; kitahara@ccs.tsukuba.ac.jp

[3] Western Australian School of Mines, Curtin University, Perth 6845, Australia; Hyongdoo.Jang@curtin.edu.au

[4] Division of Sustainable Resources Engineering, Hokkaido University, Hakodate 060-0817, Japan; kawamura@eng.hokudai.ac.jp

[*] Correspondence: zed.tungol37@gmail.com

Abstract: Fragmentation size distribution estimation is a critical process in mining operations that employ blasting. In this study, we aim to create a low-cost, efficient system for producing a scaled 3D model without the use of ground truth data, such as GCPs (Ground Control Points), for the purpose of improving fragmentation size distribution measurement using GNSS (Global Navigation Satellite System)-aided photogrammetry. However, the inherent error of GNSS data inhibits a straight-forward application in Structure-from-Motion (SfM). To overcome this, the study proposes that, by increasing the number of photos used in the SfM process, the scale error brought about by the GNSS error will proportionally decrease. Experiments indicated that constraining camera positions to locations, relative or otherwise, improved the accuracy of the generated 3D model. In further experiments, the results showed that the scale error decreased when more images from the same dataset were used. The proposed method is practical and easy to transport as it only requires a smartphone and, optionally, a separate camera. In conclusion, with some modifications to the workflow, technique, and equipment, a muckpile can be accurately recreated in scale in the digital world with the use of positional data.

Keywords: point cloud scaling; fragmentation size analysis; structure from motion

Citation: Tungol, Z.P.L.; Toriya, H.; Owada, N.; Kitahara, I.; Inagaki, F.; Saadat, M.; Jang, H.D.; Kawamura, Y. Model Scaling in Smartphone GNSS-Aided Photogrammetry for Fragmentation Size Distribution Estimation. *Minerals* **2021**, *11*, 1301. https://doi.org/10.3390/min11121301

Academic Editors: Abbas Taheri, Rajive Ganguli, Sean Dessureault and Pratt Rogers

Received: 27 July 2021
Accepted: 16 November 2021
Published: 23 November 2021

Publisher's Note: MDPI stays neutral with regard to jurisdictional claims in published maps and institutional affiliations.

1. Introduction

Fragmentation size is a key parameter to the efficiency of numerous mining operations that makes use of explosives across all of the stages of production from mine (drill, blasting, haulage, etc.) to mill (mineral processing). Several studies [1,2] going back to the late 1990s have explored this particular correlation. It is vital for companies, therefore, to monitor fragmentation size and make necessary changes to mine planning and execution. For the purpose of maintaining a consistent optimal fragmentation size, the blasting products must be monitored regularly, so that any necessary modifications to the drilling and blasting process can be made.

However, the rock is usually heterogeneous in nature, and a large amount of material is generally mined every day, so there are some difficulties in monitoring the fragmentation size distribution regularly using traditional methods. These traditional methods include manual sieving, boulder counting, and visual estimation. However, limitations on sampling and bias make these methods relatively inefficient [3]. As such, there exists a need for a quick and accessible method of rock fragmentation size distribution determination that can surmount the limitations of physical sampling and laboratory analysis. A currently used digital solution to this problem is to employ image-based particle size analysis software.

Commercial products, such as WipFrag [4], make use of images of a muckpile or orthomosaics to measure fragmentation size distribution. In addition, other 3D modelling technologies, such as lidar systems, have also been used in fragmentation measurement systems producing good results and accurate measurements without the need for scale objects.

A previous study [5] was done using laser scanning to measure the blast fragmentation in rockpiles in a mine. While specialized equipment was used to do this study, lidar is becoming increasingly accessible in recent years, with newer generation smartphone models including a lidar system. The study recognizes that this is also a good alternative to perform fragmentation measurement as they are potentially effective in an underground setting where lighting is limited, an issue that produces problems for traditional image-based photogrammetry [6].

In a previous study [7], a 3-Dimensional Fragmentation Measurement (3DFM) system was developed that makes use of 3D Photogrammetry to measure particle size distribution at accuracies greater than that of conventional methods. A theoretical visualized workflow for this particular system when applied to a mining operation is shown in Figure 1.

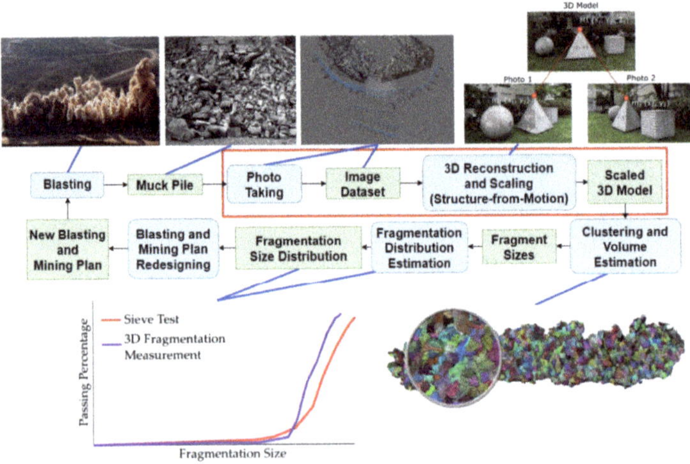

Figure 1. Whole image of a target system.

The developed system is divided into stages, utilizing multiple computational techniques in order to achieve its purpose. In a hypothetical application of the system, pictures of the muckpile from the products of blasting are taken. The sizes of muckpiles vary greatly depend on the specifications of the hauling equipment as well as the mine plan that the operation employs. In situations where the muckpile is too large or has parts that are inaccessible to photo-taking, it is possible for the system to reconstruct only a representative "slice" of the muckpile.

The images are then processed in a high-power computer by a sequence of 3D imaging techniques that will ultimately output a scaled 3D model of the muckpile in the form of a point cloud. A technique known as supervoxel clustering is then performed on the 3D model, which then undergoes supervoxel clustering in order to divide the individual fragments into segments whose dimensions have been calculated. The dimensional data can then be used in the computation of the fragment size distribution of the muckpile. Using this information, the blasting product can be judged if it is up to the expected specification. Adjustments are then made the blasting design, such as the amount and type of explosive and blasting patterns in order to achieve the required distribution.

This study focuses on the 3D model scaling aspect of this system, as highlighted with a red box in Figure 1. Specifically, the research will analyze how positional data can affect scaling error when reconstructing using Structure-from-Motion. Scaling is a critical component of fragmentation size distribution measurement using photogrammetry as this will directly determine the accuracy of the size estimation. In creating a 3D model, extrinsic data, such as ground truths, are needed to create a properly-scaled reconstruction of the scene. Traditionally, scale is resolved in photogrammetry by placing scale bars in the scene or taking a measurement of two features and then scaling the generated model using that information.

The study proposes a method that makes use of GNSS (Global Navigation Satellite System) data to create scaled 3D models without the need for post-reconstruction rescaling. GNSS positional data, and its sub-systems, such as GPS, Beidou, GLONASS, and Japan's own QZSS can be utilized. A previous study was performed with regards to using GPS in reconstruction but mostly in the context of UAV (Unmanned Aerial Vehicle) Mapping [8]. This study aims to create a system that does not need ground truth data, such as GCPs (Ground Control Points) to create a properly scaled 3D model of a muckpile. This would aid greatly in the fragmentation size distribution measurement of muckpiles using photogrammetry.

It is a known fact that inherent error exists within GNSS and its subsets, and even high-end geodetic GNSS receivers have errors in the centimeter range [9]. For this study, a smartphone was used as a GNSS receiver for the digital camera. This decision was due to the end-goal of this research, which is to be able use both image data and GNSS data from a smartphone, as this practicality can be important in a mining operation environment.

This comes at a drawback to the GNSS accuracy, as recreational grade GNSS chips, like those found in smartphones, typically have errors in the meter range [10]. To overcome this error, the study proposes to make use of an increasing number of georeferenced images to statistically decrease the scaling error of the constructed 3D model. Figure 2 shows a general overview of the proposed system for this study. Utilizing a smartphone's built-in GNSS receiver, GNSS data can be logged and sent to a camera. At the moment an image is taken, GNSS data can be embedded into the image's metadata (EXIF).

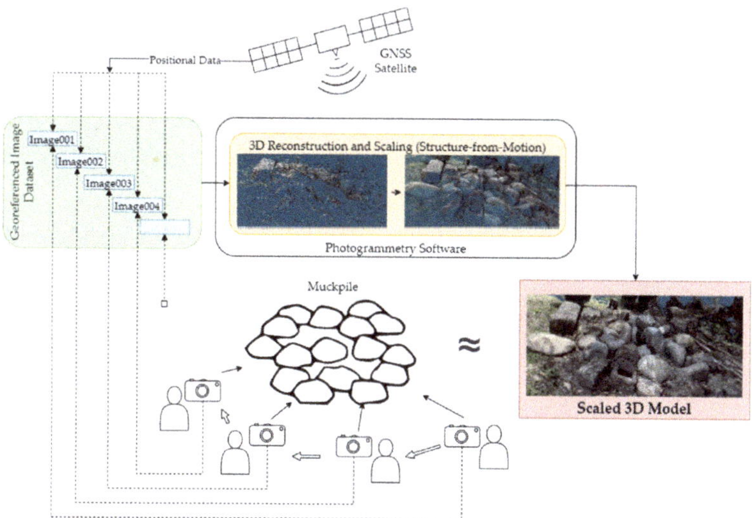

Figure 2. Workflow for the proposed method.

In a similar study [11], a Real Time Kinematic (RTK) GNSS receiver was used in conjunction with a camera for a photogrammetric survey of a geological outcrop. The method suggested in this study is a potentially cheaper alternative as it utilizes the built-in GNSS receiver in a smartphone. In a similar fashion, the method used by this study allows for greater flexibility as it is a point-and-shoot method that does not require external preparation. While this can mean that more photos will be needed to generate a 3D model, the cost-efficiency and the practicality of not having to use GCPs or physical scales can be desirable in some applications.

2. Materials and Methods

2.1. Structure-from-Motion—Multi View Stereo (SfM-MVS)

Mathematically speaking, SfM can be described as the conversion of four coordinate systems, illustrated in Figure 3:

(1)   An image pixel coordinate system, which concerns the pixels on the 2D image.
(2)   An imaging plane coordinate system, which lie on the same plane of the previous system, but whose origin is the plane's intersection with the camera's optical axis.
(3)   A camera coordinate system, which concerns a pinhole camera's point of view of the image.
(4)   A world coordinate system, which is a reference system to describe the position of the camera and the objects being taken pictures of.

$$Z_c \begin{bmatrix} u \\ v \\ 1 \end{bmatrix} = \begin{bmatrix} \frac{1}{\delta_x} & 0 & u_o \\ 0 & \frac{1}{\delta_y} & v_o \\ 0 & 0 & 1 \end{bmatrix} \begin{bmatrix} f & 0 & 0 \\ 0 & f & 0 \\ 0 & 0 & 1 \end{bmatrix} \begin{bmatrix} R & t \end{bmatrix} \begin{bmatrix} X_w \\ Y_w \\ Z_w \\ 1 \end{bmatrix} = \begin{bmatrix} f_x & 0 & u_0 \\ 0 & f_y & v_0 \\ 0 & 0 & 1 \end{bmatrix} \begin{bmatrix} R & t \end{bmatrix} \begin{bmatrix} X_w \\ Y_w \\ Z_w \\ 1 \end{bmatrix} \quad (1)$$

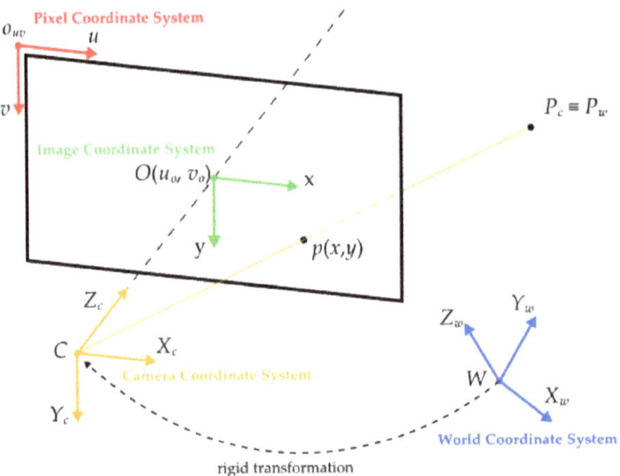

Figure 3. An illustration of the coordinate systems. Described is the conversion of world point P$_w$ to camera point P$_c$, to imaging plane coordinates (x,y), and finally pixel coordinates (u,v) [12].

The conversion of these four coordinates systems can be described by Equation (1). u and v describe the axes in the imaging planes. u_0 and v_0 are the coordinates of the origins of the imaging plane in the pixel coordinate system. δ_x and δ_y represent the physical size of each pixel in the image in the imaging plane (zoom ratio). f describes the focal length, which is the distance from the optical center of the camera to the pixel plane. $R \in \mathbb{R}^{3\times3}$ and $t \in \mathbb{R}^3$ describe the rotational and translational vectors that relate the camera and the world coordinate systems. X_w, Y_w, Z_w are the actual coordinates of a point in the world coordinate system [12].

Equation (1) represents the fact that, in order to estimate the position of a point in the real world, the external parameter matrix of the camera (i.e., R and t) needs to be measured first. Once R is known, the relative position of the object in the world coordinate system can be estimated, and once t is known, the absolute position can also be acquired as well.

Basic SfM can relatively estimate R and t. In this study, GNSS data is used as the absolute value of t. The details of this method are described in Section 2.2.

It can be inferred that there is no single 'correct' workflow or process in the conversion of 2D images into models. However, there are key processes that are present in almost all applications of the method, as shown in Figure 4. The steps are briefly described in the following sections.

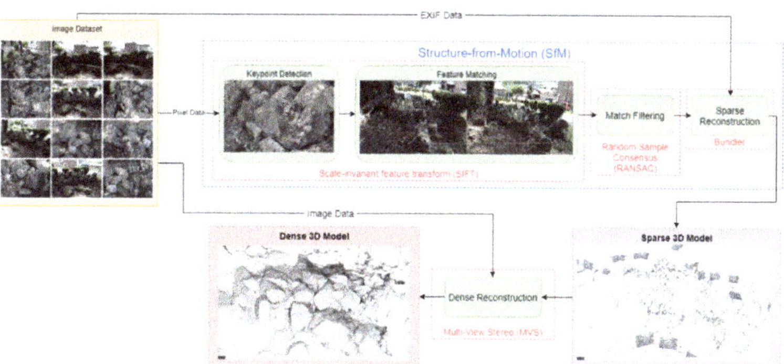

Figure 4. SfM-MVS pipeline.

2.1.1. Keypoint Detection

The initial processing step after acquiring the images is feature detection, or extraction, where possible common features (keypoints) in the individual images are identified as shown in Figure 5. It is by these features that allow the different images in the dataset to be matched at the next stage. There are several techniques that have been developed for the solution of this step [13]; however, the most widely used amongst modern SfM applications is the scale-invariant feature transform (SIFT) [14].

Figure 5. Keypoint detection on a pile of rocks.

The system recognizes feature points in the image set, which are uniform in scaling and rotation and relatively uniform to changes in lighting and 3D camera view angles. The number of keypoints that are extracted in an image relies heavily on the resolution and texture of the images themselves with high-quality, original-resolution pictures returning the most results [15].

2.1.2. Keypoint Matching

The next step is to match the keypoints and identify the correspondences between them. Matches are found by identifying a keypoint's nearest neighbor in the database. The nearest neighbor is defined as the keypoint with the least Euclidean distance for its descriptor vector, as shown in Figure 6 [14]. It is also important to note at this point that not all keypoints are guaranteed to have a good match in the dataset. It is, therefore, necessary to discard these unmatched keypoints, making use of the ratio between the Euclidean distance of the nearest neighbor with that of the second nearest at a certain minimum value as a criterion for discarding false keypoint matches [13].

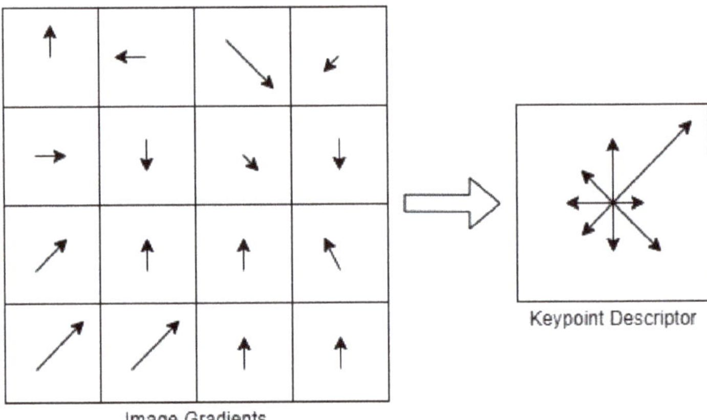

Figure 6. Graphical representation of image gradients and keypoint descriptors.

The inherent complexity of the keypoint descriptors gives rise to the need of an efficient solution to the search process, as brute-force searching for nearest neighbors proves to be computationally difficult and time-consuming. Several solutions, such as k-dimensional trees (k-d trees), best-bin first (BBF), and approximate nearest neighbor (ANN) searching, are used to solve this problem of efficiency by partitioning the data into bins that are prioritized for match searching, thus, decreasing the number of recursions needed to go through all the keypoints [13].

2.1.3. Keypoint Filtering

The third stage, also known as geometric verification or match filtering, is done to further eliminate erroneous matches. Since the initial matching is solely based on appearance, it cannot be guaranteed that the matched keypoints refer to the same point in an image (e.g., images with symmetrical or similar features) [16]. SfM then needs to verify matches by mapping keypoints across images using projective imagery. An example of this step can be illustrated by the image pair in Figure 7. The two images are of the same scene, taken at two different angles, and the keypoints found in both images are matched, as shown by colored matching tracks.

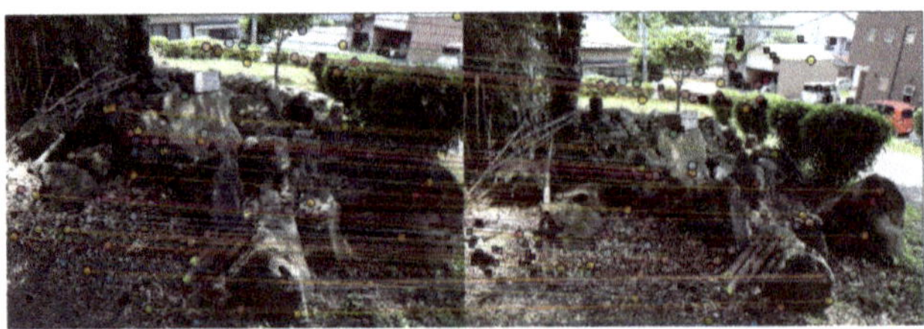

Figure 7. Keypoint matching tracks over two different views.

2.1.4. Sparse Reconstruction: Structure-from-Motion

The fourth step, which also by itself is sometimes called SfM (Structure-from-Motion), is to reconstruct the scene that was taken using 2D images into an initial sparse 3D structure. Using the verified matched keypoints, SfM aims to simultaneously reconstruct the: (a) 3D scene structure, (b) camera position and orientation (extrinsic parameters), and (c) intrinsic camera calibration parameters. The intrinsic camera parameters are defined by a camera calibration matrix that includes image scale, skew, and the principal point that is defined as the location on the image plane that intersects the optical axis.

Further intrinsic parameters are also required to resolve additional internal aberrations, such as distortion on non-pre-calibrated cameras. These intrinsic parameters are either included in the camera's image file format (e.g., EXIF) or will be resolved in additional intermediate steps. After this, a process known as bundle adjustment is used to produce sparse point-clouds [16]. This process will be described further in another part of the paper below as it is in this step that GNSS constraints will come to play in scaling the produced 3D model. A simplified illustration of this process is described by the illustration in Figure 8.

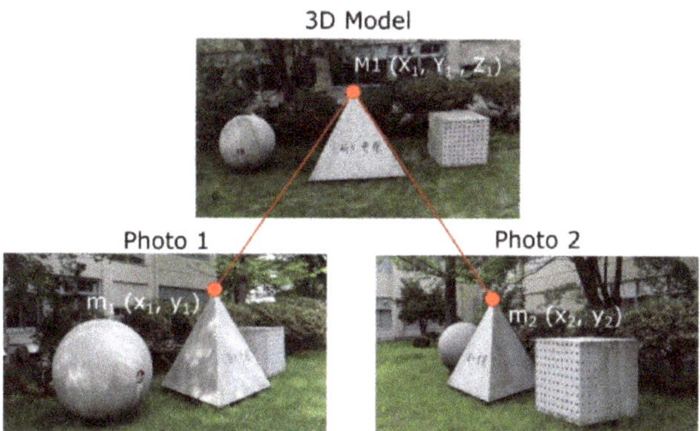

Figure 8. SfM example showing common points.

2.1.5. Dense Reconstruction: Multi View Stereo

An additional, post-processing method known as MVS (Multi-View Stereo) can be applied to the sparse 3D model from SfM in order to generate an enhanced "dense" 3D model. The final output of MVS is a complete 3D scene reconstruction from a collection of images of known intrinsic and extrinsic parameters, which is already resolved through SfM. A variety of MVS algorithms are available but recent variants called clustering views for MVS (CMVS), and patch-based MVS (PMVS) was observed to perform well against other algorithms [13].

CMVS decomposes the camera poses from bundle adjustment into manageable clusters, and PMVS is used to independently reconstruct the 3-dimensional model from these clusters [15]. Most modern MVS pipelines, including the one in the software used for this study, include features from both these variants of MVS. A comparison of the point density between sparse and dense reconstructions is illustrated in Figure 9.

Figure 9. The sparse point cloud (**a**) and dense point cloud (**b**) of a scene.

2.2. GNSS-Aided Scaling in Bundle Adjustment

In the bundle adjustment phase of SfM, the previous and imperfect solutions regarding camera positions and 3D features of the scene are refined [17]. More specifically, bundle adjustment is a non-linear minimization procedure that jointly optimizes the camera parameters and point position by minimizing the reprojection error between the image locations of observed and predicted image points. This minimization is done using nonlinear least-squares algorithms [18].

Numerous studies have been done since its inception in the 1990s regarding bundle adjustment, with most of the research going into reducing its computational burden and accelerating the problem-solving process [19]. One of such propositions is the fusion of positional data and bundle adjustment, with GNSS data being used as constraints for solving reprojection errors [20]. This concept is what this research aims to produce: accurately-scaled 3D reconstructions of muckpiles. GNSS data is used to provide position and covariance estimates for the bundle adjustment process. The nominal form of these solutions is:

$$r^M_{GNSS}(\tau) = r^M_c(\tau) + R^M_c(\tau)r^c_{GNSS} + \left(b^M_{GNSS} + d^M_{GNSS}(\tau - \tau_0)\right) \tag{2}$$

where $r^M_{GNSS}(t)$ is the position of the GNSS receiver, $r^M_c(t)$ denotes the camera position, $R^M_c(t)$ is the rotational matrix that aligns the camera and mapping space axes, and r^c_{GNSS} is the difference between the GNSS receiver and camera position. b^M_{GNSS} and d^M_{GNSS} denote bias and drift terms and are included to account for data inconsistencies and the inherent errors that exist within GNSS. A previous study [11] applied GNSS-assisted terrestrial photogrammetry to model coastal areas without the use of GCPs. With bundle adjustment being an error minimization problem with multiple factors, weights can be assigned to them, as is the case in the study's SfM workflow.

The software that was used for this study was developed with the aim of being able to perform SfM without the need for any additional intrinsic or extrinsic data (such as GNSS) aside from the image themselves. [21]. However, the software itself still allows for the importation of GNSS data from images for the purpose of constraining camera positions. Weights are assigned to this GNSS data and used in the bundle adjustment step [22]. As such, the study deems it necessary to initially prove if properly constraining camera positions will help in creating a properly scaled 3D model. A preliminary experiment was designed to test this theory, which is described in the proceeding section of this paper.

Preliminary Experiment for Validating Scaling Fundamentals

A preliminary photogrammetry experiment was performed before the main experiment to test some core concepts regarding the study, specifically the effects of known and constrained camera positions on scaling error and reconstruction quality. This small-scale experiment involved taking photos of a scene that was set-up indoors in the laboratory that consisted of a stuffed dog plush toy that was placed on the floor in such a way that it was in the middle of a grid of nine carpet panels. The panels are 50 by 50 cm in size and form a 3×3 grid measuring 150 by 150 cm in total. Figure 10 shows the general layout of the scene. The purpose of this grid is to provide a spatial reference for the camera positions when taking pictures of the scene. A detailed board was put on the middle of the scene to provide enough feature points for SfM, as initial reconstructions without the board resulted in distorted point clouds with missing parts.

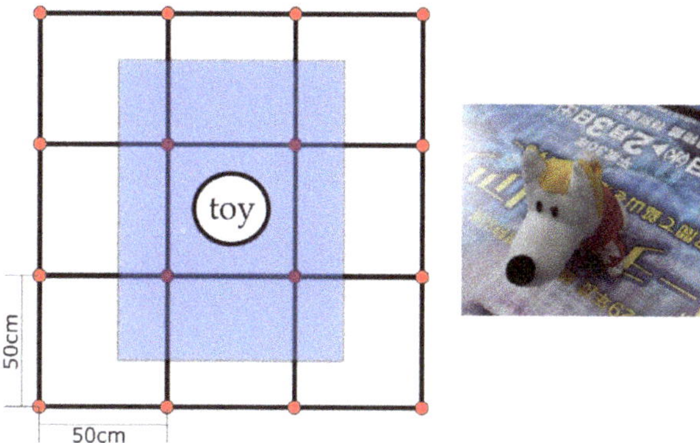

Figure 10. Grid layout of the first preliminary experiment (**left**) and a toy as the object (**right**). Red dots indicate camera positions, and the blue rectangle indicates the board that was inserted for improved feature detection.

The camera used for this preliminary and succeeding experiments was a Canon EOS R equipped with a Canon 24–105 mm lens. The f-stop was set at 4 with variable exposure times, automatic white balancing enabled, and the zoom was kept at a minimum to provide a fixed focal length, which is required for SfM. A total of 32 photos were taken, two at each of the intersection points of the grid at different heights (45 and 60 cm), with sample images shown in Figure 11. The height was maintained by mounting the camera on an adjustable tripod.

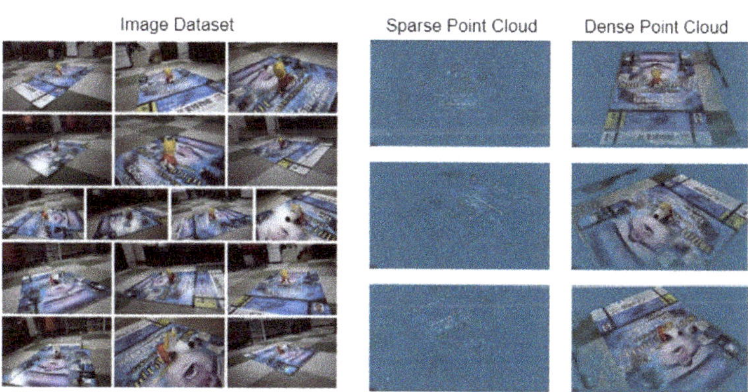

Figure 11. Data input and output of the preliminary experiment.

Along with the 50 cm spacing, this provided known relative camera positions. The captured images were then processed with a workflow that consisted of making the sparse point cloud and a dense point cloud using photogrammetry software. Creating a textured mesh was deemed unnecessary as it meant a longer processing time and larger project file size and ultimately did not contribute to analyzing the results of the experiment. After this, the camera positions were constrained to their known locations. The scaling error of the reconstructed model was then analyzed.

Without any camera constraints, the software arbitrarily designated a scale, rotation, and translation for the model, and the measurement of the dimension of the carpet panel was about 10 units (since there are no constraints applied, this value cannot be assigned a specific unit of measurement). However, upon adding constraints to the camera (at the centimeter level) by importing a file describing each image's distance from each other, the same dimension then measured at 51.2719 cm, with a difference of 1.2719 cm from the real measurement of 50 cm.

This difference was attributed to human error during the shooting process. A possible specific example is that the center of the tripod (and, by extension, the center of the camera) was used to align the camera to the grid instead of the nodal point of the camera lens. This means that the images were offset from the actual intended grid position depending on the orientation of the camera and the tripod. Despite this difference, the study still proves fundamentally that accurately constraining the camera positions to their real-world values improved the scale accuracy of the constructed 3D model.

3. GNSS-Constrained SfM on Monuments of Known Dimensions

To perform quantitative evaluation of the effects of GNSS constraints on the scaling error of 3D reconstruction using SfM, an analysis using monuments of known dimensions outside Akita University was done. The experiment aims to correlate the scaling error to the number of images used in SfM. The hypothesis of this experiment is that, as more images are used, the scaling error due to GNSS error will decrease. In this scene, the cube-shaped monument has sides measuring approximately 1 m. This dimension is used to compute the scaling error. This particular scene was chosen for this reason, in addition to the monuments being of simple 3D shapes, making analysis of measurements more accurate for the purpose of quantitative evaluation.

For this experiment, around 200 images of the scene were taken across 2 days at roughly the same time of the day, with sample images shown in Figure 12 and a map of the depicted photo taking area and the recorded camera positions in Figure 13. For this and the proceeding experiment, the camera was used freehanded without a tripod, with a Xiaomi Mi 9T Pro smartphone placed close to the camera sending GNSS data to it via Bluetooth. The dataset, as with the previous experiment, was used to create 3D reconstructions at different image numbers, with an example shown in Figure 14. The scaling error when varying number of images are used was noted and compared. For reconstruction purposes in this and the following experiments, 3DF Zephyr was used, with a setting of 50% GNSS data weight, as specified in the software's manual [22].

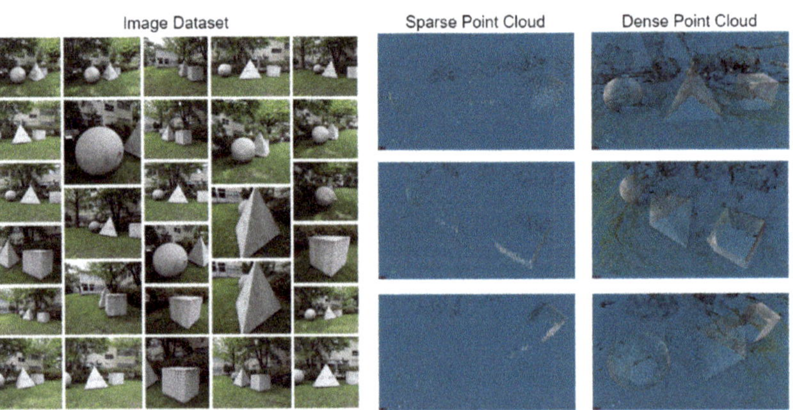

Figure 12. Data input and output of the experiment using 200 images.

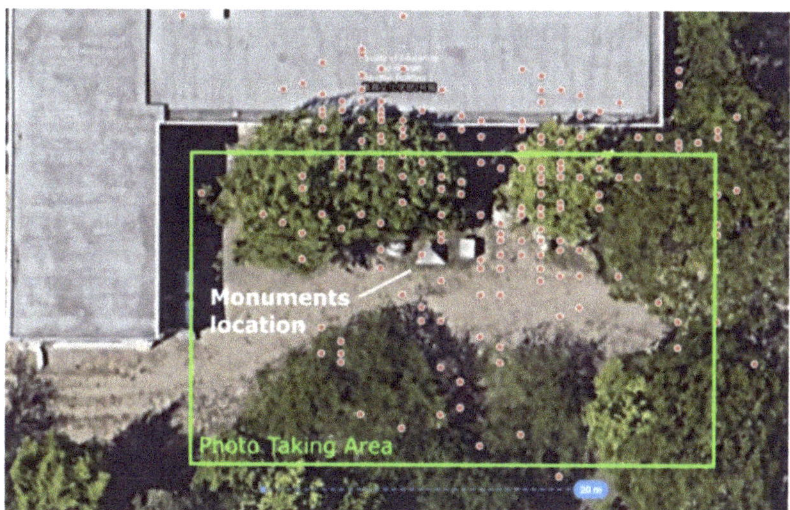

Figure 13. GNSS location of camera positions as logged by a smartphone (red dots) of the experiment on monuments of known dimensions.

Figure 14. Mesh reconstructed 3D CG model of the monuments. The third picture (bottom) shows the measured side of cube monument when using 100 images.

As shown in Table 1 and Figure 15, there is a trend that at increasing number of images used in reconstruction, the difference from the real measurement decreases. This increase in accuracy lends credence to the hypothesis that using more images for reconstruction has the tendency to lessen scale error in 3D models. Using the trendline of the data, a model with a difference from real measurement of 0.1 m (10% scaling error) can be hypothetically created if 386 (385.93) images are used.

Table 1. Results of the experiment on monuments of known dimensions.

Data	Measured (m)	Real Measurement (m)	Difference from Real Measurement (m)
50 images	2.00	1	1.00
100 images	1.97	1	0.97
150 images	1.74	1	0.74
200 images	1.61	1	0.61

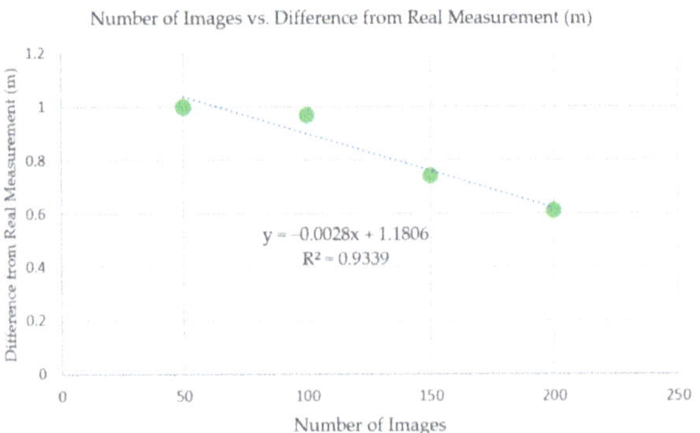

Figure 15. Graph detailing the results of the experiment.

4. Experiment on a Pseudo-Muckpile

For this test, the goal was to recreate a scene of a collection of boulder-sized rocks found at a temple site near the university, shown in Figure 16. The aim of this case study is to provide both quantitative and qualitative evaluation of the effects of GNSS constraints on the scaling error of 3D reconstruction with a subject that is a close simulation of an actual muckpile in a mining environment. The study conducted an experiment using a rock pile located near Akita University. These rocks are similar in size and shape to a muckpile, and, if the effectiveness of the method on this dataset can be confirmed, it can be assumed that the method will be equally effective on an actual muckpile in a mine site.

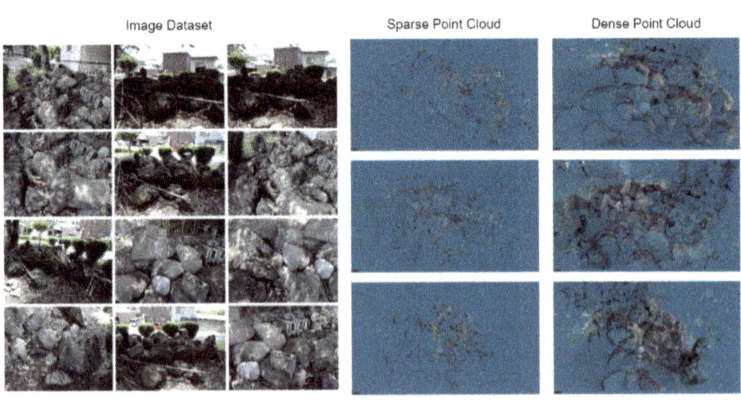

Figure 16. Data input (Set #1) and output of the experiment using 100 images.

A total of 200 photos were taken and split into two datasets (Set #1 and #2) as shown in Figures 16 and 17, with a map depicting the photo taking area and the recorded camera positions found in Figure 18 The rockpile was divided into two parts, one with bigger, angular rocks and another with smaller, rounded rocks. Both piles were around 4 m wide on their longest side and are less than a meter long. A wooden box measuring 30 by 30 by 17 cm was placed in the scene for reference, as shown in its reconstructed form in Figure 19. In addition, measurement of the big, rectangular prism-shaped rock with dimensions of 35 cm × 40 cm × 30 cm were taken for reference as well, which can be seen in its reconstructed form in Figure 20.

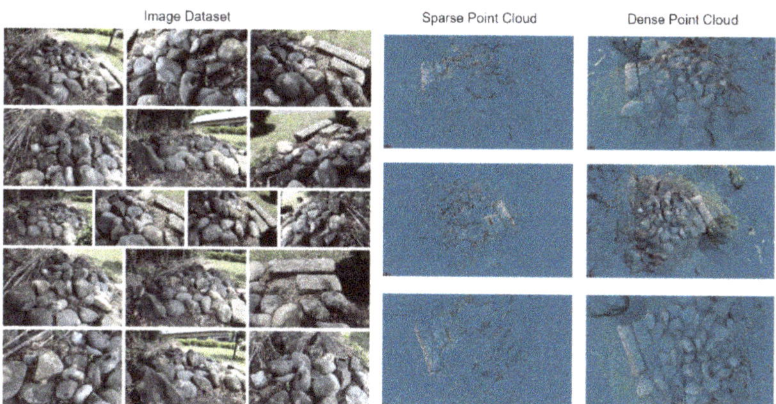

Figure 17. Data input (Set #2) and output of the experiment using 100 images.

Figure 18. GNSS location of camera positions as logged by a smartphone (red dots) of the experiment on a pseudo-muckpile.

Figure 19. Meshed 3D CG reconstruction of wooden box reference with measurement (at 100 images used).

Figure 20. Close up of Set #2 meshed 3D CG reconstruction, with measurements on long rectangular prism-shaped rock that was used as reference.

After the photos were taken, they were once more processed to produce several 3D models at different image numbers. The scaling error and the reconstruction quality was then observed in a similar fashion to the previous experiments. Since two sets of data were used for this experiment, scaling errors between using 50 images (chosen at random) and 100 images for each set were used. An additional exploratory test using 200 images using both sets was added for testing. The study's initial hypothesis, however, was that this will introduce some reconstruction errors as there are not enough images that are similar between these two scenes.

The measurement comparison is shown in Table 2. For Set #1, at 50 images used, the difference from the real measurement of the width of the box (0.3 m) was 2.6 m. At 100 images used, the difference was 1.3 m. This led to a decrease of 1.3 m in the scaling error when using 50 more images. For Set #2, at 50 images used, the difference from the real measurement of the width of the rectangular rock (1.4) was 4.8 m. At 100 images used, the difference was 4.6 m. This led to a decrease of 0.2 m in the scaling error when using 50 more images.

Table 2. Results of the experiment on a pseudo-muckpile.

Data Set and Image Count	Measured (m)	Real Measurement (m)	Difference from Real Measurement (m)
Set #1 (50 images)	2.98	0.30	2.68
Set #1 (100 images)	1.60	0.30	1.30
Set #2 (50 images)	6.11	1.40	4.71
Set #2 (100 images)	6.09	1.40	4.69
Combined Set (200 images)	0.16	0.17	0.01

For a final, investigational set using 200 images combining both previous sets, a surprising result was observed—even though there were significantly more reconstruction errors (missing parts, duplicating parts, etc.) in this particular reconstruction, the wooden box width in this reconstruction was measured at 0.167 m with a difference of 0.01 m from the real measurement. A reconstruction is shown below in Figure 21. We considered that this increase in accuracy can be attributed to not only the number of images increasing but also the general area of the scene becoming larger as it includes both pseudo-muckpiles (the effect of model size on GNSS error is discussed in a latter part of this section). However, combining the datasets also means that the scene being reconstructed is contextually different as it now includes both parts of the pseudo-muckpile.

Figure 21. 3D CG reconstruction results of the combined data sets.

From both experiments, we can see through the maps in Figures 14 and 17 the apparent GNSS drift that occurs during the photo taking. Some of the recorded camera positions are either outside the photo taking area or are in spots that are obstructed. The study recognizes that these changing boundary conditions have an effect on the results, and a separate investigation on this could provide insight for GNSS-aided photogrammetry. Aside from the inaccuracies found in GNSS, several additional factors have been considered to contribute to the drift. One of these is the effect of the partial tree cover in some of the camera positions.

A previous study [23] in a similar setting (university campus) analyzed the effect of not only partial tree cover but also nearby infrastructure on GNSS accuracy by comparing GNSS data to total station survey data. The results showed that some points were no longer suitable for GNSS positioning due to high GDOP (geometric dilution of precision), and, where it was suitable, the GNSS recorded position differed by as much as 5.7 m from the total station data. This difference is consistent with what transpired in this study's experiments, as can be seen from the maps. In a mining site, where there is usually less vegetation and obstruction, this effect should be diminished, except in situations such as benches shadowing satellites.

Another factor that can be considered is the overall scale of the pseudo-muckpile. A large majority of GNSS-aided photogrammetry applications are typically in the form of aerial imagery and mapping with a scope and scale larger than both of the terrestrial photogrammetry experiments performed in this study. A study [24] investigating the application of terrestrial photogrammetry in field geology by using SfM-MVS aided by GPS to model an outcrop that long observed scaling and rotational errors in their reconstruction. Aside from concluding that GNSS contributed highly to these model errors, they suggested that, at a larger scale, the error would be less of an issue.

In parallel to this, the study observed that the relatively small scale of the experiment area affected the data; particularly the pseudo-muckpile whose size was smaller than a muckpile that one would normally find in a mining operation. Ultimately however, the results showed that, even at this scale, incremental improvements to 3D model scaling have been made as shown in the data.

5. Conclusions

In this paper, we proposed a low-cost method of creating an accurately scaled 3D model without the use of GCPs by constraining camera positions through the use of georeferenced images as input for SfM. Monitoring fragmentation size is an important procedure in optimizing mining operations that perform blasting. In recent years, a new method that involves using 3D photogrammetry to measure fragment sizes has been developed and has the potential to surpass traditional techniques. For this particular process to be accurate, a method for properly scaling 3D model with georeferenced images using GNSS was investigated.

To validate the method, several experiments were performed. As an initial test to prove the fundamentals, an indoor scene involving a small object was recreated in 3D space using SfM with photos of the known relative positions for constraining the camera location, and good results that showed that the created 3D model had a scaling error of 1.27 cm were achieved. For the main experiment, the study took georeferenced photos of an outdoor scene with a monument of known dimensions and made several reconstructions at increasing number of images used (50, 100, 150, and 200 images, respectively). The results showed a linear pattern with an R-squared value of 0.93 in which the scaling error decreases as the number of images used increased.

Finally, an experiment was performed to verify the study's hypothesis further using a scene that included a pseudo-muckpile to simulate the usage of the proposed system for a mining operation. In a similar fashion, the results showed increasing scale accuracy with an increasing number of images used in reconstructions. Two observations can be drawn from the experimental results: (1) constraining cameras to accurate positions in SfM resulted in a properly scaled 3D model and (2) increasing the number of georeferenced images in SfM incrementally improved the scaling error of the reconstruction. These observations can help improve scale accuracy in GNSS-aided 3D fragmentation measurements. The method described in this study will be of interest when cost-efficiency and practicality are desired in a 3D fragmentation measurement system.

Author Contributions: Conceptualization, Z.P.L.T., H.T., I.K., H.D.J. and Y.K.; Data curation, Z.P.L.T.; Formal analysis, Z.P.L.T.; Funding acquisition, Y.K.; Investigation, Z.P.L.T. and H.T.; Methodology, Z.P.L.T. and H.T.; Project administration, H.T., M.S. and Y.K.; Resources, H.T.; Software, N.O. and Y.K.; Supervision, H.T., I.K., F.I., M.S., H.D.J. and Y.K.; Validation, Z.P.L.T. and H.T.; Visualization, Z.P.L.T. and H.T.; Writing—original draft, Z.P.L.T.; Writing—review & editing, Z.P.L.T., H.T., N.O., I.K., F.I. and Y.K. All authors have read and agreed to the published version of the manuscript.

Funding: This research received no external funding.

Conflicts of Interest: The authors declare no conflict of interest.

References

1. Kanchibotla, S.; Valery, W.; Morrell, S. Modeling Fines in Blast Fragmentation and Its Impact on Crushing and Grinding. In Proceedings of the Explo '99–A Conference on Rock Breaking, The Australasian Institute of Mining and Metallurgy, Kalgoorlie, Australia, 7–11 November 1999.
2. Valery, W.; Morrell, S.; Kojovic, T.; Kanchibotla, S.; Thornton, D. Modelling and Simulation Techniques Applied for Optimisation of Mine to Mill Operations and Case Studies. In Proceedings of the Conference: VI Southern Hemisphere Conference on Minerals Technology, Rio de Janeiro, Brazil, 27–31 May 2001.
3. Nefis, M.; Talhi, K. A Model Study to Measure Fragmentation by Blasting. *Min. Sci.* **2016**, *23*, 9–104. [CrossRef]
4. Palangio, T.C.; Franklin, J.A.; Maerz, N.H. WipFrag—A Breakthrough in Fragmentation Measurement. In Proceedings of the 6th High-Tech Seminar on State of the Art Blasting Technology Instrumentation, and Explosives Applications, Boston, MA, USA, 1 January 1995.
5. Onederra, I.; Thurley, M.; Catalan, A. Measuring blast fragmentation at Esperanza mine using high-resolution 3D laser scanning. *Min. Technol.* **2014**, *124*, 34–36. [CrossRef]
6. Tungol, Z.P. A Study on the Use of 3D Photogrammetry Software for Rock Fragmentation Size Distribution Measurement. Master's Thesis, Akita University, Akita, Japan, 2019.
7. Jang, H.; Kitahara, I.; Kawamura, Y.; Endo, Y.; Topal, E.; Degawa, R.; Mazara, S. Development of 3D rock fragmentation measurement system using photogrammetry. *Int. J. Min. Reclam. Environ.* **2019**, *34*, 294–305. [CrossRef]
8. Fonstad, M.A.; Dietrich, J.T.; Courville, B.C.; Jensen, J.L.; Carbonneau, P.E. Topographic structure from motion: A new development in photogrammetric measurement. *Earth Surf. Process. Landf.* **2012**, *38*, 421–430. [CrossRef]
9. Anjasmara, I.M.; Pratomo, D.G.; Ristanto, W. Accuracy Analysis of GNSS (GPS, GLONASS and BEIDOU) Obsevation for Positioning. In *Proceedings of the E3S Web of Conferences*; EDP Sciences: Bali, Indonesia, 8 May 2019; Volume 94.
10. Merry, K.; Bettinger, P. Smartphone GPS accuracy study in an urban environment. *PLoS ONE* **2019**, *14*, e0219890. [CrossRef] [PubMed]
11. Jaud, M.; Bertin, S.; Beauverger, M.; Augereau, E.; Delacourt, C. RTK GNSS-Assisted Terrestrial SfM Photogrammetry without GCP: Application to Coastal Morphodynamics Monitoring. *Remote Sens.* **2020**, *12*, 1889. [CrossRef]
12. Madali, N. Structure from Motion. Available online: https://towardsdatascience.com/structure-from-motion-311c0cb50e8d (accessed on 7 July 2021).
13. Carrivick, J.; Smith, M.; Quincey, D. *Structure from Motion in the Geosciences*; Wiley Blackwell: Chichester, UK, 2016; ISBN 9781118895849.
14. Lowe, D.G. Distinctive Image Features from Scale-Invariant Keypoints. *Int. J. Comput. Vis.* **2004**, *60*, 91–110. [CrossRef]
15. Westoby, M.J.; Brasington, J.; Glasser, N.F.; Hambrey, M.J.; Reynolds, J.M. 'Structure-from-Motion' photogrammetry: A low-cost, effective tool for geoscience applications. *Geomorphology* **2012**, *179*, 300–314. [CrossRef]
16. Schönberger, J.L.; Frahm, J. Structure-from-Motion Revisited. In Proceedings of the 2016 IEEE Conference on Computer Vision and Pattern Recognition (CVPR), Las Vegas, NV, USA, 27–30 June 2016; pp. 4104–4113.
17. Zhang, J.; Boutin, M.; Aliaga, D.G. Robust Bundle Adjustment for Structure from Motion. In Proceedings of the International Conference on Image Processing (ICIP), Atlanta, GA, USA, 8–11 October 2006; pp. 2185–2188.
18. Lourakis, M.; Argyros, A. SBA: A software package for generic sparse bundle adjustment. *ACM Trans. Math. Softw.* **2009**, *36*, 1–30. [CrossRef]
19. Triggs, B.; McLauchlan, P.F.; Hartley, R.I.; Fitzgibbon, A.W. Bundle adjustment—A modern synthesis. In *International Workshop on Vision Algorithms*; Springer: Berlin/Heidelberg, Germany, 1999; pp. 298–372.
20. Kume, H.; Taketomi, T.; Sato, T.; Yokoya, N. Extrinsic Camera Parameter Estimation Using Video Images and GPS Considering GPS Positioning Accuracy. In Proceedings of the 20th International Conference on Pattern Recognition, Istanbul, Turkey, 23–26 August 2010; pp. 3923–3926.
21. Toldo, R. Towards Automatic Acquisition of High-Level 3D Models from Images. Ph.D. Thesis, Universit'a degli Studi di Verona, Verona, Italy, May 2013.
22. Dflow 3DF Zephyr Manual. Available online: http://3dflow.net/zephyr-doc/3DF%20Zephyr%20Manual%204.500%20English.pdf (accessed on 14 June 2021).
23. Uzodinma, V.N.; Nwafor, U. Degradation of GNSS Accuracy by Multipath and Tree Canopy Distortions in a School Environment. *Asian J. Appl. Sci.* **2018**, *6*. [CrossRef]
24. Fleming, Z.; Pavlis, T. An orientation based correction method for SfM-MVS point clouds—Implications for field geology. *J. Struct. Geol.* **2018**, *113*, 76–89. [CrossRef]

Review

Advances in Blast-Induced Impact Prediction—A Review of Machine Learning Applications

Nelson K. Dumakor-Dupey [1,*], Sampurna Arya [1] and Ankit Jha [2]

[1] Department of Mining and Mineral Engineering, University of Alaska Fairbanks, Fairbanks, AK 99775, USA; snarya@alaska.edu
[2] Department of Mining Engineering and Management, South Dakota School of Mines and Technology, Rapid City, SD 57701, USA; ankitkrjha@gmail.com
* Correspondence: nkdumakordupey@alaska.edu

Abstract: Rock fragmentation in mining and construction industries is widely achieved using drilling and blasting technique. The technique remains the most effective and efficient means of breaking down rock mass into smaller pieces. However, apart from its intended purpose of rock breakage, throw, and heave, blasting operations generate adverse impacts, such as ground vibration, airblast, flyrock, fumes, and noise, that have significant operational and environmental implications on mining activities. Consequently, blast impact studies are conducted to determine an optimum blast design that can maximize the desirable impacts and minimize the undesirable ones. To achieve this objective, several blast impact estimation empirical models have been developed. However, despite being the industry benchmark, empirical model results are based on a limited number of factors affecting the outcomes of a blast. As a result, modern-day researchers are employing machine learning (ML) techniques for blast impact prediction. The ML approach can incorporate several factors affecting the outcomes of a blast, and therefore, it is preferred over empirical and other statistical methods. This paper reviews the various blast impacts and their prediction models with a focus on empirical and machine learning methods. The details of the prediction methods for various blast impacts— including their applications, advantages, and limitations—are discussed. The literature reveals that the machine learning methods are better predictors compared to the empirical models. However, we observed that presently these ML models are mainly applied in academic research.

Keywords: machine learning; blast impact; empirical model; mining; fragmentation

Citation: Dumakor-Dupey, N.K.; Arya, S.; Jha, A. Advances in Blast-Induced Impact Prediction— A Review of Machine Learning Applications. *Minerals* **2021**, *11*, 601. https://doi.org/10.3390/min11060601

Academic Editors: Rajive Ganguli, Sean Dessureault and Pratt Rogers

Received: 21 March 2021
Accepted: 28 May 2021
Published: 3 June 2021

Publisher's Note: MDPI stays neutral with regard to jurisdictional claims in published maps and institutional affiliations.

1. Introduction

Rock fragmentation in mining involves the breakage of hard rock into appropriate sizes to facilitate downstream handling and processing. Currently, the most economical and widely accepted ground fragmentation technique is drilling and blasting that involves the usage of commercial explosives (placed in blastholes) to break down a rock mass into pieces upon detonation [1–3]. The technique is also common in many civil construction projects, including the construction of tunnels, highways, subways, dams, and building demolition [4–7].

Blasting has significant environmental, operational, and cost implications, and the outcomes of a blast can impact the entire mining operation, from waste/ore transportation through beneficiation. For instance, an optimized blast fragmentation process improves excavator and dump truck production, minimizes equipment maintenance and repair costs, maximizes crusher throughput, and ultimately, minimizes operating costs [3,8,9]. There are two types of impacts for every blasting event: desirable and undesirable (see Figure 1). When an explosive detonates, it releases an enormous amount of energy in the form of gases, pressure, heat, and stress waves [10], causing the surrounding rock mass to develop cracks and get displaced. About 20–30% of the explosive energy released is utilized to fragment and throw the material [11], while the remaining 70–80% generates undesirable outcomes [12]. The undesirable outcomes include airblast/air overpressure, ground vibration, flyrock, noise, heat, fumes/dust, and backbreak. It should be noted that heat, which is a part of the undesirable outcomes, does not necessarily produce adverse effects; it is the portion of the released energy that is not fully utilized in breaking the rock mass.

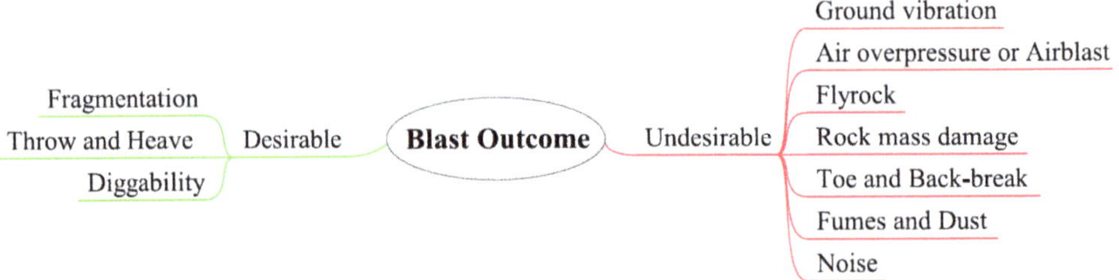

Figure 1. Desirable and undesirable outcomes of a blasting operation.

The undesirable outcomes can reach elevated levels causing discomfort to humans, a threat to human safety and health, and damage to building structures and equipment close to the blast zone. It can also affect groundwater, geological structures, and slope stability. Blasting affects groundwater when soluble substances from detonators and explosives that are not fully combusted permeate groundwater [13]. It may cause short-term turbidity and long-term changes to incumbent wells due to the expansion of fractures from loss of lateral confinement [14]. There are cases reported in the literature on groundwater contamination, including elevated nitrate levels and turbidity [15]. Blasting near cave regions can cause damages to the structural integrity of caves due to vibrations and air overpressure [16]. Incidents of frequent complaints, which, in some cases, escalate into protests against mining operations due to blast impacts, have been reported in many mining jurisdictions, including Ghana, India, Brazil, Turkey, and South Africa [17–21]. Thus, it is important to understand these phenomena and model the potential impacts of blasting activities on catchment communities.

Studies have been performed to ascertain the distance to which the adverse effects of blasting would affect the surrounding blast areas. McKenzie [22] conducted a detailed study to predict the projection range of flyrocks and suggested calculating maximum projection distance with an appropriate safety factor to establish clearance distance. The study found that the maximum flyrock distance is a function of hole diameter, shape factor, and velocity coefficient. The velocity coefficient is calculated using the scaled length of burial, which is a function of stemming length, explosive density, hole diameter, and charge length. Blanchier [23] suggested utilizing a flyrock model developed by Chiapetta et al. [24] to estimate the flyrock speed and maximum range. The model is a function of burden, linear energy of explosives, and a coefficient that expresses the probability of attaining estimated speed [23]. Richard and Moore [25] suggested using empirical formulae developed by Lundborg et al. [26] for predicting the maximum throw and projectile size of flyrock.

Generally, mining regulations prescribe blast standards to ensure that blast impacts are maintained within a certain bound. For example, in the USA, the Title 30 Code of Federal Regulations (30 CFR) specifies that flyrock shall not be cast from the blasting site: more than one-half the distance to the nearest dwelling or other occupied structure, beyond the area of control required under, or beyond the permit boundary [27]. A similar regulatory requirement exists in other mining countries. It should be noted that blast standards are established following extensive empirical and field studies based on several factors, including geology, rock type, explosive type, ground condition, wind direction, blast direction, and building types. Some of these factors (e.g., geology, rock type, and building type) vary from one location to another; therefore, the blast standard for one geological location or country may not necessarily be the same for another geological location or country. Table 1 presents a summary of blast standards for ground vibration, airblast, flyrock, and noise for the USA, Canada, and Australia.

Table 1. Blast standards for the USA, Canada, and Australia.

Blast Impact	Country		
	USA	Canada	Australia
Ground vibration	Maximum allowable PPV: 0–300 ft for PPV $\leq$ 1.25 in./s 301–5000 ft for PPV $\leq$ 1.00 in./s >5001 ft for PPV $\leq$ 0.75 in./s Frequency: 0.03 in for 1–3.5 Hz 0.75 in./s for 3.5–12 Hz 0.01 in. for 12–30 Hz 2.0 in./s for 30–100 Hz	PPV $\leq$ 12.5 mm/s measured below grade or less than 1 m above grade.	Must not exceed a PPV of 5 mm/s for nine out of any ten consecutive blasts initiated, regardless of the interval between blasts, but never over 10 mm/s for any blast.
Airblast	$\leq$0.1 Hz: peak $\leq$ 134 dB $\leq$2 Hz: peak $\leq$ 133 dB $\leq$6 Hz: peak $\leq$ 129 dB C-weighted–slow response: 105 dBC	$\leq$128 dB	Must not be more than 115 dB(lin) peak for nine out of any ten consecutive blasts initiated, regardless of the interval between blasts, but never over 120 dB(lin) peak for any blast.
Flyrock	Shall not cast: More than one-half the distance to the nearest dwelling. Beyond the area of control required under 30 CFR 816.66(c); or Beyond the permit boundary.	The blaster must take precautions for the protection of persons and property, including proper loading and stemming of holes, and where necessary, the use of cover for the blast or other effective means of controlling the blast or resultant flying material.	If debris from blasting in a surface mining operation could constitute a danger to any person or property, each responsible person at the mine must ensure that such precautions are taken as are necessary to prevent injury to persons and to minimize the risk of damage to property.
Noise	70 dBA (EPA)	$\leq$55 dBA daytime (L$_{eq\,D}$) $\leq$45 dBA at (L$_{eq\,N}$) nighttime	No worker to be exposed to noise with a level exceeding 140 dB(lin) peak

PPV is the peak particle velocity, dBA is the A-weighted decibel, dBC is the C-weighted decibel, dB(lin) or dBZ is the unweighted decibel, and EPA is the U.S. Environmental Protection Agency.

Figure 2 indicates various zones of blast influence and the potential risk to people and structures within these zones. The risk to people and equipment is highest at the innermost circle, i.e., within the immediate vicinity of the blast zone. The blast zone is a high-risk area with the highest degree of blast-induced impacts. However, the severity of the impacts reduces as they travel outward from the blast zone towards the outer perimeter, as depicted by the blast impact profile in Figure 2. The blast impacts are not confined to a single direction; they can travel radially because the explosive energy act on all points of the blasthole simultaneously [28]. However, the intensity of the associated impacts may not be the same everywhere. Figure 2 is divided into three segments (S1, S2, and S3) to illustrate the potential impact regions. Assuming the blast design is optimal in S1, then the associated undesirable effects are limited to the buffer zone, and they would be harmless even if they exceed the buffer zone. However, with the same buffer zone, increasing the explosive charge (S2) or the number of blast shots (S3) can cause undesirable effects to exceed the buffer zone, damaging structures in the concession and beyond. Usually, for a good blast operation, it is expected that the magnitude of the blast impact beyond the buffer zone will reduce below the damage threshold. In other words, blast impacts attenuate with the increasing distance. The distances between the blast zone, buffer zone, and mine concession are usually stated in the blast standards. For instance, in Ghana, the blast standard prescribes a safe distance (buffer zone) of 500 m from the blast zone. Decreasing factors, such as the quantity of explosive charge and number of blast shots, could also reduce the magnitude of blast impacts. Blast standards mandate that all employees and equipment must be cleared from the blast area to a safe location before any scheduled blast operation to prevent injury and equipment damage.

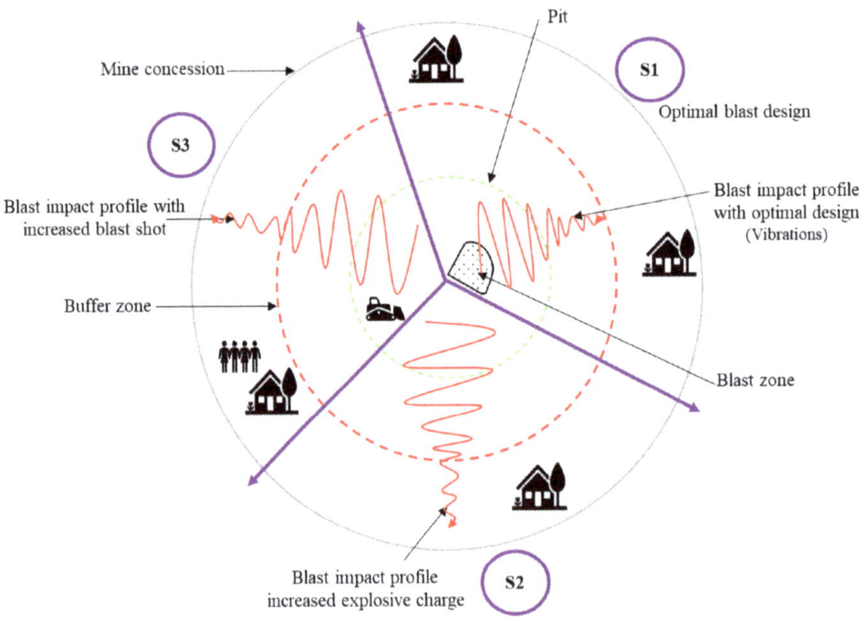

Figure 2. Blast impact zones and objects of concern. Varying blast design parameters (S1, S2, and S3) influence the magnitude and travel distance of undesirable blast effects.

Generally, there are two categories of factors that influence blast impacts: controllable and uncontrollable parameters. Controllable parameters are those that a blast engineer can modify and include the blast geometry (spacing, burden, blasthole depth, blasthole diameter, and stemming) and explosive parameters (type, density, powder factor, charge per delay/instantaneous charge, and delay time). The uncontrollable parameters include geological (rock type, discontinuities, and groundwater) and geotechnical properties (rock strength, density, etc.) of the rock formation that cannot be modified. Therefore, blasts must be designed to suit the prevailing ground conditions to generate optimal fragmentation with minimal environmental impact, fostering an excellent company–community relationship. Mining regulations are also major deciding factors in blast design, providing guidelines and blast impact threshold limits to ensure safe blast operations.

Over the years, studies have been conducted to examine blast impacts, which has led to the development of several blast impact prediction models. These models, many of which are based on empirical data, have primarily been applied in mining operations to predict and model the potential impacts of blasting. Several empirical models are in the literature for predicting blast-induced ground vibration, flyrock, dust/fumes, backbreak, and fragmentation. Though most of these models have a long history of use in the mining industry, they possess some inherent limitations, such as (1) a restriction to just two input parameters, (2) inability to concurrently predict more than one outputs, and (3) unsuitability to apply to all geological formations or mine conditions. Singh and Singh [29] noted that empirical models are analyzed datasets along specific geometries, which may or may not be favorable to understand the nonlinearity existing among various input/output parameters. Additionally, there are too many other interrelated controllable (blast geometry and explosive) and uncontrollable (geological and geotechnical) parameters, which are not incorporated in any of the available predictors [30]. In effect, the empirical models are not able to identify the nonlinear relationships, and this weakness influences the performance of these models.

A promising solution to this problem is the application of ML techniques in blast impact prediction. With the recent popularity of artificial intelligence (AI) in both academia and industry, many scholars are exploring machine learning as a robust tool to model blast impacts. In recent years, numerous scientific papers have been published in this area, and the number of new publications is ascending significantly. The wide application of ML can be attributed to its ease in handling complex engineering problems with several variables. ML is the study and application of computer algorithms to make intelligent systems that improve automatically through the experience without being explicitly programmed. It is classified as a subfield of AI, which is the science and engineering of making intelligent machines. ML applies computer algorithms to analyze and learn from data and makes decisions or predictions based on the data provided. Depending on the structure of available data being analyzed, ML models are categorized as supervised learning, unsupervised learning, or reinforcement learning [31].

In this paper, the authors performed a comprehensive review of scientific studies that applied ML techniques to predict blast impact. This paper covered a detailed examination of machine learning models for blast-induced ground vibration, flyrock, airblast, backbreak, and fragmentation. It is worth noting that most of the studies conducted in this field are related to blast-induced ground vibration.

The remainder of the paper is organized into five sections. Section 2 outlines the review methodology, followed by a description of the rock breakage mechanism in Section 3. Sections 4 and 5 discuss the empirical and ML blast impact prediction models, respectively. Section 6 presents a discussion and future trends for ML applications, while Section 7 covers the concluding remarks.

2. Methodology

This review intends to summarize the existing knowledge on the application of ML in blast-induced impact predictions and identify gaps in the current research to suggest areas for further investigation. The review scope is mainly limited to only publications related to blast-induced impacts associated with surface and underground mining and quarry operations. The primary purpose of this review was to report the current status of ML usage in predicting blast-induced impacts in mining. However, a few studies on blast impacts resulting from blasting operations in dam and tunnel construction were also considered.

Based on the stated review objective and purpose, we conducted an extensive literature search to identify relevant peer-reviewed publications indexed in major scientific research databases, such as Web of Science, Google Scholar, Scopus, and ScienceDirect. To limit the search scope, we used keywords, including "blasting", "rock fragmentation", "machine learning", "blast impacts", "ground vibration", "airblast or air overpressure", "flyrock", "backbreak", "soft computing", "neural networks", "deep learning", and "support vector machines". Boolean operators and strings were adopted to improve the search results. Another search strategy employed was snowballing (e.g., forward and backward snowballing), where the original search results led to the discovery of more papers. We screened the search results for relevance by reviewing the titles and abstracts of the publications. The published articles were required to be original, peer-reviewed, and recognized in the field.

The search scope covered research articles published from 2004 to 2020. However, a few recent articles published in early 2021 were also included. This review was mostly focused on peer-reviewed journal publications, since the intention was to rely on rigorous research addressing the subject matter. Some of the notable journals where the search results were retrieved were *Engineering with Computers, Safety Science, Environmental Earth Sciences,*

International Journal of Mining Science and Technology, Rock Mechanics and Rock Engineering, Neural Computing and Applications, and *Natural Resources Research*. From the research results, we noticed that the majority of the articles were published in *Engineering with Computers*, followed by *Natural Resources Research*, as evident in Figure 3. In a few cases, relevant papers in peer-reviewed conference proceedings and a thesis report were included.

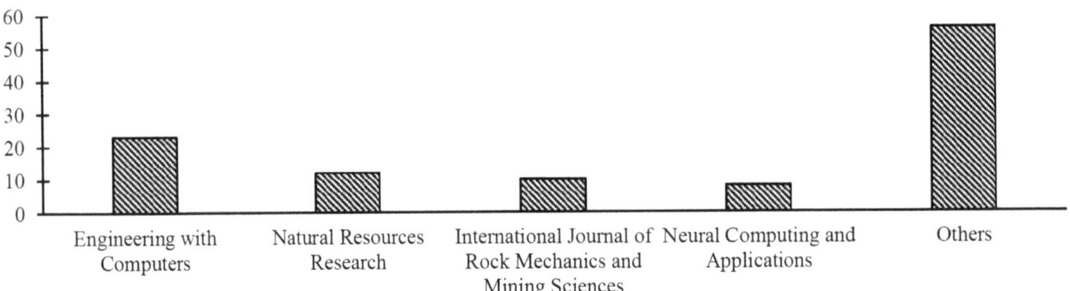

Figure 3. Journals with the most counts of publications in the machine learning application in the blast-induced impact predictions in this review.

Out of the 193 articles reviewed, approximately 112 focused on the prediction of blast-induced impacts using machine learning, while the remaining articles covered blast phenomenon and empirical prediction models. This is by no means an exhaustive list of all blast-induced impacts and ML-related articles published in this field within the period under consideration. Figure 4 illustrates the yearly distribution of publications on ML applications in blast-induced impact predictions. The distribution (Figure 4) shows an increasing trending in publications of ML techniques in this field. This positive trend can be attributed to the growing interest in ML applications in academia and the industry in recent years.

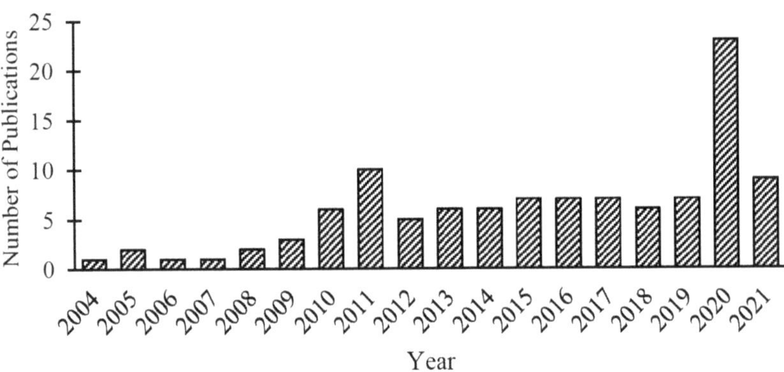

Figure 4. Publication trend for machine learning applications in blast-induced impact predictions.

Table 2 presents a summary of the number of ML applications in each blast-induced impact considered in this review. Most of the studies reviewed predicted only one blast-induced impact. It is interesting to note that a significant portion of ML applications were about ground vibrations, likely due to the drive to accurately measure and mitigate blast-induced vibration levels. Since blast-induced ground vibrations can cause structural damage to buildings, resulting in contention between mining companies and host communities, it is always prudent to ensure that the vibration levels are within the regulatory requirements. Therefore, relatively cheaper and more rapid techniques that allow the blast engineer to predict the vibration level before blasting are helpful in pre-blast planning as compared to field measurements. This may also indicate the importance placed on ground vibrations compared to other blast-induced impacts and the research efforts to improve the prediction results.

Table 2. Review statistics of blast-induced impacts predicted using machine learning.

Blast Impact	Count
Ground vibration	58
Flyrock	15
Fragmentation	13
Airblast	11
Backbreak	3
Overbreak	1
Noise	1
Ground vibration and airblast	3
Flyrock and fragmentation	3
Backbreak and fragmentation	2
Flyrock and backbreak	1
Ground vibration, airblast, and fragmentation	1

ML application in flyrock prediction has also received significant research attention, as flyrock is a potential hazard responsible for a large proportion of all blasting-related injuries and fatalities. The fragment size analysis and airblast have also received considerable attention, while backbreak and overbreak are blast-induced impacts with the least ML implementations. It is worth noting that, apart from a single impact prediction, a few studies have predicted two impacts, while one research predicted three impacts simultaneously.

3. Rock Fragmentation and Blast Impact Phenomena

The technique most commonly used for breaking rock with explosives involves drilling blastholes into a rock mass, placing explosive substances in the blastholes, initiating the fire sequence, and detonating the explosive, as illustrated in Figure 5. Upon initiation, the explosive charge detonates (i.e., an intense and rapid chemical reaction occurs), producing an enormous amount of energy in the form of gases at very high temperatures and pressure. The energy released by an explosive during a blast can be categorized into seismic, kinetic, backbreaks, heave, heat, or fragmentation energies [32]. The resulting detonation energy has the following effects: pressurizes the blasthole and fractures the vicinity rock mass, creates strong shock waves in the rock mass, which propagate as plastic and, ultimately, elastic waves and appear as a seismic wave or ground vibration, and displaces and heaves the fractured rock mass to form a muck pile that appears as kinetic energy imparted to the rock [33–36].

According to Changyou et al. [37], the theory that rock damage is a result of the coaction of the blast wave and explosive explosion is currently accepted by most scholars, as it matches the actual process of blast-induced rock breakage favorably. Nevertheless, the mechanism of rock breakage under explosive action is still being investigated, even after many decades of advancement in explosive technology for mining and civil applications. Recently, numerical modeling and simulation models have been applied to further the understanding of blasting [38–40]. Generally, the fragmentation action has been attributed to either the gases or shock waves generated or both [38,41,42].

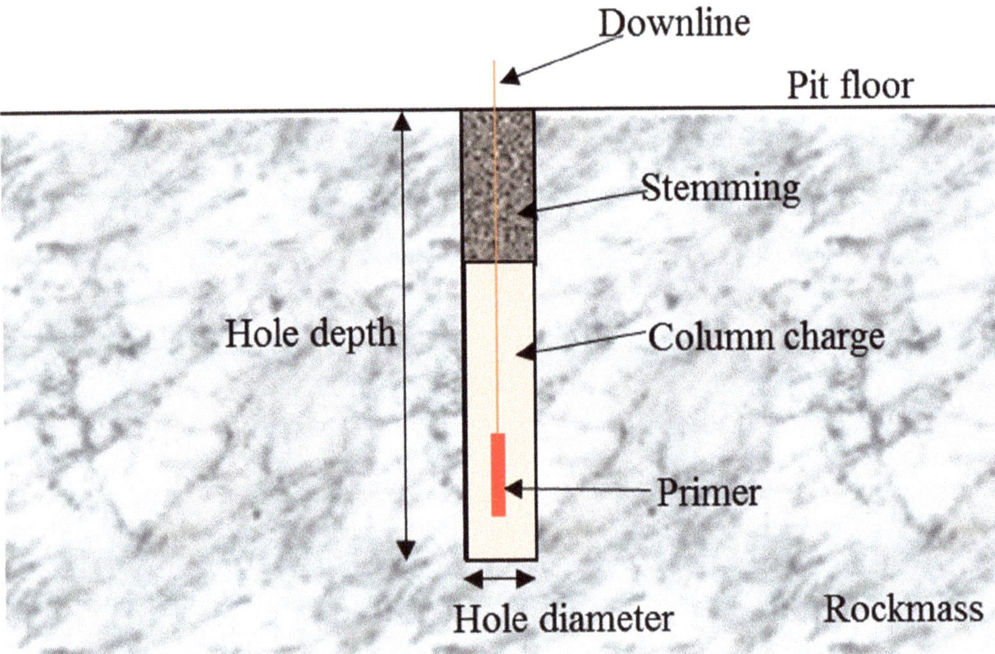

Figure 5. Schematic diagram of a charged blasthole.

The detonation waves from the explosive (with the velocity of detonation between 2000 and 7000 m/s, depending on the type of explosive) induce intense stresses in the blasthole due to the sudden acceleration of the rock mass by detonating gas pressure on the blasthole wall [35]. Bendezu et al. [28] stated that the energy released is converted into two main forms that are responsible for rock fracturing, creating new cracks and widening the already existing ones: blast-induced stress waves (dynamic load) and the overpressure of the explosive gases (quasi-static load). The strain waves transmitted to the surrounding rock sets up a wave motion in the ground. The strain energy carried out by these strain waves fragments the rock mass, resulting in different breakage mechanisms such as crushing, radial cracking, and reflection breakage in the presence of a free face. The crushed zone and radial fracture zone encompass a volume of permanently deformed rock. When the stress wave intensity diminishes to the level where no permanent deformation occurs in the rock mass (i.e., beyond the fragmentation zone), strain waves propagate through the medium as elastic waves, oscillating the particles through which they travel. These waves in the elastic zone are known as ground vibrations, which closely conform to viscoelastic behavior. The wave motion spreads concentrically from the blast point in all directions and attenuates as it travels farther from the origin through the rock medium.

The fragmentation action does not exhaust all the explosive energy; some portion of it is transformed into ground vibration, airblast, and flyrock. Bendezu et al. [28] pointed out that there is no clear indication about the amount of energy converted into stress wave energy; how much is available as high-pressure gases; and how much is lost to other sources, such as ground vibration, air blast, heat, and smoke/dust. The energy distribution depends on the type of explosive. However, some studies have reported that approximately 20–30% of the explosive energy is utilized to fragment and throw the rock mass, while the remaining 70–80% goes toward the generation of other blast-induced impacts [11]. Even though ground vibrations attenuate exponentially with distance, the large quantity of explosives used means that ground vibrations can still be high enough to cause damage to buildings and other structures by causing dynamic stresses that exceed the material's strength [35]. The blast phenomena and the mechanisms of ground vibrations, airblast, flyrock, and fragmentation have been well-documented. Figure 6 depicts a blast event with its associated vibrations and undesirable effects, such as flyrock, ground vibrations, and airblast.

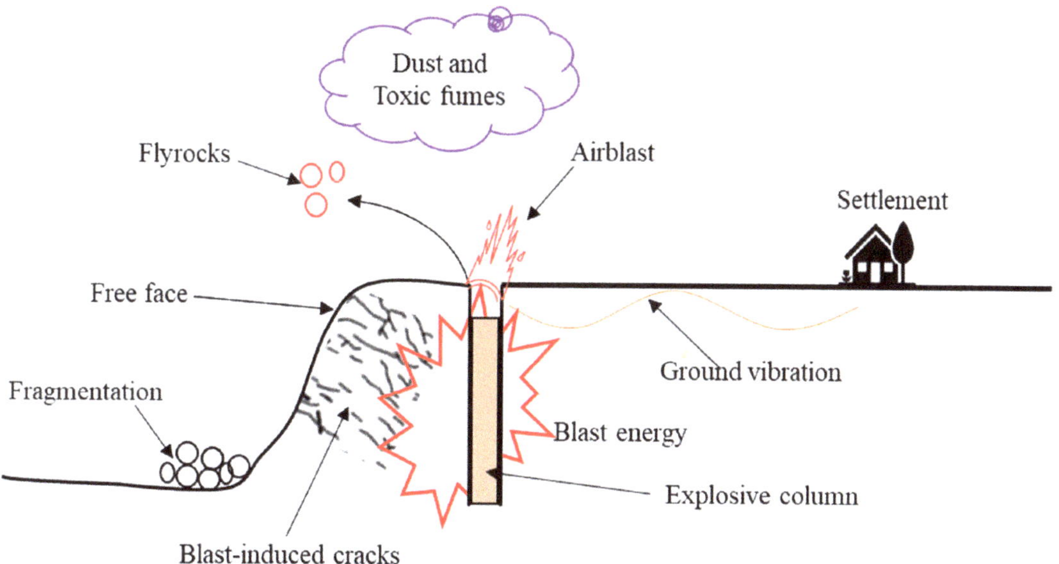

Figure 6. Blast waves and impacts.

4. Empirical Models

Empirical blast impact prediction models are established following rigorous and extensive field studies; data collection; and site observations of several blast parameters, including blast geometry, the geology of the area, rock type, blast direction, wind direction, the location of building structures relative to a blast zone, etc. The empirical models are based on two main factors: (1) the maximum charge per delay and (2) the distance from the blast face to the monitoring point. The models are generally mine-specific due to the heterogeneity of geological formations and variations in site conditions from one location to another. To apply empirical models for site-specific predictions of blast impacts, the models are calibrated using field measurements and established site constants. Tables 3–5 are summaries of some empirical models for predicting blast-induced ground vibrations, airblast/air overpressure, and flyrock, respectively. The models presented in these tables are not exhaustive, and references can be made to Murmu et al. [12] and Kumar et al. [43] for a more comprehensive list, particularly for blast-induced ground vibrations.

Table 3. Empirical models for predicting blast-induced ground vibrations.

Prediction Model	Equation	Reference
USBM	$PPV = k(D/\sqrt{Q})^{-\beta}$	[44]
Langefors–Kihlstrom	$PPV = k(Q^{1/2}/D^{3/4})^{\beta}$	[45]
General predictor	$PPV = k \times D^{-\beta} \times Q^{A}$	[43]
Ambraseys–Hendron	$PPV = (D/\sqrt[3]{Q})^{-\beta}$	[46]
Indian Standard	$PPV = k(Q/D^{2/3})^{\beta}$	[46]
Ghosh–Daemen 1	$PPV = k(D/\sqrt{Q})^{-\beta} \times e^{-\alpha \times D}$	[43]
Ghosh–Daemen 2	$PPV = k(D/\sqrt[3]{Q})^{-\beta} \times e^{-\alpha \times D}$	[43]
Gupta et al.	$PPV = k(D/\sqrt[3]{Q})^{-\beta} \times e^{-\alpha \times (D/Q)}$	[43]
CMRI predictor	$PPV = n + k(D/\sqrt{Q})^{-1}$	[43]
Rai–Singh	$PPV = k \times D^{-\beta} \times Q^{A} \times e^{-\alpha \times D}$	[47]

PPV is the peak particle velocity (mm/s), D is the distance between the blast face to the monitoring point (m), and Q is the cooperating charge (kg). The values k and β are the site-specific constants (coefficients) obtained through a linear regression model by plotting the graph between the PPV versus scaled distance (SD) on a log–log scale [48].

Table 4. Empirical models for predicting the airblast or air overpressure.

Prediction Model	Equation	Reference
USBM	$P = \beta_1 \times (D/Q^{0.33})^{\beta_2}$	[49]
NAASRA	$P = 140\sqrt[3]{Q/200}/d \; (kPa)$	[50]
Ollofson; Persson et al.	$P = 0.7 \times Q^{1/3}/D \, (mbar)$	[51]
Holmberg-Persson	$P = k \times 0.7 \times Q^{1/3}/D \; (mbar)$	[51]
Mckenzie	$P = 165 - 24 \log D/Q^{1/3} \; (dB)$	[52]

P is the airblast or overpressure, Q is the mass of the explosive charge (kg), D is the distance from the charge (m) to the monitoring point, and H and β are the site factors.

Table 5. Empirical models for predicting the flyrock.

Prediction Model	Equation	Reference
Lundborg et al.	$L_m = 260 \times d^{2/3} \; T_b = 0.1 \times d^{2/3}$	[53]
Chiapetta et al.	$R_1 = V_0 \times (2\sin 2\theta \, /g)$ $R_2 = V_0 \times \cos(V_0 \sin\theta + 2V_0 \sin\theta + 2gH)/g$	[34]
Gupta	$L = 155.2 \times D^{-1.37}$	[54]

L_m is the flyrock range (m), d is the blasthole diameter (inch), T_b is the flyrock fragment size (m), L is the ratio of the length stemming the column to burden, D is the distance traveled by the flyrock (m), R_1 is the distance traveled (m) by the rock along a horizontal line at the original elevation of the rock on the face, R_2 is the total distance traveled (m) by a fragment ejected from the blast, accounting for its height above the pit floor, V_0 is the initial velocity of the flyrock, θ is the angle of departure with the horizontal, and g is the gravitational constant.

5. Machine Learning Models

AI refers to a branch of computer science concerned with building smart machines capable of performing tasks that typically require human intelligence [55]. AI techniques have been increasing steadily in many engineering fields, including image processing [56], mineral exploration [57], and mine planning [58,59]. Simeone [60] believes that the widespread use of data-driven AI methods is motivated by the successes of ML-based pattern recognition tools. ML is a branch of AI that systematically applies algorithms to synthesize the underlying relationships among data and information [61]. ML focuses on the application of computer algorithms to process large amounts of data, detect patterns or regularities in data, and improve their performance based on experience [62,63]. Such applications may offer more understanding about a system and can be used to predict or modify the future behavior of the system. Given sufficient input data and a sequence of instructions (algorithms), a computer can perform the desired task of predicting an output. Algorithms for some desired tasks can be developed easily using traditional programming (TP), and a computer will be able to execute them following all the steps required to solve the problem without learning. However, for more advanced tasks (e.g., prediction of consumer behavior or natural occurrences), it can be challenging for a human to manually create the needed algorithms. In practice, it can turn out to be more effective to help the machine develop its model rather than having human programmers specify every needed step [64–66]. It may be impossible to develop an explicit program of such an advanced system, but the ML models provide good and useful approximations. Unlike TP, ML automates the process of learning a model (program) that captures and subsequently predicts the relationship between the input and output variables in a dataset by searching through a set of possible prediction models that best defines the relationship between the variables [67]. A good prediction model must be able to predict events that are not in the current data, i.e., it must generalize well.

Samuel [68] described ML as the "field of study that gives computers the ability to learn without being explicitly programmed". Alpaydin [65] also defined ML as programming computers to optimize a performance criterion using example data or experience. In other words, given a sufficient dataset (e.g., historical blast monitoring data), an ML algorithm can identify patterns; predict blast impact values (e.g., PPV, frequency, flyrock, fragment size, etc.); and improve the previous predictions as more data are made available. Once programmed, the algorithm can learn from the data and improve the learning experience with little human interference. The algorithm synthesizes the various independent variables, such as hole diameter, hole depth, blast size, spacing, burden, stemming height, explosives blasted per delay, and distance between the blast zone and measuring point, with weights that depict their influence on the dependent variable. A generalized ML implementation procedure is presented in Figure 7. The first step in the ML model development cycle (Problem definition) deals with an understanding of the problem, characterizing it, and eliciting required knowledge in acquiring the relevant data. The second step (Data collection) is the collection of all relevant and comprehensive data, followed by data preparation and feature extraction. Next, the data is divided into training, validation, and testing sets based on a predefined ratio (Data partition). Following that, an ML model is selected, trained, validated, and tested using the partitioned datasets (Train model). Here, the programmer can try different algorithms and compare their performances.

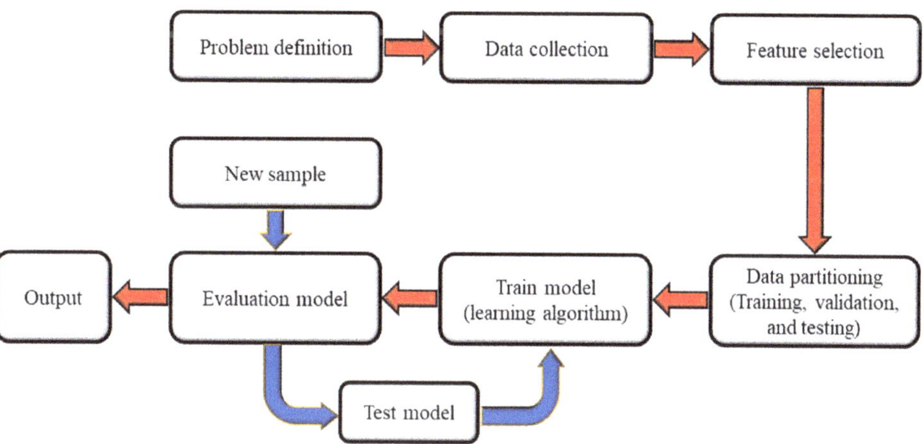

Figure 7. Machine learning implementation stages.

Model evaluation involves the usage of some metrics or a combination of metrics to measure the objective performance of the selected ML model (Evaluate model). The model parameters can be revised (hyperparameter-tuned) until a satisfactory performance is achieved; then, it is adopted for prediction. A few of the statistical criteria used to evaluate the performance of ML models include the mean absolute error (MAE), root mean square error (RMSE), correlation coefficient (R), and determination coefficient (R^2).

The ML methods that have been employed in blast impact prediction are the artificial neural network (ANN), support vector machine (SVM), random forest (RF), gaussian processes (GP), and fuzzy theory sets. These models have been successfully applied in evaluating various blast impacts. The ANN is a computational network presenting a simplified abstraction of the human brain. Conceptually, this computational network mimics the operations of biological neural networks to recognize existing relationships in a set of data. It consists of layers of interconnected nodes that represent artificial neurons. The layers are categorized into three divisions: input layer (receives the raw data), hidden layer (process the raw data), and output layer (processed data). The number of layers and neurons (topology) in a network determines the structure of a neural network or network architecture [66]. Figure 8 depicts an ANN architecture for predicting maximum flyrock distance. The model comprises one input layer with seven neurons, two hidden layers with eight and seven neurons, respectively, and one output layer with one neuron.

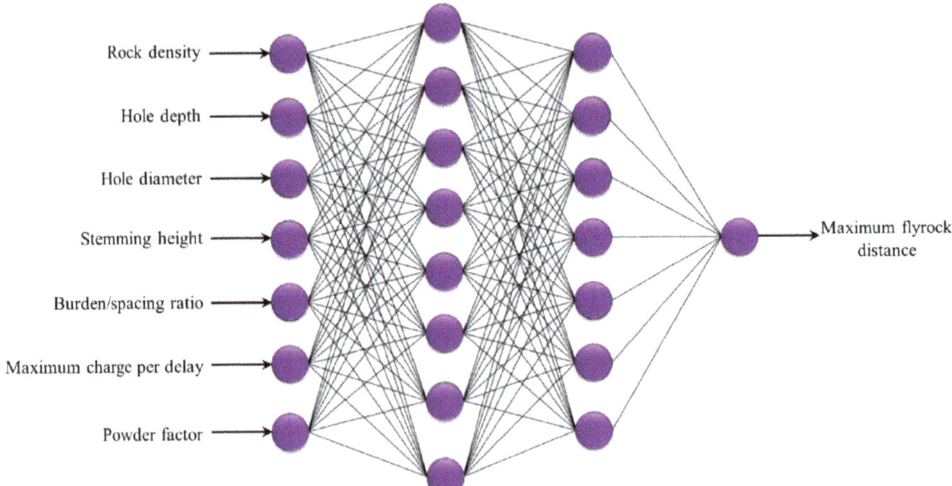

Figure 8. General structure of an ANN model to forecast the maximum flyrock distance.

SVM is an ML algorithm based on the structural risk minimization principle [69,70]. The algorithm uses the concept of decision planes that utilize decision boundaries to optimally separate data into different categories [69]. SVM can solve classification, regression, and outlier detection problems, and when it is applied to regression problems, it is called a support vector machine (SVR). The process of training an SVM decision function involves identifying a reproducible hyperplane that maximizes the distance (i.e., the "margin") between the support vectors of both class labels, and thus, the optimal hyperplane is that which "maximizes the margin" between the classes [71].

RF is a supervised learning algorithm consisting of multiple independent decision trees (DT) that are trained independently on a random subset of data [72,73]. It is an ensemble method that uses bagging (bootstrapping and aggregation) to train several DTs in parallel (i.e., uncorrelated forest of trees) whose prediction by committee is more accurate than that of any individual trees [73,74]. RF can solve both classification and regression problems.

GP is a "collection of random variables, any finite number of which have (consistent) joint Gaussian distributions" [75]. It is characterized by mean and covariance functions.

GPs are attractive because of their flexible nonparametric natures and computational simplicity, and they are designed to solve regression and probabilistic classification problems.

The fuzzy set theory uses natural language to formulate a mathematical model of vague qualitative or quantitative data by attributing a degree to which a certain object belongs to a set [76,77]. The model is based on the generalization of the classical concepts of the set and its characteristic functions. Fuzzy sets and fuzzy logic are an extension of classical set theory and built around the central concept of a fuzzy set or membership function [78]. The model provides a natural way of dealing with problems in which the source of imprecision inhabits a precise definition of class membership criteria [76]. Fuzzy set theory has been shown to cope well with the complexity of complicated and ill-defined systems flexibly and reliably [79].

The following subsections review the application of these algorithms to blast impact prediction problems. There is extensive documentation in the literature regarding the assumptions, mathematical computations, and architecture of these techniques; thus, this paper focused largely on their application.

5.1. Ground Vibration

Several ML models, including the ANN, RF, SVM, and logistic regression, have been employed in predicting and modeling blast-induced ground vibrations. Currently, ground vibrations are, by far, the most studied blast impact for many ML applications. The prediction procedure involves the selection of input parameters, a training model, and predicting the outcome. The input parameters can vary from two to as many as possible, depending on the strength of the algorithm and the computing resources available. Different studies have considered different sets of influential factors in predicting the ground vibrations and designed varying ANN architectures to ensure the accuracy of these predictions. Some of these studies only considered as few as two parameters, while others considered as many as 13 parameters to predict the blast-induced ground vibrations [80]. In fact, due to the complexity of the blast phenomenon and the many factors involved, it has been a challenge to identify the specific influential factors. Nevertheless, studies have considered explosive characteristics, blast design parameters, geological conditions, and rock mass properties as the major factors influencing blast-induced ground vibrations. Among the main factors, the distances between the blast zone and monitoring point, maximum charge per delay, velocity of detonation, blasthole depth, burden, spacing, stemming height, powder factor, rock-quality designation (RQD) and p-wave velocity were the most common factors in estimating blast-induced ground vibrations. Due to the limitations of the parameters and datasets, studies have tried to change the number of hidden layers and the hidden neurons to ensure the accuracy of their predictions [81]. For instance, Amnieh et al. [82] designed an ANN model with four hidden layers (hidden neurons in each layer: 20-17-15-10) and four influential parameters that showed better performances in predicting the PPV for a problem with 25 datasets.

Most scholarly articles applied an ANN, particularly the feed-forward back-propagation neural network (BPNN), for the prediction of blast-induced ground vibrations [29,32,48,80–89]. We present a review of some of these papers in this section. BPNN is a strong modeling technique for input/output pattern identification problems and is a commonly used ANN, often applied to solve nonlinear problems. The calculation process of BPNN is divided into two steps: forward calculation and backward propagation. The connection weights and bias values are adjusted by gradient descent algorithms. The weights of the interneuron connections are adjusted according to the difference between the predicted and the actual network outputs [81]. Normally, closer mapping is required to obtain more satisfactory model performance [90], and it is recommended that the numeric values of the pertinent parameters be normalized in a range of 0 to 1 to achieve a reasonable solution [46].

Singh et al. [91] used the ANN technique for the prediction of p-wave velocity and anisotropy, taking chemical composition and other physicomechanical properties of rocks as the input parameters. Due to data limitation, the leaving-one-out cross-validation method was used, and the network had three layers with six inputs, five hidden neurons, and two output neurons. Using the Bayesian regulation, overfitting of the data was mitigated, and the network was trained with 1500 training epochs, resulting in a high correlation coefficient and low mean absolute percentage error between the predicted and observed values, respectively. Khandelwal and Singh [80] used a BPNN consisting of three layers to predict PPV and its corresponding frequency based on the rock mass mechanical, explosive, and blast design properties. Khandelwal and Singh [92] evaluated and predicted blast-induced ground vibrations and frequencies by incorporating the rock properties, blast design, and explosive parameters into an ANN. Mohamed [93] determined the effect of varying the number of input parameters (blast variables) on the performance of a neural network for ground vibration prediction. Khandelwal et al. [94] incorporated the explosive charges per delay and blast monitoring distance to evaluate and predict ground vibrations using an ANN. With an optimum architecture of 4-10-5-1, Monjezi et al. [95] compared the performances of a BPNN model with empirical predictors and a regression analysis. The comparison revealed that the most influential parameter was the distance between the blast zone and the monitoring point, while the least effective parameter was stemming the height.

Other types of ANN applied in the prediction of blast-induced ground vibrations include GRNN, quantile regression neural network (QRNN), wavelet neural network (WNN), hybrid neural fuzzy inference system (HYFIS), adaptive neuro-fuzzy inference system (ANFIS), and group method of data handling (GMDH). Arthur et al. [96] estimated blast-induced ground vibrations by comparing five ANNs (WNN, BPNN, RBFNN, GRNN, and GMDH) and four empirical models (Indian Standard, the United State Bureau of Mines, Ambrasey-Hendron, and Langefors and Kilhstrom). The study revealed that WNN with a single hidden layer and three wavelons produced highly satisfactory results compared to the benchmark methods of BPNN and RBFNN. Xue and Yang [97] also predicted blast-induced ground vibrations and frequencies by incorporating rock properties, blast design, and explosive parameters using the general regression neural network (GRNN) technique. The GRNN model provided excellent predictions with a high degree of correlation when compared with multivariate regression analysis (MVRA). Nguyen et al. [98] argued that MLP recorded the most accurate prediction over BRNN and HYFIS. They also observed that not all ANN models (e.g., HYFIS) are useful for blast impact predictions in open-pit mines, depending on the input parameters and training algorithms.

Generally, ANN-based models are better predictors with superior performances compared to empirical models when it comes to predicting blast-induced ground vibration levels. However, this is not to say that ANN results are always accurate and are without challenges. ANN algorithms also have some weaknesses, such as overfitting [99], long training times, and falling easily into the local minimum [81]. According to Dreiseitl and Ohno-Machado [99], ANN models are more flexible and, thus, more susceptible to overfitting. This usually occurs when the ANN model begins "to memorize the training set instead of learning them and consequently loses the ability to generalize" [48]. The methods proposed for resolving it include early stopping, noise injection, cross-validation, Bayesian regularization, and the optimization approximation algorithm [48,100,101]. Paneiro et al. [102] employed bilevel optimization to avoid overfitting and reduce the complexity of an ANN-based ground vibration model. The authors concluded that the improved ANN model offered a much higher generalization ability than traditional and other ANN models applied to ground vibration predictions. Piotrowski and Napiorkowski [100] also cautioned that the ANN architecture should be kept relatively simple, as complex models are much more prone to overfitting. Dreiseitl and Ohno-Machado advised that, in constructing the model, the network size can be restricted by decreasing the number of variables and hidden neurons and by pruning the network after training. Alternatively, one can require the model output to be sufficiently smooth through regularization [99].

Studies have integrated ANN with other soft computing techniques, such as data mining and feature selection algorithms, to improve the accuracy and robustness of ANN-based ground vibration models. In some instances, preprocessing of the raw data involves data mining to find relationships and patterns in the raw data. For example, before training the ANN model, Amiri et al. [103] applied itemset mining (IM) to identify patterns and extract frequently occurring sets of items in a database. Based on the extracted knowledge, association rules were formed that helped select the best instance for training the neural network model. The proposed itemset mining and neural networks (IM–NN) model showed superior prediction results compared to the classical ANN.

To overcome the limitations associated with ANN in predicting blast-induced ground vibrations, studies have also applied other ML algorithms that are without these shortcomings. Some of the algorithms applied included SVM [104–111], relevance vector regression [112], particle swarm optimization [113,114] Bayesian network and random forest [108], Gaussian process regression [115], classification and regression trees, chi-square automatic interaction detection, random forest [1,116,117], hybrid artificial bee colony algorithm [118], fuzzy Delphi method and hybrid ANN-based systems [119], cuckoo search algorithm [120], extreme learning machine [121], extreme gradient boosting (XGBoost) [122], and the firefly algorithm [123–126].

5.2. Airblast

Airblast or air overpressure are among the undesirable effects of blasting operations. They are explosion-induced large shock waves that are refracted horizontally by density variations in the atmosphere. The atmospheric pressure waves of airblasts consist of a high audible frequency and subaudible low-frequency sound [50,127]. Airblasts can impact structures close to the blast zone by rattling windows and the roofing materials.

Several scholarly studies have attempted to predict airblasts based on some identified influential factors, such as the maximum explosive charge per delay, burden, spacing, stemming, wind direction, temperature, and distance from the blast zone to the monitoring point. There are empirical models (see Table 2) for predicting airblasts, in addition to more recent applications of machine learning techniques, such as the ANN, support vector regression, particle swarm optimization, and adaptive neuro-fuzzy inference system.

Khandelwal and Singh [128] attempted to predict airblasts using an ANN by incorporating the maximum charge per delay and distance between the blast zone and the monitoring point and demonstrated that the neural network model yields better predictions when compared to a generalized equation and conventional statistical relations. Mohamed [129] predicted airblasts using the fuzzy inference system and ANN. Comparing the results of these methods with the values obtained by a regression analysis and measured field data, Mohamed asserted that the neural network and fuzzy models had accurate predictions compared to the regression analysis. Khandelwal and Kankar [130] predicted airblasts using SVM and compared the values with the results of the generalized predictor equation. They showed that the predicted values of airblasts by SVM were much closer to the actual values as compared to the predicted values by the predictor equation. Nguyen and Bui [72] developed and combined five ANN models with an RF algorithm to form an ANN-RF model to predict blast-induced air overpressure. The input variables of the model included the maximum explosive charge capacity, monitoring distance, vertical distance, powder factor, burden, spacing, and length of stemming. The results indicate that the proposed ANN-RF model was a superior model to the empirical technique, ANN, and RF models.

Mohamad et al. [131] employed the empirical, ANN, and a hybrid model of the genetic algorithm (GA-ANN) to estimate airblasts based on a maximum charge per delay and the distance from the blast face input parameters. The results show that the GA-ANN technique can provide a higher performance in predicting airblasts compared to the ANN and empirical models. The superior performance of GA-ANN in airblast prediction was also reported by Armaghani et al. [132]. They compared it with the ANN, USBM, and MLR models and observed that, with a coefficient of determination of 0.965, GA-ANN was a better airblast predictor than the other models implemented. Hajihassani et al. [133]

developed a hybrid airblast model where the particle swarm optimization (PSO) algorithm was used to train ANNs instead of the backpropagation algorithm. Using nine input parameters, the proposed model had a correlation coefficient of 0.94, suggesting a superior predictive strength compared to empirical models. AminShokravi et al. [134] evaluated the acceptability and reliability of three PSO-based airblast models (the PSO-linear, PSO-power, and PSO-quadratic models) and found that the PSO-linear model showed a higher predictive ability than the PSO-power, PSO-quadratic, ANN, and USBM models.

Armaghani et al. [135] also optimized an ANN with an imperialist competitive algorithm for airblast prediction. They also developed conventional ANN models to compare the results with the new model. The results demonstrated that the proposed model could predict airblasts more accurately than the other presented techniques. Nguyen et al. [136] investigated the feasibility of three ensemble machine learning algorithms, including the gradient boosting machine (GBM), random forest (RF), and Cubist, for predicting airblasts in open-pit mines. The ensemble model results were compared with those of an empirical model. Their findings revealed that the ensemble models yielded more precise accuracy than those of the empirical model. Of the ensemble models, the Cubist model provided a better performance than those of the RF and GBM models. Besides, they also indicated that the explosive charge capacity, spacing, stemming, monitoring distance, and air humidity were the most important inputs for the airblast predictive models using AI.

5.3. Flyrock

Flyrock is a loose rock fragment ejected from blasting processes that can travel over long distances away from the zone of influence of the blast. The Institute of Makers of Explosives (IME) defines flyrock as the rock propelled beyond the blast area by the force of an explosion [137]. According to Amini et al. [138], there are three mechanisms via which flyrock can occur (Figure 9): riffling, catering, and face bursting. Riffling occurs when the stemming material is insufficient, causing blast gases to stream up the blast hole along the path of least resistance, resulting in stemming ejection and, sometimes, ejection of the collar rock. Catering is due to the venting of gasses through the stemming region (i.e., blasthole collar), which usually contains a weakened layer due to the previous blasting from the bench above. Face bursting occurs when explosive charges are adjacent to the major geological structures or zones of weakness, allowing high-pressure gases to jet along the weakness zones [138].

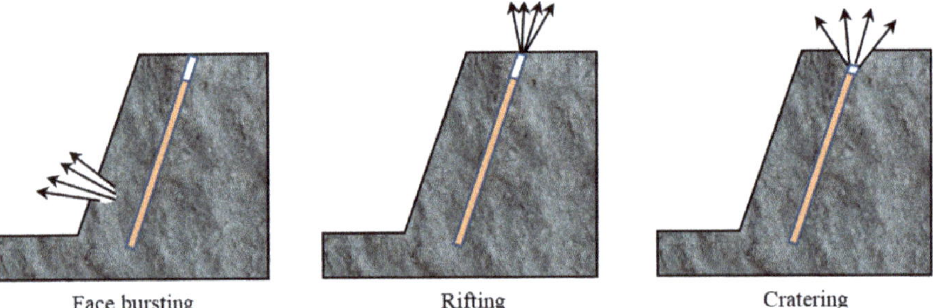

Figure 9. Main categories of flyrock in open-pit mines.

Flyrock has the potential to cause serious damage to the properties or cause injuries and fatalities in communities located close to a blast zone. As a result, researchers have made efforts to develop empirical models to predict and help mitigate flyrock. Equations have also been formulated based on Newton's law of motion with two possible solutions: an approximate numerical solution and the application of the Runge-Kutta algorithm of the fourth order to predict the maximum throw of flyrock fragments and estimate safe distances [139]. More recently, ML has proven to be a useful tool with surging applications in predicting flyrock. Amini et al. [138] tested the capability of SVM in flyrock prediction of a copper mine. Comparing the obtained results of the SVMs with those of an ANN, they concluded that the SVM model was faster and more precise than the ANN model in predicting flyrock. A new combination (FA-ANN) can be used as a powerful and practical technique in predicting the flyrock distance before blasting operations. Li et al. [140] selected the most important factor for flyrock predictions using the fuzzy Delphi method and developed a firefly algorithm (FA) and ANN model to estimate the flyrock distance. They observed that the FA-ANN model provided the best optimization of the weights and biases and recorded the lowest network error compared to the other ANN-based models.

Manoj and Monjezi [141] also analyzed flyrock predictions using the support vector machine and multivariate regression analysis. They found that the SVM results were more accurate than those of the multivariate regression analysis. Rad et al. [142] also conducted a similar study, comparing least squares support vector machines (LS-SVM) and support vector regression (SVR), and based on the performances of the two models, they concluded that the LS-SVM model was more useful than the SVR model in the estimation of blast-induced flyrock. A sensitivity analysis of the model showed that the powder factor and rock density were the most effective parameters on flyrock. Hasanipanah et al. [143] also developed a flyrock prediction equation based on particle swarm optimization (PSO) in quarry operations. For comparison purposes, multiple linear regression (MLR) was also used. Five effective parameters (burden, spacing, stemming, rock density, and powder factor) were used as the input parameters, while flyrock was considered as the output parameter. The results revealed that the proposed PSO equation was more reliable than MLR in predicting flyrock. Based on the sensitivity analysis results, it was also found that the rock density was the most effective parameter on flyrock in the studied cases.

Recently, Lu et al. [144] presented two machine learning models, including the extreme learning machine (ELM) and outlier robust ELM (ORELM), for predicting flyrock. To construct and verify the proposed ELM and ORELM models, a database including 82 datasets collected from three granite quarry sites was used. Additionally, the ANN and multiple regression models were used for comparison. The results showed that both the ELM and ORELM models performed satisfactorily, and their performances were far better compared to the performances of the ANN and multiple regression models. Armaghani et al. [145] estimated the flyrock distance using three machine learning methods: principal component regression (PCR), support vector regression (SVR), and multivariate adaptive regression splines (MARS). The SVR model showed a better performance in predicting the flyrock distance compared to the other proposed models. Further, the SVR model was optimized by gray wolf optimization (GWO), resulting in a 4% decrease in flyrock distance. The authors asserted that the SVR prediction model can be used to accurately predict the flyrock distance and properly establish the blast safety zone. An ELM was also optimized using the biogeography-based optimization (BBO) algorithm to form a hybrid flyrock prediction model [146]. Compared to the particle swarm optimization (PSO-ELM) and ELM models, the BBO-ELM proved to be a powerful model for predicting flyrock, with a superior performance. Dehghani et al. [147] used the gene expression programming (GEP) model and cuckoo optimization algorithm to predict and minimize the flyrock range. In this study, the burden, spacing, stemming, charge length, and powder factor were used as the input parameters in the GEP model; then, the equation from the GEP was used as a cost function for minimizing flyrock by the cuckoo optimization algorithm. They concluded that the GEP model showed a good performance in predicting blast-induced flyrock using the blast design parameters, and the cuckoo algorithm reduced the maximum flyrock distance relative to the values obtained from the initial blast designs. This study also revealed the powder factor as the input parameter sensitivity in the analysis and, hence, the most effective parameter on the flyrock phenomenon.

6. Discussion and Future Trends

The impacts of blasting operations have significant effects on mining in varied ways, from mineral processing to environmental sustainability. Undesirable blast impacts, such as ground vibration, airblast, and flyrock, pose severe risks, including human irritation, structural damage, injury, and even fatalities to receptor communities if a blast is not conducted properly [148]. In other words, blast results could increase a mine's operating cost and community complaints, which can escalate to contention between the management and the community if not addressed early. Blast-induced ground vibrations, which are measured in the PPV, are, by far, the most studied blast impact; consequently, most of the blast impact models focus on this area. The popularity of ground vibrations in this field can be attributed to the fact that ground motions accompanying blast events cannot be avoided, and they often result in community complaints. It is one of the major concerns in mining with stringent environmental standards, and a slight breach or incompliance with the rules could impede production and deteriorate the cordial relationship (i.e., social license) between a mining company and a host community. For example, the La Arena gold mine in Peru owned by Tahoe Resources Inc. had to suspend operations temporarily following a protest by some community members demanding compensation for unspecified damage caused by dust and vibrations from blasting at the mine [149]. Given increasing concerns about the environmental impacts of mining, it is now more crucial than ever to ensure that blasting operations are conducted with greater precision. The goal of every blast engineer is to conduct a blast that produces optimal fragmentation, good heave, and minimal backbreak with minimal ground vibration, airblast, flyrock, and fumes. Thus, blast impact studies are vital to determine the most appropriate blast design that would optimize the desirable effects and minimize the undesirable ones. Blasting is a complex phenomenon, and many factors influence its resulting impacts. Different methods based on numerical, empirical, and, more recently, machine learning have been developed for predicting blast impacts.

Several factors affect blast impacts. As highlighted by Yan et al. [81], some common parameters identified to influence blast impacts include the burden, spacing, free face, charge structure, delays, blasthole dimension, charge parameters, stemming, and geological conditions. It is often difficult to incorporate all the influential parameters in the blast impact model, so the practice is to identify the important parameters peculiar to the problem being addressed. Additionally, due to the heterogeneity of geological formations [150], there will be variations in the site conditions (e.g., rock strength and discontinuities) from one mine to another. Therefore, the prevailing local situation, mine plan, and environmental standards must be considered when formulating a blast impact model. The parameter selections are therefore very important, and they have a significant influence on the predictive powers of a blast impact model. Indeed, a blast impact model is as powerful and accurate as the set of parameters employed in developing the model. Studies expend significant resources in deciding which parameters should be included in a model.

Even though empirical blast models are formulated following extensive field experiments and data collection on various blast impact parameters, only a few parameters are considered in the final model. Empirical models for predicting the PPV, for example, are built using mainly the maximum charge per delay, the distance between blast zone and monitoring point, and the geological conditions, which are accounted for as site-specific constants [81]. Similar parameters are used in estimating airblasts and flyrock. The limited number of parameters could result in inaccurate predictions. Cognizant of the limitations of the empirical models, Monjezi et al. [151] modified the United State Bureau of Mines (USBM) model by incorporating the effect of water in addition to the charge per delay and distance from the blast face to develop a new predictive model based on gene expression programming (GEP). They observed that the proposed model was able to predict blast-induced ground vibrations more accurately than the other developed techniques. Nevertheless, empirical models remain the most widely used blast impact predictive tools in the mining industry. This wide usage could be attributed to their computational simplicity and reasonable prediction results. Statistical blast impact models such as those used by Hudaverdi [152] also consider only the blast design parameters and consider them as ratios instead, using their actual values. Despite the wide application of the conventional blast impact models, they possess inherent inefficiencies as a result of their inability to accommodate more relevant parameters affecting the outcome of a blast.

In addressing this challenge, researchers have employed ML techniques to estimate blast impacts. These are computer models that can accommodate several input variables and deduce the relationships between them to predict an output. Considering the numerous parameters involved in estimating blast impacts, ML has proven to be a formidable tool in this area. Besides establishing complex relationships, machine learning tools are also efficient in feature selection. Again, the literature has shown that, compared to the conventional blast impact models, the ML approach is more robust and yields better prediction results. For example, Bayat et al. [125] minimized the blast-induced ground vibrations by decreasing the PPV to 17 mm/s (60%) using an ANN combined with a FA. A burden of 3.1 m, spacing of 3.9 m, and charge per delay of 247 kg were reported as the optimized blast design parameters. Similarly, the authors of [153] employed gene expression programming (GEP) and the cuckoo optimization algorithm (COA) to optimize the blast patterns in an iron mine, resulting in a considerable reduction in the PPV values (55.33%). Armaghani et al. [145] achieved a 4% decrease in the minimum flyrock distance by using SVR in a quarry operation. Table 6 summarizes some of the ML techniques used to predict blast-induced impacts. The summary includes predicted impacts, techniques that are usually compared with ML, the prediction parameters, the number of datasets, and the ML model performance measure (coefficient of determination).

Table 6. Summary of the ML-based blast-induced impact prediction models.

ML Method	Other Models	Operation	Parameter	Dataset	Impact	Performance (R^2)	Reference
ANN	USBM, Langefors–Kihlstrom, Ambraseys–Hendron, Bureau of Indian Standard, CMRI predictor	Coal mine	Q, D	130	Ground vibration	0.919	[94]
ANN	MVR	Coal mine	Q, D, HD, HZ, B, ST CH, BI, E, V, PV, VOD, ED	150	Ground vibration	0.9994	[80]
SVM	USBM, Ambraseys–Hendron, Davies et al., Indian Standard	Dam construction	Q, D	80	Ground vibration	0.957	[105]
GA-ANN	ANN, USBM, and MLR	Quarry	Q, D	97	Airblast	0.965	[132]
PSO	MLR	Quarry	S, B, ST, PF, RD	76	Flyrock	0.966	[143]
PSO–ANN	ICA, GA	Quarry	HD, HZ, BS, Q, PF	262	Flyrock	0.943	[154]
ANN	MVR	Copper mine	B, S, Q, PF, ST, HD, NR, BH	135	Fragmentation	0.94	[155]

VOD is the velocity of detonation, Q is the maximum charge per delay, D is the distance from the blasting face, B is the burden, S is spacing, ST is stemming, HD is the hole diameter, HZ is the hole depth, CH is the charge length, BI is the blastability index, E is the Young's modulus, V is Poisson's ratio, PV is the P-wave velocity, ED is the explosive density, RD is the rock density, PF is the powder factor, BS is the burden-to-spacing ratio, NR is the number of rows, and BH is the bench height.

The most common machine learning methods used for blast impact prediction are the ANN, SVM, and PSO (Table 6). Hybrid models were also developed by combining some of these algorithms. Among these algorithms, the artificial neural network remains the most popular, with wide implementation in ground vibrations [29,80,85], airblasts [98], flyrock [95,156–158], fragmentation [155,159–162], backbreak analyses [159,160,163–165], and noise [166]. We observed that these ML techniques were generally employed to predict blast-induced impacts, just like the empirical models, and not necessarily to improve or reduce the impacts. The performances of the models were judged based on a set of statistical metrics, including the mean absolute error (MAE), root mean square error (RMSE), correlation coefficient (R), and coefficient of determination (R^2), which only showed the prediction strength of the ML techniques compared to the other models. A summary of the ML-based blast impact prediction models and common parameters is presented in Figure 10.

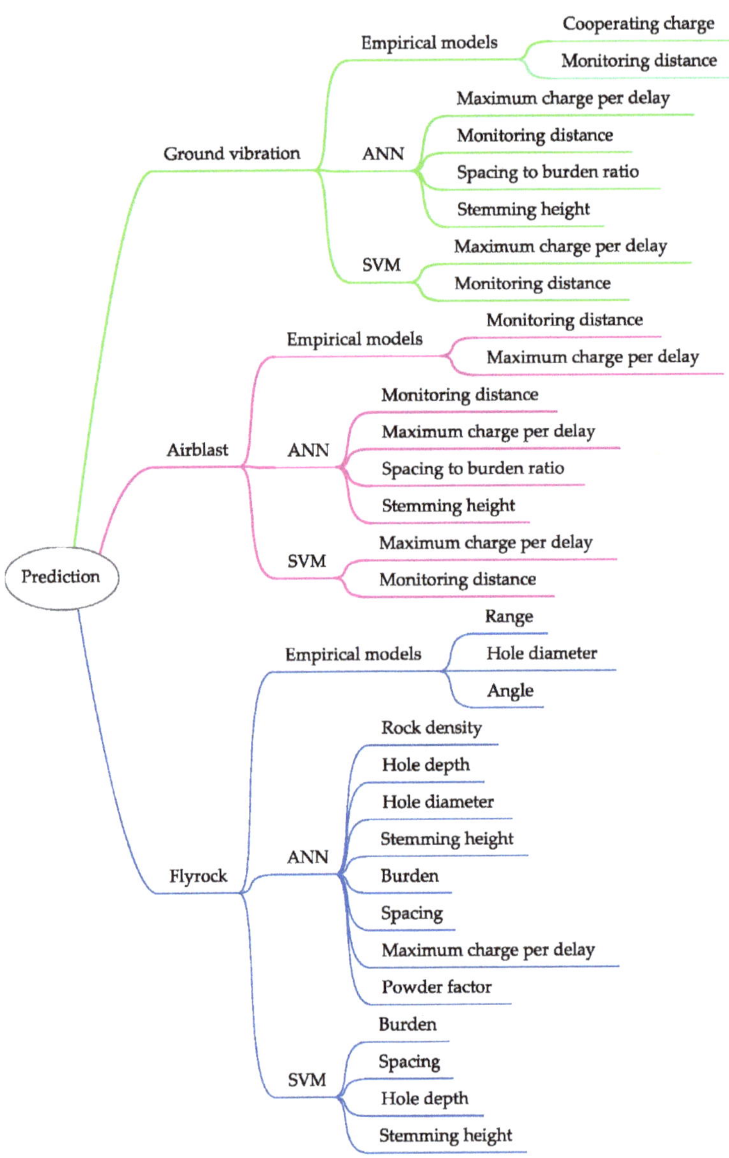

Figure 10. Predominate parameters for predicting blast impacts.

In implementing the machine learning algorithms, feature selection is considered the first step and is usually achieved using the principal component analysis (PCA). The PCA identifies the principal independent variables and eliminates irrelevant ones [153], and it is one inherent feature of the classification and regression tree (CART) algorithm, as applied by Hasanipanah et al. [167], in predicting ground vibrations. The selected features are synthesized in the chosen machine learning algorithm to estimate the blast impact. Currently, there seems to be consensus backing ANN as a suitable blast impact predictor. However, studies have also highlighted some limitations of the ANN, including a long training period and the possibility of easily falling into the local minimum [81]. Thus, the ANN is combined with other algorithms to optimize and improve the accuracy of predicting blast impacts.

This paper discussed mostly the undesirable impacts of blasting and how machine learning models have been employed to predict these impacts. However, another aspect of blasting is the desired outcomes in terms of fragmentation and heave. Mining companies and quarries desire to produce fragment sizes that can be mucked easily and directly fed into a crusher without the need for secondary blasting. At times, there are too many fine or oversized boulders. The blast input parameters are altered to control the fragment sizes. There are empirical equations [168–172] for predicting the fragment size distribution as well. Examples of empirical models for predicting blast-induced fragment distribution include Kuz-Ram models, Julius Kruttschnitt Mineral Research Centre (JKMRC) models, the Bond comminution method, and the Swebrec function. Images of a blast muck pile can be analyzed using digital image processing software such as Split-Desktop® and WipFrag 3 to determine the particle size distribution of the fragmented rock.

Additionally, attempts are being made by researchers to introduce new and improved fragment distribution models, leveraging on the advances gained in computer power in recent years. Studies such as An et al. [173], Tao et al. [174], and Yi et al. [123] have utilized numerical modeling techniques and image processing to predict fragment size distributions. One merit of the numerical approach is that it allows the researcher to simulate a series of fragment size distribution scenarios under various blast configurations and fracture patterns [174]. It is worth noting that ML applications in this area are also gaining interest in the scientific community. Generally, the process involves the provision of a set of input data (e.g., blast design parameters and muck pile image), which is processed by the ML model to generate a rock fragment size profile (Figure 11). The ML techniques being applied for evaluating the fragment size distribution are different from those used in the prediction of ground vibrations, airblasts, and flyrock. These new techniques are deep learning, a subset of ML. Deep learning naturally takes advantage of automatically discovering and extracting features and patterns from large datasets combined with modeling structures capable of capturing highly complex behaviors [175]. Examples of deep learning algorithms include convolutional neural networks (CNN), recurrent neural networks (RNNs), long short-term memory networks (LSTMs), stacked auto-encoders, deep Boltzmann machine (DBM), and deep belief networks (DBN). These algorithms have tremendously improved image classification, object detection, and natural language processing in many fields. Recent applications of deep learning in blasting include the prediction of flyrock [157], rock fragment distribution [176], and classification of mine seismic events, among others. Further, we observed that most common ML algorithms for blast-induced fragment size predictions include the ANN [159–162], SVM [104,177], PCA [177], fuzzy inference system [178–180], adaptive neuro-fuzzy inference system [177,181,182], bee colony algorithm [162], PSO [183,184], ant colony optimization [185], and gaussian process regression [186]. The ML-based fragment size prediction models performed significantly better than the empirical models [187].

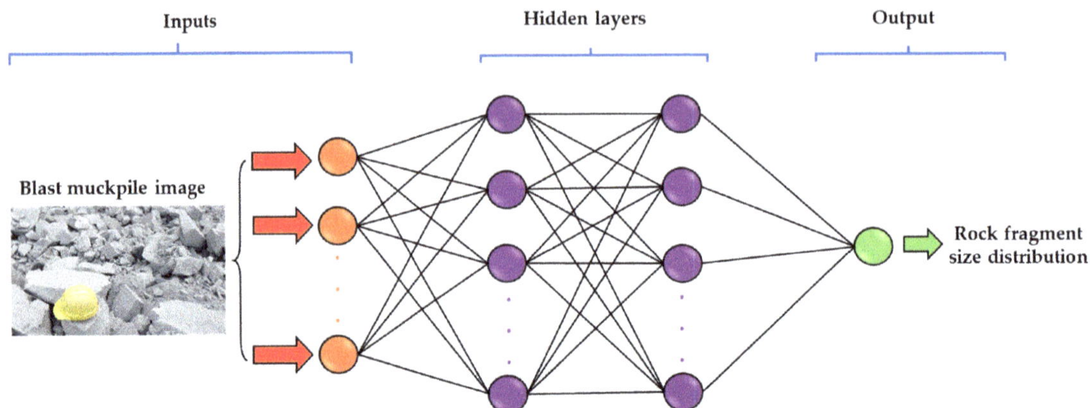

Figure 11. Schematic diagram of an ML model for predicting rock fragment size distributions.

From the literature, many of the proposed models could predict only one blast impact. Only a few models were developed to predict ground vibrations and airblasts [188,189], backbreak and rock fragmentation [160,162], and flyrock and rock fragmentation [190,191]. Meanwhile, all the blast impacts occur concurrently and are equally influenced by similar blast parameters and geological conditions. Currently, only one study (a master of a science thesis report) has been able to develop an integrated prediction model for rock fragmentation, ground vibrations, and airblasts using an ANN with 7-13-3 architecture [8]. The input parameters were the charge per delay, distance from the blast zone to the monitoring point, hole depth, stemming length, hole diameter, powder factor, and spacing-to-burden ratio, while rock fragmentation, ground vibrations, and airblasts were the corresponding output parameters. The ANN model proved to be more effective with improved fragmentation and minimal blast impacts compared to the empirical equations and multivariate regression. An integrated model of this kind saves resources and allows the blast engineer to examine the influence of the input parameters on the blast outcome in one attempt. Therefore, a more holistic and robust ML-based blast impact model should integrate all the blast impacts, both desirable and undesirable. An improved ML model development can be connecting the input (most influential blast design parameters) to the output (blast outcomes). Subsequently, with sufficient training of the ML model using an adequate dataset, the blast outcomes can be predicted before the actual blast event that would inform further modification of the input parameters to achieve the desired outcome. Compared to the other blast impacts, ML applications for blast-induced dust/fume and noise prediction have not received intensive research attention. From the existing blast features, ML models can be developed to estimate noise level and dust/fume volume and direction.

Nowadays, with automation and the internet of things (IoT), mining companies can receive real-time information on drill operations, including high-resolution rock images and ground conditions. Similarly, several measurements, such as blast images and videos, vibration results, fragment distribution, plume movement, and loading and crushing performances, can be obtained during and after a blast. With the availability of such large datasets combined with improvements in algorithms and computing power, we foresee a field-wide implementation of big data analytics coupled with deep learning applications to integrate all the aspects of mine operations, from exploration to reclamation, leading to more efficient and accurate decision-making in the industry. These applications will automatically learn from the result of each drilling and blasting operation and analyze how the parameters such as the drill pattern, hole deviation, ground condition, timing, and powder factor contribute to the resulting fragmentation and heave, material handling, and crushing performance. In fact, unlike most traditional ML algorithms applied in this field, deep learning algorithms would automatically discover, extract, and optimize the blast-induced features without human intervention. Deep learning could overcome some of the deficiencies in traditional data-driven methods as more data becomes available. Deep learning models can also make it possible for researchers to predict all blast-induced impacts simultaneously. Integrating these applications into the current systems will form part of the ongoing efforts to improve mine-to-mill processes and automate mining processes.

It is essential to mention that the foundation of a functional ML model rests upon a rich dataset. The quality, size, and partition of the dataset used in implementing ML influence the model's performance in accuracy and generalization. Thus, the application of various AI methods, including ML and deep learning, requires a reasonably large dataset to work properly. Without an adequate dataset, the model's usefulness and potential can be undermined or negated completely. Generally, it is widely accepted within the research community that AI demands an enormous dataset, and a too-little dataset will yield poor results. However, what constitutes an adequate dataset size is not clearly defined, as the amount of data required depends on different factors, such as the problem definition, model complexity, and algorithm type [192]. Fortunately, renowned researchers working in AI within the mining industry have put forth their experience in modeling problems relating to the mining and mineral industry and recommended good practices, especially when modeling with a sparse dataset.

Ganguli et al. [193] provided good practices regarding AI implementation in mining. They recommended a thorough understanding of the modeling process before implementation and advised caution when using business intelligence tools and software products. Their recommendation also included the random splitting of a dataset into training, testing, and validation subsets and achieving similar characteristics among the three subsets, irrespective of the data partition. Further, they suggested that the training subset should contain the highest and lowest values, and samples should be assigned to the training subset first, followed by validation and testing, during data grouping/segmentation. Moreover, the best data collection and processing practices should be observed during model development to ensure the dataset is of high quality, sufficient, and representative of the population.

7. Conclusions

A blast impact is a complex phenomenon with numerous influential factors that must be incorporated into blast impact prediction models to predict accurate results. However, the industry-accepted empirical models lack the computational capacity to accommodate all the influential factors. Thus, these models may not be accurate in their predictions. The importance of achieving accurate predictions is well-known, as it informs proper blast design and helps allay doubts about compliance with the established blast standards. Recent advances in computer power have ushered in soft computing tools that can address some of the limitations of the empirical models used in blast engineering. ML algorithms are powerful tools for solving both linear and nonlinear complex mining problems with several influential factors. ML algorithms, such as the ANN, SVM, and CART, can take several variables and predict blast impacts with high levels of accuracy. These models are promising tools for optimizing the blast parameters and blast outcomes to increase the production efficiency while reducing the costs. The models' predictive powers could also be improved by synthesizing with other algorithms.

Future models could focus on developing a one-shop model that could estimate all the blast impacts, perhaps using deep learning, instead of predicting a single impact such as ground vibrations or airblasts. Additionally, these new models should incorporate the geological variability and consider datasets from different mine sites or operations to develop a more holistic model. The models should be user-friendly and devoid of complex mathematical language so that industry practitioners can easily implement them.

Author Contributions: Conceptualization, N.K.D.-D. and S.A.; writing—original draft preparation, N.K.D.-D.; and writing—review and editing, S.A. and A.J. All authors have read and agreed to the published version of the manuscript.

Funding: This research received no external funding.

Data Availability Statement: Not applicable.

Conflicts of Interest: The authors declare no conflict of interest.

References

1. Yu, Z.; Shi, X.; Zhou, J.; Chen, X.; Qiu, X. Effective Assessment of Blast-Induced Ground Vibration Using an Optimized Random Forest Model Based on a Harris Hawks Optimization Algorithm. *Appl. Sci.* **2020**, *10*, 1403. [CrossRef]
2. Bayat, P.; Monjezi, M.; Mehrdanesh, A.; Khandelwal, M. Blasting pattern optimization using gene expression programming and grasshopper optimization algorithm to minimise blast-induced ground vibrations. *Eng. Comput.* **2021**, 1–10. [CrossRef]
3. Abbaspour, H.; Drebenstedt, C.; Badroddin, M.; Maghaminik, A. Optimized design of drilling and blasting operations in open pit mines under technical and economic uncertainties by system dynamic modelling. *Int. J. Min. Sci. Technol.* **2018**, *28*, 839–848. [CrossRef]
4. Zou, B.; Luo, Z.; Wang, J.; Hu, L. Development and Application of an Intelligent Evaluation and Control Platform for Tunnel Smooth Blasting. *Geofluids* **2021**, *2021*, 6669794. [CrossRef]
5. Koopialipoor, M.; Armaghani, D.J.; Haghighi, M.; Ghaleini, E.N. A neuro-genetic predictive model to approximate overbreak induced by drilling and blasting operation in tunnels. *Bull. Int. Assoc. Eng. Geol.* **2019**, *78*, 981–990. [CrossRef]
6. Ocak, I.; Bilgin, N. Comparative studies on the performance of a roadheader, impact hammer and drilling and blasting method in the excavation of metro station tunnels in Istanbul. *Tunn. Undergr. Space Technol.* **2010**, *25*, 181–187. [CrossRef]
7. Singh, R.K.; Sawmliana, C.; Hembram, P. Time-constrained demolition of a concrete cofferdam using controlled blasting. *Innov. Infrastruct. Solut.* **2020**, *6*, 20. [CrossRef]
8. Tille, R.N. Artificial Neural Network Approach to Predict Blast-Induced Ground Vibration, Airblast and Rock Fragmentation. Master's Thesis, Missouri University of Science and Technology, Rolla, MO, USA, 2016.
9. Bilim, N.; Çelik, A.; Kekeç, B. A study in cost analysis of aggregate production as depending on drilling and blasting design. *J. Afr. Earth Sci.* **2017**, *134*, 564–572. [CrossRef]
10. Trivedi, R.; Singh, T.N.; Gupta, N. Prediction of Blast-Induced Flyrock in Opencast Mines Using ANN and ANFIS. *Geotech. Geol. Eng.* **2015**, *33*, 875–891. [CrossRef]
11. Monjezi, M.; Khoshalan, H.A.; Varjani, A.Y. Prediction of flyrock and backbreak in open pit blasting operation: A neuro-genetic approach. *Arab. J. Geosci.* **2012**, *5*, 441–448. [CrossRef]
12. Murmu, S.; Maheshwari, P.; Verma, H.K. Empirical and probabilistic analysis of blast-induced ground vibrations. *Int. J. Rock Mech. Min. Sci.* **2018**, *103*, 267–274. [CrossRef]
13. Karnen, B. *Rock Blasting and Water Quality Measures that Can Be Taken to Protect Water Quality and Mitigate Impacts*; New Hampshire Department of Environmental Services: Concord, NH, USA, 2010.
14. Hawkins, J. Impacts of Blasting on Domestic Water Wells. In *Workshop on Mountaintop Mining Effects on Groundwater*; OSMRE: Washington DC, USA, 2000; pp. 1–10.
15. Pelham, K.; Lane, D.; Smerenkanicz, J.R.; Miller, W. A Proactive Approach To Limit Potential Impacts from Blasting to Drinking Water Supply Wells. In Proceedings of the 60th Highway Geology Symposium, Windham, NH, USA, 29 September–1 October 2009; pp. 1–19.
16. Birch, W.J.; Hosein, S.; Tompkin, S. Blasting in proximity to a world heritage site—A success story. In Proceedings of the 16th Extractive Industry Geology Conference, Portsmouth, UK, 8–11 September 2010; pp. 139–145.
17. Varris, P.; Thorpe, M. Community concerns and input for open pit closure in a West African urban setting. In Proceedings of the Seventh International Conference on Mine Closure, Brisbane, Australia, 25–27 September 2012.
18. Bansah, K.J.; Kansake, B.A.; Dumakor-Dupey, N.K. Baseline Structural Assessment: Mechanism for Mitigating Potential Conflicts Due to Blast Vibration. In Proceedings of the 4th UMaT Biennial International Mining and Mineral Conference, Tarkwa, Ghana, 3–6 August 2016; pp. 42–48.
19. Torres, V.N.; Silveira, L.G.; Lopes, P.; Lima, H. Assessing and controlling of bench blasting-induced vibrations to minimize impacts to a neighboring community. *J. Clean. Prod.* **2018**, *187*, 514–524. [CrossRef]
20. Agrawal, H.; Mishra, A.K. An innovative technique of simplified signature hole analysis for prediction of blast-induced ground vibration of multi-hole/production blast: An empirical analysis. *Nat. Hazards* **2020**, *100*, 111–132. [CrossRef]
21. Karadogan, A.; Kahriman, A.; Ozer, U. A new damage criteria norm for blast-induced ground vibrations in Turkey. *Arab. J. Geosci.* **2013**, *7*, 1617–1626. [CrossRef]
22. McKenzie, C.K. Flyrock range and fragment size prediction. In Proceedings of the 35th Annual Conference on Explosives and Blasting Technique, Denver, CO, USA, 8 – 11 February 2009; Volume 2.

23. Blanchier, A. Quantification of the levels of risk of flyrock. In *Rock Fragmentation by Blasting, Proceedings of the 10th International Symposium on Rock Fragmentation by Blasting, Fragblast 10, New Delhi, India,26–29 November 2012*; CRC Press: Leiden, The Netherlands, 2013; pp. 549–553.
24. Chiapetta, R.F.; Borg, D.G. Increasing productivity through field control and high-speed photography. In Proceedings of the 1st Int. Symp. on Rock Fragmentation by Blasting, Lulea, Sweden, 23–26 August 1983; pp. 301–331.
25. Richards, A.; Moore, A. Flyrock control-by chance or design. In Proceedings of the Annual Conference on Explosives and Blasting Technique, New Orleans, LA, USA, 1–4 February 2004; Volume 1, pp. 335–348.
26. Lundborg, N. *The Probability of Flyrock*; SveDeFo: Stockholm; Sweden, 1981.
27. Bajpayee, T.S.; Verakis, H.C.; Lobb, T.E. Blasting safety–revisiting site security. In Proceedings of the 31st Annual Conference of Explosives and Blasting Technique, Orlando, FL, USA, 6–9 February 2005; pp. 1–13. Available online: https://www.cdc.gov/niosh/mining/userfiles/works/pdfs/bsrss.pdf (accessed on 5 January 2021).
28. Bendezu, M.; Romanel, C.; Roehl, D. Finite element analysis of blast-induced fracture propagation in hard rocks. *Comput. Struct.* **2017**, *182*, 1–13. [CrossRef]
29. Singh, T.N.; Singh, V. An intelligent approach to prediction and control ground vibration in mines. *Geotech. Geol. Eng.* **2005**, *23*, 249–262. [CrossRef]
30. Singh, T.N.; Kanchan, R.; Verma, A.K. Prediction of Blast Induced Ground Vibration and Frequency Using an Artificial Intelligent Technique. *Noise Vib. Worldw.* **2004**, *35*, 7–15. [CrossRef]
31. Rebala, G.; Ravi, A.; Churiwala, S. *An Introduction to Machine Learning*; Springer Science and Business Media LLC: Berlin, Germany, 2019.
32. Das, A.; Sinha, S.; Ganguly, S. Development of a blast-induced vibration prediction model using an artificial neural network. *J. South. Afr. Inst. Min. Metall.* **2019**, *119*, 187–200. [CrossRef]
33. Sanchidrián, J.A.; Segarra, P.; López, L.M. Energy components in rock blasting. *Int. J. Rock Mech. Min. Sci.* **2007**, *44*, 130–147. [CrossRef]
34. Raina, A.K.; Murthy, V.M.S.R.; Soni, A.K. Flyrock in bench blasting: A comprehensive review. *Bull. Int. Assoc. Eng. Geol.* **2014**, *73*, 1199–1209. [CrossRef]
35. Khandelwal, M.; Singh, T. Evaluation of blast-induced ground vibration predictors. *Soil Dyn. Earthq. Eng.* **2007**, *27*, 116–125. [CrossRef]
36. Jimeno, C.L.; Jimeno, E.L.; Carcedo, F.J.A.; de Ramiro, Y.V. *Drilling and Blasting of Rocks*; CRC Press: Boca Raton, FL, USA, 2017.
37. Changyou, L.; Jingxuan, Y.; Bin, Y. Rock-breaking mechanism and experimental analysis of confined blasting of borehole surrounding rock. *Int. J. Min. Sci. Technol.* **2017**, *27*, 795–801. [CrossRef]
38. Cho, S.H.; Kaneko, K. Influence of the applied pressure waveform on the dynamic fracture processes in rock. *Int. J. Rock Mech. Min. Sci.* **2004**, *41*, 771–784. [CrossRef]
39. Gui, Y.; Zhao, Z.; Zhou, H.; Goh, A.; Jayasinghe, L. Numerical Simulation of Rock Blasting Induced Free Field Vibration. *Procedia Eng.* **2017**, *191*, 451–457. [CrossRef]
40. Cho, S.; Nakamura, Y.; Mohanty, B.; Yang, H.; Kaneko, K. Numerical study of fracture plane control in laboratory-scale blasting. *Eng. Fract. Mech.* **2008**, *75*, 3966–3984. [CrossRef]
41. Fourney, W.L. 2—Mechanisms of Rock Fragmentation by Blasting. In *Excavation, Support and Monitoring*; Hudson, J.A., Ed.; Pergamon: Oxford, UK, 1993; pp. 39–69.
42. Bhandari, S. *Operations, Engineering Rock Blasting*; A.A. Balkema: Rotterdam, The Netherlands, 1997.
43. Kumar, R.; Choudhury, D.; Bhargava, K. Determination of blast-induced ground vibration equations for rocks using mechanical and geological properties. *J. Rock Mech. Geotech. Eng.* **2016**, *8*, 341–349. [CrossRef]
44. Davies, B.I.; Farmer, P.A. Engineer, and Undefined 1964, Ground Vibration from Shallow Sub-Surface Blasts. Available online: https://trid.trb.org/view/140358 (accessed on 17 October 2020).
45. Ragam, P.; Nimaje, D.S. Evaluation and prediction of blast-induced peak particle velocity using artificial neural network: A case study. *Noise Vib. Worldw.* **2018**, *49*, 111–119. [CrossRef]
46. Monjezi, M.; Ghafurikalajahi, M.; Bahrami, A. Prediction of blast-induced ground vibration using artificial neural networks. *Tunn. Undergr. Space Technol.* **2011**, *26*, 46–50. [CrossRef]
47. Rai, R.; Singh, T.N. A new predictor for ground vibration prediction and its comparison with other predictors. *Indian J. Eng. Mater. Sci.* **2004**, *11*, 178–184.
48. Saadat, M.; Khandelwal, M.; Monjezi, M. An ANN-based approach to predict blast-induced ground vibration of Gol-E-Gohar iron ore mine, Iran. *J. Rock Mech. Geotech. Eng.* **2014**, *6*, 67–76. [CrossRef]
49. Siskind, D.E.; Stagg, M.S.; Kopp, J.W.; Dowding, C.H. *Structure Response and Damage Produced by Ground Vibration from Surface Mine Blasting*; US Department of the Interior, Bureau of Mines: New York, NY, USA, 1980.
50. Hasanipanah, M.; Armaghani, D.J.; Khamesi, H.; Amnieh, H.B.; Ghoraba, S. Several non-linear models in estimating air-overpressure resulting from mine blasting. *Eng. Comput.* **2016**, *32*, 441–455. [CrossRef]
51. Bansah, N.K.; Arko-Gyimah, K.J.; Kansake, K.; Dumakor-Dupey, B.A. Mitigating Blast Vibration Impact. In Proceedings of the 4th UMaT Biennial International Mining and Mineral Conference, Tarkwa, Ghana, 3–6 August 2016; pp. 30–36.
52. Nguyen, H.; Bui, X.-N. Soft computing models for predicting blast-induced air over-pressure: A novel artificial intelligence approach. *Appl. Soft Comput.* **2020**, *92*, 106292. [CrossRef]

53. Armaghani, D.J.; Mohamad, E.T.; Hajihassani, M.; Abad, S.V.A.N.K.; Marto, A.; Moghaddam, M.R. Evaluation and prediction of flyrock resulting from blasting operations using empirical and computational methods. *Eng. Comput.* **2016**, *32*, 109–121. [CrossRef]
54. Gupta, R.N. *Surface Blasting and Its Impact on Environment. Impact of Mining on Environment*; Ashish Publishing House: New Delhi, India, 1980; pp. 23–24.
55. Russell, S.; Norvig, P. *Artificial Intelligence: A Modern Approach*, 3rd ed.; Prentice Hall: Upper Saddle River, NJ, USA, 2010.
56. Wiley, V.; Lucas, T. Computer Vision and Image Processing: A Paper Review. *Int. J. Artif. Intell. Res.* **2018**, *2*, 22–31. [CrossRef]
57. Tuşa, L.; Kern, M.; Khodadadzadeh, M.; Blannin, R.; Gloaguen, R.; Gutzmer, J. Evaluating the performance of hyperspectral short-wave infrared sensors for the pre-sorting of complex ores using machine learning methods. *Miner. Eng.* **2020**, *146*, 106150. [CrossRef]
58. Guo, H.; Nguyen, H.; Vu, D.-A.; Bui, X.-N. Forecasting mining capital cost for open-pit mining projects based on artificial neural network approach. *Resour. Policy* **2019**, 101474. [CrossRef]
59. Zhang, H.; Nguyen, H.; Bui, X.-N.; Nguyen-Thoi, T.; Bui, T.-T.; Nguyen, N.; Vu, D.-A.; Mahesh, V.; Moayedi, H. Developing a novel artificial intelligence model to estimate the capital cost of mining projects using deep neural network-based ant colony optimization algorithm. *Resour. Policy* **2020**, *66*, 101604. [CrossRef]
60. Simeone, O. A Very Brief Introduction to Machine Learning With Applications to Communication Systems. *IEEE Trans. Cogn. Commun. Netw.* **2018**, *4*, 648–664. [CrossRef]
61. Awad, M.; Khanna, R. *Efficient Learning Machines*; Apress: Berkeley, CA, USA, 2015.
62. Jakhar, D.; Kaur, I. Artificial intelligence, machine learning and deep learning: Definitions and differences. *Clin. Exp. Dermatol.* **2020**, *45*, 131–132. [CrossRef]
63. Vieira, S.; Pinaya, W.; Mechelli, A. Introduction to machine learning. In *Machine Learning*; Elsevier BV: Amsterdam, The Netherlands, 2020; pp. 1–20.
64. Kubat, M. *An Introduction to Machine Learning*; Springer: Cham, Switzerland, 2017.
65. Alpaydin, E. *Introduction to Machine Learning Ethem Alpaydin*, 3rd ed.; MIT Press: Cambridge, MA, USA, 2014.
66. Shanmuganathan, S. Artificial neural network modelling: An introduction. In *Studies in Computational Intelligence*; Springer: Cham, Switzerland, 2016.
67. Kelleher, J.D.; Mac Namee, B.; D'arcy, A. *Fundamentals of Machine Learning for Predictive Data Analytics: Algorithms, Worked Examples, and Case Studies*; MIT Press: Cambridge, MA, USA, 2015.
68. Samuel, A.L. Some Studies in Machine Learning. *IBM J. Res. Dev.* **1959**, *3*, 210–229. [CrossRef]
69. Tong, J.C.; Ranganathan, S. Computational T cell vaccine design. In *Computer-Aided Vaccine Design*; Elsevier: Amsterdam, The Netherlands, 2013; pp. 59–86.
70. Nguyen, H.; Bui, X.-N.; Choi, Y.; Lee, C.W.; Armaghani, D.J. A Novel Combination of Whale Optimization Algorithm and Support Vector Machine with Different Kernel Functions for Prediction of Blasting-Induced Fly-Rock in Quarry Mines. *Nat. Resour. Res.* **2021**, *30*, 191–207. [CrossRef]
71. Pisner, D.A.; Schnyer, D.M. *Chapter 6—Support Vector Machine. Machine Learning: Methods and Applications to Brain Disorders*; Mechelli, A., Vieira, S.B.T.-M.L., Eds.; Academic Press: Cambridge, MA, USA, 2020; pp. 101–121.
72. Nguyen, H.; Bui, X.-N. Predicting Blast-Induced Air Overpressure: A Robust Artificial Intelligence System Based on Artificial Neural Networks and Random Forest. *Nat. Resour. Res.* **2019**, *28*, 893–907. [CrossRef]
73. Misra, S.; Li, H. *Chapter 9—Noninvasive Fracture Characterization Based on the Classification of Sonic Wave Travel Times. Machine Learning for Subsurface Characterization*; Misra, S., Li, H., He, J., Eds.; Gulf Professional Publishing: Huston, TX, USA, 2020; pp. 243–287.
74. Bui, X.-N.; Nguyen, H.; Le, H.-A.; Bui, H.-B.; Do, N.-H. Prediction of Blast-induced Air Over-pressure in Open-Pit Mine: Assessment of Different Artificial Intelligence Techniques. *Nat. Resour. Res.* **2019**, *29*, 571–591. [CrossRef]
75. Rasmussen, C.E. Gaussian Processes in Machine Learning. In *Summer School on Machine Learning*; Springer: Cham, Switzerland, 2004; pp. 63–71. [CrossRef]
76. Zimmermann, H.-J. Fuzzy set theory. *Wiley Interdiscip. Rev. Comput. Stat.* **2010**, *2*, 317–332. [CrossRef]
77. Gupta, M.; Kiszka, J. *Fuzzy Sets, Fuzzy Logic, and Fuzzy Systems*; Elsevier BV: Amsterdam, The Netherlands, 2003; pp. 355–367.
78. Fişne, A.; Kuzu, C.; Hüdaverdi, T. Prediction of environmental impacts of quarry blasting operation using fuzzy logic. *Environ. Monit. Assess.* **2010**, *174*, 461–470. [CrossRef] [PubMed]
79. Monjezi, M.; Rezaei, M.; Yazdian, A. Prediction of backbreak in open-pit blasting using fuzzy set theory. *Expert Syst. Appl.* **2010**, *37*, 2637–2643. [CrossRef]
80. Khandelwal, M.; Singh, T. Prediction of blast induced ground vibrations and frequency in opencast mine: A neural network approach. *J. Sound Vib.* **2006**, *289*, 711–725. [CrossRef]
81. Yan, Y.; Hou, X.; Fei, H. Review of predicting the blast-induced ground vibrations to reduce impacts on ambient urban communities. *J. Clean. Prod.* **2020**, *260*, 121135. [CrossRef]
82. Amnieh, H.B.; Mozdianfard, M.; Siamaki, A. Predicting of blasting vibrations in Sarcheshmeh copper mine by neural network. *Saf. Sci.* **2010**, *48*, 319–325. [CrossRef]
83. Rajabi, A.M.; Vafaee, A. Prediction of blast-induced ground vibration using empirical models and artificial neural network (Bakhtiari Dam access tunnel, as a case study). *J. Vib. Control.* **2020**, *26*, 520–531. [CrossRef]

84. Dehghani, H.; Ataee-Pour, M. Development of a model to predict peak particle velocity in a blasting operation. *Int. J. Rock Mech. Min. Sci.* **2011**, *48*, 51–58. [CrossRef]
85. Taqieddin, S.A.; Ash, R.; Smith, N.; Brinkmann, J. Effects of some blast design parameters on ground vibrations at short scaled distances. *Min. Sci. Technol.* **1991**, *12*, 167–178. [CrossRef]
86. Rathore, S.; Jain, S.; Parik, S. *Comparison of two near-field blast vibration estimation models: A theoretical study. Rock Fragmentation by Blasting*; CRC Press: Leiden, The Netherlands, 2012; pp. 485–492.
87. Álvarez-Vigil, A.; González-Nicieza, C.; Gayarre, F.L.; Álvarez-Fernández, M. Predicting blasting propagation velocity and vibration frequency using artificial neural networks. *Int. J. Rock Mech. Min. Sci.* **2012**, *55*, 108–116. [CrossRef]
88. Görgülü, K.; Arpaz, E.; Demirci, A.; Koçaslan, A.; Dilmaç, M.K.; Yüksek, A.G. Investigation of blast-induced ground vibrations in the Tülü boron open pit mine. *Bull. Int. Assoc. Eng. Geol.* **2013**, *72*, 555–564. [CrossRef]
89. Lawal, A.I.; Idris, M.A. An artificial neural network-based mathematical model for the prediction of blast-induced ground vibrations. *Int. J. Environ. Stud.* **2019**, *77*, 318–334. [CrossRef]
90. Monjezi, M.; Bahrami, A.; Varjani, A.Y. Simultaneous prediction of fragmentation and flyrock in blasting operation using artificial neural networks. *Int. J. Rock Mech. Min. Sci.* **2010**, *47*, 476–480. [CrossRef]
91. Singh, T.N.; Kanchan, R.; Saigal, K.; Verma, A.K. Prediction of p-wave velocity and anisotropic property of rock using artificial neural network technique. *J. Sci. Ind. Res.* **2004**, *63*, 32–38.
92. Khandelwal, M.; Singh, T. Prediction of blast-induced ground vibration using artificial neural network. *Int. J. Rock Mech. Min. Sci.* **2009**, *46*, 1214–1222. [CrossRef]
93. Mohamed, M.T. Artificial neural network for prediction and control of blasting vibrations in Assiut (Egypt) limestone quarry. *Int. J. Rock Mech. Min. Sci.* **2009**, *46*, 426–431. [CrossRef]
94. Khandelwal, M.; Kumar, D.L.; Yellishetty, M. Application of soft computing to predict blast-induced ground vibration. *Eng. Comput.* **2009**, *27*, 117–125. [CrossRef]
95. Monjezi, M.; Bahrami, A.; Varjani, A.Y.; Sayadi, A.R. Prediction and controlling of flyrock in blasting operation using artificial neural network. *Arab. J. Geosci.* **2009**, *4*, 421–425. [CrossRef]
96. Arthur, C.K.; Temeng, V.A.; Ziggah, Y.Y. Soft computing-based technique as a predictive tool to estimate blast-induced ground vibration. *J. Sustain. Min.* **2019**, *18*, 287–296. [CrossRef]
97. Xue, X.; Yang, X. Predicting blast-induced ground vibration using general regression neural network. *J. Vib. Control.* **2013**, *20*, 1512–1519. [CrossRef]
98. Nguyen, H.; Bui, X.-N.; Bui, H.-B.; Mai, N.-L. A comparative study of artificial neural networks in predicting blast-induced air-blast overpressure at Deo Nai open-pit coal mine, Vietnam. *Neural Comput. Appl.* **2020**, *32*, 3939–3955. [CrossRef]
99. Dreiseitl, S.; Ohno-Machado, L. Logistic regression and artificial neural network classification models: A methodology review. *J. Biomed. Inform.* **2002**, *35*, 352–359. [CrossRef]
100. Piotrowski, A.P.; Napiorkowski, J. A comparison of methods to avoid overfitting in neural networks training in the case of catchment runoff modelling. *J. Hydrol.* **2013**, *476*, 97–111. [CrossRef]
101. Qiu, Y.; Zhou, J.; Khandelwal, M.; Yang, H.; Yang, P.; Li, C. Performance evaluation of hybrid WOA-XGBoost, GWO-XGBoost and BO-XGBoost models to predict blast-induced ground vibration. *Eng. Comput.* **2021**, 1–18. [CrossRef]
102. Paneiro, G.; Durão, F.O.; Silva, M.C.E.; Bernardo, P.A. Neural network approach based on a bilevel optimization for the prediction of underground blast-induced ground vibration amplitudes. *Neural Comput. Appl.* **2019**, *32*, 5975–5987. [CrossRef]
103. Amiri, M.; Hasanipanah, M.; Amnieh, H.B. Predicting ground vibration induced by rock blasting using a novel hybrid of neural network and itemset mining. *Neural Comput. Appl.* **2020**, *32*, 14681–14699. [CrossRef]
104. Shi, X.-Z.; Zhou, J.; Wu, B.-B.; Huang, D.; Wei, W. Support vector machines approach to mean particle size of rock fragmentation due to bench blasting prediction. *Trans. Nonferrous Met. Soc. China* **2012**, *22*, 432–441. [CrossRef]
105. Hasanipanah, M.; Monjezi, M.; Shahnazar, A.; Armaghani, D.J.; Farazmand, A. Feasibility of indirect determination of blast induced ground vibration based on support vector machine. *Measurement* **2015**, *75*, 289–297. [CrossRef]
106. Dindarloo, S. Peak particle velocity prediction using support vector machines: A surface blasting case study. *J. South. Afr. Inst. Min. Metall.* **2015**, *115*, 637–643. [CrossRef]
107. Khandelwal, M. Evaluation and prediction of blast-induced ground vibration using support vector machine. *Int. J. Rock Mech. Min. Sci.* **2010**, *47*, 509–516. [CrossRef]
108. Khandelwal, M. Blast-induced ground vibration prediction using support vector machine. *Eng. Comput.* **2011**, *27*, 193–200. [CrossRef]
109. Verma, A.K.; Singh, T.N. Comparative study of cognitive systems for ground vibration measurements. *Neural Comput. Appl.* **2012**, *22*, 341–350. [CrossRef]
110. Temeng, V.A.; Arthur, C.K.; Ziggah, Y.Y. Suitability assessment of different vector machine regression techniques for blast-induced ground vibration prediction in Ghana. *Model. Earth Syst. Environ.* **2021**, 1–13. [CrossRef]
111. Mohammadnejad, M.; Gholami, R.; Ramezanzadeh, A.; Jalali, M.E. Prediction of blast-induced vibrations in limestone quarries using Support Vector Machine. *J. Vib. Control.* **2012**, *18*, 1322–1329. [CrossRef]
112. Fattahi, H.; Hasanipanah, M. Prediction of Blast-Induced Ground Vibration in a Mine Using Relevance Vector Regression Optimized by Metaheuristic Algorithms. *Nat. Resour. Res.* **2021**, *30*, 1849–1863. [CrossRef]

113. Hasanipanah, M.; Naderi, R.; Kashir, J.; Noorani, S.A.; Qaleh, A.Z.A. Prediction of blast-produced ground vibration using particle swarm optimization. *Eng. Comput.* **2017**, *33*, 173–179. [CrossRef]
114. Shahnazar, A.; Rad, H.N.; Hasanipanah, M.; Tahir, M.M.; Armaghani, D.J.; Ghoroqi, M. A new developed approach for the prediction of ground vibration using a hybrid PSO-optimized ANFIS-based model. *Environ. Earth Sci.* **2017**, *76*, 527. [CrossRef]
115. Arthur, C.K.; Temeng, V.A.; Ziggah, Y.Y. Novel approach to predicting blast-induced ground vibration using Gaussian process regression. *Eng. Comput.* **2020**, *36*, 29–42. [CrossRef]
116. Zhou, J.; Asteris, P.G.; Armaghani, D.J.; Pham, B.T. Prediction of ground vibration induced by blasting operations through the use of the Bayesian Network and random forest models. *Soil Dyn. Earthq. Eng.* **2020**, *139*, 106390. [CrossRef]
117. Zhang, H.; Zhou, J.; Armaghani, D.J.; Tahir, M.M.; Pham, B.T.; Van Huynh, V. A Combination of Feature Selection and Random Forest Techniques to Solve a Problem Related to Blast-Induced Ground Vibration. *Appl. Sci.* **2020**, *10*, 869. [CrossRef]
118. Taheri, K.; Hasanipanah, M.; Golzar, S.B.; Majid, M.Z.A. A hybrid artificial bee colony algorithm-artificial neural network for forecasting the blast-produced ground vibration. *Eng. Comput.* **2016**, *33*, 689–700. [CrossRef]
119. Huang, J.; Koopialipoor, M.; Armaghani, D.J. A combination of fuzzy Delphi method and hybrid ANN-based systems to forecast ground vibration resulting from blasting. *Sci. Rep.* **2020**, *10*, 19397. [CrossRef]
120. Fouladgar, N.; Hasanipanah, M.; Amnieh, H.B. Application of cuckoo search algorithm to estimate peak particle velocity in mine blasting. *Eng. Comput.* **2016**, *33*, 181–189. [CrossRef]
121. Arthur, C.K.; Temeng, V.A.; Ziggah, Y.Y. A Self-adaptive differential evolutionary extreme learning machine (SaDE-ELM): A novel approach to blast-induced ground vibration prediction. *SN Appl. Sci.* **2020**, *2*, 1845. [CrossRef]
122. Ding, Z.; Nguyen, H.; Bui, X.-N.; Zhou, J.; Moayedi, H. Computational Intelligence Model for Estimating Intensity of Blast-Induced Ground Vibration in a Mine Based on Imperialist Competitive and Extreme Gradient Boosting Algorithms. *Nat. Resour. Res.* **2020**, *29*, 751–769. [CrossRef]
123. Shang, Y.; Nguyen, H.; Bui, X.-N.; Tran, Q.-H.; Moayedi, H. A Novel Artificial Intelligence Approach to Predict Blast-Induced Ground Vibration in Open-Pit Mines Based on the Firefly Algorithm and Artificial Neural Network. *Nat. Resour. Res.* **2020**, *29*, 723–737. [CrossRef]
124. Ding, X.; Hasanipanah, M.; Rad, H.N.; Zhou, W. Predicting the blast-induced vibration velocity using a bagged support vector regression optimized with firefly algorithm. *Eng. Comput.* **2020**, 1–12. [CrossRef]
125. Bayat, P.; Monjezi, M.; Rezakhah, M.; Armaghani, D.J. Artificial Neural Network and Firefly Algorithm for Estimation and Minimization of Ground Vibration Induced by Blasting in a Mine. *Nat. Resour. Res.* **2020**, *29*, 4121–4132. [CrossRef]
126. Chen, W.; Hasanipanah, M.; Rad, H.N.; Armaghani, D.J.; Tahir, M.M. A new design of evolutionary hybrid optimization of SVR model in predicting the blast-induced ground vibration. *Eng. Comput.* **2021**, *37*, 1455–1471. [CrossRef]
127. Hasanipanah, M.; Shahnazar, A.; Amnieh, H.B.; Armaghani, D.J. Prediction of air-overpressure caused by mine blasting using a new hybrid PSO–SVR model. *Eng. Comput.* **2017**, *33*, 23–31. [CrossRef]
128. Khandelwal, M.; Singh, T.N. Prediction of Blast Induced Air Overpressure in Opencast Mine. *Noise Vib. Worldw.* **2005**, *36*, 7–16. [CrossRef]
129. Mohamed, M.T. Performance of fuzzy logic and artificial neural network in prediction of ground and air vibrations. *Int. J. Rock Mech. Min. Sci.* **2011**, *48*, 845–851. [CrossRef]
130. Khandelwal, M.; Kankar, P.K. Prediction of blast-induced air overpressure using support vector machine. *Arab. J. Geosci.* **2009**, *4*, 427–433. [CrossRef]
131. Mohamad, E.T.; Armaghani, D.J.; Hasanipanah, M.; Murlidhar, B.R.; Alel, M.N.A. Estimation of air-overpressure produced by blasting operation through a neuro-genetic technique. *Environ. Earth Sci.* **2016**, *75*, 174. [CrossRef]
132. Armaghani, D.J.; Hasanipanah, M.; Mahdiyar, A.; Majid, M.Z.A.; Amnieh, H.B.; Tahir, M.M.D. Airblast prediction through a hybrid genetic algorithm-ANN model. *Neural Comput. Appl.* **2018**, *29*, 619–629. [CrossRef]
133. Hajihassani, M.; Armaghani, D.J.; Sohaei, H.; Mohamad, E.T.; Marto, A. Prediction of airblast-overpressure induced by blasting using a hybrid artificial neural network and particle swarm optimization. *Appl. Acoust.* **2014**, *80*, 57–67. [CrossRef]
134. AminShokravi, A.; Eskandar, H.; Derakhsh, A.M.; Rad, H.N.; Ghanadi, A. The potential application of particle swarm optimization algorithm for forecasting the air-overpressure induced by mine blasting. *Eng. Comput.* **2018**, *34*, 277–285. [CrossRef]
135. Armaghani, D.J.; Hajihassani, M.; Marto, A.; Faradonbeh, R.S.; Mohamad, E.T. Prediction of blast-induced air overpressure: A hybrid AI-based predictive model. *Environ. Monit. Assess.* **2015**, *187*, 666. [CrossRef]
136. Nguyen, H.; Bui, X.-N.; Tran, Q.-H.; Van Hoa, P.; Nguyen, D.-A.; Hoang, B.B.; Le, Q.-T.; Do, N.-H.; Bao, T.D.; Bui, H.-B.; et al. A comparative study of empirical and ensemble machine learning algorithms in predicting air over-pressure in open-pit coal mine. *Acta Geophys.* **2020**, *68*, 325–336. [CrossRef]
137. The Institute of Makers of Explosives. *Glossary of Commercial Explosives Industry Terms*; IME: Washington, DC, USA, 1997.
138. Amini, H.; Gholami, R.; Monjezi, M.; Torabi, S.R.; Zadhesh, J. Evaluation of flyrock phenomenon due to blasting operation by support vector machine. *Neural Comput. Appl.* **2011**, *21*, 2077–2085. [CrossRef]
139. Stojadinovic, S.; Pantovic, R.; Zikic, M. Prediction of flyrock trajectories for forensic applications using ballistic flight equations. *Int. J. Rock Mech. Min. Sci.* **2011**, *48*, 1086–1094. [CrossRef]
140. Li, D.; Koopialipoor, M.; Armaghani, D.J. A Combination of Fuzzy Delphi Method and ANN-based Models to Investigate Factors of Flyrock Induced by Mine Blasting. *Nat. Resour. Res.* **2021**, *30*, 1905–1924. [CrossRef]

141. Manoj, K.; Monjezi, M. Prediction of flyrock in open pit blasting operation using machine learning method. *Int. J. Min. Sci. Technol.* **2013**, *23*, 313–316. [CrossRef]
142. Armaghani, D.J.; Hasanipanah, M.; Amnieh, H.B.; Mohamad, E.T. Feasibility of ICA in approximating ground vibration resulting from mine blasting. *Neural Comput. Appl.* **2018**, *29*, 457–465. [CrossRef]
143. Hasanipanah, M.; Armaghani, D.J.; Amnieh, H.B.; Majid, M.Z.A.; Tahir, M.M.D. Application of PSO to develop a powerful equation for prediction of flyrock due to blasting. *Neural Comput. Appl.* **2017**, *28*, 1043–1050. [CrossRef]
144. Lu, X.; Hasanipanah, M.; Brindhadevi, K.; Amnieh, H.B.; Khalafi, S. ORELM: A Novel Machine Learning Approach for Prediction of Flyrock in Mine Blasting. *Nat. Resour. Res.* **2020**, *29*, 641–654. [CrossRef]
145. Armaghani, D.J.; Koopialipoor, M.; Bahri, M.; Hasanipanah, M.; Tahir, M.M. A SVR-GWO technique to minimize flyrock distance resulting from blasting. *Bull. Int. Assoc. Eng. Geol.* **2020**, *79*, 1–17. [CrossRef]
146. Murlidhar, B.R.; Kumar, D.; Armaghani, D.J.; Mohamad, E.T.; Roy, B.; Pham, B.T. A Novel Intelligent ELM-BBO Technique for Predicting Distance of Mine Blasting-Induced Flyrock. *Nat. Resour. Res.* **2020**, *29*, 4103–4120. [CrossRef]
147. Dehghani, H.; Pourzafar, M.; Zadeh, M.A. Prediction and minimization of blast-induced flyrock using gene expression programming and cuckoo optimization algorithm. *Environ. Earth Sci.* **2021**, *80*, 12. [CrossRef]
148. Vasović, D.; Kostić, S.; Ravilić, M.; Trajković, S. Environmental impact of blasting at Drenovac limestone quarry (Serbia). *Environ. Earth Sci.* **2014**, *72*, 3915–3928. [CrossRef]
149. Taylor, S. Tahoe Suspends Mining at Peru Operation after Protest. Reuters. 2018. Available online: https://www.reuters.com/article/us-tahoe-resources-protest-peru/tahoe-suspends-mining-at-peru-operation-after-protest-idUSKCN1LG21G (accessed on 14 January 2021).
150. Jha, A.; Rajagopal, S.; Sahu, R.; Purushotham, T. Detection of Geological Features using Aerial Image Analysis and Machine Learning. In Proceedings of the 46th Annual Conference on Explosives & Blasting Technique, Denver, CO, USA, 26–29 January 2020; pp. 1–11.
151. Monjezi, M.; Baghestani, M.; Faradonbeh, R.S.; Saghand, M.P.; Armaghani, D.J. Modification and prediction of blast-induced ground vibrations based on both empirical and computational techniques. *Eng. Comput.* **2016**, *32*, 717–728. [CrossRef]
152. Hudaverdi, T. Application of multivariate analysis for prediction of blast-induced ground vibrations. *Soil Dyn. Earthq. Eng.* **2012**, *43*, 300–308. [CrossRef]
153. Faradonbeh, R.S.; Monjezi, M. Prediction and minimization of blast-induced ground vibration using two robust meta-heuristic algorithms. *Eng. Comput.* **2017**, *33*, 835–851. [CrossRef]
154. Koopialipoor, M.; Fallah, A.; Armaghani, D.J.; Azizi, A.; Mohamad, E.T. Three hybrid intelligent models in estimating flyrock distance resulting from blasting. *Eng. Comput.* **2019**, *35*, 243–256. [CrossRef]
155. Monjezi, M.; Mohamadi, H.A.; Barati, B.; Khandelwal, M. Application of soft computing in predicting rock fragmentation to reduce environmental blasting side effects. *Arab. J. Geosci.* **2012**, *7*, 505–511. [CrossRef]
156. Monjezi, M.; Mehrdanesh, A.; Malek, A.; Khandelwal, M. Evaluation of effect of blast design parameters on flyrock using artificial neural networks. *Neural Comput. Appl.* **2013**, *23*, 349–356. [CrossRef]
157. Guo, H.; Zhou, J.; Koopialipoor, M.; Armaghani, D.J.; Tahir, M.M. Deep neural network and whale optimization algorithm to assess flyrock induced by blasting. *Eng. Comput.* **2021**, *37*, 173–186. [CrossRef]
158. Trivedi, R.; Singh, T.; Raina, A. Prediction of blast-induced flyrock in Indian limestone mines using neural networks. *J. Rock Mech. Geotech. Eng.* **2014**, *6*, 447–454. [CrossRef]
159. Bahrami, A.; Monjezi, M.; Goshtasbi, K.; Ghazvinian, A. Prediction of rock fragmentation due to blasting using artificial neural network. *Eng. Comput.* **2010**, *27*, 177–181. [CrossRef]
160. Sayadi, A.R.; Monjezi, M.; Talebi, N.; Khandelwal, M. A comparative study on the application of various artificial neural networks to simultaneous prediction of rock fragmentation and backbreak. *J. Rock Mech. Geotech. Eng.* **2013**, *5*, 318–324. [CrossRef]
161. Enayatollahi, I.; Bazzazi, A.A.; Asadi, A. Comparison Between Neural Networks and Multiple Regression Analysis to Predict Rock Fragmentation in Open-Pit Mines. *Rock Mech. Rock Eng.* **2014**, *47*, 799–807. [CrossRef]
162. Ebrahimi, E.; Monjezi, M.; Khalesi, M.R.; Armaghani, D.J. Prediction and optimization of back-break and rock fragmentation using an artificial neural network and a bee colony algorithm. *Bull. Int. Assoc. Eng. Geol.* **2016**, *75*, 27–36. [CrossRef]
163. Monjezi, M.; Amiri, H.; Farrokhi, A.; Goshtasbi, K. Prediction of Rock Fragmentation Due to Blasting in Sarcheshmeh Copper Mine Using Artificial Neural Networks. *Geotech. Geol. Eng.* **2010**, *28*, 423–430. [CrossRef]
164. Kumar, S.; Mishra, A.K.; Choudhary, B.S. Prediction of back break in blasting using random decision trees. *Eng. Comput.* **2021**, 1–7. [CrossRef]
165. Monjezi, H.D.M. Evaluation of effect of blasting pattern parameters on back break using neural networks. *Int. J. Rock Mech. Min. Sci.* **2008**, *45*, 1446–1453. [CrossRef]
166. Temeng, V.A.; Ziggah, Y.Y.; Arthur, C.K. Blast-Induced Noise Level Prediction Model Based on Brain Inspired Emotional Neural Network. *J. Sustain. Min.* **2021**, *20*, 28–39. [CrossRef]
167. Hasanipanah, M.; Faradonbeh, R.S.; Amnieh, H.B.; Armaghani, D.J.; Monjezi, M. Forecasting blast-induced ground vibration developing a CART model. *Eng. Comput.* **2017**, *33*, 307–316. [CrossRef]
168. Ouchterlony, F.; Sanchidrián, J. A review of development of better prediction equations for blast fragmentation. *J. Rock Mech. Geotech. Eng.* **2019**, *11*, 1094–1109. [CrossRef]

169. Cunningham, C.V.B. The Kuz-Ram fragmentation model—20 years on. In *Brighton Conference proceedings*; European Federation of Explosives Engineers: Brighton, UK, 2005.
170. Ouchterlony, F. The Swebrec© function: Linking fragmentation by blasting and crushing. *Min. Technol.* **2005**, *114*, 29–44. [CrossRef]
171. Spathis, A.T. Formulae and techniques for assessing features of blast-induced fragmentation distributions. In *Fragblast 9, Proceedings of the 9th International Symposium on Rock Fragmentation by Blasting, Granada, Spain, 13–17 August 2009*; Taylor & Francis Group: London, UK, 2010.
172. Lawal, A.I. A new modification to the Kuz-Ram model using the fragment size predicted by image analysis. *Int. J. Rock Mech. Min. Sci.* **2021**, *138*, 104595. [CrossRef]
173. An, H.; Song, Y.; Liu, H.; Han, H. Combined Finite-Discrete Element Modelling of Dynamic Rock Fracture and Fragmentation during Mining Production Process by Blast. *Shock. Vib.* **2021**, *2021*, 6622926. [CrossRef]
174. Tao, J.; Yang, X.-G.; Li, H.-T.; Zhou, J.-W.; Qi, S.-C.; Lu, G.-D. Numerical investigation of blast-induced rock fragmentation. *Comput. Geotech.* **2020**, *128*, 103846. [CrossRef]
175. Fu, Y.; Aldrich, C. Deep Learning in Mining and Mineral Processing Operations: A Review. *IFAC-PapersOnLine* **2020**, *53*, 11920–11925. [CrossRef]
176. Andersson, T.; Thurley, M.; Carlson, J. A machine vision system for estimation of size distributions by weight of limestone particles. *Miner. Eng.* **2012**, *25*, 38–46. [CrossRef]
177. Esmaeili, M.; Salimi, A.; Drebenstedt, C.; Abbaszadeh, M.; Bazzazi, A.A. Application of PCA, SVR, and ANFIS for modeling of rock fragmentation. *Arab. J. Geosci.* **2015**, *8*, 6881–6893. [CrossRef]
178. Dunford, J. *Control and Prediction of Blast Fragmentation and its Impact on Comminution*; University of Exeter: Exeter, UK, 2016.
179. Monjezi, M.; Rezaei, M.; Varjani, A.Y. Prediction of rock fragmentation due to blasting in Gol-E-Gohar iron mine using fuzzy logic. *Int. J. Rock Mech. Min. Sci.* **2009**, *46*, 1273–1280. [CrossRef]
180. Shams, S.; Monjezi, M.; Majd, V.J.; Armaghani, D.J. Application of fuzzy inference system for prediction of rock fragmentation induced by blasting. *Arab. J. Geosci.* **2015**, *8*, 10819–10832. [CrossRef]
181. Zhou, J.; Li, C.; Arslan, C.A.; Hasanipanah, M.; Amnieh, H.B. Performance evaluation of hybrid FFA-ANFIS and GA-ANFIS models to predict particle size distribution of a muck-pile after blasting. *Eng. Comput.* **2021**, *37*, 265–274. [CrossRef]
182. Iphar, M.; Yavuz, M.; Ak, H. Prediction of ground vibrations resulting from the blasting operations in an open-pit mine by adaptive neuro-fuzzy inference system. *Environ. Earth Sci.* **2007**, *56*, 97–107. [CrossRef]
183. Hasanipanah, M.; Amnieh, H.B.; Arab, H.; Zamzam, M.S. Feasibility of PSO–ANFIS model to estimate rock fragmentation produced by mine blasting. *Neural Comput. Appl.* **2018**, *30*, 1015–1024. [CrossRef]
184. Sayevand, K.; Arab, H. A fresh view on particle swarm optimization to develop a precise model for predicting rock fragmentation. *Eng. Comput.* **2019**, *36*, 533–550. [CrossRef]
185. Zhang, S.; Bui, X.-N.; Trung, N.-T.; Nguyen, H.; Bui, H.-B. Prediction of Rock Size Distribution in Mine Bench Blasting Using a Novel Ant Colony Optimization-Based Boosted Regression Tree Technique. *Nat. Resour. Res.* **2019**, *29*, 867–886. [CrossRef]
186. Gao, W.; Karbasi, M.; Hasanipanah, M.; Zhang, X.; Guo, J. Developing GPR model for forecasting the rock fragmentation in surface mines. *Eng. Comput.* **2018**, *34*, 339–345. [CrossRef]
187. Xie, C.; Nguyen, H.; Bui, X.-N.; Choi, Y.; Zhou, J.; Nguyen-Trang, T. Predicting rock size distribution in mine blasting using various novel soft computing models based on meta-heuristics and machine learning algorithms. *Geosci. Front.* **2021**, *12*, 101108. [CrossRef]
188. Amiri, M.; Amnieh, H.B.; Hasanipanah, M.; Khanli, L.M. A new combination of artificial neural network and K-nearest neighbors models to predict blast-induced ground vibration and air-overpressure. *Eng. Comput.* **2016**, *32*, 631–644. [CrossRef]
189. Hajihassani, M.; Armaghani, D.J.; Monjezi, M.; Mohamad, E.T.; Marto, A. Blast-induced air and ground vibration prediction: A particle swarm optimization-based artificial neural network approach. *Environ. Earth Sci.* **2015**, *74*, 2799–2817. [CrossRef]
190. Asl, P.F.; Monjezi, M.; Hamidi, J.K.; Armaghani, D.J. Optimization of flyrock and rock fragmentation in the Tajareh limestone mine using metaheuristics method of firefly algorithm. *Eng. Comput.* **2018**, *34*, 241–251. [CrossRef]
191. Trivedi, R.; Singh, T.; Raina, A. Simultaneous prediction of blast-induced flyrock and fragmentation in opencast limestone mines using back propagation neural network. *Int. J. Min. Miner. Eng.* **2016**, *7*, 237. [CrossRef]
192. Ajiboye, A.R.; Abdullah-Arshah, R.; Qin, H.; Isah-Kebbe, H. Evaluating the effect of dataset size on predictive model using supervised learning technique. *Int. J. Softw. Eng. Comput. Syst.* **2015**, *1*, 75–84. [CrossRef]
193. Ganguli, R.; Dagdelen, K.; Grygiel, E. System Engineering. In *SME Mining Engineering Handbook*; Darling, P., Ed.; Society for Mining Metallurgy: Englewood, CO, USA, 2011; Volume 1, pp. 839–853.

 *minerals*

Article

Study of the Influence of Non-Deposit Locations in Data-Driven Mineral Prospectivity Mapping: A Case Study on the Iskut Project in Northwestern British Columbia, Canada

Alix Lachaud [1], Adam Marcus [2], Slobodan Vučetić [3] and Ilija Mišković [1,*,†]

1   Norman B. Keevil Institute of Mining Engineering, Faculty of Applied Science, University of British Columbia, Vancouver, BC V6T 1Z4, Canada; alix.lachaud@alumni.ubc.ca
2   Seabridge Gold Inc., Toronto, ON M5A 1E1, Canada; marcus@seabridgegold.net
3   Department of Computer and Information Sciences, Temple University, Philadelphia, PA 19122, USA; vucetic@temple.edu
*   Correspondence: eli.miskovic@ubc.ca
†   Current address: 517-6350 Stores Road, Vancouver, BC V6T 1Z4, Canada.

Abstract: The accuracy of data-driven predictive mineral prospectivity models relies heavily on the training datasets used. These models are usually trained using data for "known" deposit locations as well as "non-deposit" locations that are based on randomly generated point patterns. In this study, data related to the Seabridge Gold Inc Iskut project, an epithermal Au deposit in northwestern British Columbia (BC), Canada, are used to test the utility of data-driven mineral prospectivity modeling. The input spatial dataset is comprised mostly of publicly available data. Data for 18 vein and epithermal Au known mineral occurrences (KMO) are obtained from the BC Geological Survey's MINFILE repository and selected as training deposit locations. A total of eleven sets of non-deposit locations (NDL) were also created, including one set of selected non-prospective KMO for Au deposits from the MINFILE and ten sets of random point patterns. Given the scale of this study, most of the KMO recorded on the property are of the epithermal deposit type. Hence, they could not be used as a selection criterion. Data-driven mineral potential models are generated using the random forest (RF) algorithm and trained on multiple data sets. The comparison of RF models demonstrated that using non-prospective KMO generates more accurate predictions than the random point pattern. The produced mineral prospectivity maps delineated multiple areas with higher discovery potential, which matched viable targets for the Au-Cu epithermal-porphyry system identified through previous Seabridge Gold Inc. (Toronto, ON, Canada) field reconnaissance and drilling programs.

Keywords: mineral prospectivity mapping; random forest algorithm; machine learning; epithermal gold; unstructured data

check for
updates

Citation: Lachaud, A.; Adam, M.; Vucetic, S.; Mišković, I. Study of the Influence of Non-Deposit Locations in Data-Driven Mineral Prospectivity Mapping: A Case Study on the Iskut Project in Northwestern British Columbia, Canada. *Minerals* **2021**, *11*, 597. https://doi.org/10.3390/min11060597

Academic Editor: Amin Beiranvand Pour

Received: 15 April 2021
Accepted: 28 May 2021
Published: 1 June 2021

Publisher's Note: MDPI stays neutral with regard to jurisdictional claims in published maps and institutional affiliations.

1. Introduction

With new mineral deposits becoming more challenging to find, geoscientists have focused on development of novel methods to assist with mineral deposit discovery. Development of the geographic information system (GIS) technology, improved computing power, and application of data-driven methods, such as machine learning, are enabling the evolution of quantitative methods of geoscientific data analysis, including mineral potential mapping (MPM) [1,2]. For instance, in 2018, Goldcorp Inc., Vancouver, BC, Canada (now part of the Newmont Corporation) and IBM announced a partnership with a goal to utilize the IBM Watson supercomputer and its artificial intelligence (AI) framework to aid mineral targeting at the Red Lake Mines in northwestern Ontario, Canada.

MPM consists of combining multiple layers of geoscience data into a map identifying areas favorable for mineral exploration. The process can be summarized into five main steps: definition of the exploration model for the type of deposit sought, selection of the geospatial dataset to be used, processing of data, creation of the predictor maps, integration of the predictor maps to create a predictive model [1,2].

There are three main categories of modeling methods using GIS-environment: knowledge-driven, data-driven and hybrid.

The knowledge-driven approach requires "experts" to assign weights and to assess the relative importance of each evidential layer as they relate to the specific exploration model being used. This method is subjective but has the advantage of being well suited for greenfield areas with missing or scarce data and where few deposits are known [3]. Examples of knowledge-driven approach include Boolean logic [1,2,4], index overlays [1,2,4,5], fuzzy logic [6–10] and evidential belief [2,6].

The data-driven approach uses the spatial relationship between the geospatial features and known mineral deposits (training set) to estimate the model parameters. It is well suited for well-established mining areas where a large number of known mineral deposits is available to quantify the spatial associations with evidential features and to guarantee the performance and robustness of the model. Methods such as weights of evidence [4,7,11–13], logistic regression [11,12], neural networks [11,14,15], support vector machine [15–18] and random forest [3,13,19–22] are examples of data-driven approaches.

Since 2000, the number of publications on data-driven methods for MPM has largely increased, especially articles using machine learning algorithms (MLA) [23]. In recent years, the random forest (RF) algorithm has proven to offer a new approach to MPM. Contrarily to other MLA, like artificial neural networks or support vector machine, RF is not a 'black-box' algorithm, meaning that the inner workings of the algorithm are known and can even be represented (i.e., decision trees). It is also simpler to parameterize, more stable and computationally light [15,22,24,25].

For any data-driven method, the training dataset should contain a sufficient number of samples to train a given model, and studies showed that RF can be accurate even with a small training set (i.e., less than 20 deposit locations) [19,20,24,26,27]. Moreover, the training dataset should be balanced, meaning that the dataset must contain an equal number of deposits and non-deposit, to avoid results to be biased for one class or the other [19]. Training deposit locations are usually discovered deposits or known mineral occurrences (KMO) in the study area for the commodity and the type of deposit sought. On the other hand, non-deposit locations are usually generated by random points following specific criteria [19,20,24,25,27] or random locations in lithologies considered unprospective [3,15,22,26].

In this paper, the relative influence of the non-deposit locations in the training dataset is assessed by the accuracy of the RF model to MPM. MPM models using the RF algorithm were created with different training data set. We generated ten sets of the random point pattern using a three criteria selection that we compared with KMO that were categorized as non-prospective and distal to every deposit location. In a broader sense, this study is testing whether non-prospective KMO (for one commodity but can be of similar deposit type) should be preferably used instead of randomly generated locations in MPM, when such data set is available.

2. RF Algorithm

The RF algorithm is a collection of decision tree classifiers trained to increase their diversity and reduce generalization error of the aggregate classifier made of the individual trees [28]:

$$\{h(x, \theta_k), k = 1, \dots \},\tag{1}$$

where the $\{\theta_k\}$ are independent identically distributed random vectors, and each tree casts a unit vote for the most popular class at input x. A RF can be composed of either classification or regression trees.

The algorithm uses a modified version of the bagging (or bootstrap aggregating) technique to create an ensemble of n_{tree} decision trees [29]. This technique increases the diversity of the trees. In the bagging process, each tree is trained on 2/3 of the input samples. The training set is sampled randomly from the original dataset with replacement (i.e., no deletion of the data selected from the original dataset for the generation of the next subset). In other words, to grow a tree, the input data can be used more than once or not at all. This allows the RF to be more stable and robust to outliers in the input data set, as well as increasing prediction accuracy [28].

The remaining 1/3 of the training samples are referred to as out-of-bag (OOB) samples. The OOB samples can be used to evaluate performance, removing the need for cross-validation. The resulting OOB error is an unbiased estimate of the generalization error and converges as the number of trees increases; thus, RF does not over-fit the data [28].

Each tree is grown on a random subset of m_{try} features selected from the input evidential features. This increases the diversity of trees within the forest and reduces the correlations between the trees. The RF algorithm does not apply pruning on the grown trees. The output of the RF is calculated differently depending on the type of decision trees. For regression trees, the output is the average of the predictions from all the trees, whereas for classification trees, the output is the majority vote of all the trees. A simplified diagram of the RF algorithm is presented in Figure 1.

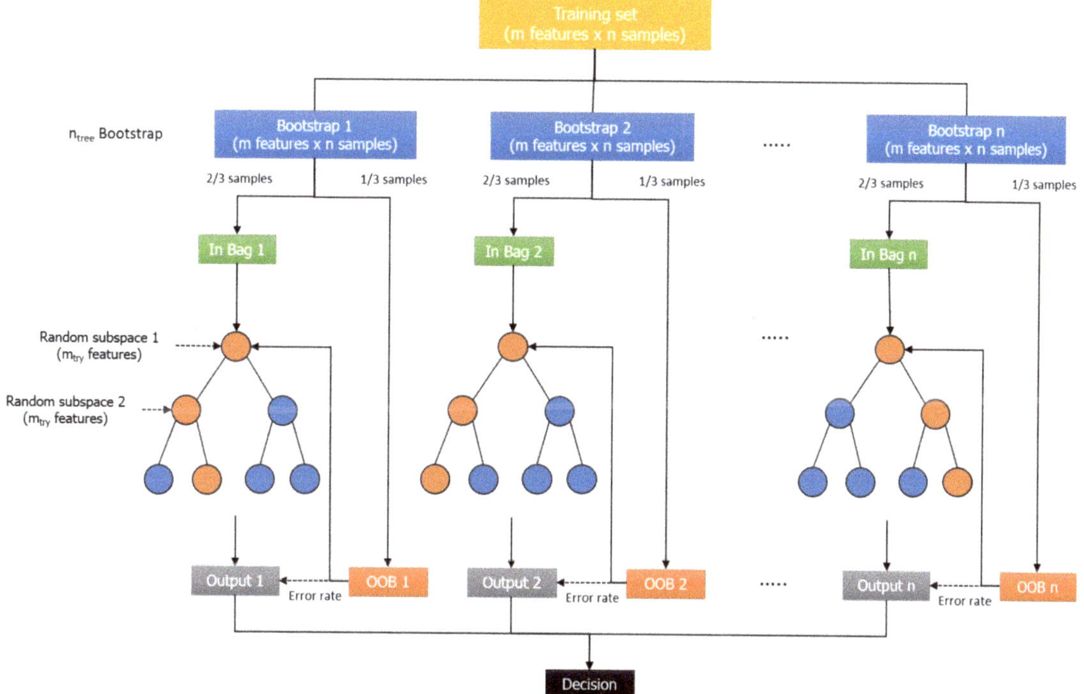

Figure 1. Workflow of the random forest algorithm.

The RF algorithm tries to maximize purity of the tree grown by making each child nodes 'purer' than the parent node. The tree impurity $I(T)$ is defined by [30]:

$$I(T) = \sum_{t \in T} I(t), \qquad (2)$$

where $I(t) = i(t)p(t)$ with $p(t)$ an impurity function and $i(t)$ the node impurity function.

The decision tree search through all candidate splits to find the optimal split to reduce $I(T)$ or, equivalently, maximizes the information gain [30]:

$$\Delta I(s, t) = I(t) - I(t_L) - I(t_R), \qquad (3)$$

where $I(t)$ is the impurity of the parent node, $I(t_L)$ is the impurity of the left leaf and $I(t_R)$ is the impurity of the right leaf.

The Gini Index (I_G) is the impurity function employed in this study. It is defined as [30]:

$$I_G = 1 - \sum_{j=1}^{m} p_j^2, \tag{4}$$

where p_j is the proportion of samples that belongs to class m for a particular node and m the number of classes. The decision tree splitting criterion is based on choosing the attribute with the lowest Gini impurity index (I_G).

The importance of each feature can be evaluated by the RF algorithm. To measure the importance of the k-th feature, the values of the k-th feature are permuted among the training data while keeping the rest constant. The OOB error estimation is used to measure the decrease in accuracy [28].

3. Study Area: Iskut Project

The Iskut project has undergone mineral exploration since the early 1900s. Since the discovery of the Snip Mine in 1964 (Skeena Resources: https://www.skeenaresources.com/projects/snip; accessed on 21 October 2019), the property has had relatively systematic exploration, which has led to the discovery of the Johnny Mountain Mine and the definition of the Bronson Slope deposit. These discoveries, in conjunction with surface anomalies across the property, have seen over C\$38 million spent on exploration looking for analogs to these deposits. Seabridge Gold acquired the property in 2016 and has since undertaken exploration for porphyry and epithermal deposits on the property.

3.1. Geological Setting

The Iskut property is located in northwestern British Columbia (BC), in the metallogenically important Stewart-Iskut River area also known as the "Golden Triangle.

The Iskut Project lies on the western margin of the Stikine Terrane (Stikinia). Three distinct units of the Stikine Terrane ranging in age from Upper Paleozoic to Middle Jurassic were recognized in the area (Figure 2). The oldest rocks are Upper Paleozoic metamorphosed and deformed limestone, clastic sedimentary rocks, and polymodal volcanic rocks of the Stikine Assemblage [31,32]. Two groups of the Mesozoic arc-related strata are present in the area: Late Triassic folded marine volcanic and sedimentary arc-related strata with some degree of alteration and low-grade metamorphism of the Stuhini Group and the Early to Middle Jurassic subaerial and submarine volcanic and sedimentary rocks of the Hazelton Group [31]. The two units are separated by a regional angular unconformity. The Bowser Lake basin sedimentary rocks unconformably overly the Jurassic strata and cover mineral deposits to the East of the study area. Quaternary basalts and local volcanism are observed to the North and Northeast of the property along the Iskut river.

There are three major intrusive suites mapped on the property. Two of these are major regional metallogenic events that occur in Stikinia during the late Triassic and over an extended period from Early to Middle Jurassic [31,33]. The Stikine Plutonic Suite emerged from a pulse of arc growth in the Late Triassic (221–236 Ma; [33]) and is coeval with the Stuhini Group strata [31,33]. This intrusion is coincident with emplacement of Cu-Au-Ag enriched pluton, an important metallogenic event within the Cordillera [31,33].

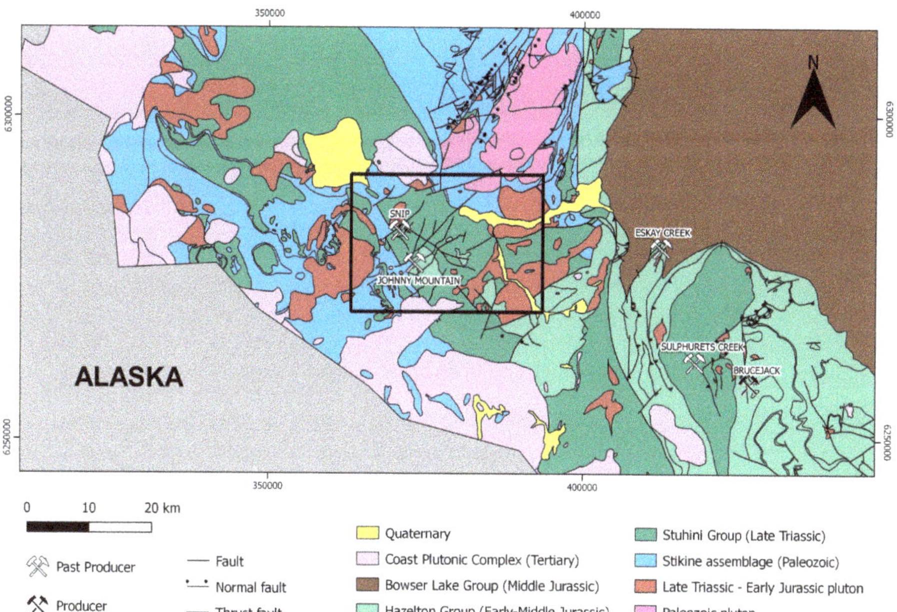

Figure 2. Regional geology of the Iskut River area with the study area outlined in black from BC digital geology data repository.

Two magmatic events occurred between Early to Middle Jurassic and are associated with two different mineralization styles. The first event, in Early Jurassic (187–195 Ma; [33]), is associated with porphyry deposits (e.g., Kerr-Sulphurets-Mitchell, Galore Creek and Red Chris) and epithermal gold-veins (e.g., Brucejack) that tend to occur in lower Hazelton Group volcanic rocks [31,33]. The second event is associated with exhalative mineralization (Eskay Camp deposits) and characterized by bimodal volcanic rocks coeval with the upper volcanic sequence of the Hazelton Group [31]. These form as exhalative deposits within a deep marine oceanic crust setting.

3.2. Au Mineralization

The Iskut project lies in the metallogenic rich "Golden Triangle" area. Two formerly producing mines and a well defined mining project are located in the study area: the Johnny Mountain mine, a vein hosted gold deposit with a production of 92,300 oz gold, 145,000 oz silver, 2,270,000 lbs copper, the Snip mine, a shear-vein gold deposit with a production of 1,032,000 oz gold, 390,000 oz silver, 550,000 lbs copper and the copper-gold Bronson Slope deposit (190Mt@0.36 g/t Au, 0.122% Cu) (Richards, 2005, unpublished).

Mineralization in the area is principally shear-hosted gold and base metal veins deposits like the Snip, Johnny Mountain (Stonehouse) or Inel deposits. They are associated with brittle-ductile deformation and porphyritic stock and intrusion of the Early Jurassic Texas Creek Plutonic Suite [31,33]. The structural style may be host-rock dependent: mineralized shear-veins are hosted by a clastic sequence of the Stuhini Group at the Snip and the Inel deposits, while dilatant quartz-sulfide veins are hosted by Jurassic coarse volcanic and intrusive rocks at the Stonehouse deposit (but are also present at the Snip and the Inel deposits) [34]. From a comparison between the Stonehouse and the Snip deposit, Rhys [35] suggests that intrusion, semi-brittle deformation, and a mineralizing hydrothermal system were closely related temporally and genetically and that gold was deposited during the formation of the vein and not by later mineralization or remobilization.

3.3. Conceptual Exploration Model

Since epithermal ore deposits are formed by tectonic scale earth process systems that concentrate hydrothermal fluids, the exploration model presented below accounts for processes critical to the formation of that type of deposit: source-pathway-trap (physical and/or chemical)-preservation. Based on the mineral system approach to exploration targeting, Seabridge Gold defines the conceptual exploration model at a district-scale for epithermal Au deposits at the Iskut project with the following criteria:

- Proximity to mineralized porphyritic intrusions;
- Proximity to faults;
- Presence of hydrothermal alteration zones;
- Geochemical enrichment in gold and associated pathfinder elements;
- Viability of host rock.

The most probable source of gold-bearing fluids in the area are the porphyry intrusives of Texas Creek Plutonic Suite and Stikine Plutonic Suite. Mineralization is synchronic with brittle-ductile deformation (i.e., faulting, folding, and shearing) characterized by mineralized dikes and veins having similar orientation as tectonic structures. Faults and fractures can serve two functions; they can be the both conduits taken by metal-rich fluids and physical traps to those fluids. Lithologies from the Stuhini Group and the Hazelton Group are the host lithologies of the known deposits in the area (e.g., Stonehouse, Inel, Bronson slope). They can be considered chemical traps, due to a change in RedOx conditions or geochemical assemblage for instance, as well as physical traps, because of a change in density for example. Hydrothermal alteration and geochemical enrichment in gold and associated pathfinder elements are evidence of chemical traps as the reaction of the mineralized hydrothermal fluids with wall rock.

4. Methods

4.1. Spatial Data Input

4.1.1. Target Variable

For deposit location, we selected 18 vein and epithermal gold deposit locations (i.e., past producers and prospects) from the KMO depository (https://catalogue.data.gov.bc.ca/dataset/minfile-mineral-occurrence-database), (accessed on 25 March 2019) a public online data repository hosted at Data Catalogue by the Government of British Columbia) (MINFILE) by the BC Geological Survey (See Appendix A).

For NDL, three selection criteria are used. First, all sets should have an equal number of NDL to that of the deposit locations. Second, the NDL should be located far from any known deposit to ensure different geospatial characteristics to nearby deposits [2,19]. The third criteria depend on the nature of the data: random locations or KMO.

Point pattern analysis was applied to define a buffer distance from every deposit location. That buffer represents the distance beyond which there is a 100% probability of finding another deposit from any deposit. In the study area given the selected deposits, that distance is 8000 m (Figure 3). However, this length is too restrictive for this study, as it would exclude more than half of the study area. Instead, a buffer distance of 2000 m, representing a 78% probability of finding a neighboring deposit from that distance. Hence, NDL are to be selected in the study area excluding a 2000 m buffer from every deposit location (Figure 4).

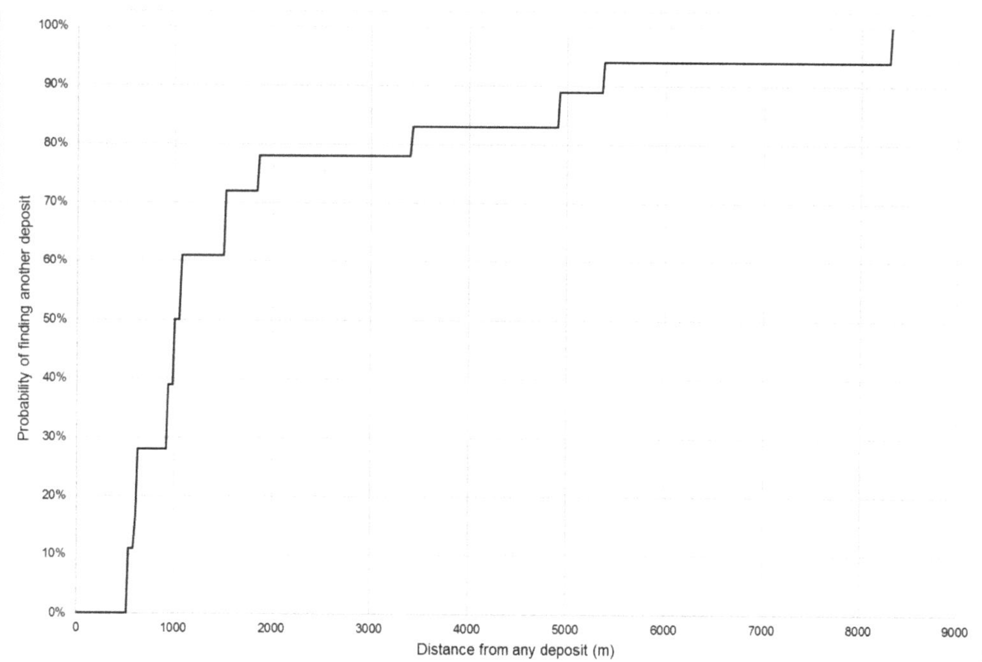

Figure 3. Result of point pattern analysis showing probability of finding another Au deposit from any given deposit for different distances.

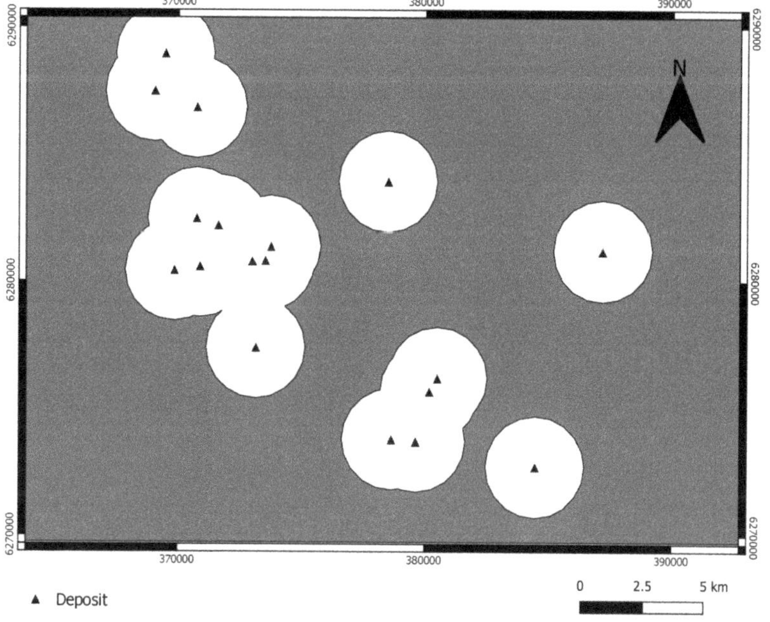

Figure 4. Location of deposits showing the buffer zone (in grey).

Unlike deposit locations, which are 'rare' events and tend to cluster, the NDL should be distributed randomly through the study area as they should not be representative of any particular geological process [2]. Therefore, we generated ten sets of the random point pattern containing 18 independent points, that we generated using the *rpoint* function from the *spatstat* (v1.61-0) library in R software (R version 3.5.3 (Great Truth), released on 11 March 2019. Retrieved from R project website: (available online: https://www.r-project.org/; accessed on 18 March 2019). The number of sets generated was chosen arbitrarily so that the different models could be easily plotted and compared. The number of points was chosen to be equal to the number of deposit locations, to obtain a balanced dataset (i.e., equal number of deposit to non-deposit location).

The second set of NDL constitutes of selected KMO from the MINFILE depository that were considered non-prospective for the Au deposit. Every occurrence that has gold listed as one of their first three listed commodities was discarded. A total of 19 locations were left after this selection process (see Appendix A). However, due to the scale of the study area (i.e., project-scale), the majority of the KMO in the area are related either to an epithermal or a porphyry system. Therefore, the deposit type could not be used as a selection criterion as it would have been too restrictive (i.e., not enough samples to conduct this study). Thus, some selected non-prospective KMO can be pre or post-dating the selected prospective KMO and have a similar geospatial signature (e.g., similar pathways, similar traps). It is assumed that the source of fluids is different, which would explain the difference in metals association between the prospective and non-prospective KMO (i.e., with or without Au).

Each sample in our training sets is attributed to a binary variable such as 1 s for prospective locations and 0 s for non-prospective location.

4.1.2. Predictor Maps

The geospatial dataset for this study is selected based on availability of the data and its usefulness to be used as proxy for our conceptual exploration model.

Geological data were derived from 1:50,000 British Columbia digital geology data compilation that was last updated on 5 April 2018 [36] (original dataset related to this article can be found at https://catalogue.data.gov.bc.ca/dataset/bedrock-geology, a public online data repository hosted at Data Catalogue by the Government of British Columbia; accessed on 25 March 2019). In the study area, Au-hydrothermal deposits are strongly correlated with intrusions from the Late Triassic-Early Jurassic and structurally controlled [34]. Therefore, we created predictor maps of distances to Texas Creek Plutonic Suite and Stikine Plutonic Suite intrusions and distance to fault traces at 500 m intervals. Moreover, reactive lithologies can act as a chemical trap for Au deposition. As such, reactive lithologies of the Stuhini Group and the Hazelton Group were categorized as 'favorable' host-rock while the other lithologies present in the study area were categorized as 'non-favorable' host-rock.

The geochemical data comes from a compiled database of soil samples from historic geochemical surveys conducted by private companies from 1981 to 2011. The elements analyzed and the analytical methods used in each survey varied and were not always adequately reported. Exploration efforts focused on areas surrounding known prospects, thus do not cover the entire study area. Only the twelve most present elements in the database were kept for further analysis: Ag, Au, As, Ba, Co, Cu, Fe, Mn, Mo, Pb, Sb, Zn. Values below the lower detection limit were replace by half the detection limit. No imputation was performed on the missing data. Hence, our geochemical dataset contains some missing data. The dataset was transformed using centered log-ratio (clr) [37]. Then, principal component analysis (PCA) is applied to the transformed dataset. Principal components can be interpreted as describing separate geological processes (i.e., differentiation, alteration, mineralization, weathering) [38]. In this study, only PC1 and PC2 were kept as the variation between the loadings of the different elements decreased as the number of principal components increased. They account for 34% and 14% of the variance respectively. Based on the loadings of the different elements (Table 1), PC1 and PC2 seem to represent potential metal associations (e.g., Ag-Au-Sb), and enrichment or depletion. The clr-transformed

Au values and the PC1 and PC2 were interpolated using the Inverse Distance Weighted (IDW) algorithm.

Table 1. Rotated component matrix of principal component analysis of clr-transformed soil samples. Significant loadings (bolded values) are based on the absolute threshold of value 0.3.

	Ag	Au	As	Ba	Co	Cu	Fe	Mn	Mo	Pb	Sb	Zn
PC1	**−0.41**	**−0.39**	0.21	0.28	0.04	0.12	−0.26	**0.35**	−0.24	0.25	−0.29	**0.39**
PC2	0.09	0.28	0	0.21	0.1	**−0.38**	**−0.38**	**0.40**	−0.15	**−0.46**	−0.29	**−0.30**

The geophysical data consists of magnetic data only, as it was the only available data in the area with a relatively small spatial resolution (200 m × 200 m) and was downloaded from the Canadian Aeromagnetic data base. Only processed data was available for download, and we chose to use the first vertical derivative as it is useful to enhance near-surface structure.

For remote sensing data, we used ASTER Level 1 Precision Terrain Corrected Registered At-Sensor Radiance (L1T) data, with a spatial resolution of 15 m, 30 m, and 90 m for the NIR, SWIR and TIR bands respectively. The SWIR and TIR band images were re-sampled to the VNIR band images resolution. The project study area is densely vegetated in the valleys but vegetation becomes scarce with altitude. The area is covered by snow from October to April with presence of multiple glaciers. In order to maximize bedrock exposure, we selected scenes acquired in summer to minimize snow cover and with minimum cloud cover (Figure 5). The ASTER data was atmospherically corrected and converted to relative reflectance using the Semi-Automatic Classification Plugin in QGIS software.

Figure 5. False color ASTER image derived from Band 2, Band 4 and Band 7 as RGB color combination showing the glacier and water (blue), vegetation (green) and exposed bedrock and sediments (yellow).

Hydrothermal alteration can effectively be mapped with ASTER data [39–41]. Band ratio (BR) and relative absorption band depth (RBD) methods are useful to enhance the absorption feature of some characteristic alteration minerals [42,43]. In this study, the following ratios to map the main alteration present in a hydrothermal system are used [44,45]:

$$Argilic = \frac{b4 + b6}{b5} \tag{5}$$

$$Ironoxides = \frac{b4}{b2} \tag{6}$$

$$Phyllic = \frac{b5 + b7}{b6} \tag{7}$$

$$Propylitic = \frac{b7 + b9}{b8} \tag{8}$$

$$Silica = \frac{b11}{b12} \tag{9}$$

A total of twelve predictor maps are generated to map the mineral potential of epithermal Au in the study area (Figures 6 and 7).

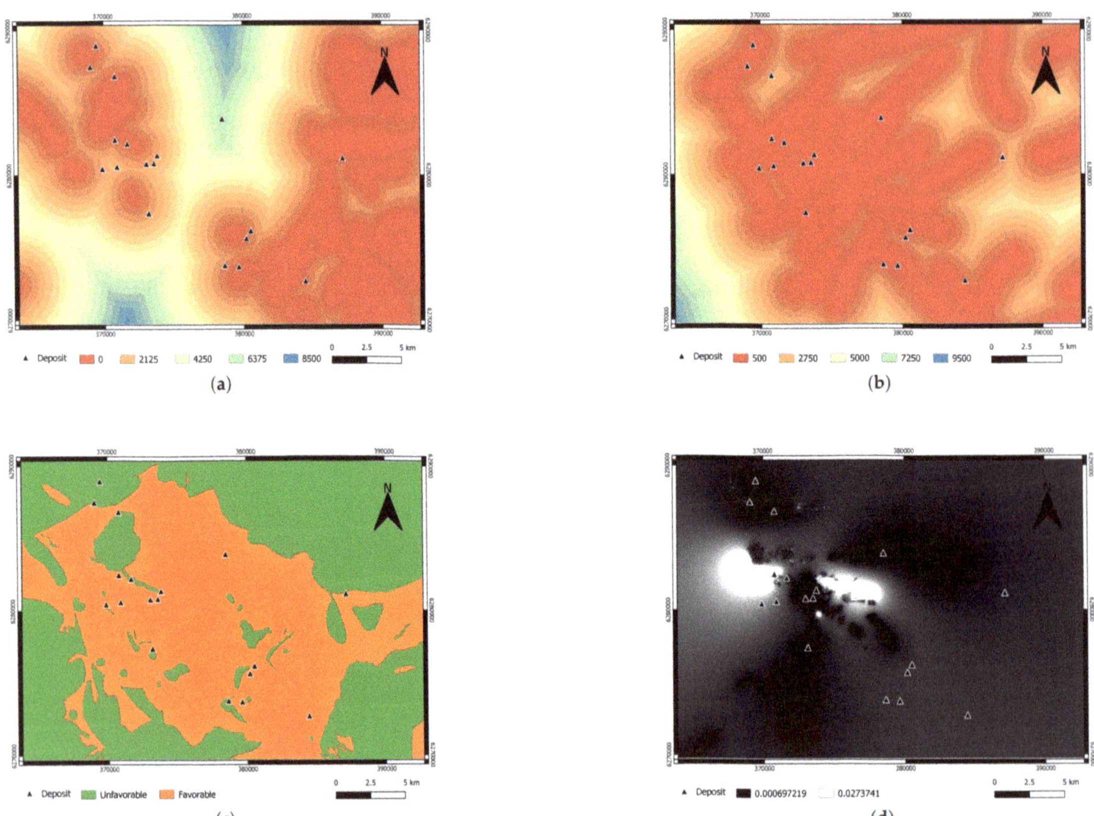

(a)

(b)

(c)

(d)

Figure 6. *Cont.*

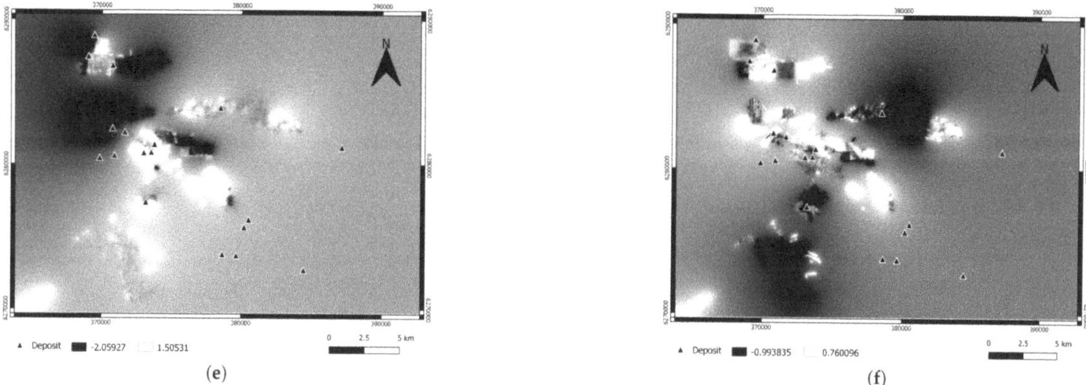

Figure 6. Predictor maps of (**a**) distance to intrusions, (**b**) distance to fault, (**c**) favorable host-rock, (**d**) Au geochemical anomaly, (**e**) PC1, and (**f**) PC2.

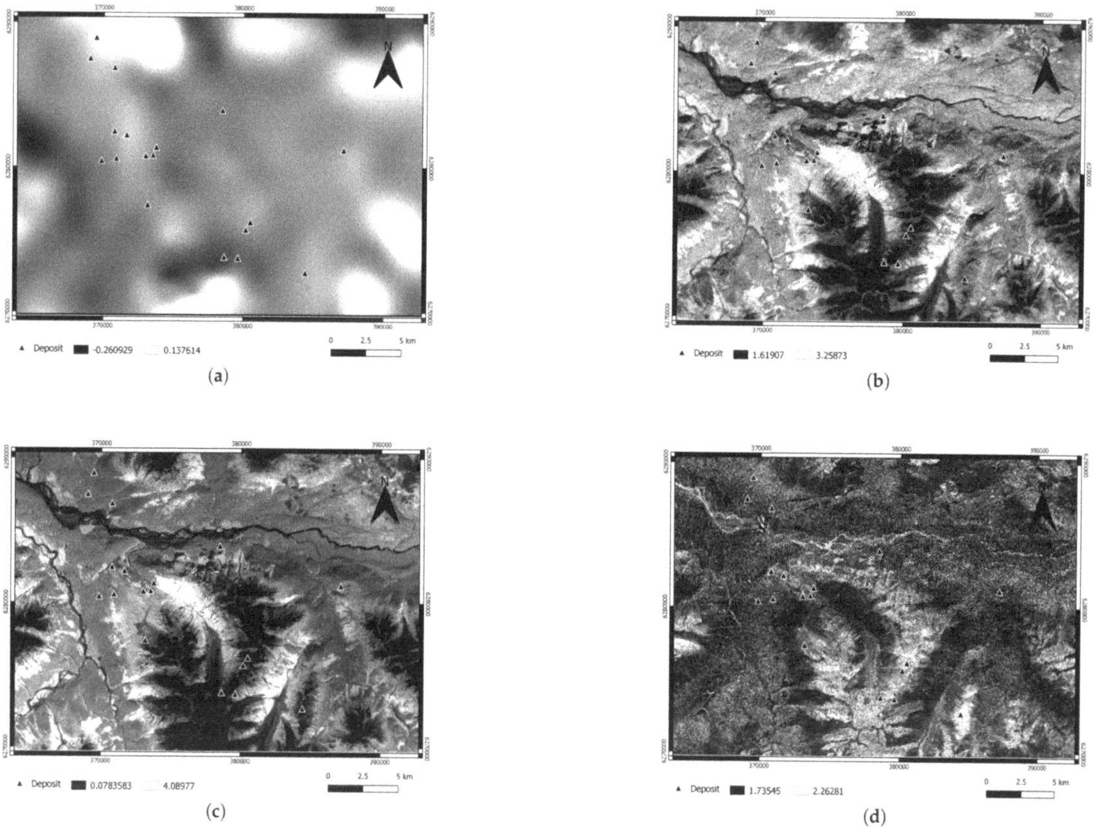

Figure 7. *Cont.*

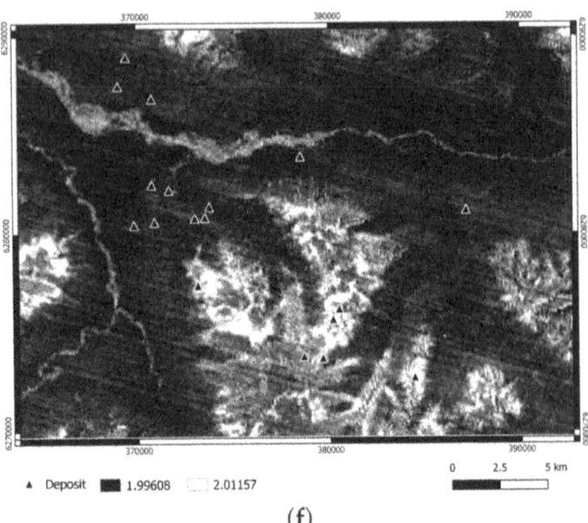

(e)　　　　　　　　　　　　　　　　　　(f)

Figure 7. Predictor maps of (**a**) magnetic first vertical derivative, (**b**) argillic alteration, (**c**) iron oxides alteration, (**d**) phyllic alteration, (**e**) propylitic alteration and (**f**) silica alteration.

4.1.3. Cell Size

Using the methodology laid out by Carranza [46], the unit cell size is determined using point pattern analysis. First, the higher limit of a set of suitable cell size is determined by finding the distance where the likelihood of finding deposits located next to one another is null. That distance is 515 m. The lower limit depends on the predictor map with the smallest scale and can be estimated using the following formula [47]:

$$x = MS \times 0.00025,　　　　　　(10)$$

where MS is the map scale factor. In this study, the predictor maps were derived from 1:50,000 geological map and ASTER images (resolution of 15 m, 30 m, 90 m for the NIR, SWIR and TIR bands respectively). Thus, 12.5 m is the lower limit. The most suitable cell-size can be determined by fitting straight lines to the log-log plot of the rate of increase of the ratio $[N(D)] : [N(T) - N(D)]$ based on a cell-size to the next coarser cell-size, with $[N(T)]$ being the total number of cells and $[N(D)]$ being the number of cells containing one deposit [46]. The most suitable cell size for our study area is defined by the intersection of the two straight lines fitted to the log-log plot (Figure 8); thus, we selected a 50 m cell size.

4.2. RF Algorithm Parameters

As presented in Section 2, RF requires only two essential parameters: k and m_{try}. The k parameter represents the number of trees in the ensemble and m_{try} is the number of input features selected to do the splitting at each node of a tree. In this study, various values of k (from 500 to 5000) and selected $k = 1000$ as the generalization error started to converge from $k \geq 1000$. The lowest RMSE on the OOB samples was used to select the optimal value of m_{try}.

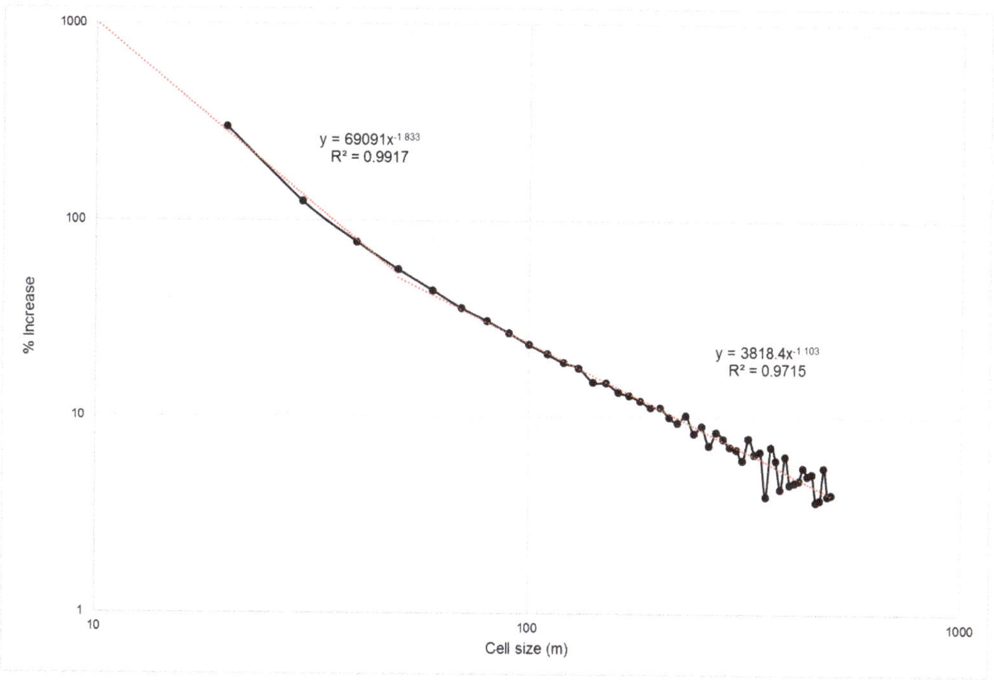

Figure 8. $Log - log$ plot of rates (in %) of increase in the ratio $[N(D)] : [N(T)N(D)]$ as a function of unit cell size.

4.3. Model Evaluation

After optimization of the RF parameter, the performance of the best-fit models was comprehensively evaluated using confusion matrix, indices of predictive accuracy, and success-rate curves.

A series of statistical indices are calculated from the confusion matrix and permit to evaluate the predictive performance of the trained model:

$$Sensitivity = \frac{TP}{TP + FN} \tag{11}$$

$$Specificity = \frac{TN}{TN + FP} \tag{12}$$

$$Accuracy = \frac{TP + TN}{TP + TN + FP + FN} \tag{13}$$

$$Kappa = \frac{(TP + TN) - \frac{(TP+FN)(TP+FP)+(FP+TN)(FN+TN)}{(TP+FP+TN+FN)}}{(TP + FP + TN + FN) - \frac{(TP+FN)(TP+FP)+(FP+TN)(FN+TN)}{(TP+FP+TN+FN)}}, \tag{14}$$

where TP is True Positive, TN is True Negative, FN is False Negative, and FP is False Positive.

The kappa index measures the fit between the predictor maps and the training dataset [48], the sensitivity and specificity indicate whether the deposit cells or non-deposit cells are correctly classified to their corresponding class respectively. Success-rate curves can be employed to evaluate the overall performance of the models. The curve is generated by calculating the percentage of correctly delineated training deposit in a prospective area for a given threshold with an increment of 5-percentile of the likelihood value [22].

5. Results

5.1. Relative Importance of Predictor Maps

As explained in Section 2, the RF algorithm measures the importance of each predictor maps which provides insights on the best proxies for mineralization in the study area (Figure 9).

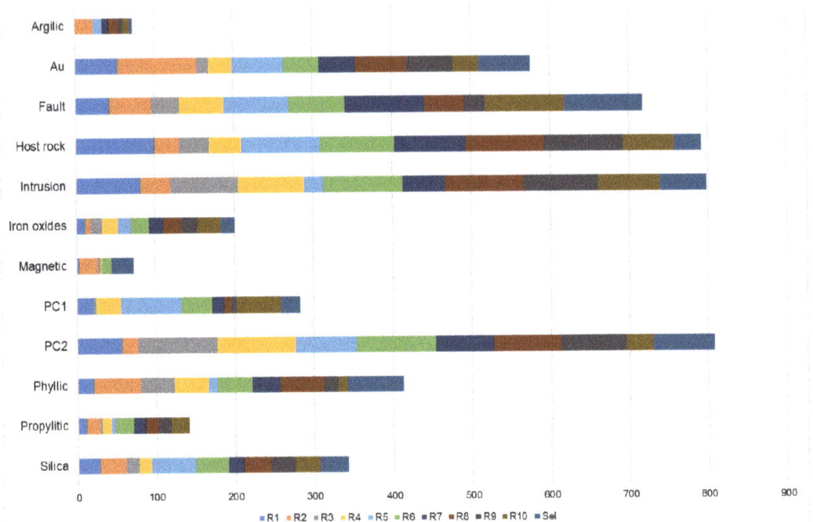

Figure 9. Sum of the relative importance percentage of each predictor map from models generated with different set of random point pattern non-deposit location (R1 to R10) and selected non-prospective KMO (Sel).

Overall, the five most important predictors are PC2, intrusion, host rock, fault, and Au in respective order from higher to lower cumulated percentage. These features represent the principal components of any epithermal Au exploration: source, pathway, physical trap, and chemical trap.

The source of the mineralized hydrothermal fluids is represented by the proximity to porphyritic intrusive bodies of Late Triassic to Early Jurassic age. The faults correspond to the pathway taken by gold-bearing fluids and they can also act as physical traps for those fluids. Lithologies from the Stuhini and Hazelton Group can act as a chemical trap for mineralized fluids. Geochemical anomalies of Au and other pathfinder elements, represented by Au and PC2 predictor maps, are also proxies of a chemical trap.

Among the different hydrothermal alteration mapped using the ASTER images, the phyllic and silica alteration have the highest cumulated percentage.

5.2. Predictive Accuracy of the Model

The predicted values range between 0 and 1 and symbolize the probability of occurrence of Au mineral deposit. A threshold of 0.5 was used to classify predictions and to calculate the statistical indices in Tables 2 and 3. Cells with a value higher than the threshold are considered prospective, whereas values below the threshold are considered non-prospective.

Table 2. Accuracy of models generated with different sets of random point pattern NDL (R1 to R10).

	R1	R2	R3	R4	R5	R6	R7	R8	R9	R10
Accuracy	88%	67%	72%	78%	75%	81%	83%	83%	81%	72%
Kappa	78%	33%	44%	56%	50%	61%	67%	67%	61%	44%
Sensitivity	94%	61%	67%	78%	78%	78%	89%	83%	78%	67%
Specificity	83%	72%	78%	78%	72%	83%	78%	83%	83%	78%

Table 3. Accuracy of models generated with selected non-prospective KMO NDL and the mean and standard deviation of accuracy indices of models generated with different sets of random point pattern NDL.

	Selected	Mean	Sd
Accuracy	84%	78%	6%
Kappa	67%	56%	13%
Sensitivity	79%	77%	10%
Specificity	89%	79%	4%

The average accuracy of the random models is lower than the selected model, with 78% and 84%, respectively. The kappa values of 56% and 67% of the averaged random models and the selected model respectively indicate that both models have a moderate fit between the predictor maps and the training datasets [48].

Of all the random models, only R1 yields better results than the selected model, but overall, the selected model is more accurate than the random models. Although the results indicate that all models can capture the spatial relationship between the predictor maps and the training datasets, the selected model is the most accurate.

5.3. Performance of RF Modelling

In the previous section, the predictive accuracy of each model is reviewed. However, the models are not evaluated in a spatial context. For that purpose, success-rate curves, describing the performance of the RF modeling based on the resulting predictive maps, are used (Figure 10).

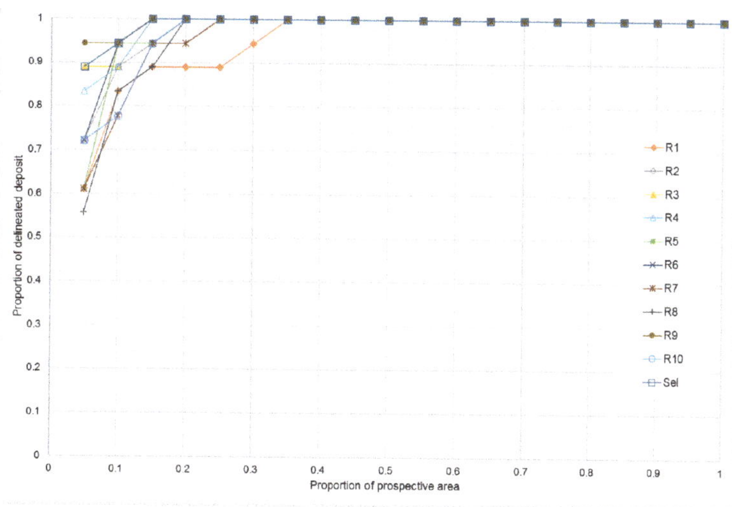

Figure 10. Success-rate curves of predictive map of Au prospectivity obtained by using training set with random non-deposit location (R1–R10) and with selected non-prospective KMO.

Of the eleven mineral prospectivity maps, the worst performing model is the R1 model which requires 35% of the study area to delineate all of the training deposits. In contrast, the best performing maps capture all of the training deposits in 15% of the study area. Those maps are obtained with the Sel, R6 and R9 models (Figures 11–13).

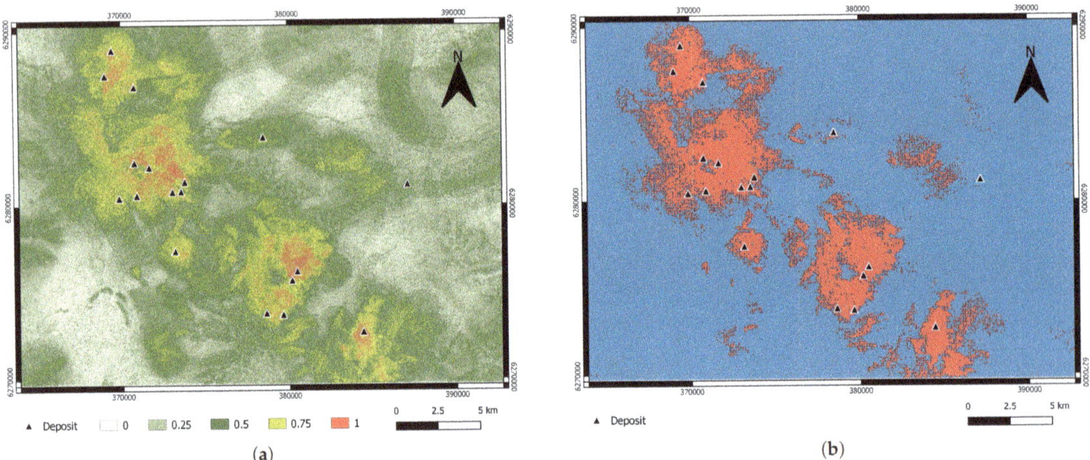

Figure 11. Mineral prospectivity map with Sel model (**a**) and delineated prospective area (**b**).

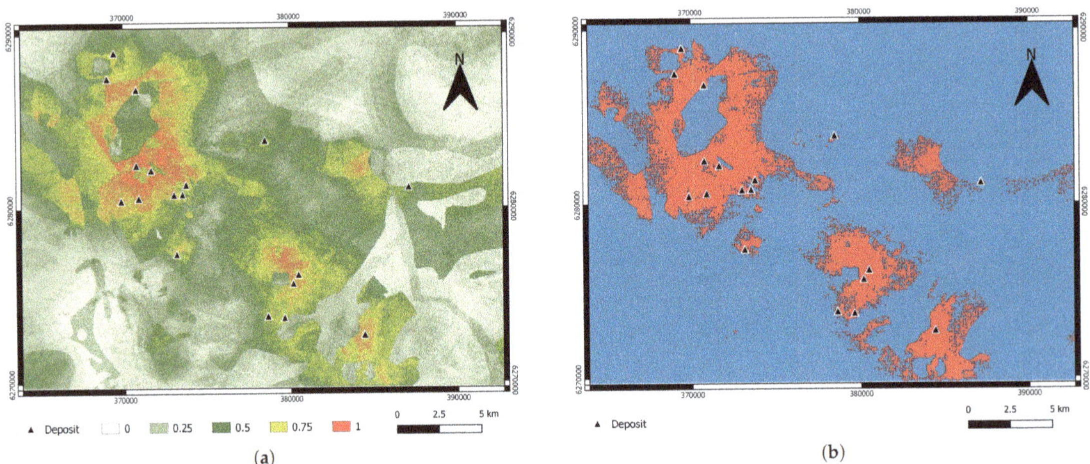

Figure 12. Mineral prospectivity map with R6 model (**a**) and delineated prospective area (**b**).

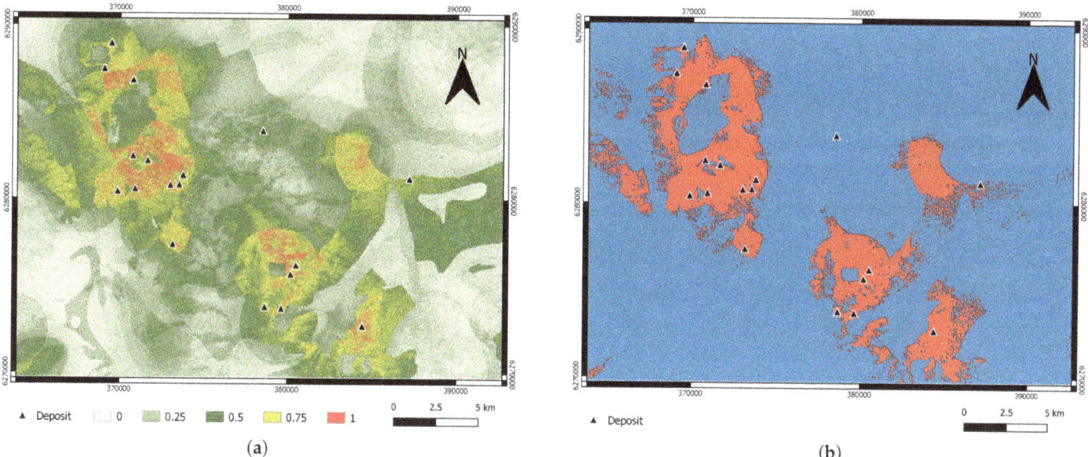

Figure 13. Mineral prospectivity map with R9 model (**a**) and delineated prospective area (**b**).

6. Conclusions

In this study, different training sets using selected non-prospective KMO and ten sets of randomly generated non-deposit location are compared. The type of deposit could not be used as a selection criterion to select the KMO because the area is an extensive epithermal system. Hence, some of the non-prospective KMO can have a similar geospatial signature as our deposit training data set and introduce bias.

Across all the models, the five predictor maps with the highest cumulated relative importance are representative of the source, pathway, and both physical and chemical traps of an epithermal Au deposit. The different model predictive accuracy are compared and it is found that the model using selected non-prospective KMO has higher accuracy than the average accuracy of the models using random NDL with 84% and 78%, respectively. Thus, using the listed commodities as a discriminant to select non-prospective KMO is enough to have an accurate resulting model.

The predictive maps are evaluated using success-rate curves. The best mineral prospectivity maps are obtained with the R6, R9,and Sel training sets that capture 100% training deposits in 15% of the prospective area. Therefore, when available, it is recommended using KMO classified as 'non-prospective' for the commodity sought rather than the randomly generated non-deposit training set. However, for the larger study area were a diversity of deposit types is present, it is recommended to use the commodities and the deposit type as selection criteria to strengthen the predictive accuracy of the model further.

As found in this study, MPM using RF can be used in early stages of an exploration project when only public data are available. By analyzing and interpreting the response of the target variable to a set of predictor variables, RF is very similar to other knowledge-guided data-driven methods such as evidential belief and weights of evidence modeling. RF can also impute missing values, both continuous and categorical data, particularly when handling heterogeneous datasets, which is the case in this study. This yields an out-of-bag imputation error estimate without the need of a test set or elaborate cross-validation. These characteristics make RF a non-black-box exploration method, which is more suitable for mineral prospectivity modeling than other currently used machine learning approaches.

Author Contributions: Conceptualization, A.L., I.M. and A.M.; methodology, A.L. and I.M.; validation, A.L. and I.M.; formal analysis, A.L.; investigation, A.L.; data curation, A.L.; writing—original draft preparation, A.L.; writing—review and editing, A.L., A.M., S.V. and I.M.; supervision, I.M. All authors have read and agreed to the published version of the manuscript.

Funding: This research received no external funding.

Appendix A

Table A1. Deposit locations from the mineral occurrences depository (MINFILE). The coordinates are in Universal Transverse Mercator (UTM) coordinate system (NAD83/zone 9N).

MINFILE #	NAME	STATUS	DEPOSIT TYPE	UTM NORTH	UTM EAST
104B 077	BRONSON SLOPE	Developed Prospect	High-sulfidation epithermal	6,282,211	371,642
104B 089	SNIP NORTH - EAST ZONE	Prospect	Massive sulphide Cu-Pb-Zn	6,286,850	370,775
104B 107	JOHNNY MOUNTAIN	Past Producer	Subaqueous hot spring Ag-Au	6,277,401	373,149
104B 113	INEL	Developed Prospect	Massive sulphide Cu-Zn	6,275,679	380,178
104B 116	TAMI (BLUE RIBBON)	Prospect	Alkalic porphyry Cu-Au	6,272,714	384,430
104B 138	KHYBER PASS	Prospect	Massive sulphide Cu-Zn	6,273,715	379,627
104B 204	WARATAH 6	Prospect	Au pyrrhotite veins	6,283,926	378,489
104B 250	SNIP	Past Producer	Au pyrrhotite veins	6,282,486	370,764
104B 264	C3 (REG)	Prospect	Au pyrrhotite veins	6,280,600	370,900
104B 300	BRONSON	Prospect	Au pyrrhotite veins	6,281,374	373,763
104B 356	GORGE	Prospect	Vein	6,287,500	369,050
104B 357	GREGOR	Prospect	Unspecified	6,288,962	369,467
104B 537	MYSTERY	Prospect	Au pyrrhotite veins	6,281,200	387,150
104B 557	AK	Prospect	Subaqueous hot spring Ag-Au	6,276,200	380,500
104B 563	CE CONTACT	Prospect	Au pyrrhotite veins	6,280,800	373,000
104B 567	SMC	Prospect	Massive sulphide Cu-Pb-Zn	6,280,450	369,850
104B 571	CE	Prospect	Au pyrrhotite veins	6,280,829	373,529
104B 685	KHYBER WEST	Prospect	Unspecified	6,273,802	378,627

Table A2. Selected non-deposit location from the mineral occurrences depository (MINFILE). The coordinates are in Universal Transverse Mercator (UTM) coordinate system (NAD83/zone 9N).

MINFILE #	NAME	STATUS	DEPOSIT TYPE	UTM NORTH	UTM EAST
104B 005	CRAIG RIVER	Showing	Cu skarn	6,2761,77	366,697
104B 205	HANDEL	Showing	Polymetallic veins	6,281,905	376,693
104B 206	WOLVERINE	Showing	Polymetallic veins Ag-Pb-Zn	6,277,250	377,150
104B 256	WOLVERINE (INEL)	Showing	Cu skarn	6,277,063	383,766
104B 268	HANGOVER TRENCH	Showing	Polymetallic veins Ag-Pb-Zn	6,275,185	369738
104B 272	DAN 2	Showing	Polymetallic veins Ag-Pb-Zn	6,271,824	375,475
104B 292	GIM (ZONE 1)	Showing	Polymetallic veins Ag-Pb-Zn	6,281,770	383,605
104B 305	MILL	Showing	Porphyry Cu-Mo-Au	6,272,879	363,417
104B 306	NORTH CREEK	Showing	Polymetallic veins Ag-Pb-Zn	6,275,031	368,709
104B 324	IAN 4	Showing	Cu-Ag quartz veins	6,286,725	379,485
104B 326	CAM 9	Showing	Cu skarn	6,279,635	391,709
104B 327	CAM SOUTH	Showing	Polymetallic veins Ag-Pb-Zn	6,279,579	392,696
104B 331	IAN 8	Showing	Cu skarn	6,286,038	383,655
104B 362	KIRK MAGNETITE	Showing	Fe skarn	6,276,565	389,635
104B 368	ELMER	Showing	Fe skarn	6,275,780	391,286
104B 377	ROCK AND ROLL	Developed Prospect	Massive sulphide Cu-Zn	6,288,261	363,286
104B 416	IAN 6 SOUTH	Showing	Massive sulphide Cu-Pb-Zn	6,286,900	382,200
104B 500	KRL-FORREST	Showing	Vein	6,288,950	393,400
104B 536	ANDY	Showing	Pb-Zn skarn	6,278,300	385,825

References

1. Bonham-Carter, G.F. *Geographic Information Systems for Geoscientists: Modelling with GIS*; Pergamon (Elsevier Science Ltd.): Oxford, UK, 1994.
2. Carranza, E.J.M. Geochemical Anomaly and Mineral Prospectivity Mapping in GIS. In *Handbook of Exploration and Environmental Geochemistry*, 2009th ed.; Elsevier Science: Amsterdam, The Netherlands, 2008; Volume 11.
3. Harris, J.; Grunsky, E.; Behnia, P.; Corrigan, D. Data- and knowledge-driven mineral prospectivity maps for Canada's North. *Ore Geol. Rev.* **2015**, *71*, 788–803. [CrossRef]
4. Harris, J.; Wilkinson, L.; Heather, K.; Fumerton, S.; Bernier, M.A.; Ayer, J.; Dahn, R. Application of GIS Processing Techniques for Producing Mineral Prospectivity Maps—A Case Study: Mesothermal Au in the Swayze Greenstone Belt, Ontario, Canada. *Nat. Resour. Res.* **2001**, *34*, 91–124. [CrossRef]
5. Yousefi, M.; Carranza, E.J.M. Geometric average of spatial evidence data layers: A GIS-based multi-criteria decision-making approach to mineral prospectivity mapping. *Comput. Geosci.* **2015**, *83*, 72–79. [CrossRef]

6. Abedi, M.; Mostafavi Kashani, S.B.; Norouzi, G.H.; Yousefi, M. A deposit scale mineral prospectivity analysis: A comparison of various knowledge-driven approaches for porphyry copper targeting in Seridune, Iran. *J. Afr. Earth Sci.* **2017**, *128*, 127–146. [CrossRef]
7. Joly, A.; Porwal, A.; McCuaig, T.C. Exploration targeting for orogenic gold deposits in the Granites-Tanami Orogen: Mineral system analysis, targeting model and prospectivity analysis. *Ore Geol. Rev.* **2012**, *48*, 349–383. [CrossRef]
8. Porwal, A.; Das, R.D.; Chaudhary, B.; Gonzalez-Alvarez, I.; Kreuzer, O. Fuzzy inference systems for prospectivity modeling of mineral systems and a case-study for prospectivity mapping of surficial Uranium in Yeelirrie Area, Western Australia. *Ore Geol. Rev.* **2015**, *71*, 839–852. [CrossRef]
9. Yousefi, M.; Carranza, E.J.M. Fuzzification of continuous-value spatial evidence for mineral prospectivity mapping. *Comput. Geosci.* **2015**, *74*, 97–109. [CrossRef]
10. Yousefi, M.; Nykänen, V. Data-driven logistic-based weighting of geochemical and geological evidence layers in mineral prospectivity mapping. *J. Geochem. Explor.* **2016**, *164*, 94–106. [CrossRef]
11. Harris, D.; Zurcher, L.; Stanley, M.; Marlow, J.; Pan, G. A Comparative Analysis of Favorability Mappings by Weights of Evidence, Probabilistic Neural Networks, Discriminant Analysis, and Logistic Regression. *Nat. Resour. Res.* **2003**, *12*, 241–255.:NARR.0000007804.27450.e8. [CrossRef]
12. Porwal, A.; González-Álvarez, I.; Markwitz, V.; McCuaig, T.; Mamuse, A. Weights-of-evidence and logistic regression modeling of magmatic nickel sulfide prospectivity in the Yilgarn Craton, Western Australia. *Ore Geol. Rev.* **2010**, *38*, 184–196. [CrossRef]
13. Zhang, Z.; Zuo, R.; Xiong, Y. A comparative study of fuzzy weights of evidence and random forests for mapping mineral prospectivity for skarn-type Fe deposits in the southwestern Fujian metallogenic belt, China. *Sci. China Earth Sci.* **2016**, *59*, 556–572. [CrossRef]
14. Brown, W.M.; Gedeon, T.D.; Groves, D.I.; Barnes, R.G. Artificial neural networks: A new method for mineral prospectivity mapping. *Aust. J. Earth Sci.* **2000**, *47*, 757–770. [CrossRef]
15. Rodriguez-Galiano, V.F.; Sanchez-Castillo, M.; Chica-Olmo, M.; Chica-Rivas, M. Machine learning predictive models for mineral prospectivity: An evaluation of neural networks, random forest, regression trees and support vector machines. *Ore Geol. Rev.* **2015**, *71*, 804–818. [CrossRef]
16. Abedi, M.; Norouzi, G.H.; Bahroudi, A. Support vector machine for multi-classification of mineral prospectivity areas. *Comput. Geosci.* **2012**, *46*, 272–283. [CrossRef]
17. Geranian, H.; Tabatabaei, S.H.; Asadi, H.H.; Carranza, E.J.M. Application of Discriminant Analysis and Support Vector Machine in Mapping Gold Potential Areas for Further Drilling in the Sari-Gunay Gold Deposit, NW Iran. *Nat. Resour. Res.* **2016**, *25*, 145–159. [CrossRef]
18. Shabankareh, M.; Hezarkhani, A. Application of support vector machines for copper potential mapping in Kerman region, Iran. *J. Afr. Earth Sci.* **2017**, *128*, 116–126. [CrossRef]
19. Carranza, E.J.M.; Laborte, A.G. Data-driven predictive mapping of gold prospectivity, Baguio district, Philippines: Application of Random Forests algorithm. *Ore Geol. Rev.* **2015**, *71*, 777–787. [CrossRef]
20. Carranza, E.J.M.; Laborte, A.G. Data-Driven Predictive Modeling of Mineral Prospectivity Using Random Forests: A Case Study in Catanduanes Island (Philippines). *Nat. Resour. Res.* **2016**, *25*, 35–50. [CrossRef]
21. Hariharan, S.; Tirodkar, S.; Porwal, A.; Bhattacharya, A.; Joly, A. Random Forest-Based Prospectivity Modelling of Greenfield Terrains Using Sparse Deposit Data: An Example from the Tanami Region, Western Australia. *Nat. Resour. Res.* **2017**, *26*, 489–507. [CrossRef]
22. Rodriguez-Galiano, V.F.; Chica-Olmo, M.; Chica-Rivas, M. Predictive modelling of gold potential with the integration of multisource information based on random forest: A case study on the Rodalquilar area, Southern Spain. *Int. J. Geogr. Inf. Sci.* **2014**, *28*, 1336–1354. [CrossRef]
23. Carranza, E.J.M. Natural Resources Research Publications on Geochemical Anomaly and Mineral Potential Mapping, and Introduction to the Special Issue of Papers in These Fields. *Nat. Resour. Res.* **2017**, *26*, 379–410. [CrossRef]
24. Parsa, M.; Maghsoudi, A.; Yousefi, M. Spatial analyses of exploration evidence data to model skarn-type copper prospectivity in the Varzaghan district, NW Iran. *Ore Geol. Rev.* **2018**, *92*, 97–112. [CrossRef]
25. Sun, T.; Chen, F.; Zhong, L.; Liu, W.; Wang, Y. GIS-based mineral prospectivity mapping using machine learning methods: A case study from Tongling ore district, eastern China. *Ore Geol. Rev.* **2019**, *109*, 26–49. [CrossRef]
26. McKay, G.; Harris, J. Comparison of the Data-Driven Random Forests Model and a Knowledge-Driven Method for Mineral Prospectivity Mapping: A Case Study for Gold Deposits Around the Huritz Group and Nueltin Suite, Nunavut, Canada. *Nat. Resour. Res.* **2016**, *25*, 125–143. [CrossRef]
27. Zhang, S.; Xiao, K.; Carranza, E.J.M.; Yang, F. Maximum Entropy and Random Forest Modeling of Mineral Potential: Analysis of Gold Prospectivity in the Hezuo–Meiwu District, West Qinling Orogen, China. *Nat. Resour. Res.* **2019**, *28*, 645–664. [CrossRef]
28. Breiman, L. Random Forests. *Mach. Learn.* **2001**, 5–32. [CrossRef]
29. Breiman, L. Bagging predictors. *Mach. Learn.* **1996**, *24*, 123–140. [CrossRef]
30. Breiman, L.; Friedman, J.H.; Olshen, R.A.; Stone, C.J. *Classification And Regression Trees*; Taylor & Francis Group: Boca Raton, FL, USA, 1984.

31. Macdonald, A.J.; Lewis, P.D.; Thompson, J.F.H.; Nadaraju, G.; Bartsch, R.; Bridge, D.J.; Rhys, D.A.; Roth, T.; Kaip, A.; Godwin, C.I.; et al. Metallogeny of an Early to Middle Jurassic arc, Iskut River area, northwestern British Columbia. *Econ. Geol.* **1996**, *91*, 1098–1114. [CrossRef]

32. Nelson, J.; Kyba, J. Structural and stratigraphic control of porphyry and related mineralization in the Treaty Glacier KSM Brucejack Stewart trend of western Stikinia. *Br. Columbia Geol. Surv. Pap.* **2014**, *2013*, 111–140.

33. Logan, J.M.; Mihalynuk, M.G. Tectonic Controls on Early Mesozoic Paired Alkaline Porphyry Deposit Belts (Cu-Au Ag-Pt-Pd-Mo) Within the Canadian Cordillera. *Econ. Geol.* **2014**, *109*, 827–858. [CrossRef]

34. Rhys, D.A. Geology of the Snip Mine and Its Relationship to the Magmatic and Deformational History of the Johnny Mountain Area, Northwestern Bristish Columbia. Master's Thesis, University of British Columbia, Vancouver, BC, Canada, 1993.

35. Rhys, D.A. *Geology of the Stonehouse Gold Deposit (Johnny Mountain Gold Mine) and Exploration Implications*; Technical Report; Spirit Bear Minerals Ltd.: Vancouver, BC, Canada, 1994.

36. Cui, Y.; Miller, D.; Schiarizza, P.; Diakow, L.J. British Columbia Digital Geology. In *Open File 2017-8, British Columbia Ministry of Energy, Mines and Petroleum Resources*; British Columbia Geological Survey: Victoria, BC, Canada, 2017.

37. Aitchison, J. *The Statistical Analysis of Compositional Data*; Chapman and Hall: London, UK, 1986.

38. Grunsky, E.C. The interpretation of geochemical survey data. *Geochem. Explor. Environ. Anal.* **2010**, *10*, 27–74. [CrossRef]

39. Crósta, A.P.; De Souza Filho, C.R.; Azevedo, F.; Brodie, C. Targeting key alteration minerals in epithermal deposits in Patagonia, Argentina, using ASTER imagery and principal component analysis. *Int. J. Remote Sens.* **2003**, *24*, 4233–4240. [CrossRef]

40. Di Tommaso, I.; Rubinstein, N. Hydrothermal alteration mapping using ASTER data in the Infiernillo porphyry deposit, Argentina. *Ore Geol. Rev.* **2007**, *32*, 275–290. [CrossRef]

41. Moore, F.; Rastmanesh, F.; Asadi, H.; Modabberi, S. Mapping mineralogical alteration using principal-component analysis and matched filter processing in the Takab area, north-west Iran, from ASTER data. *Int. J. Remote Sens.* **2008**, *29*, 2851–2867. [CrossRef]

42. Crowley, J.K.; Brickey, D.W.; Rowan, L.C. Airborne imaging spectrometer data of the Ruby Mountains, Montana: Mineral discrimination using relative absorption band-depth images. *Remote Sens. Environ.* **1989**, *29*, 121–134. [CrossRef]

43. Rowan, L.C.; Mars, J.C. Lithologic mapping in the Mountain Pass, California area using Advanced Spaceborne Thermal Emission and Reflection Radiometer (ASTER) data. *Remote Sens. Environ.* **2003**, *84*, 350–366. [CrossRef]

44. Pour, A.B.; Hashim, M. Identification of hydrothermal alteration minerals for exploring of porphyry copper deposit using ASTER data, SE Iran. *J. Asian Earth Sci.* **2011**, *42*, 1309–1323. [CrossRef]

45. Rajendran, S.; Nasir, S. Characterization of ASTER spectral bands for mapping of alteration zones of volcanogenic massive sulphide deposits. *Ore Geol. Rev.* **2017**, *88*, 317–335. [CrossRef]

46. Carranza, E.J.M. Objective selection of suitable unit cell size in data-driven modeling of mineral prospectivity. *Comput. Geosci.* **2009**, *35*, 2032–2046. [CrossRef]

47. Hengl, T. Finding the right pixel size. *Comput. Geosci.* **2006**, *32*, 1283–1298. [CrossRef]

48. Landis, J.R.; Koch, G.G. The Measurement of Observer Agreement for Categorical Data. *Biometrics* **1977**, *33*, 159. [CrossRef] [PubMed]

MDPI
St. Alban-Anlage 66
4052 Basel
Switzerland
Tel. +41 61 683 77 34
Fax +41 61 302 89 18
www.mdpi.com

Minerals Editorial Office
E-mail: minerals@mdpi.com
www.mdpi.com/journal/minerals